Cell and Tissue Ultrastructure

A Functional Perspective

Patricia C. Cross, Ph.D., Associate Professor

K. Lynne Mercer, M.Sc.

Department of Cell Biology
Stanford University School of Medicine

W. H. Freeman and Company
New York

Library of Congress Cataloging-in-Publication Data

Cross, Patricia C.
Cell and tissue ultrastructure : a functional perspective / Patricia C. Cross, K. Lynne Mercer.
p. cm.
Includes bibliographical references and index.
ISBN 0-7167-7033-4
1. Histology. 2. Ultrastructure (Biology) I. Mercer, K. Lynne. II. Title.
QM551.C75 1993
611'.0189—DC20 93-10479
CIP

Printed in the United States of America

Third printing, 1998

For my sons, Jake and Wyatt,
who enrich my life and works

Patricia C. Cross

To my courageous and loving parents, and to
Larry, for the inspiration they have provided
and for all they have done

K. Lynne Mercer

Contents

PREFACE

Cell and Tissue Ultrastructure: A Functional Perspective describes the ultrastructure of most cells in the body and how this structure relates to function. The book was designed to combine the clarity of an atlas with the perspective of an up-to-date cell biology text. An atlas format is maintained throughout the book, with electron micrographs on right-hand pages and corresponding text on facing pages. For all Chapters except Cell (Chapter One), a special diagram on each facing page clarifies the orientation of the electron micrograph. The diagram includes (1) a shaded rectangle within which the structures visible in the corresponding electron micrograph are directly illustrated to scale and (2) a continuation of the drawing beyond the shaded rectangle that positions the high-magnification micrograph in its (idealized) light microscope context. The diagram thus leads the student from a familiar low-magnification perspective to ultrastructure viewed at a magnification as great as 100,000×.

Throughout the book the text is divided into basic sections describing fundamental structure and function and enrichment sections (indented and in smaller print) containing more detailed information. Course directors can use this distinction to separate required and optional material. For students interested in exploring beyond the information in the enrichment sections, we have included a relatively extensive reference list at the end of the book. Whenever possible this list provides reviews that lend perspective and provide "stepping stones" to current research and the primary sources.

With the exciting growth in the field of cell biology and the increasing merger of basic research with clinical medicine, cell ultrastructure has progressed from a peripheral to a central role in the study of cells and tissues. It is at the ultrastructural level that students' interest peaks and that we, as teachers, are best able to relate structure to function. It is from the ultrastructure and related function of individual cells that we build our entire understanding of organs and organ systems. The membrane projections holding digestive enzymes in the small intestine provide striking contrast to the organized contractile assembly of striated muscles. It becomes apparent that, as different as various cells are, they are unified by essential processes reflected in their common organelle complement. There are many examples of the same organelle carrying out a similar function in different cells, although frequently with unique outcomes. Pumps held in the polarized infolded membranes of the osteoclast secrete acid that dissolves and thus remodels bone, whereas pumps in the infolded membranes of the stomach parietal cell secrete acid that destroys bacteria and thus protects the gastrointestinal lining. The smooth endoplasmic reticulum in the olfactory region removes unwanted compounds, thus clearing the sensory region and maintaining its sensitivity; in the liver, a similar enzyme complex in this same organelle functions to detoxify drugs.

Micrographs used in this book represent mammalian tissues except where indicated. With the exception of those credited to other people, the micrographs were made by Lynne Mercer for use in the histology course for first-year medical students at Stanford. Publishing this book was in part a response to the demand of our students, who have used the micrographs and text as a bridge between the lecture material and their observations with the light microscope. They have found that this overview provides a broad foundation of knowledge in cell biology and physiology that carries over to many of their other courses. We would appreciate hearing from readers of our book about ways they find it useful and ways it could be improved.

ACKNOWLEDGMENTS

We would like to thank Dr. Sylvia Friedberg, who, with Lynne Mercer, assembled the original electron micrographs in 1974 and 1975. It was this initial collection of micrographs, expanded over the years, that led to the development of this book. We are indebted to the following researchers who provided specific micrographs, as cited in the text—A. S. Breathnach, Edward P. Gogol, A. J. Hudspeth, Richard A. Jacobs, Lawrence H. Mathers, A. L. Olins, D. E. Olins, Anna C. Spudich, James A. Spudich, Nigel Unwin, Luciano Zamboni— and to the many people who donated tissue for this project. We would also like to thank the following reviewers for offering valuable comments: Lawrence Eng, Linda Giudice, Aaron Hseuh, Roger Kornberg, U. Jack McMahan, Roy Maffly, Jack Moriarity, Peter Parham, Edwin Rock, James Spudich, Eric Stark, Lubert Stryer, Richard Tsien, and Conrad Vial. We extend our sincere appreciation to Joanne Tisch for typing the references.

Patricia C. Cross
K. Lynne Mercer
July 1993

CELL

Interphase Nucleus

Interphase cells have nuclei that are defined by a nuclear envelope (arrowheads, micrograph 1). Within the nucleus, the parts of chromosomes that are densely packed are visible as heavily stained areas of **heterochromatin** (arrows, micrograph 1). The more extended parts of chromosomes, the **euchromatin,** are usually not visible in routine electron micrographs. Both heterochromatin and euchromatin consist of DNA and its associated proteins.

The primary level of chromosome packing can be observed after nuclear disruption and dilution. Following this treatment a "string of beads" arrangement is revealed (inset, courtesy of Dr. A. L. Olins and Dr. D. E. Olins). Each bead is a **nucleosome** consisting of a core particle of eight histones wrapped by DNA. Further packing of nucleosomes into 30-nm thick heterochromatin filaments appears to depend upon specific interactions with histone H1, which is outside the nucleosome core. One possible secondary packing arrangement, or **solenoid,** is shown in the diagram. The 30-nm filaments are frequently packed together in dense areas next to the nuclear envelope.

The DNA in euchromatin is accessible for transcription. In contrast, the DNA in the highly compacted heterochromatin is transcriptionally inactive.

Two classes of heterochromatin exist.

1. Constitutive heterochromatin consists of a large amount of the eukaryotic genome that is never involved in transcription, including the highly repeated centromeric DNA, which seems to coordinate mitosis but does not code for proteins.

2. Facultative heterochromatin consists of DNA that is not being transcribed at a particular time. Micrograph 1 illustrates how the amount of facultative heterochromatin varies with cellular activity and cell type. The euchromatic nucleus of the lymphoblast (A) is characteristic of cells particularly active in the synthesis of proteins required for division. More heterochromatin distinguishes its precursor, the less active small lymphocyte (B), prior to its activation by antigen.

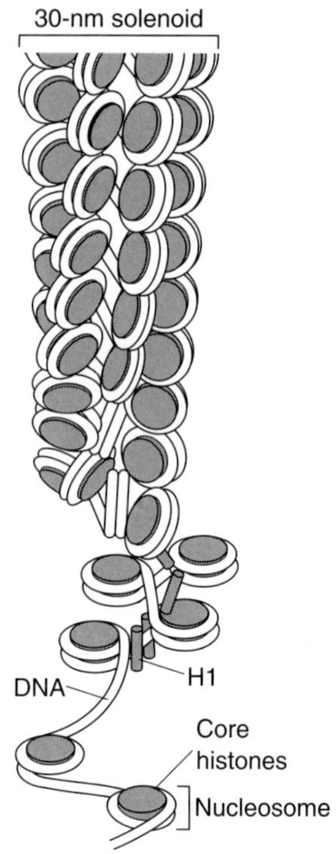

Modified from R. D. Kornberg and A. Klug, *Sci. Amer. 244:* 52 (1981).

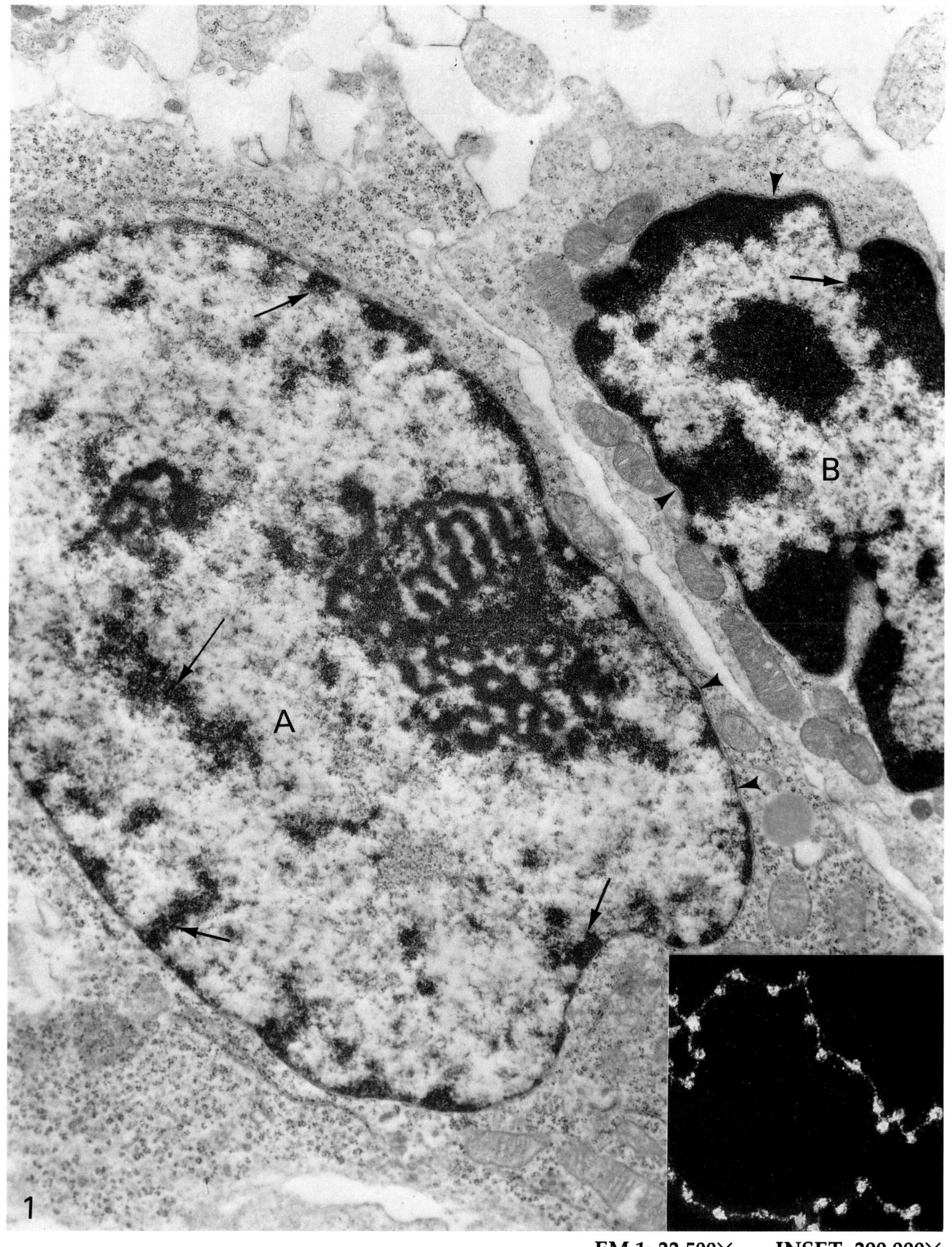

EM 1: 22,500× INSET: 290,000×

Mitotic Chromosomes

At the onset of **cell division,** as transcription comes to an end, more euchromatin is packed into heterochromatin. Heterochromatin undergoes further packing as mitosis progresses, resulting in a clear definition of **chromosomes** (ch, micrograph); 50,000 μm of DNA is condensed into a chromosome 5 μm long. The 30-nm solenoids are positioned in chromosomes in hierarchical arrangements of either loop domains or helical structures. Nonhistone proteins form a scaffold important in these topological conversions of chromosomes during mitosis. Microtubules (arrowheads, micrograph) of the spindle bind to and direct chromosome movements.

In the telophase cell represented in this micrograph, the **nuclear envelope** (arrows), dispersed throughout the cytoplasm during most of mitosis as small vesicles (100–200 nm), has begun to re-form on the surface of chromosomes. Unique proteins associated with the nuclear envelope, the lamins, appear to be important attachment sites for chromosomes. The association between the nuclear envelope, the lamins, and the chromosomes occurs at specific DNA sequences and may position the chromosomes in the nucleus in a defined way.

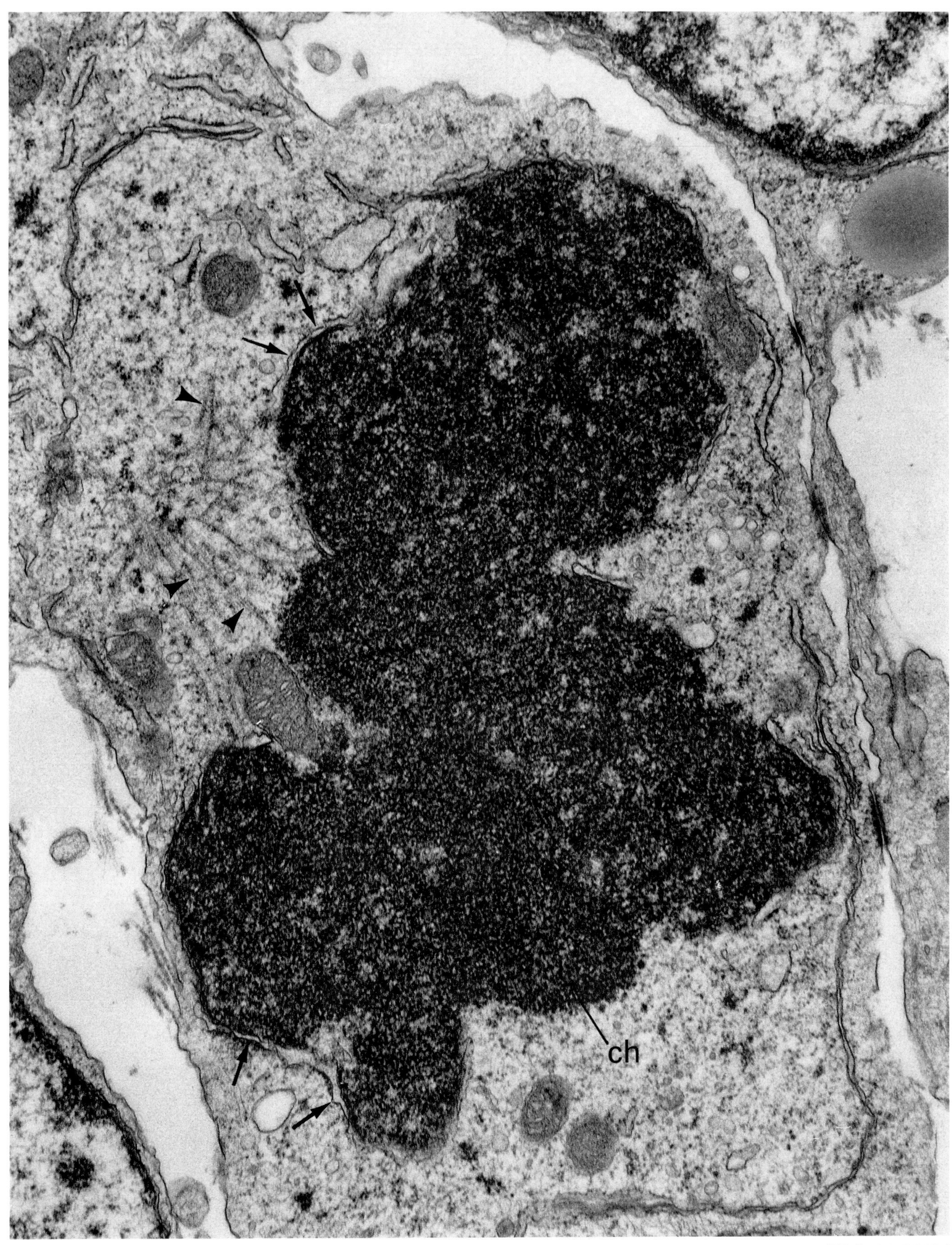

EM: 31,000×

NUCLEOLUS

Nucleoli (brackets, micrographs 1 and 2) are the sites of ribosome production and are therefore prominent nuclear structures in cells that are synthesizing proteins. Within each nucleolus, ribosomal RNA (except for the 5S component) is synthesized on many copies of DNA in one or more pale-staining areas called **nucleolar organizing regions** (nor, micrograph 2). The processing of this RNA and its association with proteins occurs initially in the nucleolar organizing regions, next in the **fibrillar region** (f), and finally in the **granular region** (g), as shown in micrographs 1 and 2. During this sequence of ribosomal subunit development, a 45S precursor of rRNA is cleaved and modified to form the large and small ribosomal subunits. The subunits, structurally close to completion but not yet functional, leave the nucleus via nuclear pores. Even though the pores are not resolved in micrograph 1, their presence is inferred at regions devoid of heterochromatin (arrows).

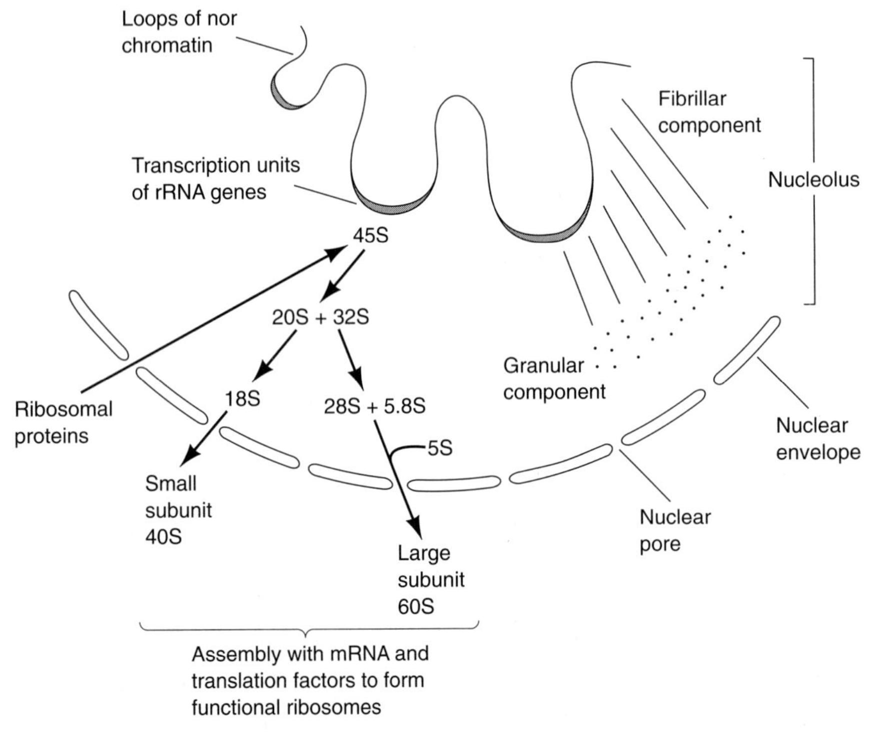

Modified from J. Sommerville, *Nature 318:* 410 (1985).

In many cells a relatively large amount of heterochromatin (hc, micrographs 1 and 2) is observed in close association with the nucleolus. This chromatin does not participate in nucleolus formation. In many female somatic cells it represents the inactivated X chromosome. Sex chromatin is not always found adjacent to the nucleolus; in oral mucosa it is associated with the nuclear envelope as a Barr body, and in neutrophils it occupies a small lobe of the nucleus commonly referred to as the drumstick.

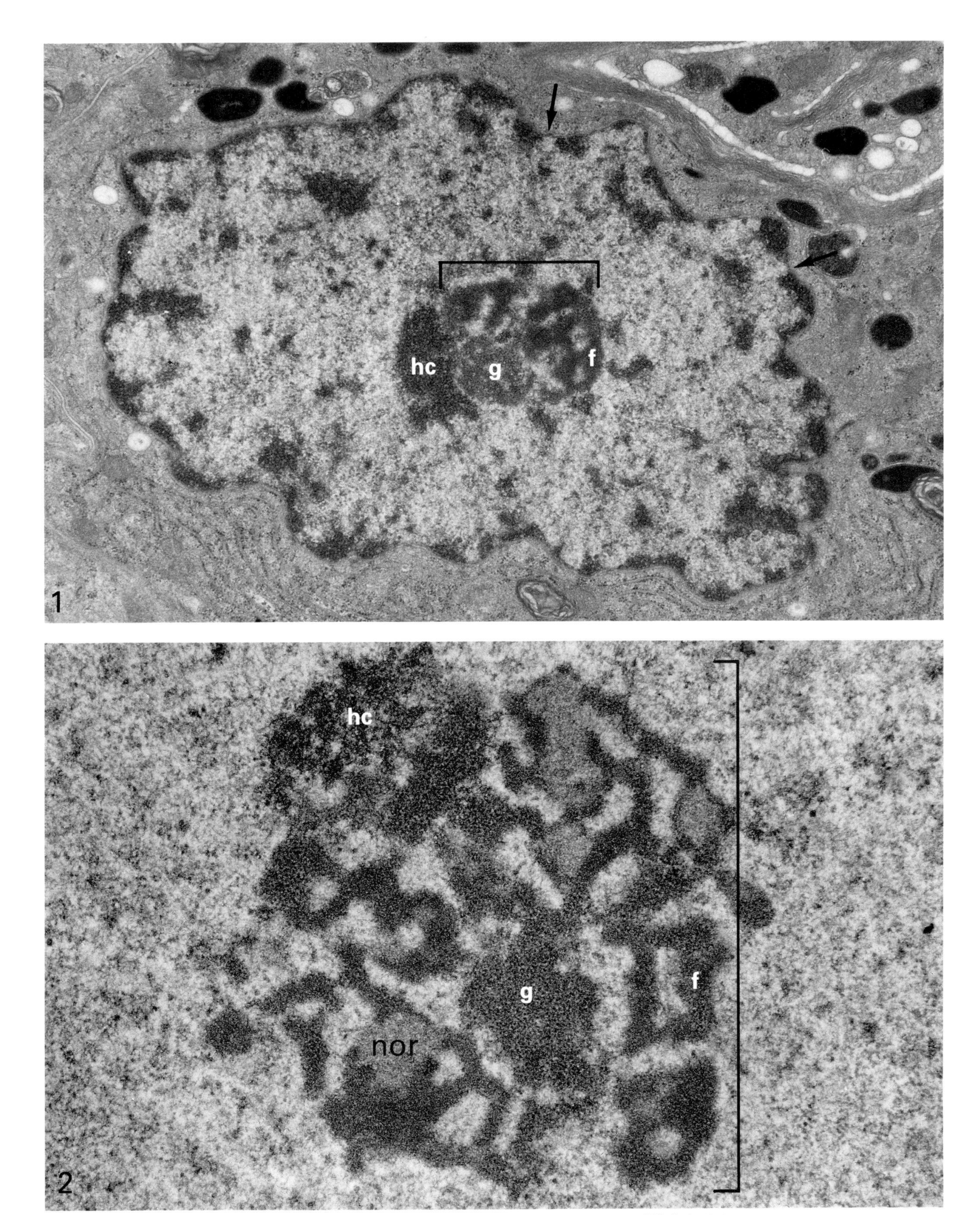

EM 1: 3,840× EM 2: 33,800×

NUCLEAR ENVELOPE

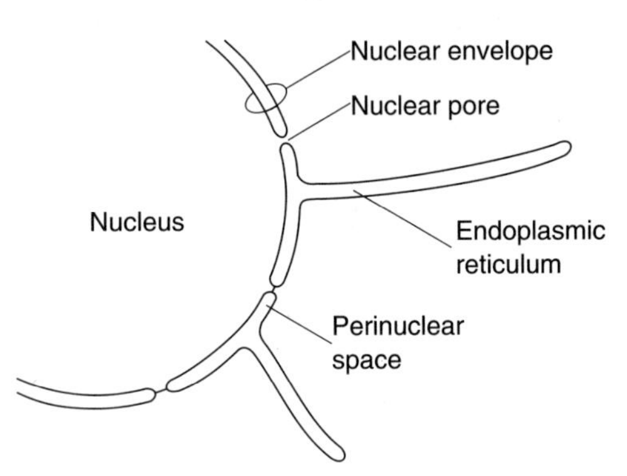

The **nuclear envelope** (shown in micrographs 1, 2 and 3) surrounds and functionally separates the interphase nucleus (N) from the cytoplasm (C). It consists of two parallel unit membranes enclosing a perinuclear space. The outer membrane of the nuclear envelope is continuous with the endoplasmic reticulum (arrowheads, micrograph 1). In many protein-secreting cells, such as the antibody-producing plasma cell, this outer membrane is identical to rough endoplasmic reticulum, with attached ribosomes and a perinuclear space swollen with protein secretion (arrows, micrograph 2). The proteins pass to the cisternae of the rough endoplasmic reticulum via direct connections such as the one shown in micrograph 1.

The structure and function of the inner and outer membranes of the nuclear envelope are different. In many cell types the nuclear side is thickened by a fibrous layer, the **lamina** (arrows, micrograph 3). The lamina is composed of three proteins, A, B, and C lamins, which help maintain the integrity of the nucleus. The strength provided by the lamins comes from alpha helical sections that form coiled-coil structures similar to those found in intermediate filaments of the cytoplasm. Lamins disassemble during the cell cycle as the nuclear envelope breaks down at prophase and reassemble around the telophase chromosomes as the nuclear envelope reforms.

Lamins are associated with disease (antibodies to lamin B have been found in the sera of certain patients with systemic lupus erythematosus, a connective tissue autoimmune disease), development (lamin B is synthesized early in cell development, whereas lamins A and C appear only at later stages of differentiation), and the communication between the nucleus and cytoplasm (lamins interact on one side with chromatin and on the other with cytoplasmic intermediate filaments).

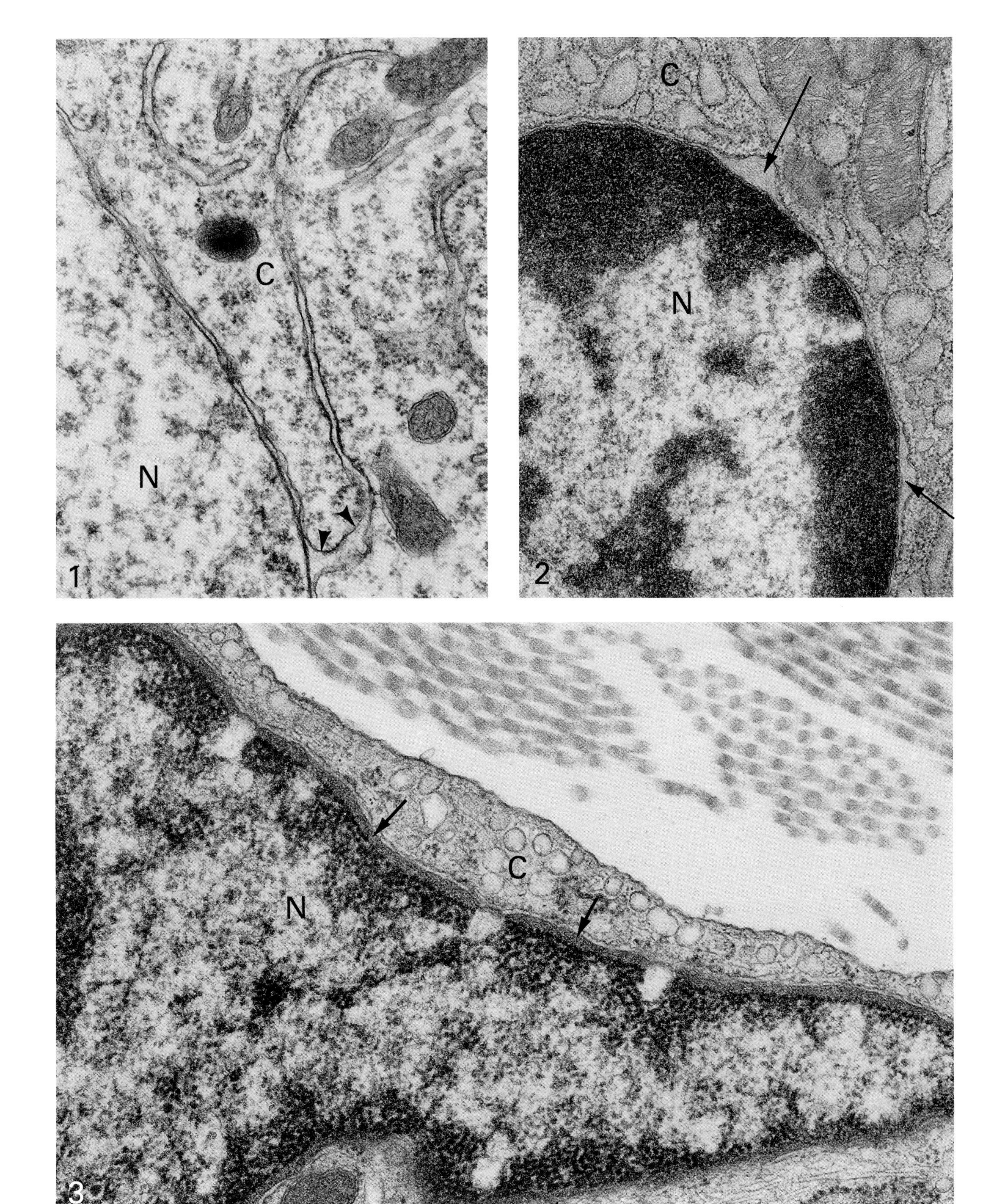

EM 1: 46,000× EM 2: 34,000× EM 3: 51,000×

Nuclear pore complexes interrupt the nuclear envelope at a number of sites (arrows, micrograph 1). The nucleoplasm adjacent to each complex is typically free of heterochromatin and lamins. At the margins of the pore complex there is continuity of the inner and outer membrane of the nuclear envelope.

Nuclear pores serve as pathways for traffic between the nucleus and the cytoplasm. In most sections, as in micrograph 1, pores appear to be completely closed. Even in instances in which the section does seem to pass through a pore opening, centrally placed "plugs" have been observed. These "plugs," particularly well seen in the negatively stained preparation in the inset (courtesy of Dr. Nigel Unwin), represent material in transit (e.g., ribosomes heading to the cytoplasm; DNA polymerase to the nucleus) and a system of proteins, the central transporter, that regulates this exchange. A current model of this transporter shows a double iris diaphragm with eightfold symmetry that opens and closes as substances are translocated. When the transporter is "closed," a 90-Å pore remains, which allows passive diffusion of small molecules. Transport of larger molecules is energy dependent and involves opening of the central pore of the transporter complex.

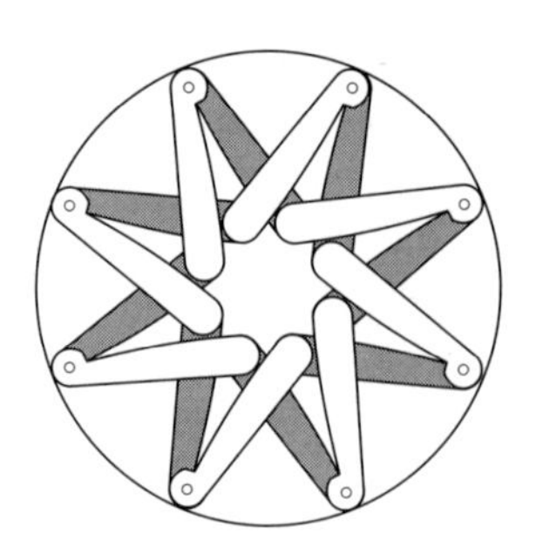

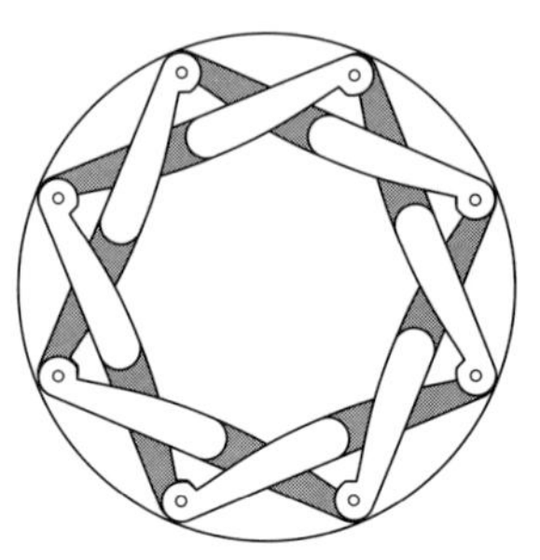

Modified from C. W. Akey, *Biophys. J. 58:* 341 (1990).

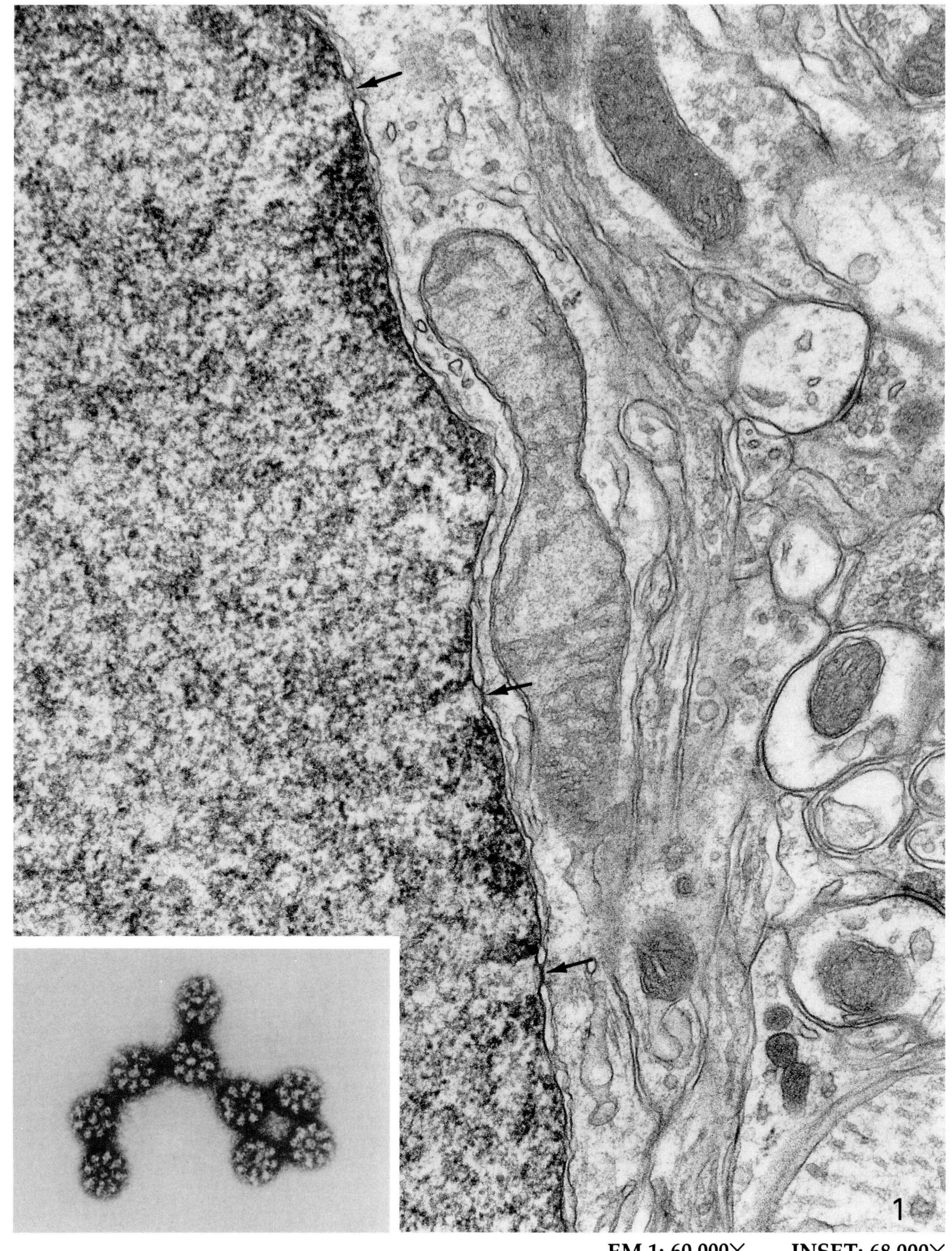

EM 1: 60,000× INSET: 68,000×

NUCLEAR PORE COMPLEX: Structure

Part of a nucleus is sectioned tangentially in the micrograph, revealing **nuclear pore complexes** *en face* (arrows). As in this fibroblast, pore complexes occupy a considerable portion of the nuclear envelope in most cells, averaging approximately 4,000 per nucleus. The number has been shown to vary with general metabolic activity and even more specifically with hormonal cycles and aging. Pores may be evenly distributed or clustered and arranged in regular or irregular patterns.

Nuclear envelopes have been isolated and the pore complexes examined using image processing to determine their molecular structure. Surrounding the central particle (transporter and material in transit) are spokes and two coaxial rings, one facing the nucleus and one the cytoplasm, each with an eightfold symmetry. The rings are tightly attached to the nuclear envelope, but only one protein is an integral membrane protein. Many of the pore complex proteins are glycosylated, and the removal or masking of the glycosylated regions reduces nuclear/cytoplasmic transport.

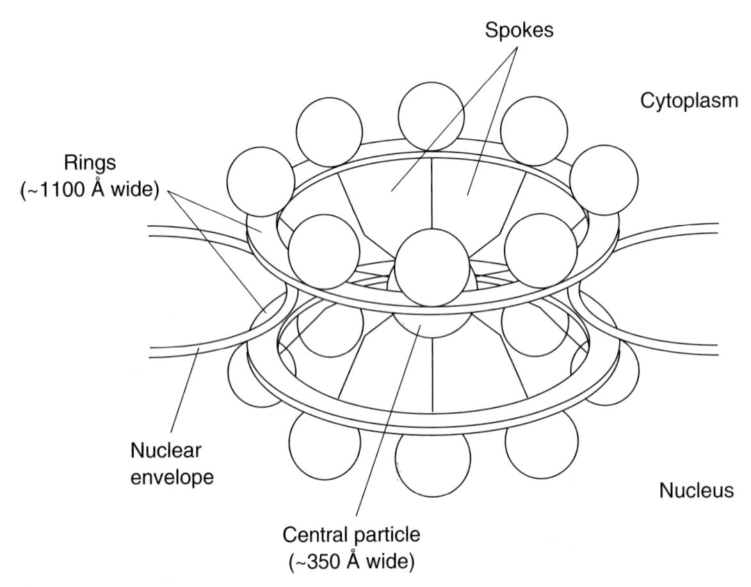

Courtesy of Dr. N. Unwin and Dr. R. Milligan.

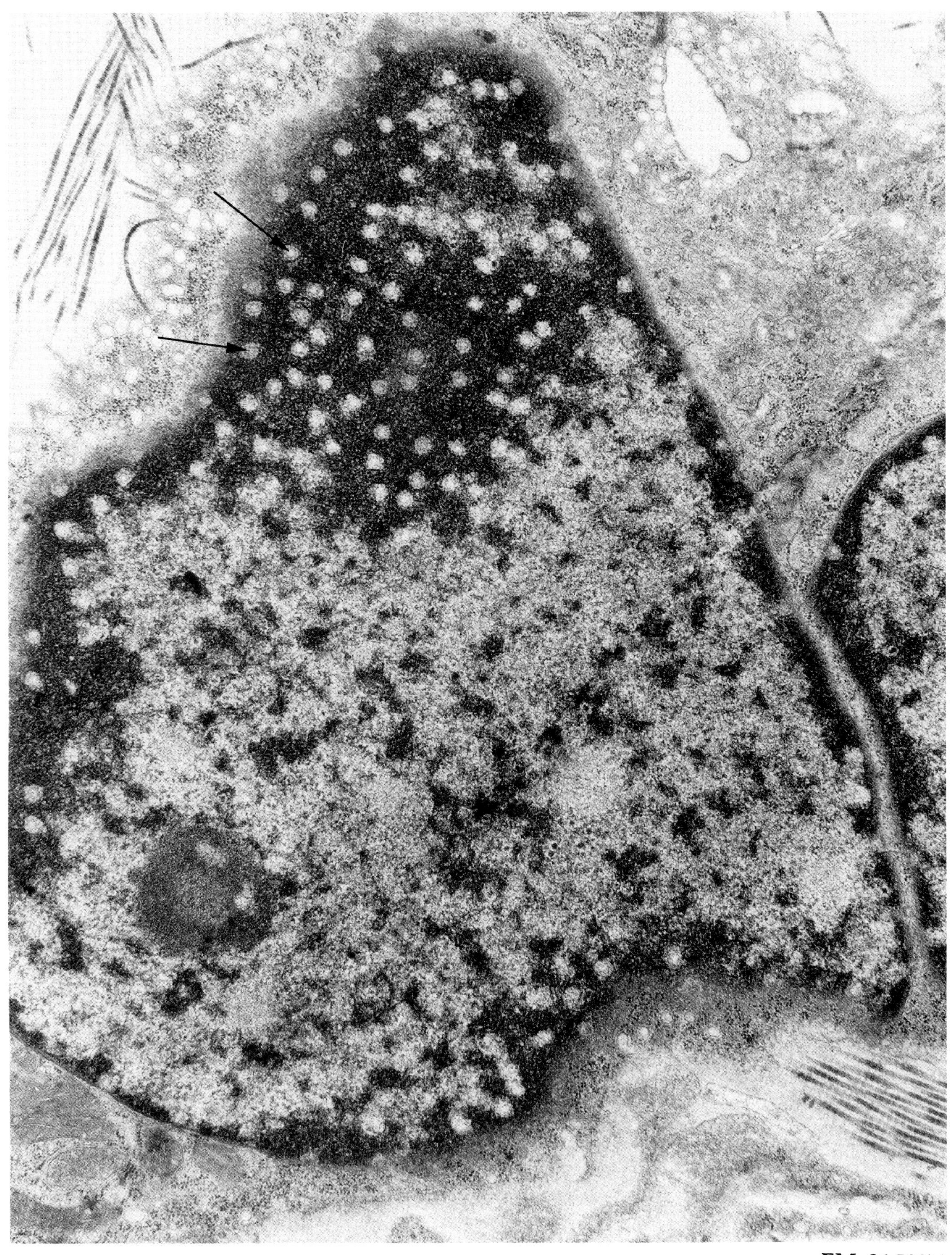

EM: 24,500×

RIBOSOMES

Ribosomes, particles consisting of two-thirds RNA and one-third protein, are the molecular assemblies where all protein synthesis occurs. In micrographs 1 and 2, mature ribosomes, 30 nm in their largest dimension, can be distinguished (arrows). Each ribosome is made up of a large (60 S) and a small (40 S) subunit, which play unique, intricate roles in protein synthesis.

Mature ribosomes are always attached to mRNA and are either initiating or in the process of protein synthesis. Typically a single mRNA attaches to more than one ribosome, creating clusters with each ribosome synthesizing the same protein. These groups, known as **polyribosomes** (or polysomes), are seen particularly well in micrograph 2 (circles). Polyribosomes are found either free in the cytosol or attached to the endoplasmic reticulum (ER), a continuous labyrinth of membrane-enclosed cisternae (arrowheads, micrographs 1 and 2) in the cytoplasm. Isolated ribosomes seen in micrographs are a part of polysome complexes that are not in the plane of the section.

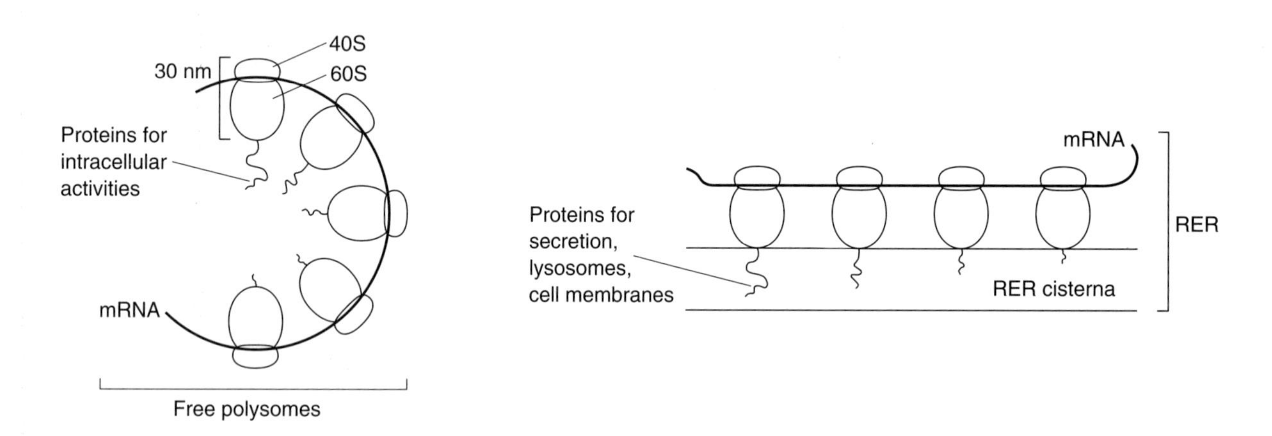

Free and attached ribosomes have the same structure; however, the types of proteins synthesized in the two locations are different. In general, proteins synthesized on polyribosomes free in the cytoplasm are used for activities within the cell, such as cytoplasmic filament formation, while proteins synthesized on polysomes attached to the ER enter the cisternae of the ER, travel to the Golgi apparatus, and are destined for secretion, lysosomes, or plasma, ER, or Golgi membranes. All protein synthesis begins on mRNA/polysome complexes free in the cytoplasm. Proteins are directed to different locations (e.g., nucleus, mitochondria, ER) by specific "signal sequences." Proteins directed to the ER attach to a receptor, and the protein is translocated, as it is being synthesized, into either the ER membrane or the cisternae. Evidence suggests that translocation into the cisternae occurs through protein-conducting channels in the membrane that are opened by the signal peptides.

Endoplasmic reticulum with attached ribosomes is referred to as **rough endoplasmic reticulum** (rough ER). The degree of development of the rough ER as well as its form varies between different cell types, as can be seen by comparing the enzyme-producing pancreatic cell in micrograph 1 and the nerve cell in micrograph 2. In cells that are extremely active in the synthesis of protein for export, such as the pancreatic cells, the rough ER is packed in stacks of flattened cisternae.

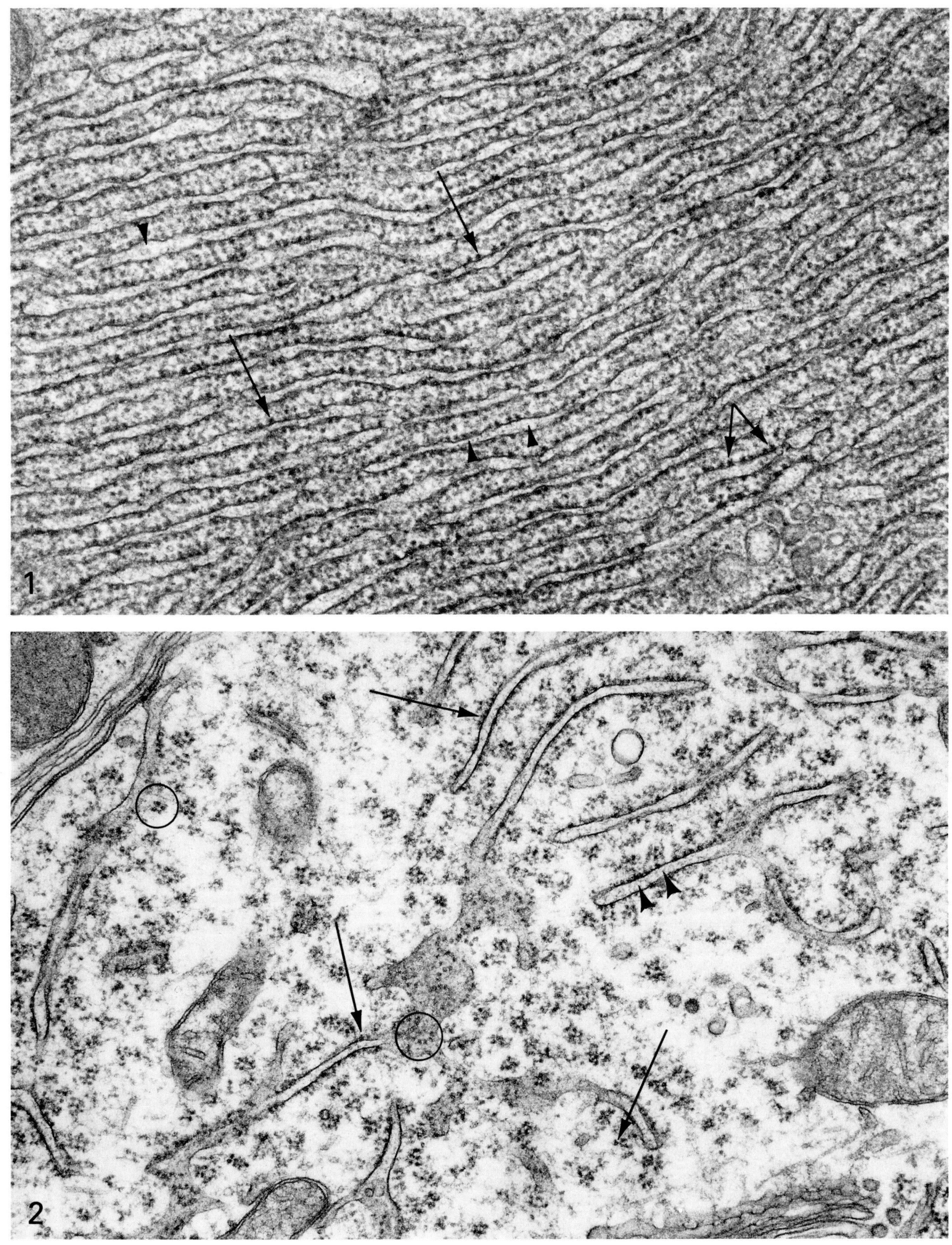

EM 1: 60,000× EM 2: 50,000×

Golgi Apparatus

Proteins synthesized on the rough ER are sequestered in transport vesicles and travel to the **Golgi,** where they are structurally modified and sorted according to their destination. In addition to directing proteins to secretory vesicles, lysosomes, or particular membranes, the Golgi in polarized epithelial cells carry out a more refined sorting to apical versus basolateral plasma membrane domains.

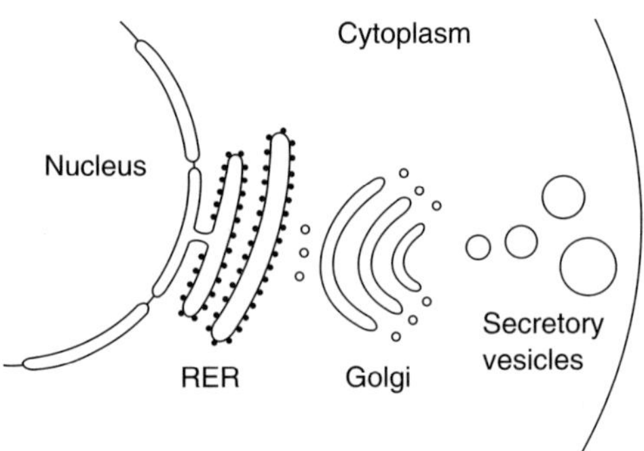

Each Golgi apparatus is made up of **cisternae** and related **vesicles** and is easily recognized on electron micrographs by its unique crescent form with a convex **(cis, entry)** and a concave **(trans, exit)** surface (micrograph 1). The Golgi functions in a highly ordered sequential manner with separate biochemical events occurring initially in the cis, then in the medial, and finally in the trans cisternae. Specialized vesicles transport the protein not only to and from the Golgi complex but also between the cis, medial, and trans regions. Many of these small vesicles (50 – 100 nm) have a fuzzy coat (arrows, micrograph 1) important in the surface changes associated with membrane budding. In some Golgi-associated vesicles this protein is **clathrin,** a coating protein also used in receptor-mediated endocytosis.

Some proteins are glycosylated in the Golgi, but many proteins coming from the rough ER have already been glycosylated. In these cases the Golgi "fine-tunes" the carbohydrate portion in a variety of ways including the addition and removal of residues, phosphorylation, and sulfation. These modifications can determine the destination of a particular glycoprotein. For example, the phosphorylation of a mannose residue at the 6 carbon position tags proteins destined for lysosomes. The mannose is phosphorylated in the cis region of the Golgi and, when the enzyme reaches the trans region, it binds to mannose-6-phosphate receptors and is shuttled into vesicles destined for lysosomes.

Golgi, present in all cells, are most well developed in cells specialized for secretion. In these cells, secretory vesicles (arrowheads, micrograph 1) are formed on the trans surface. Usually the protein secretion becomes concentrated within secretory vesicles and appears more electron dense than it does in the cisternae of the rough ER or Golgi; in some cells, such as the β cells of the pancreas (micrograph 2), the secretion (insulin in this case) actually crystallizes. A crystalline pattern is evident in the vesicles indicated by the arrows.

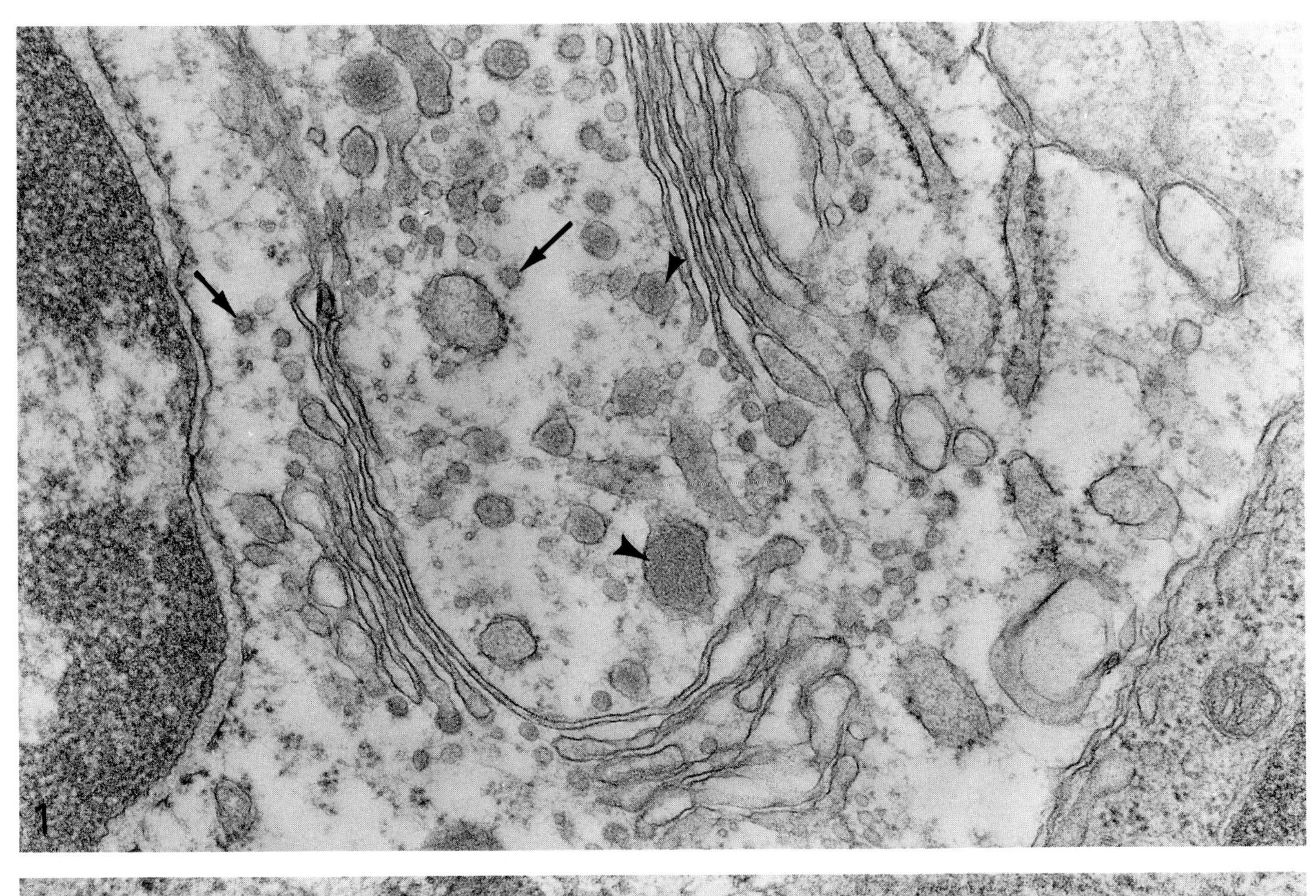

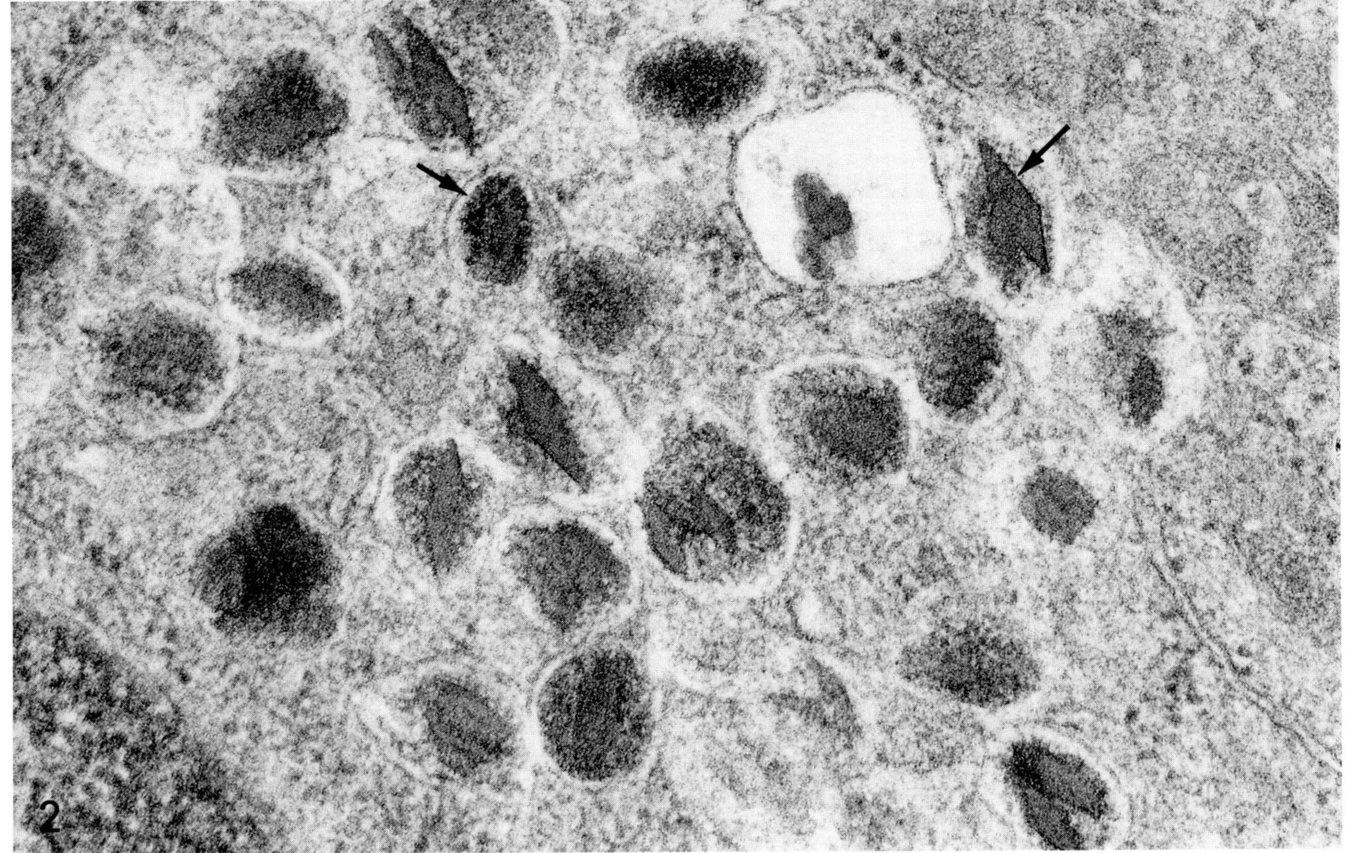

EM 1: 53,000× EM 2: 108,000×

LYSOSOMES: Receptor-Mediated Endocytosis

Lysosomes are the digestive system of the cell. In their simplest form as **primary lysosomes** (arrow, micrograph 1), they are homogeneous, dense, membrane-bound organelles packed with acid hydrolases capable of breaking down polymers of all types. The low pH required for the hydrolase activities (below pH 5) is maintained by a membrane ATP-dependent hydrogen ion pump.

A major function of lysosomes is to process material entering from the extracellular environment before it is released into the cytoplasm. **Receptor-mediated endocytosis** is the most common way for proteins to enter cells. Initially proteins bind to receptors on the plasma membrane in pits that are coated on the cytoplasmic surface with clathrin (arrows, micrograph 2), a protein needed for vesicle budding. After a vesicle is formed, the clathrin coat rapidly disassembles, resulting in naked endosomes. Endosomes, like lysosomes, have an acidic pH, which in many cases acts to separate ligands from their receptors. The receptor is then recycled to the cell surface, and the free ligand is transported in the endosome to a lysosome for final processing.

Cholesterol, carried in low-density lipoprotein (LDL) particles in blood, is taken up by cells as needed via receptor-mediated endocytosis. Defects in the synthesis of the LDL receptor can lead to excess cholesterol in the circulation, resulting in a predisposition to vascular disease.

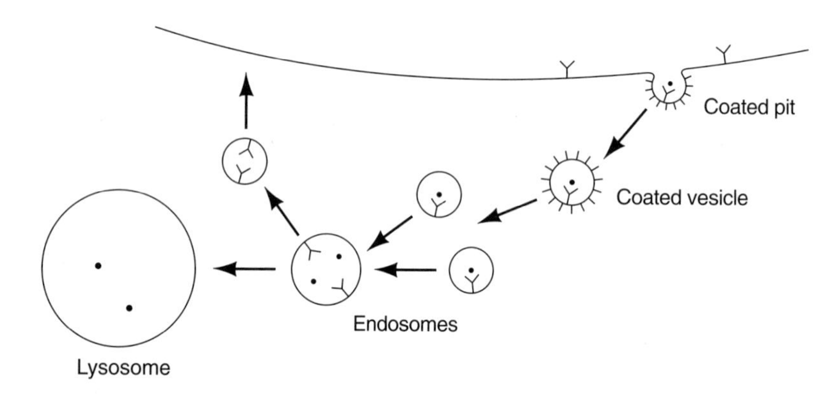

Lysosomes and endosomes play unique roles in different organs. In the kidney proximal tubule cell in micrograph 3, the **coated vesicles** (arrows) collect peptides and small proteins that have been reabsorbed after escaping the filtration barrier in the glomerulus. Transport to lysosomes will result in their breakdown to amino acids that will either be utilized by the cell or released to the circulation.

The endocytotic pathway, so important to the survival of cells, is also traveled by many toxins and viruses bent on destruction. The low pH in lysosomes is essential for the fusion of the viral and lysosomal membranes and release of the virus into the cytosol. Drugs that increase lysosomal pH reduce this type of viral infection.

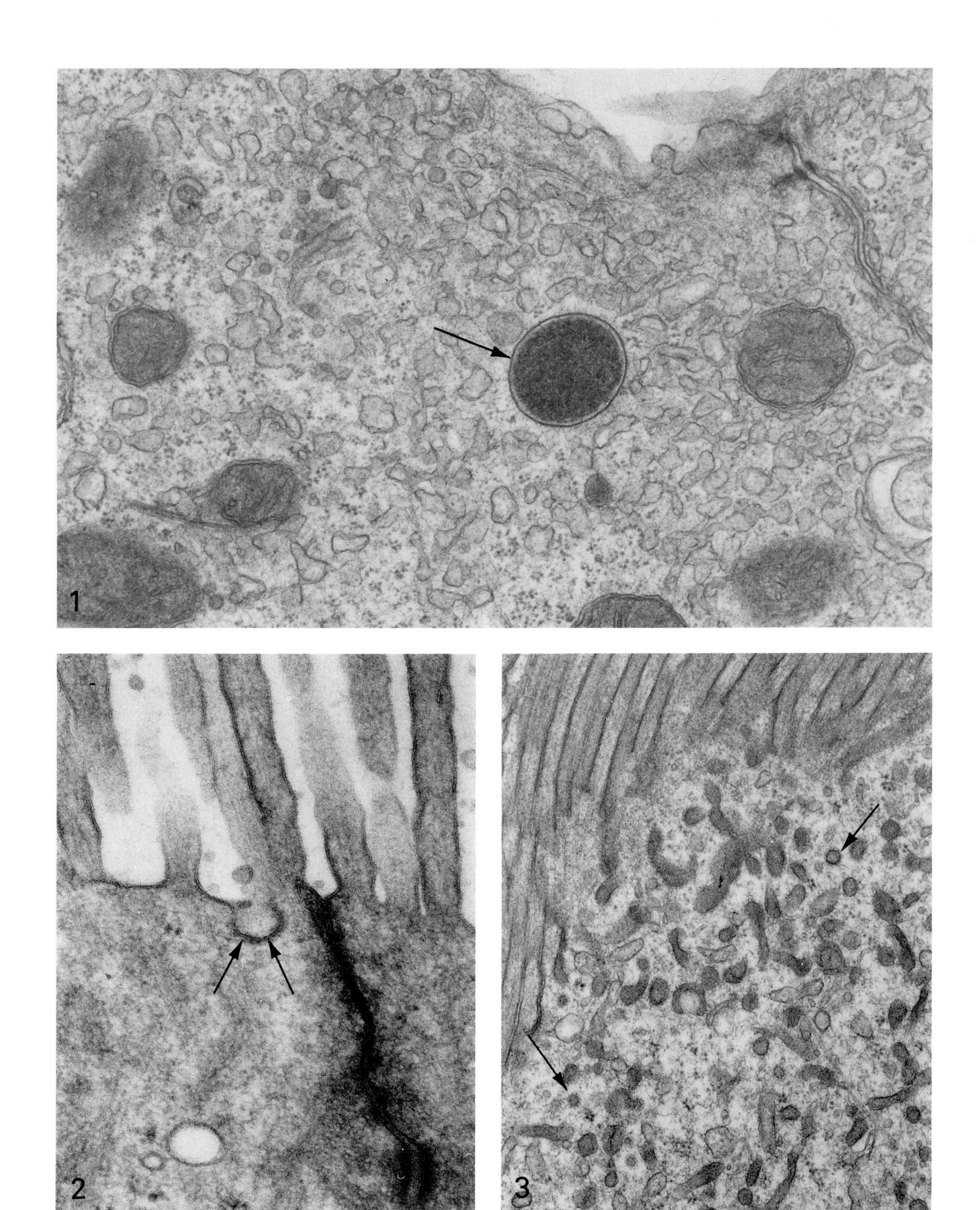

EM 1: 51,800× EM 2: 92,400× EM 3: 36,000×

Lysosomes are particularly well developed in phagocytes such as macrophages. The **macrophage** in micrograph 1 is from the thymus and has ingested four lymphocytes, which are in various stages of digestion. Phagocytosis, a type of endocytosis, includes the formation of a vesicle, in this case a phagosome, which then fuses with a lysosome. Once primary lysosomes have fused with phagosomes and are in the process of digestion, they are called **secondary lysosomes** and are identified easily when they are large and heterogeneous (arrows, micrograph 1). Small **primary lysosomes** are shown in contrast (arrowheads, micrograph 1).

The lysosomes of phagocytes contain acid hydrolases and function in digestion as they do in other cells, but in addition they have the capacity to kill invading organisms and thus play an added critical role in host defense. Molecules produced in the initial immune response, such as antibody and complement, coat the foreign bacteria and cells and also bind to special phagocyte receptors. This binding is essential to the initiation of the events of phagocytosis and triggers not only the formation of the phagosome and increased synthesis of lysosomal enzymes, but also the **respiratory burst,** which is the central event in killing. This burst is an increase in nonmitochondrial O_2 consumption following the activation of a unique membrane-bound reduced nicotinamide adenine dinucleotide phosphate (NADPH) oxidase. In contrast to cytochrome oxidase in mitochondria, this oxidase only partially reduces oxygen, leading to the formation of highly reactive microbicidal oxygen species (O_2^-, H_2O_2, $HO\cdot$, HOCl).

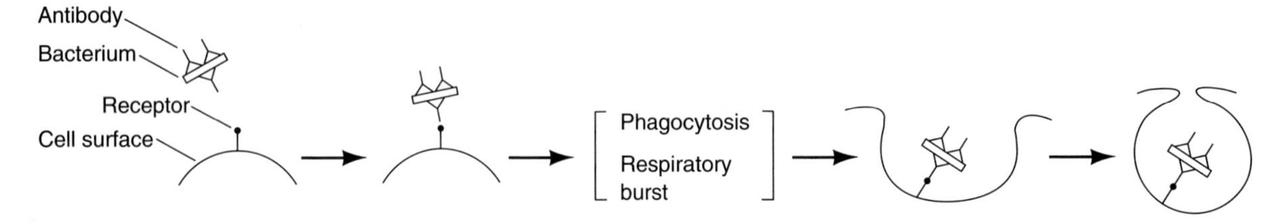

Intracellular digestion results in either complete breakdown of phagocytosed material or indigestible residues that persist in the form of heterogeneous **residual bodies** (outlined by arrows in micrographs 2 and 3). These bodies are variable in form and include a brownish pigment that is referred to by light microscopists as lipochrome or lipofuscin. Residual bodies sometimes consist of tightly packed membranes, reflecting the difficulty lysosomal enzymes have with lipid digestion. The residual body in micrograph 3 from a kidney cell is the end result of autophagy rather than phagocytosis. The swirls of lipid are the remains of the membranes of aged organelles. All cells use lysosomes for the removal of cellular components as they undergo routine turnover.

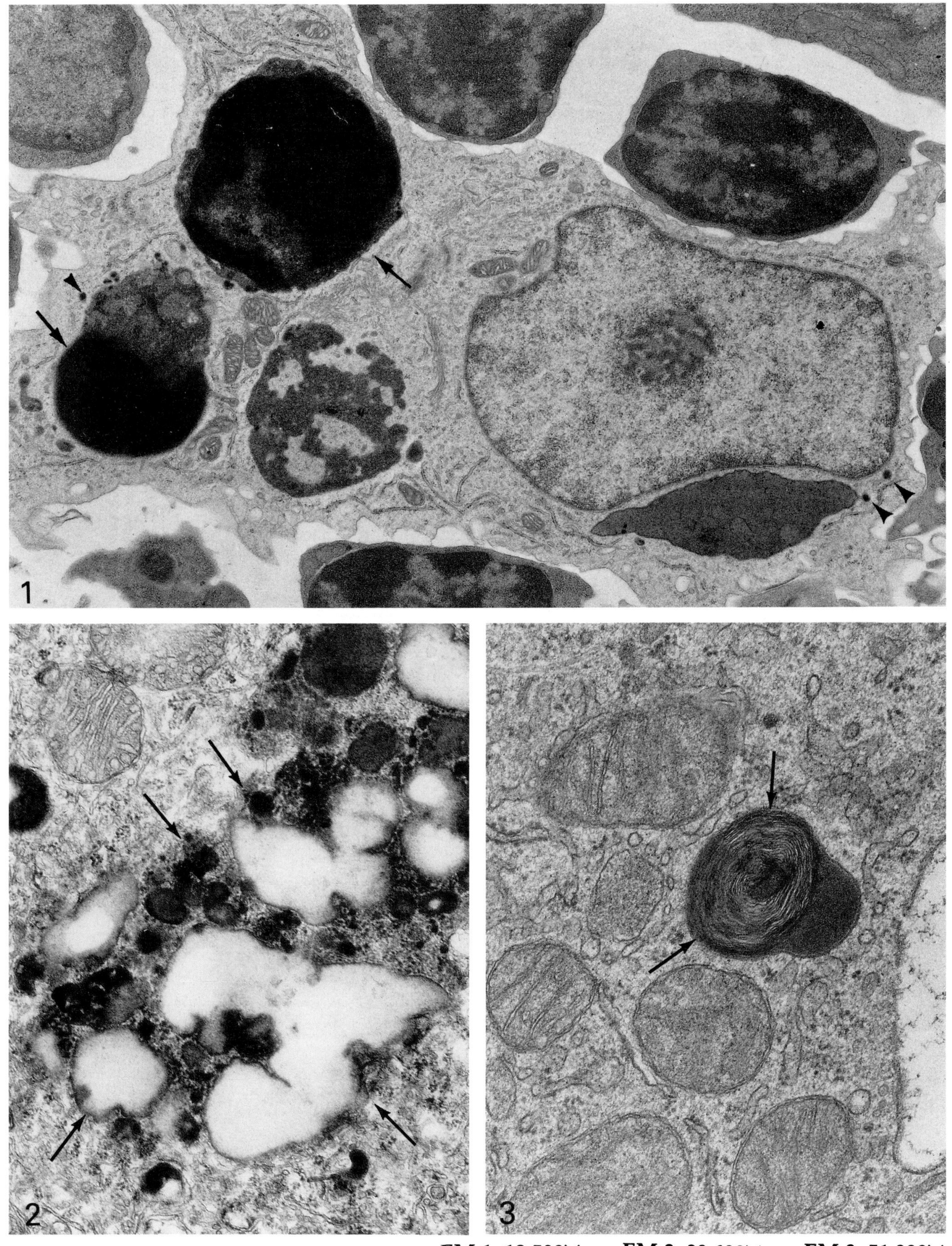

EM 1: 13,500× EM 2: 30,600× EM 3: 51,000×

Smooth Endoplasmic Reticulum

Smooth endoplasmic reticulum (smooth ER), even though directly continuous with rough ER, has a completely different form and function. Instead of being primarily arranged in cisternae as in the case of rough ER, smooth ER consists of a ribosome-free network of fine tubules seen longitudinally in micrograph 1 (arrows) and in cross section in micrograph 2 (arrows). The biochemical reactions mediated by smooth ER membranes are diverse, as illustrated by the examples described below, ranging from the synthesis of steroid hormones to the detoxification of drugs.

Steroid hormone synthesis is a result of a series of reactions catalyzed by enzymes in the smooth ER and mitochondria, with intermediates shuttling back and forth between these two organelles. The synthesis of the precursor, cholesterol, and the final steps in the elaboration of the steroids occur in the smooth ER. Frequently, in steroid-producing cells such as the Leydig cell in micrograph 1, large lipid droplets of cholesterol are intimately surrounded by smooth ER. Even though cells synthesizing steroid hormones do not store the final hormone (testosterone in this case), they frequently store the cholesterol precursor.

Another major function of SER is **detoxification.** In the liver cell, smooth ER alters the biological activities of many exogenous compounds and is thus a major means of protection against chemical insult. Certain drugs are inactivated by hydroxylation in the presence of cytochrome P-450, a mixed-function oxidase associated with the smooth ER. The hydroxylation increases the drugs' solubility and thus facilitates their excretion by the kidney. The liver cell in micrograph 2 is from an animal treated with daily doses of phenobarbital. Such treatment induces a fourfold to fivefold increase in liver cytochrome P-450 activity, which reflects de novo synthesis of the enzyme. At a morphological level this induction is observed as a significant proliferation of the smooth ER. With increased smooth ER and cytochrome P-450, phenobarbital is inactivated more rapidly and a tolerance develops to both this drug and other similar drugs metabolized by this system.

The hepatic hydroxylation system is significant in another important respect: many potentially carcinogenic compounds are changed by hydroxylation into active carcinogens.

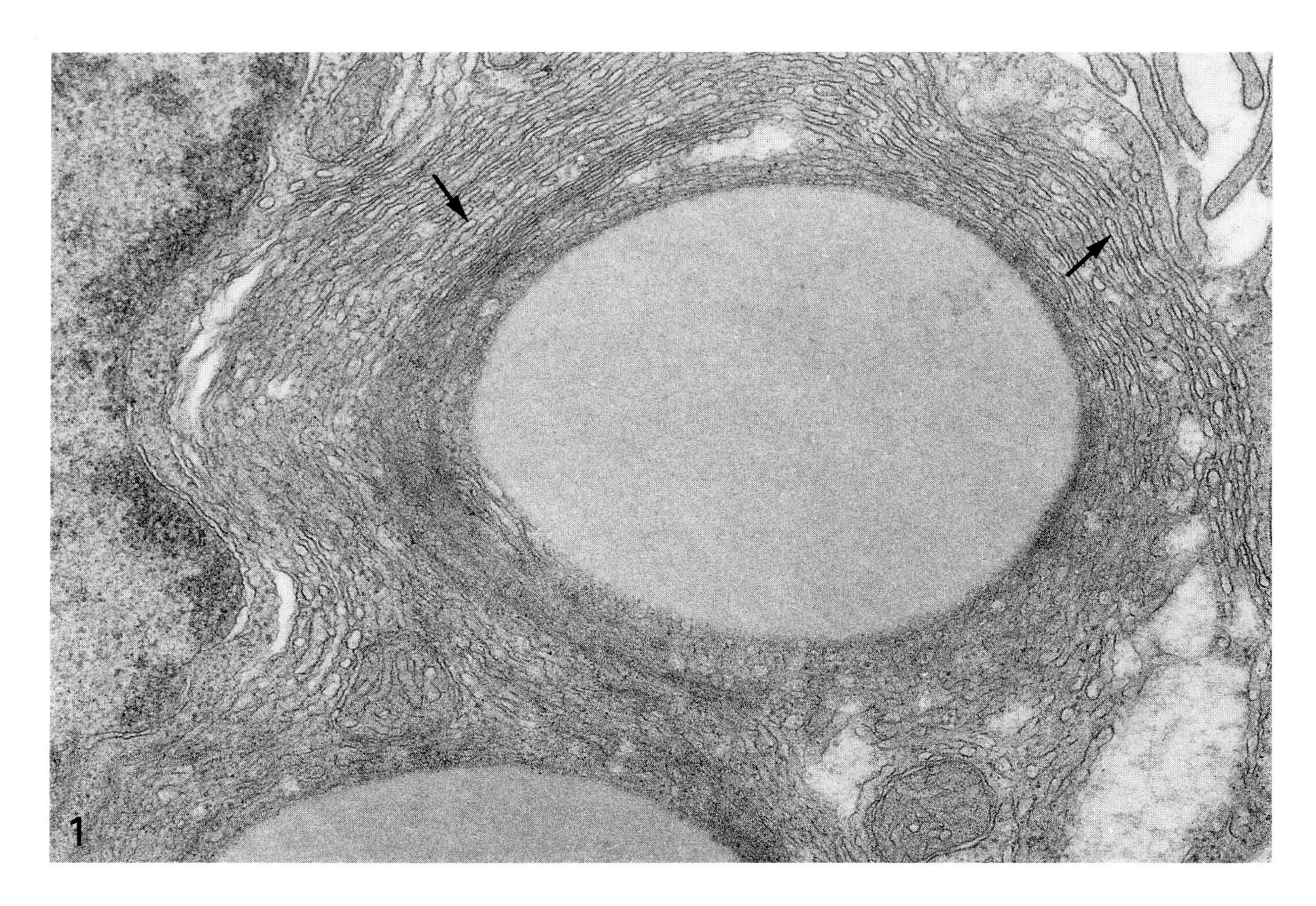

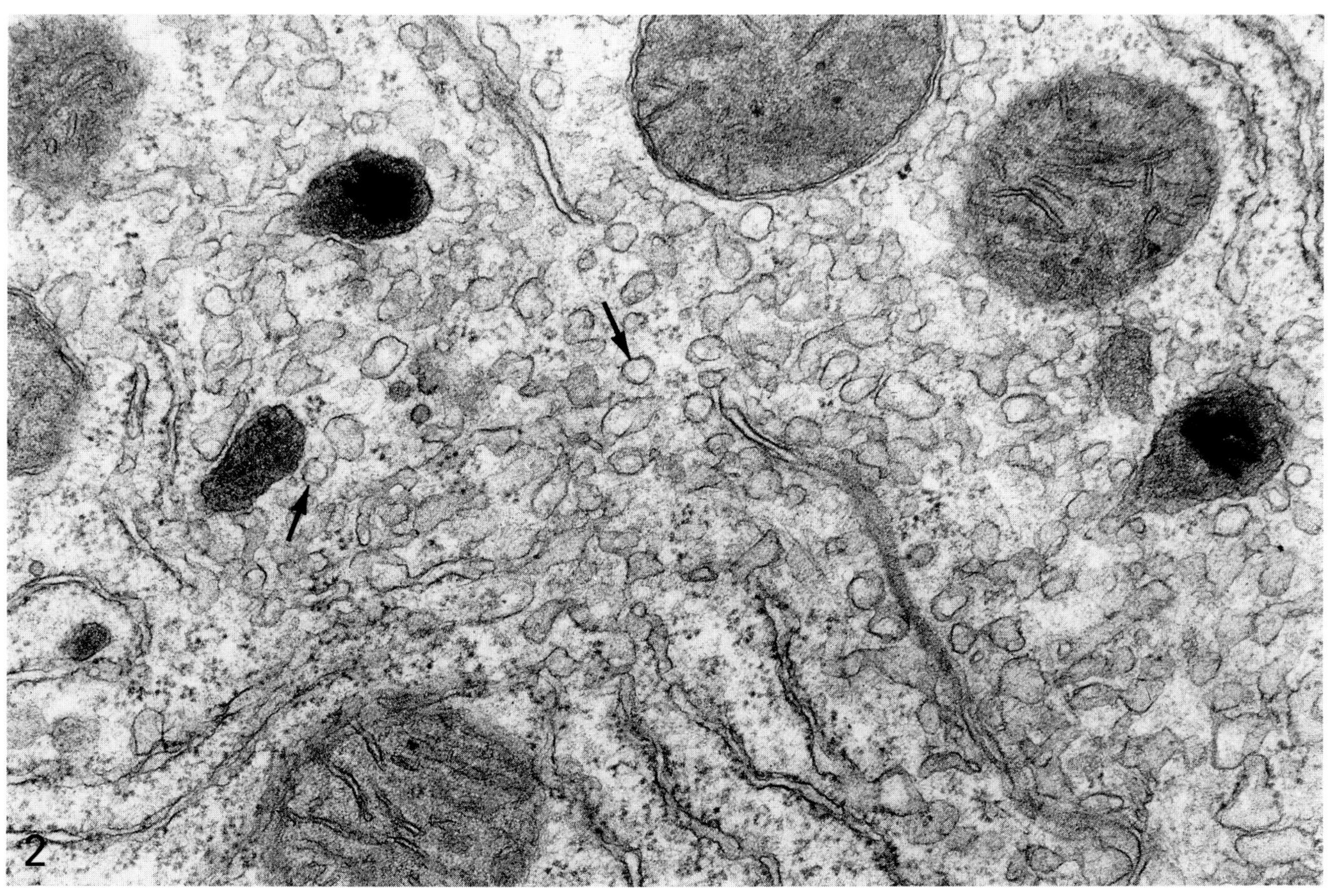

EM 1: 35,000× **EM 2: 50,500×**

PEROXISOMES

Peroxisomes (arrows) are recognized in many species by the presence of a central **crystalline core** of urate oxidase, as seen in the peroxisomes from rat liver (micrograph 1) and rat brain (inset). In humans, this crystalline core is not present since urate, a product of purine breakdown, is excreted as such rather than being degraded further prior to excretion. Therefore, transmission electron micrographs do not distinguish between lysosomes and peroxisomes in human tissue. Both are present as membrane-bound organelles packed with electron-dense enzymes. Histochemistry clearly demonstrates, however, that lysosomal and peroxisomal enzyme content are different.

Oxidases are the most prevalent and characteristic components of peroxisomes. They catalyze many diverse reactions, including the oxidation (to acetyl-CoA) of very long chain saturated fatty acids that are not handled well by mitochondria. In the liver cell (micrograph 1) the activity of various oxidases can account for as much as 20% of oxygen consumption. Many oxidative reactions in peroxisomes lead to the formation of H_2O_2. This relatively toxic byproduct is removed by peroxisomal catalase, thus protecting the cell from potential damage. The detoxification of alcohol is frequently coupled to this action of catalase.

Peroxisomes are particularly abundant in the liver. This may reflect a role in cholesterol metabolism and gluconeogenesis in this organ. In other locations, their enzyme content reflects still other functions, such as (1) the synthesis of plasmalogens, phospholipids abundant in central nervous system myelin, and (2) the synthesis of complex lipids by sebaceous glands in the skin.

The importance of peroxisomal activity to normal cell metabolism becomes obvious by looking at the severe effects of diseases in which these organelles are missing. Patients with Zellweger syndrome, an autosomal recessive disease, lack peroxisomes, accumulate very long chain fatty acids, and die in early childhood, due in part to a deficiency of plasmalogens, which leads to neurological disorders.

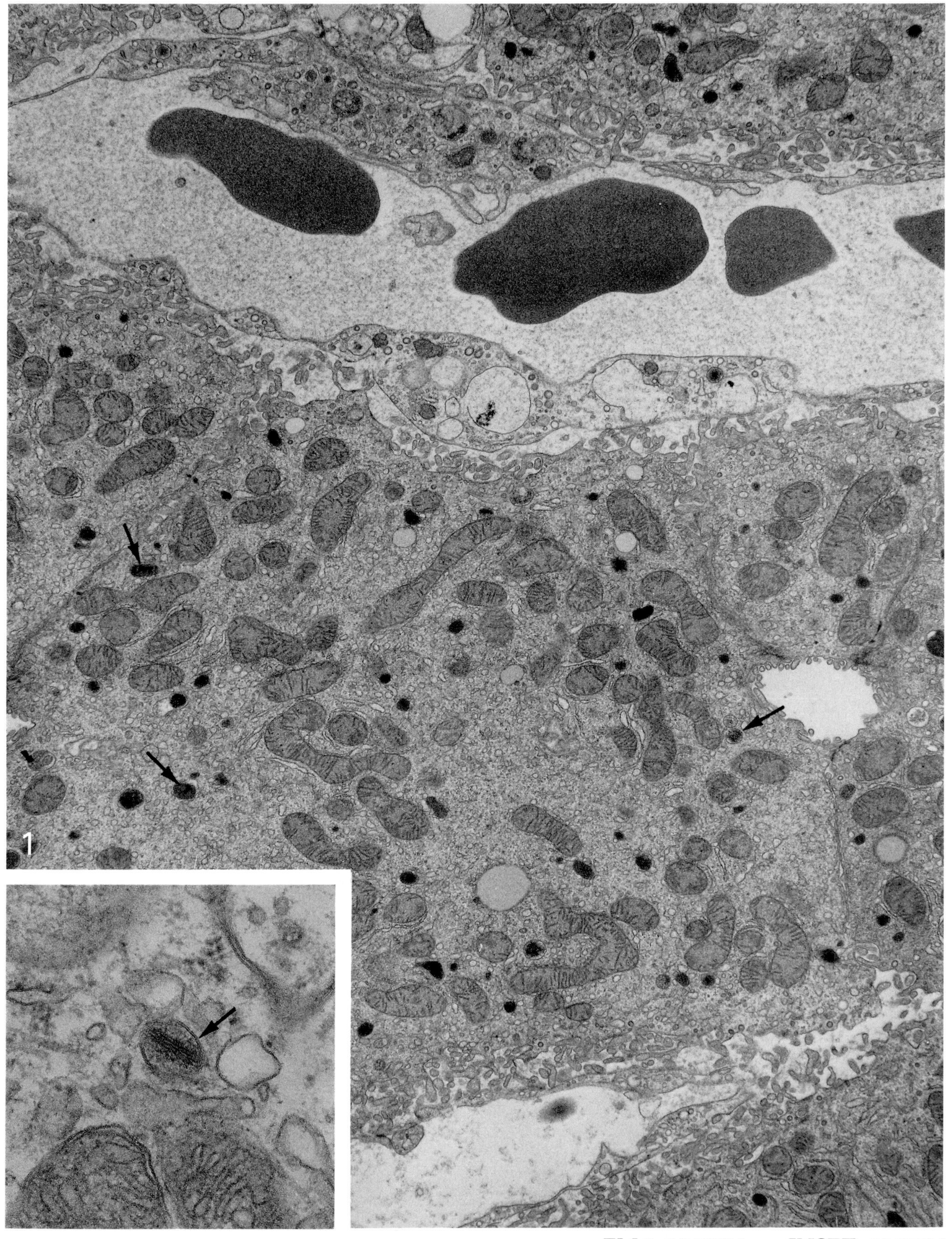

EM 1: 10,200× INSET: 66,875×

MITOCHONDRIA

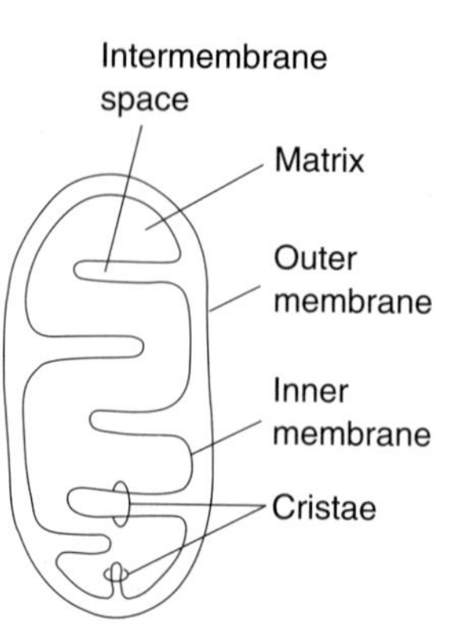

Mitochondria provide cells with energy in the form of ATP. They are defined by two limiting membranes. The inner membrane forms folds or invaginations called cristae, which project into the interior of the organelle. **Cristae** may be shelflike, as in micrograph 1, or tubular, as in the steroid-secreting cell seen in micrograph 2. Careful examination of the mitochondrion in micrograph 1 reveals the continuity (arrowheads) between the cristae and the inner membrane. The size and shape of mitochondria vary considerably (compare the three micrographs); even within one cell, mitochondria move, change shape, divide, and fuse.

In the **matrix** of the mitochondria, pyruvate and fatty acids are converted to acetyl-CoA, which is fed into the tricarboxylic acid cycle. The resulting reduced nicotinamide adenine dinucleotide (NADH) and flavin adenine dinucleotide ($FADH_2$) become the main source of electrons for the respiratory chain enzymes, which are a part of the **inner membrane.** The electron-transfer reactions of the respiratory chain enzymes are coupled to the formation of an electrochemical proton gradient across the inner membrane, which drives ATP formation. The inner mitochondrial membrane has one of the highest concentrations of proteins of any cell membrane. In addition to the respiratory chain and ATP synthase enzymes, membrane proteins include channels that regulate metabolite exchange. Very specific permeability characteristics of the inner membrane are essential in order to maintain the electrochemical gradient. In contrast, the outer membrane allows most small molecules ($<10{,}000$ daltons) to enter the intermembrane space.

Dark **matrix granules** (arrows) are scattered in the relatively dense matrix of the mitochondria in micrograph 3. Evidence suggests that these represent sites of accumulation of divalent cations in a nonionized form. Mitochondria may sequester calcium in this form and contribute to the extremely low intracellular concentration of this ion.

Even though not resolved in these micrographs, the mitochondrial matrix also contains DNA, ribosomes, and other components needed for localized protein synthesis. The mitochondrial DNA genome, in addition to encoding ribosomal and transfer RNAs, encodes for certain cytochrome subunits of the electron transport chain. Most mitochondrial proteins, however, are encoded by nuclear DNA and synthesized on cytoplasmic polysomes. These proteins contain a signal sequence tagging them for transport into mitochondria. Both mitochondria and cytoplasmic protein synthesis are necessary for the formation of new mitochondria from preexisting mitochondria.

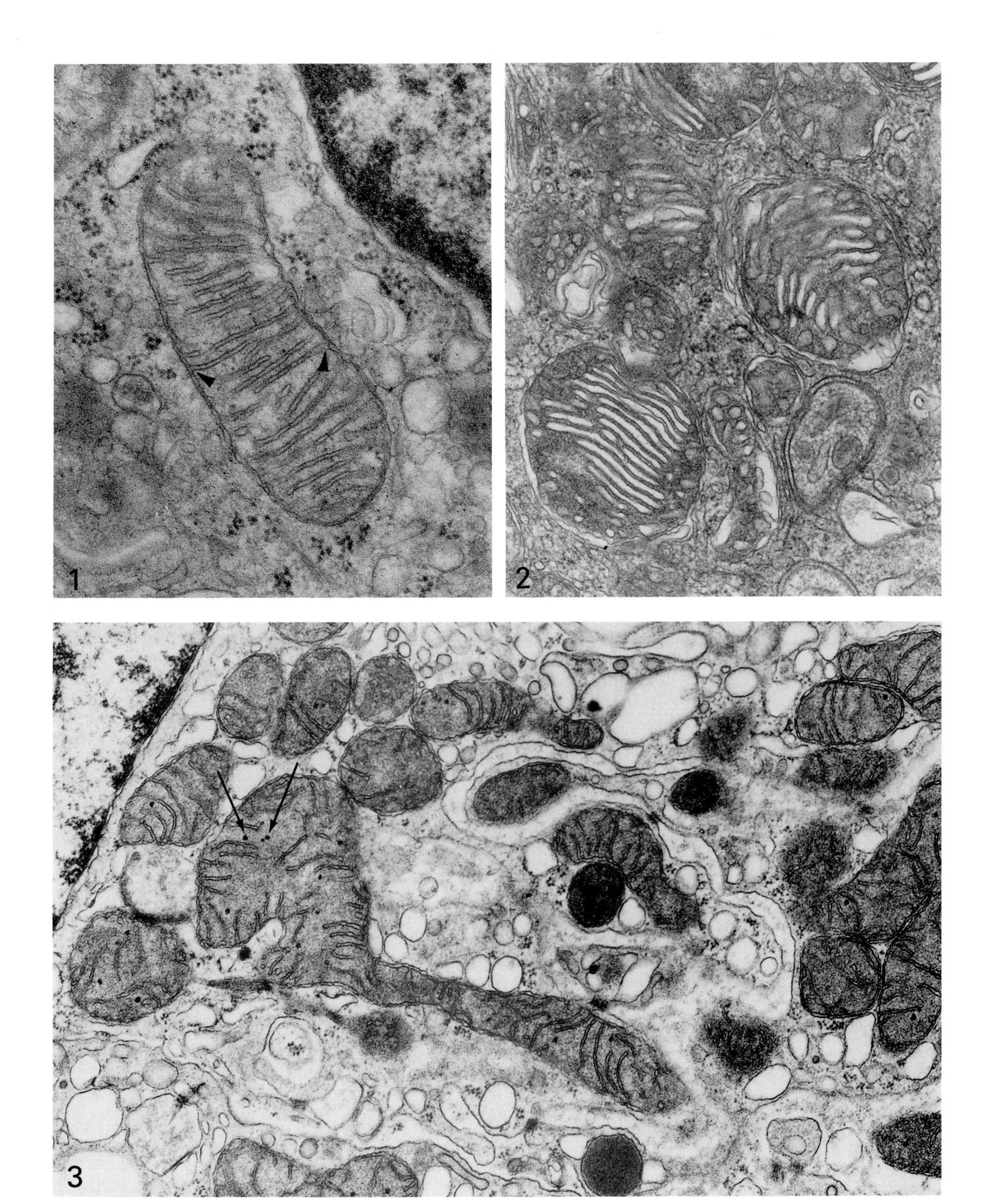

EM 1: 60,000× EM 2: 28,600× EM 3: 31,200×

Glycogen and Lipids

Glycogen, a storage form of **glucose,** is found predominantly in liver and muscle cells as cytoplasmic granules ranging in size from 10 to 40 nm. In liver cells, as shown in micrograph 1, these granules group together to form **rosettes** (arrows), frequently in close proximity to smooth endoplasmic reticulum (arrowheads). Liver smooth ER contains glucose-6-phosphatase, an enzyme that removes phosphate, permitting glucose to leave the cell and enter the circulation for distribution to other tissues and organs as a major energy source.

Triacylglycerols, a storage form of fatty acids, are stored as isolated non-membrane-bound lipid droplets. In micrograph 2 of a liver cell, a **lipid droplet** (l) is wrapped by a mitochondrion, reflecting the intimate functional relationship between these two structures. Fatty acids released from the lipid-droplet triacylglycerols are first combined with coenzyme A (CoA) in the outer mitochondrial membrane. They are then transported across the inner mitochondrial membrane into the matrix, where they are oxidized to acetyl-CoA, which enters the tricarboxylic acid (TCA) cycle.

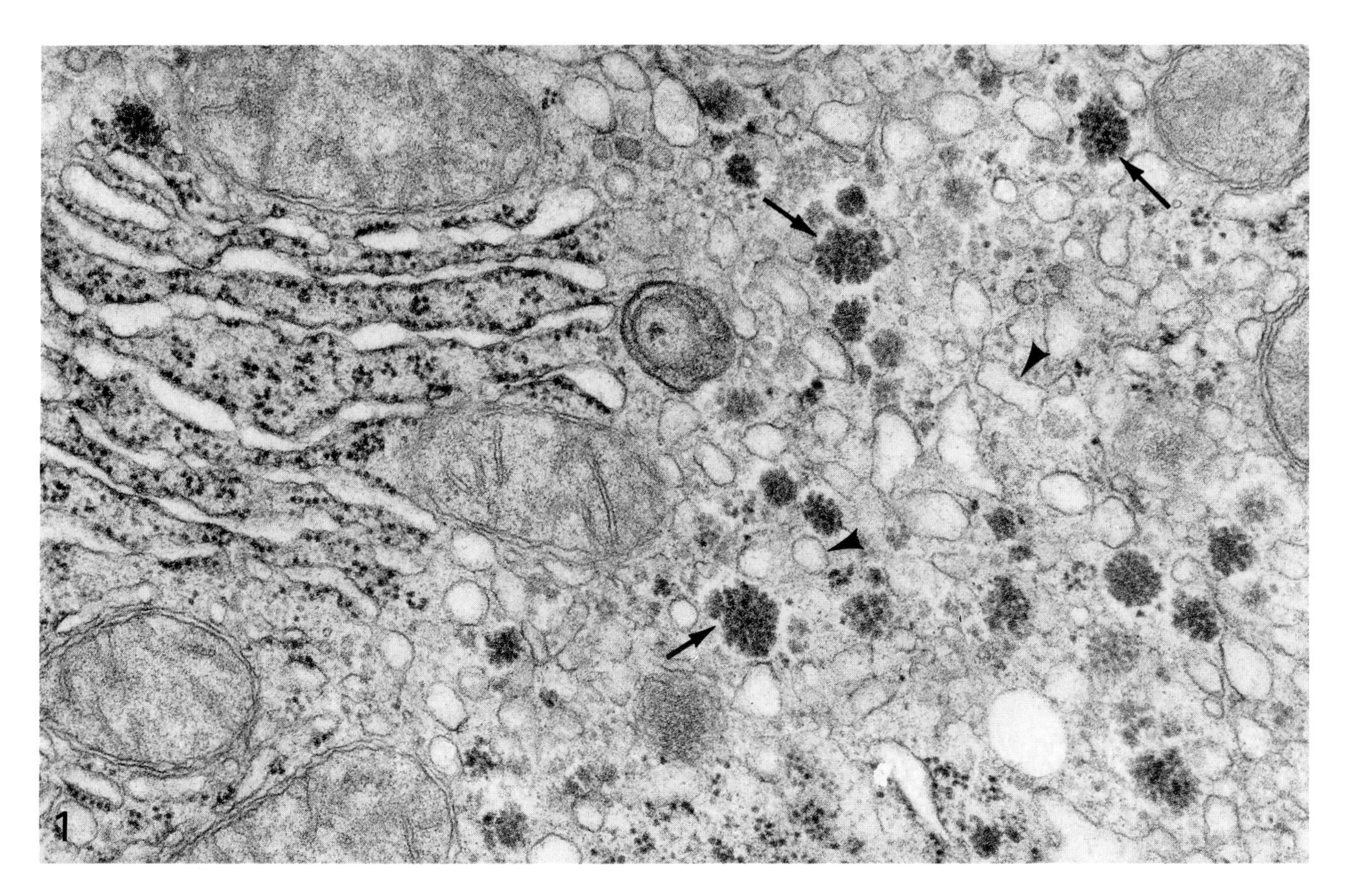

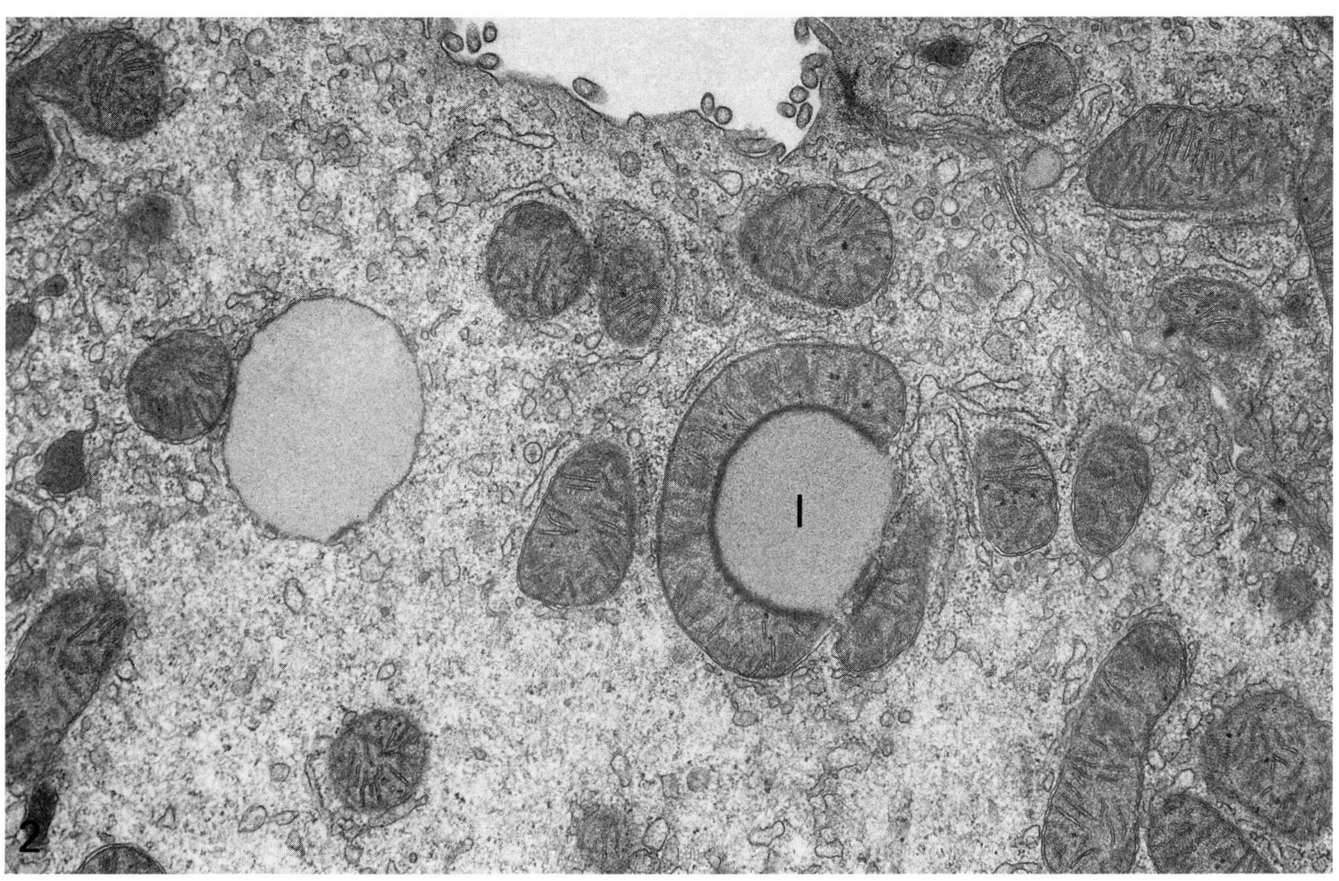

EM 1: 65,600× EM 2: 20,500×

Actin filaments, ubiquitous cytoskeletal proteins, are important in the structural support and movement of cells. Each actin filament (F-actin) consists of two strands of globular subunits (G-actin) wrapped around each other to form a polarized unit. The globular substructure of actin filaments, not seen in routine thin sections, is clearly illustrated in the negative-stained preparation of isolated filaments in micrograph 1 (courtesy of Dr. J. Spudich; see References, Uyemura and Spudich). At the ionic condition within cells, most actin is in the filamentous form.

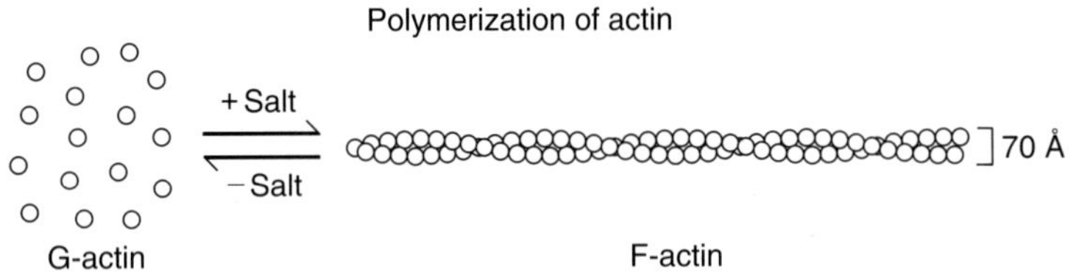

Polymerization of actin

+ Salt

− Salt

70 Å

G-actin

F-actin

Courtesy of Dr. J. Spudich.

Actin filaments within all cell types bind to a variety of accessory proteins, including those that prevent or enhance lengthening, those that break filaments, and those that bind filaments together and to other structures, particularly membranes. In the smooth muscle cells in micrograph 2, actin filaments, evident throughout the muscle cytoplasm (M), are part of contractile machinery that includes other proteins such as myosin, tropomyosin, and calmodulin (a calcium-binding enzyme regulator).

In the intestinal lining cells in micrograph 3, actin is concentrated in the luminal side of the cell as a component of the terminal web (arrowheads) and as the core (arrows) of cell surface projections, microvilli. Filamentous actin and its associated proteins (Epithelium, page 60) provide structural support throughout the short lifetime of these cells.

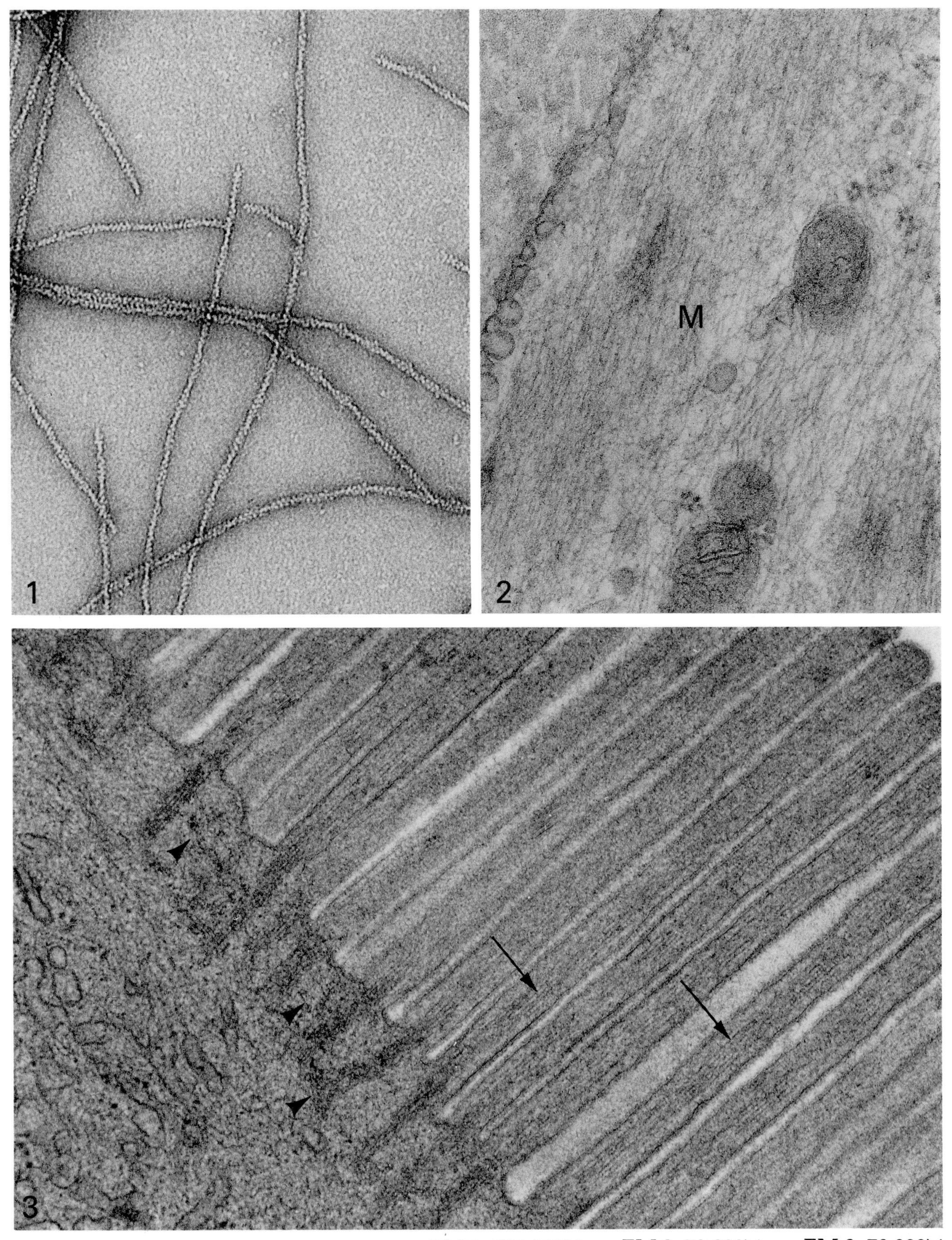

EM 1: 225,000× EM 2: 72,000× EM 3: 70,000×

All cells are involved in **migration** at some point in their development or as a part of their mature function. Both the attachment and movement phases of migration are mediated by actin filaments. Actin is well suited to a role in cell migration since it can change form relatively rapidly.

When fibroblasts attach to a surface, membrane receptors at the site of attachment trigger a cascade of events. The network of actin that extends into the processes, such as the filopodia (arrows) in micrograph 1, associates with the cell membrane at defined areas called **focal adhesion plaques,** which are rich in many actin-binding proteins such as vinculin and α-actinin. Cross-links occur between actin filaments to form rigid **stress fibers.** In addition to these intracellular events, surface attachment in the plaque region involves the association of integrins, unique integral membrane proteins, with fibronectin, a large extracellular glycoprotein. An intact sequence from intracellular actin to extracellular fibronectin is necessary for adhesion.

Actin is essential to the **movement** phase of migration as well as the attachment phase. Cytochalasin B, a fungal alkaloid known to inhibit actin polymerization, stops fibroblast movement. What changes the actin filament undergoes during movement and how (if at all) myosin is involved are topics of current investigation. Myosin is found associated with the leading-edge membrane in migrating cells.

Classical stress fibers are not formed in very active cells, such as amoebae, even though when they attach to a surface, actin does aggregate and bind to the cell membrane. Aggregated actin (arrowheads) is evident at low (micrograph 2) and high (micrograph 3) magnifications in migrating Dictyostelium treated to remove most cellular structures. Actin aggregation may relate to the change from the sol to gel state in pseudopods. (Micrographs reproduced courtesy of Dr. A. Spudich.)

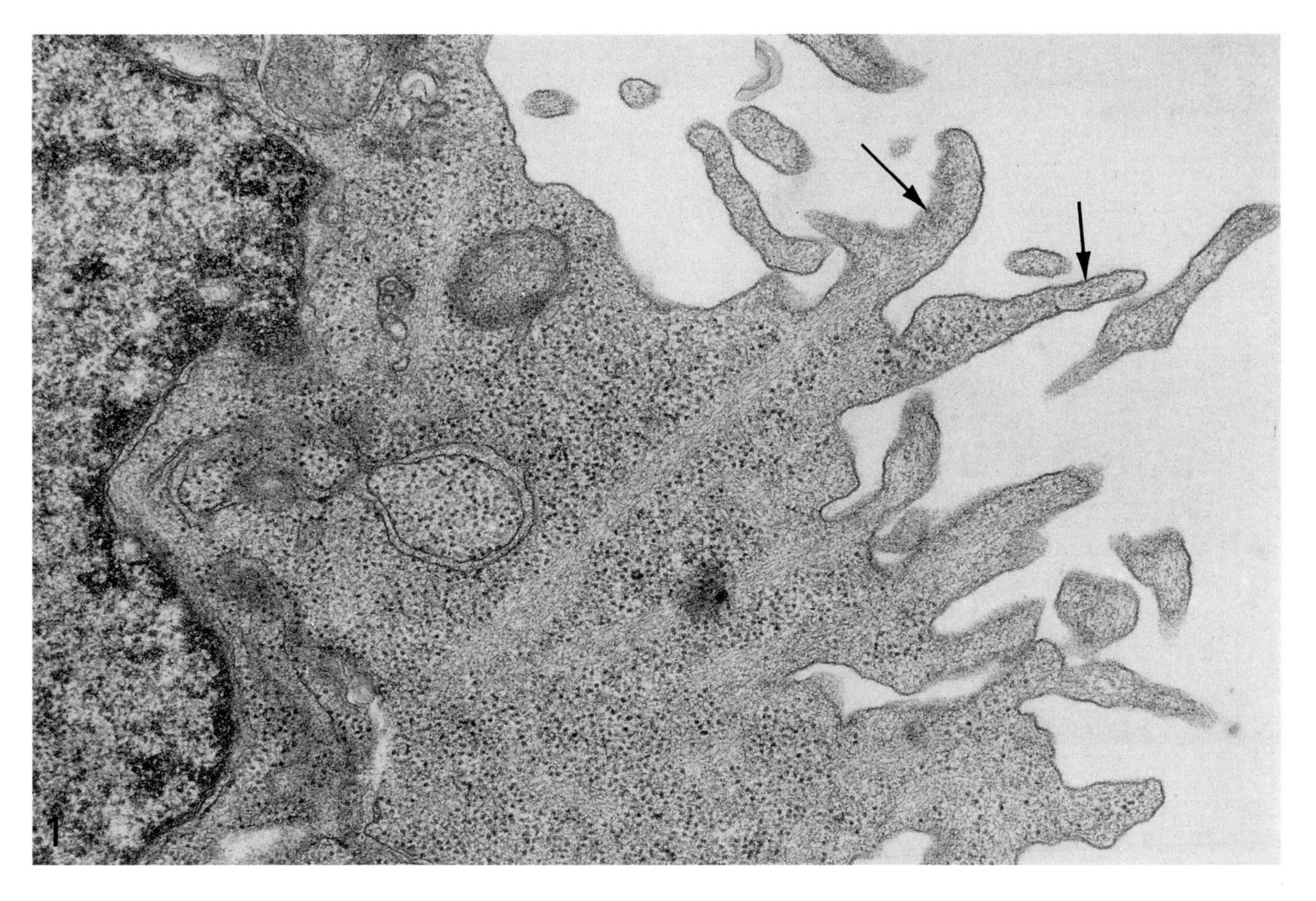

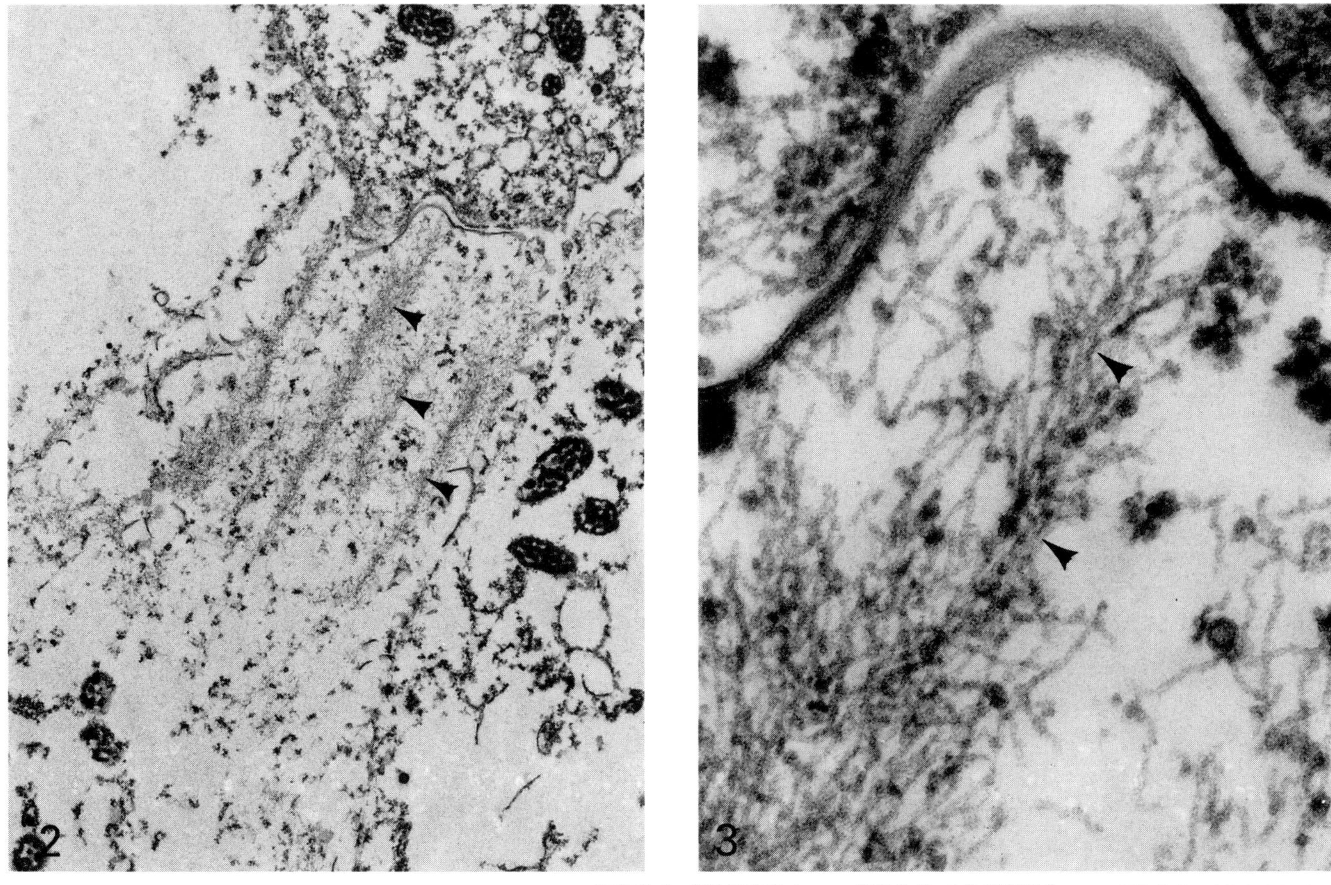

EM 1: 37,500× EM 2: 19,500× EM 3: 126,000×

CYTOSKELETON: Microtubules, Spindle Apparatus

Microtubules, like actin filaments, are highly conserved, polarized, cytoskeletal elements important in cell support and movement. Also, like actin, the elongated form is in equilibrium with globular subunits. Within each microtubule the subunits (α and β **tubulin**) are lined up in longitudinal rows to form 13 protofilaments. Protofilaments are helically arranged as hollow cylinders, usually several microns long, with a diameter of 20 to 25 nm. Microtubules have a slow-growing end (−) attached to an organizing region and a fast-growing end (+) farthest from the cell center.

Microtubules are the major component of the **spindle apparatus.** In the dividing cell shown in micrograph 1, microtubules (arrowheads) radiate out from the centrosome or cell center. The centrosome includes a pair of **centrioles** (curved arrows, micrograph 1), each a cylinder consisting of nine triplets of parallel microtubules (seen in cross section in micrograph 3), and closely associated electron-dense regions, the **microtubule organizing centers** (MTOCs, arrows in micrograph 2). The actual nucleation of microtubules occurs at the MTOCs and not the centriole, but the centriole does seem essential for functioning of the MTOCs.

Spindle and centriole microtubules have the same fundamental structure but differ in at least two respects: (1) spindle but not centriole microtubules are sensitive to colchicine, a drug that prevents the addition of tubulin subunits; (2) centriole and spindle microtubules have different microtubule-associated proteins (MAPs).

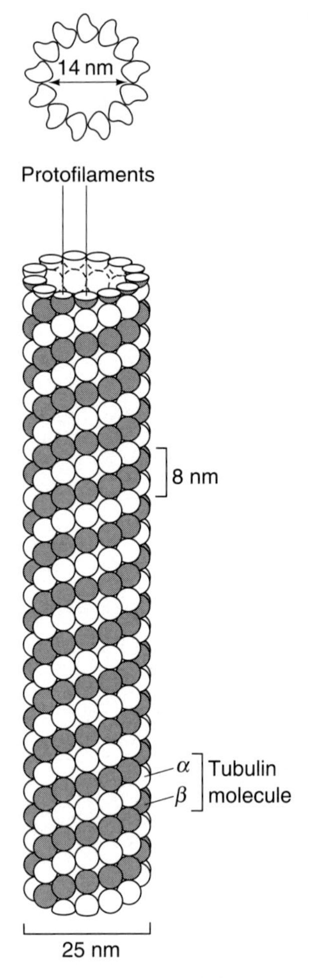

Modified from B. Alberts et al., *Molecular Biology of the Cell,* Garland, New York, 1989.

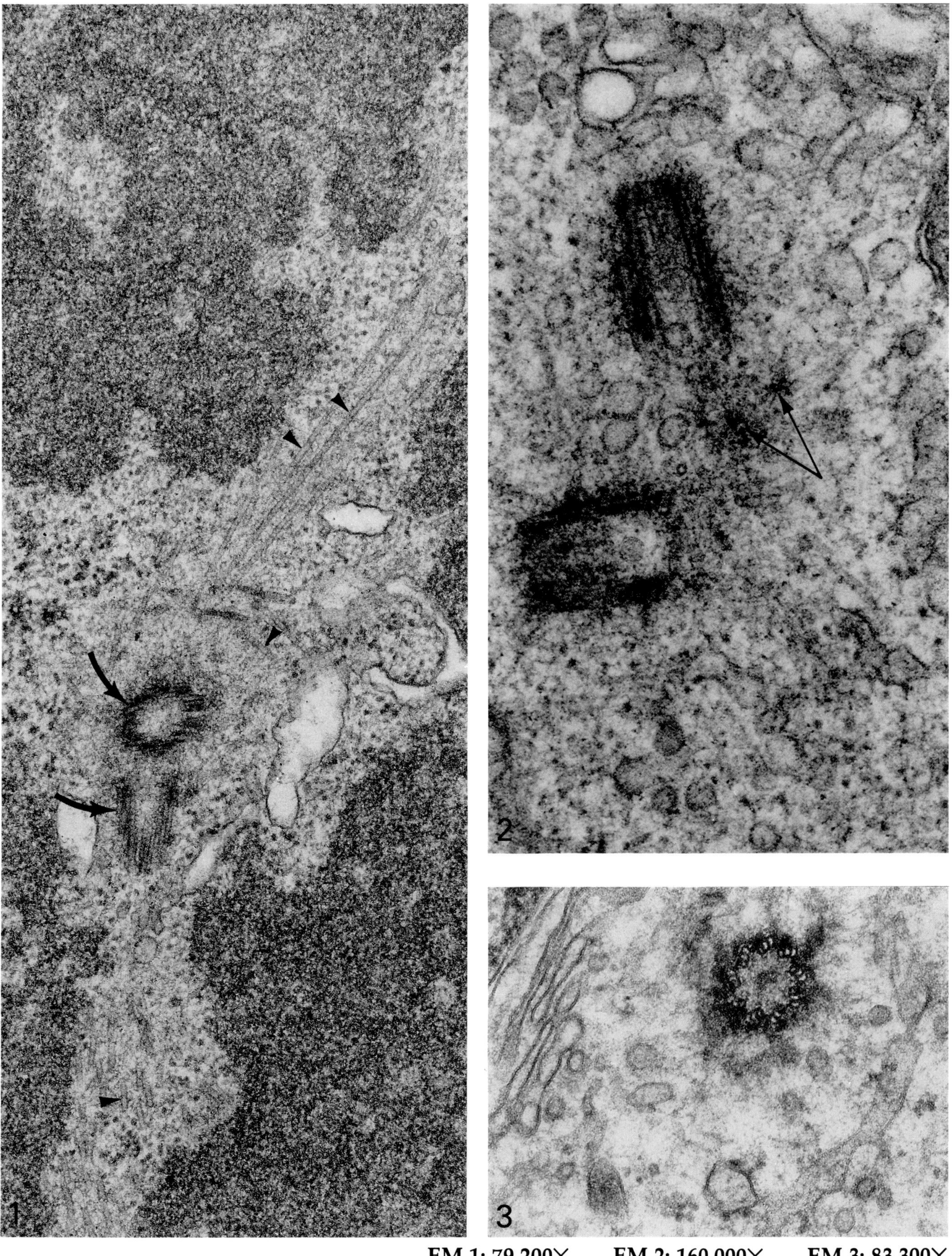

EM 1: 79,200× **EM 2: 160,000×** **EM 3: 83,300×**

The alignment of chromosomes at the metaphase plate (micrograph, frog intestinal tract) and subsequent separation at anaphase depend upon complex microtubule dynamics. Prior to mitosis, each centriole duplicates, forming a daughter at right angles. As each centriole pair moves to opposite poles of the cell, **spindle microtubules** grow out from the cell center in all directions. Some attach to the kinetochores, specialized regions of each sister chromatid in the constricted centromeric region. Other microtubules extend from the poles to the midregions of the spindle where they overlap. Some microtubules do not orient specifically with the chromosomes but instead are relatively short and radiate in all directions from each pole. All three types of spindle microtubules, **kinetochore, polar,** and **astral,** are formed initially at the MTOCs of the centrosome.

During cell division, spindle microtubules are polarized with the minus end at the poles and the plus end at the kinetochore and the region of overlap. The plus sites at the kinetochore are the primary regions of the depolymerization and polymerization that are responsible for the movement of chromosomes during the alignment at the metaphase plate and the movement to the poles.

At the same time that the chromosomes are moving toward the poles (anaphase A), the entire spindle is lengthening as the poles become farther apart (anaphase B). Anaphase B involves a polymerization at the plus region of polar fiber overlay and a sliding between adjacent microtubules.

Just as the myosin motor moves actin, other protein motors move microtubules. The direction of the force relates to the polarity of the microtubules. Motors, closely related to those that drive cilia movement (dyneins) and organelle transport to nerve terminals (kinesins), are essential to the spindle dynamics of mitosis.

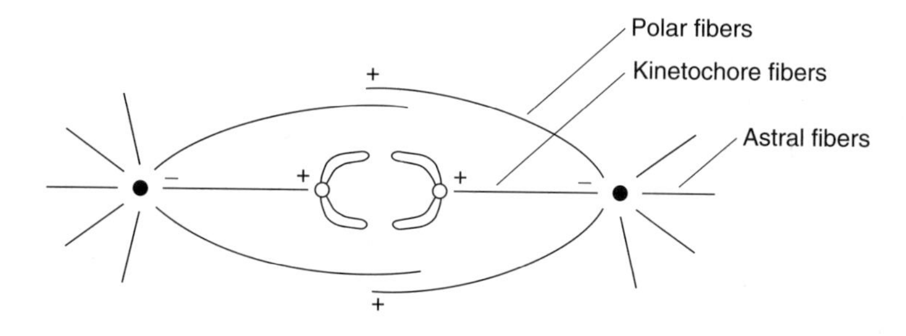

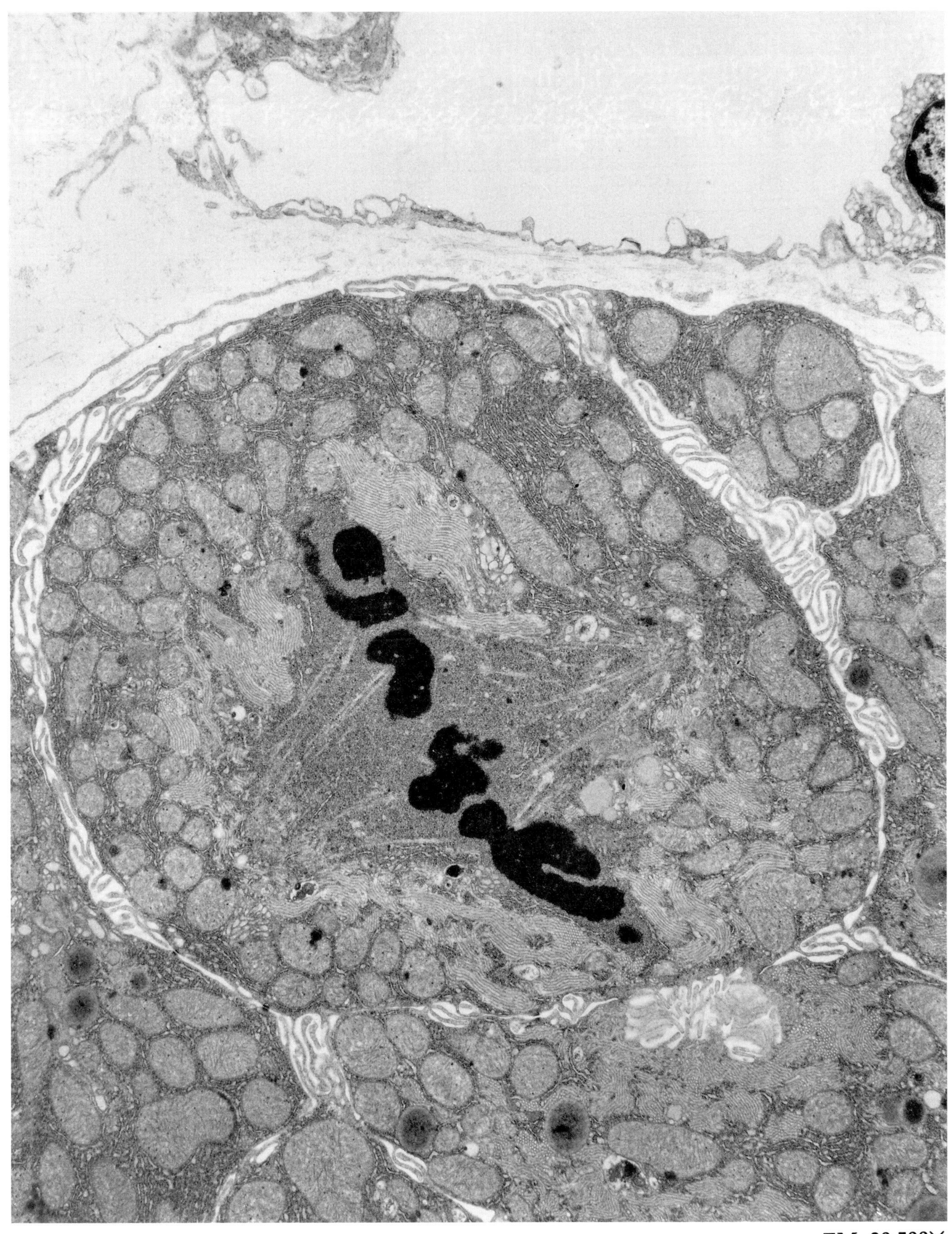

EM: 30,500×

As **chromosomes** separate during anaphase, actin and myosin polymerize in the form of a **contractile ring** at the future site of the cleavage furrow. Contraction will result in the final division of the cell during telophase. This is the most well documented instance of an actin/myosin contraction mechanism in nonmuscle cells.

In the **telophase** shown in the micrograph, the contractile ring has narrowed to produce a deep cleavage furrow and the reconstitution of the nuclear membrane (arrowheads) is well under way. The completion of cytokinesis is temporarily prevented by the presence of a spindle remnant. The slender cytoplasmic bridge is filled with spindle microtubules held together by dense amorphous material known as the **midbody** (arrows). Immunological techniques demonstrate that components of the midbody are also characteristic of the midzone of the spindle during other stages of mitosis. Since this is the region where the polar spindle fibers overlap, it has been suggested that these proteins are involved in spindle dynamics.

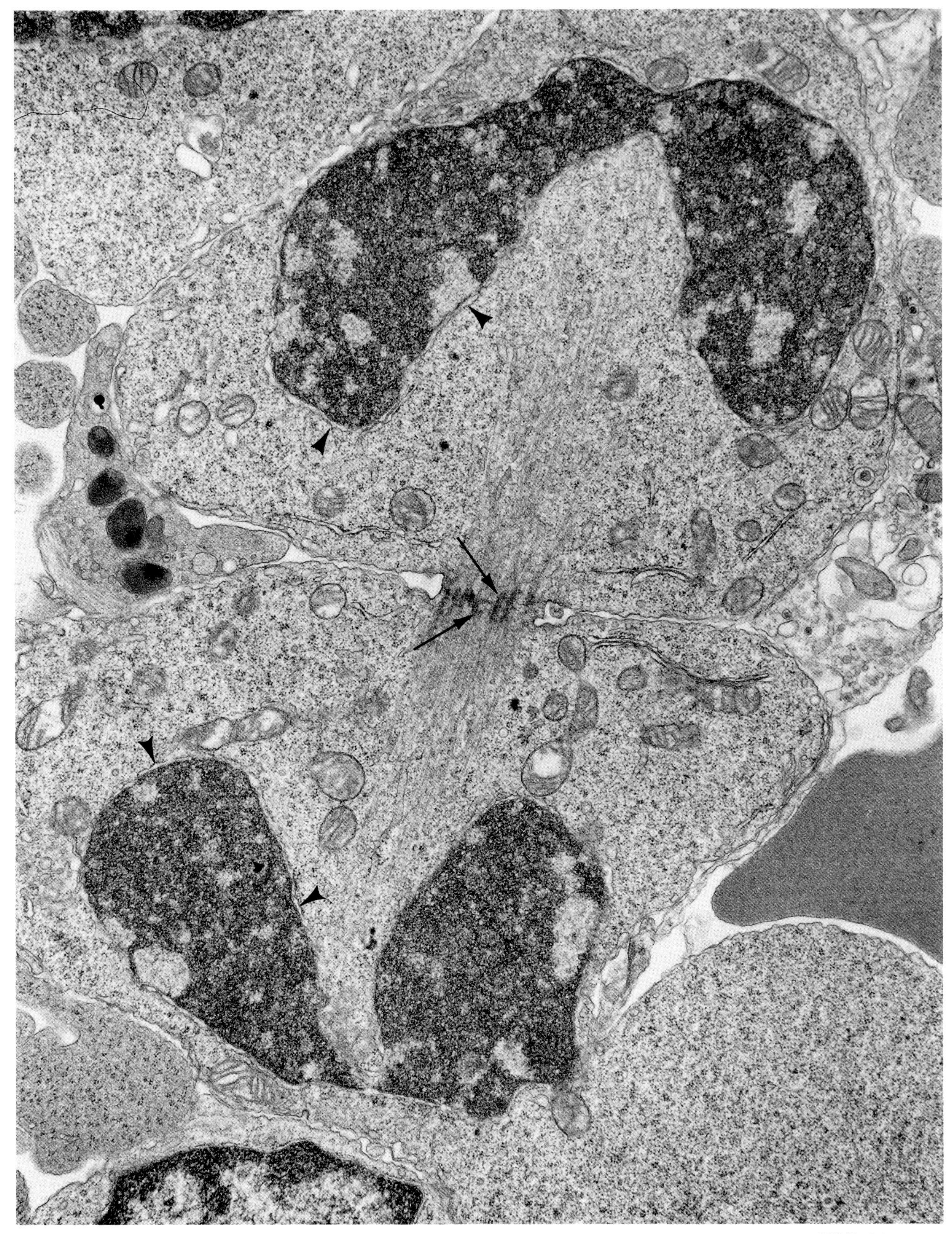

EM: 26,400×

The spindle apparatus of mitosis is just one site of microtubule accumulation. These cytoskeletal elements (arrows) are found in all interphase cells, such as the fibroblast in micrograph 1, and even in fully differentiated cells that do not subsequently divide, such as the neurons shown in micrograph 2. In addition, microtubules are the central support and machinery for cilia and flagella (see Epithelium, page 56).

The coordinated **movement of entire cells** depends upon microtubules. During fibroblast migration microtubules distribute, along with actin, to the leading edge of the cell, where they may stabilize the contacts necessary for actin stress fiber formation. Interrupting microtubule formation with colchicine results in slower and less directed types of cell movement.

Movement of components **within cells** also depends upon microtubules. In axons, crossbridges of the protein kinesin connect organelles to microtubules and move them in an anterograde direction (away from the cell center), while retrograde movement is effected by dynein. Individual microtubules do not seem to be particularly selective with regard to the type of organelle transported or the direction of transport; mitochondria and small vesicles travel in both directions on the same microtubule.

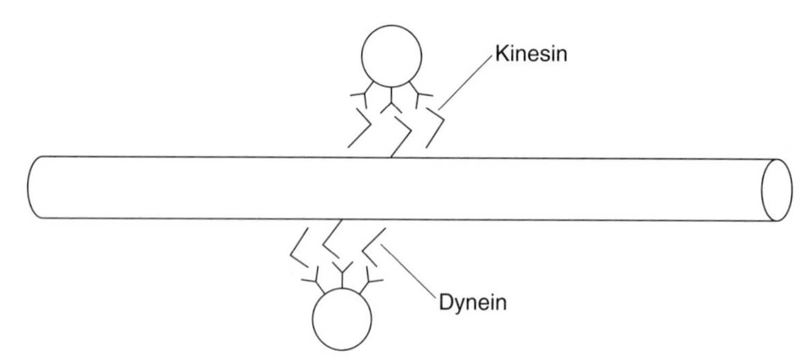

Modified from a drawing provided by Dr. Ronald Vale.

Microtubules in neurons function in structural support as well as transport; however, much of the axonal support is provided by **neurofilaments** (arrowheads, micrograph 2), fibrous proteins that are part of the third major cytoskeletal group, intermediate-sized filaments. Neurofilaments are 10 nm in diameter and consist of three polypeptides (70, 140, and 210 kd) that are unique to neurons. Neurofilaments and microtubules continually move down the axon at a much slower rate than do the organelles.

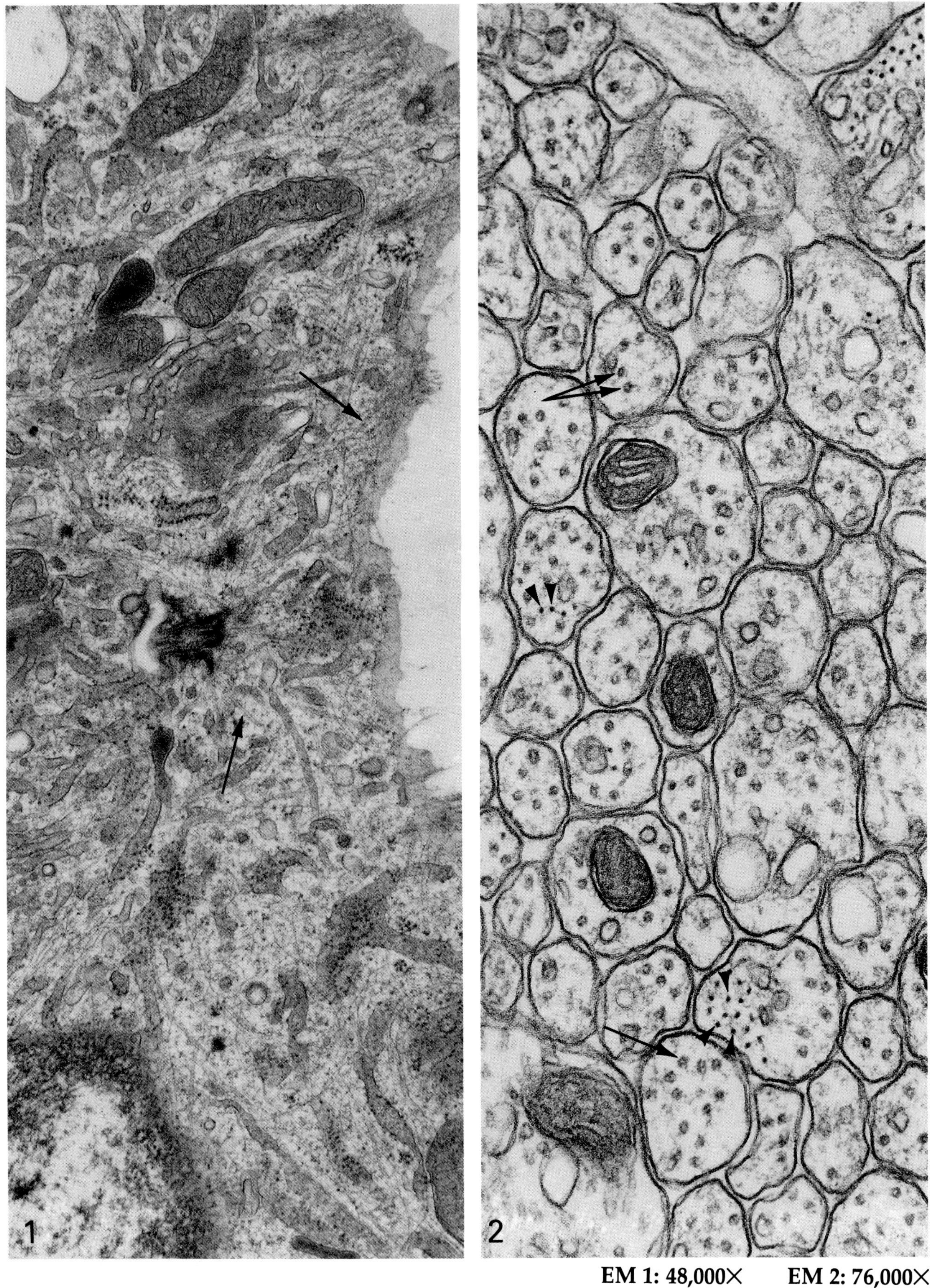

EM 1: 48,000× EM 2: 76,000×

CYTOSKELETON: Intermediate Filaments

Cytoplasmic **intermediate filaments** are divided into five classes: epithelial keratins, muscle desmin, fibroblast vimentin, glial filaments, and neurofilaments. All of these classes are composed of proteins, encoded by the same gene family, that are 10 nm in diameter, **fibrous, insoluble,** and **relatively stable.** The α-helical regions of the subunits within the 10-nm filaments wind around each other to form coiled-coils. It is generally assumed, due to the stability and arrangement of intermediate filaments, that all classes function, like the epithelial keratins, to provide support and strength.

The urinary bladder epithelial cell in micrograph 1 contains numerous keratin filaments (f). These filaments, also referred to as **tonofilaments,** crisscross through the cell and anchor at the plasma membrane at several spotlike junctions called desmosomes (arrows). Other proteins cross the intercellular space, and this continuum throughout and between the cells distributes stress during mechanical distortion. This is particularly important to accommodate the stretching and relaxing associated with changes in bladder size.

In fibrous astrocytes in the white matter of the brain and spinal cord, highly ordered intermediate filaments (arrowheads, micrograph 2) pack the cell processes to the near exclusion of other cellular components. Short projections between these filaments can be seen within the circle. These may represent cross-links between filaments that have been shown to occur in the helical regions.

Astrocyte intermediate filaments are made up of **glial fibrillary acidic protein** (GFAP). Following neuronal damage, astrocytes hypertrophy, with increased synthesis of GFAP filaments. The resulting astroglial scarring may interfere with axon regeneration.

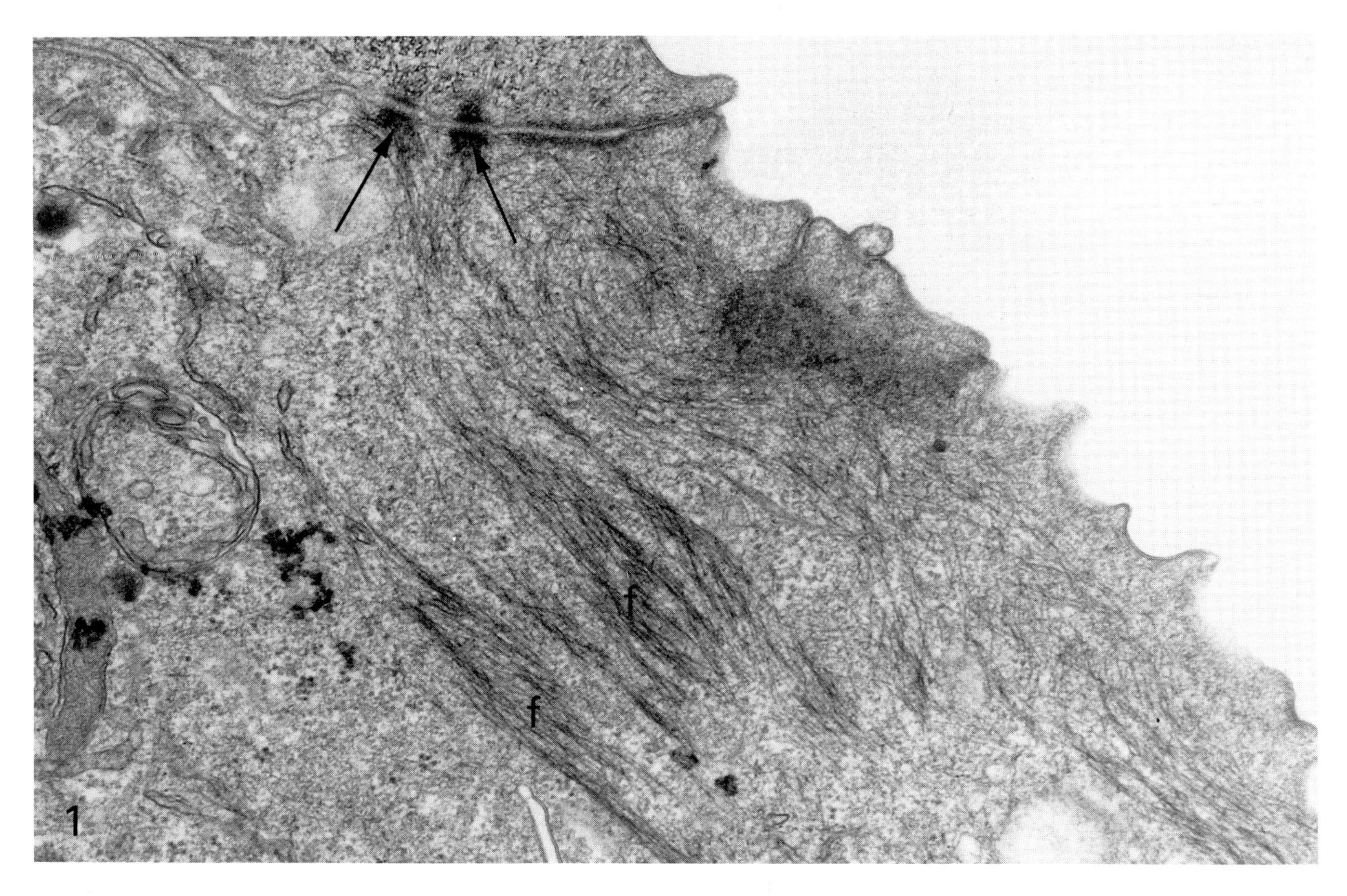

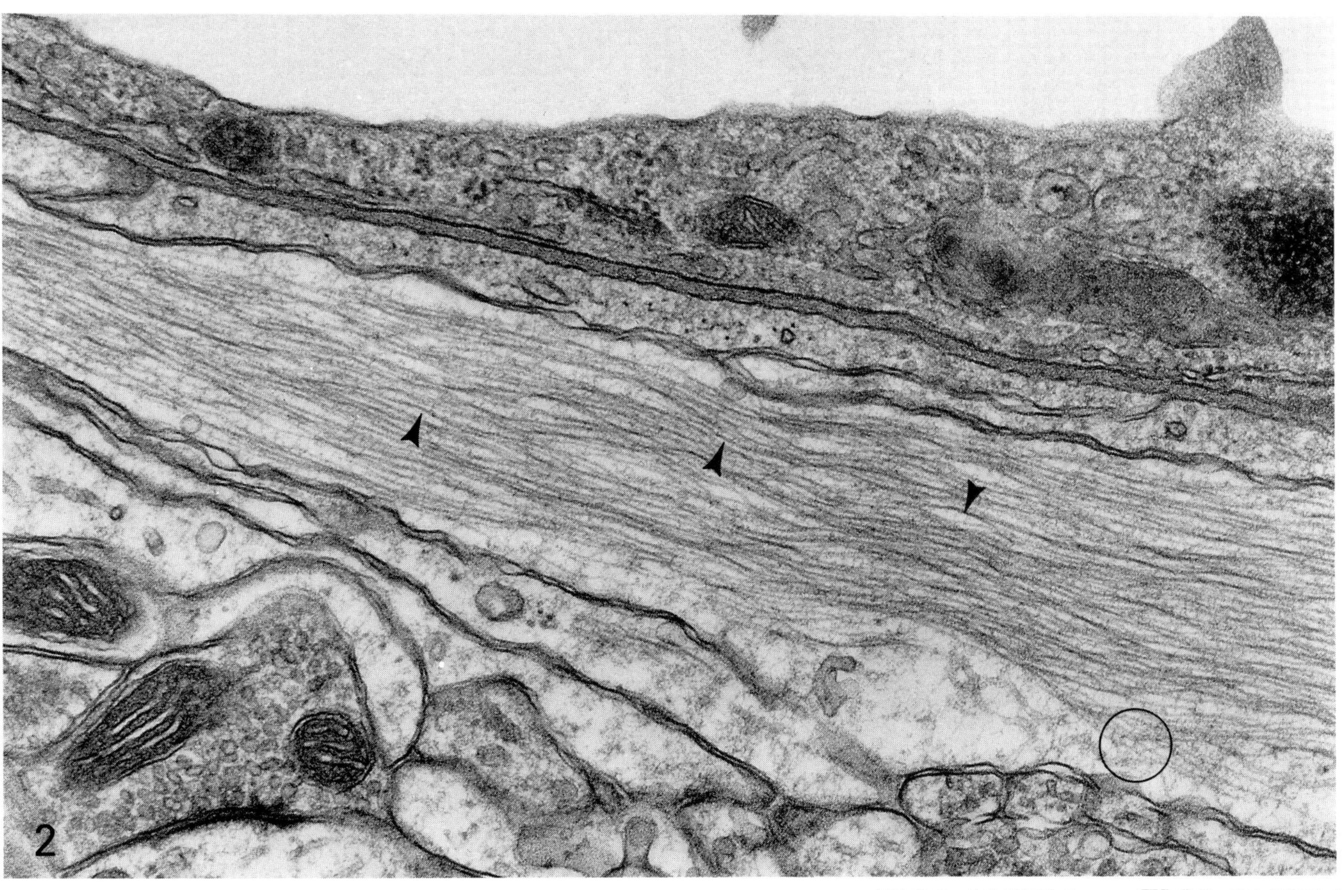

EM 1: 46,800× **EM 2: 51,000×**

EPITHELIUM

Overview

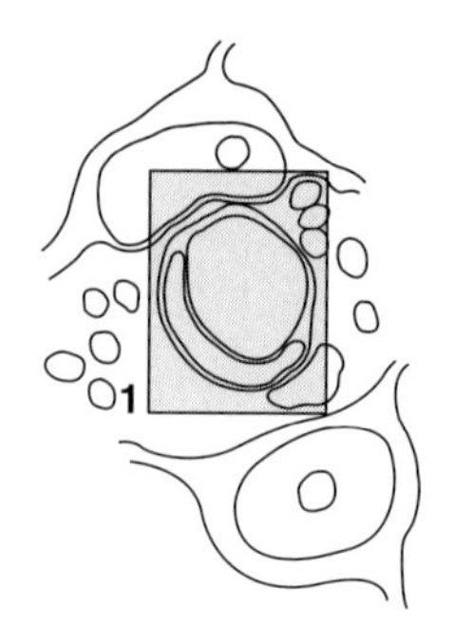

Epithelial tissue is composed of cells that work closely together as glands or as sheets that separate a lumen or space from underlying tissue. Internally, epithelial sheets line most surfaces, including the lumen of blood vessels (micrograph 1), kidney tubules (micrograph 2), and the gastrointestinal tract (micrograph 3). Externally, epithelium covers the entire body as the epidermis of skin (micrograph 4).

Epithelial sheets maintain a distinct **polarity** that relates to their function and morphology. In simple (single-layered) epithelium, which may be squamous, or flat (micrograph 1), cuboidal (micrograph 2), or columnar (micrograph 3), polarity is classically defined by distinctly different apical (adjacent to the lumen) and baso (abutting underlying tissue)-lateral (adjacent to neighboring cells) surfaces.

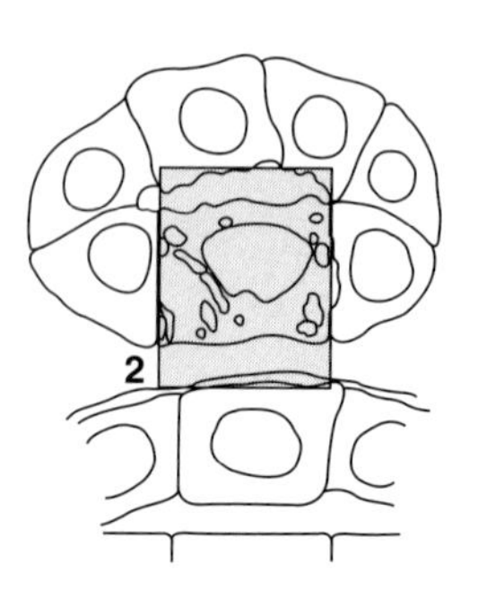

Apical specializations are best illustrated in the epithelium lining the intestine (micrograph 3). In goblet cells (G), granules (arrows) are concentrated close to the apical surface where they will release their contents; in enterocytes (E), microvilli (arrowheads) projecting from the apical surface increase the surface area for the absorption and digestion of nutrients. **Basolateral specializations** are most apparent in the distal tubule of the kidney (micrograph 2). Mitochondria (m) concentrated in the basal cytoplasm and the adjacent infolded basolateral membrane (arrowheads) reflect the active ion transport that occurs in this region. A well-defined basal lamina (arrows) attaches the cell to the underlying tissue.

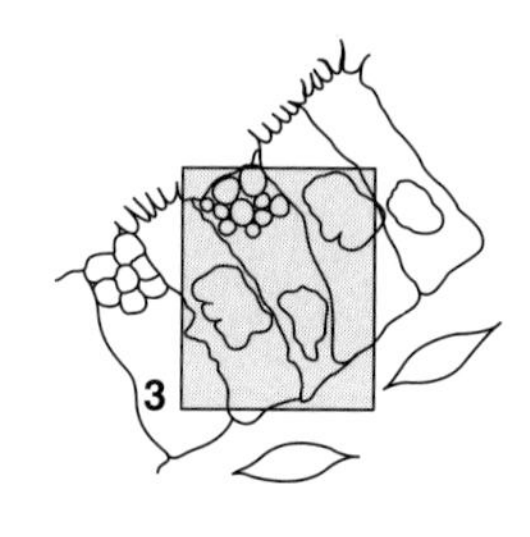

In contrast to the examples above, polarity is not obvious in the ultrastructure of the simple squamous lining of the brain capillary (micrograph 1). The thinness of the lining reflects its primary function in rapid exchange of substances between blood and tissue. This apparent simplicity, however, is misleading, since this epithelium is specialized to function as a major barrier in the brain. Certain membrane proteins control mechanisms that exclude potentially harmful substances from crossing into the brain and interfering with neuron activity. Other membrane proteins regulate the supply of essential nutrients to this tissue. The selectivity of the barrier and transport processes depends upon the placement of these proteins in specific apical versus basolateral positions.

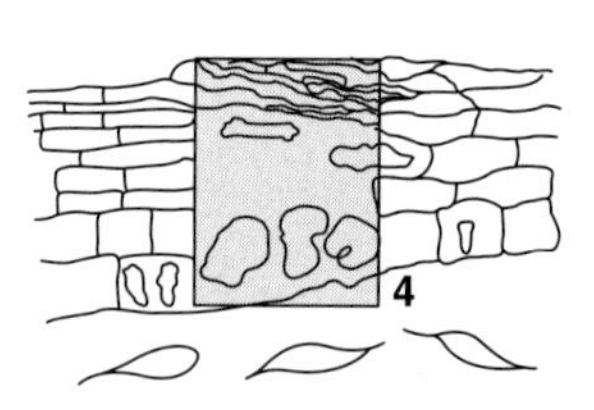

Whereas the polarity in simple epithelium refers to differences between parts of the same cell, in stratified epithelium it is often defined by differences between whole cells. In skin (micrograph 4) the basal cells (B) are undifferentiated and specialized for division whereas the apical cells (A) are highly differentiated "dead" packages of protein that, together with a lipid "mortar," prevent water loss and penetration by environmental insults.

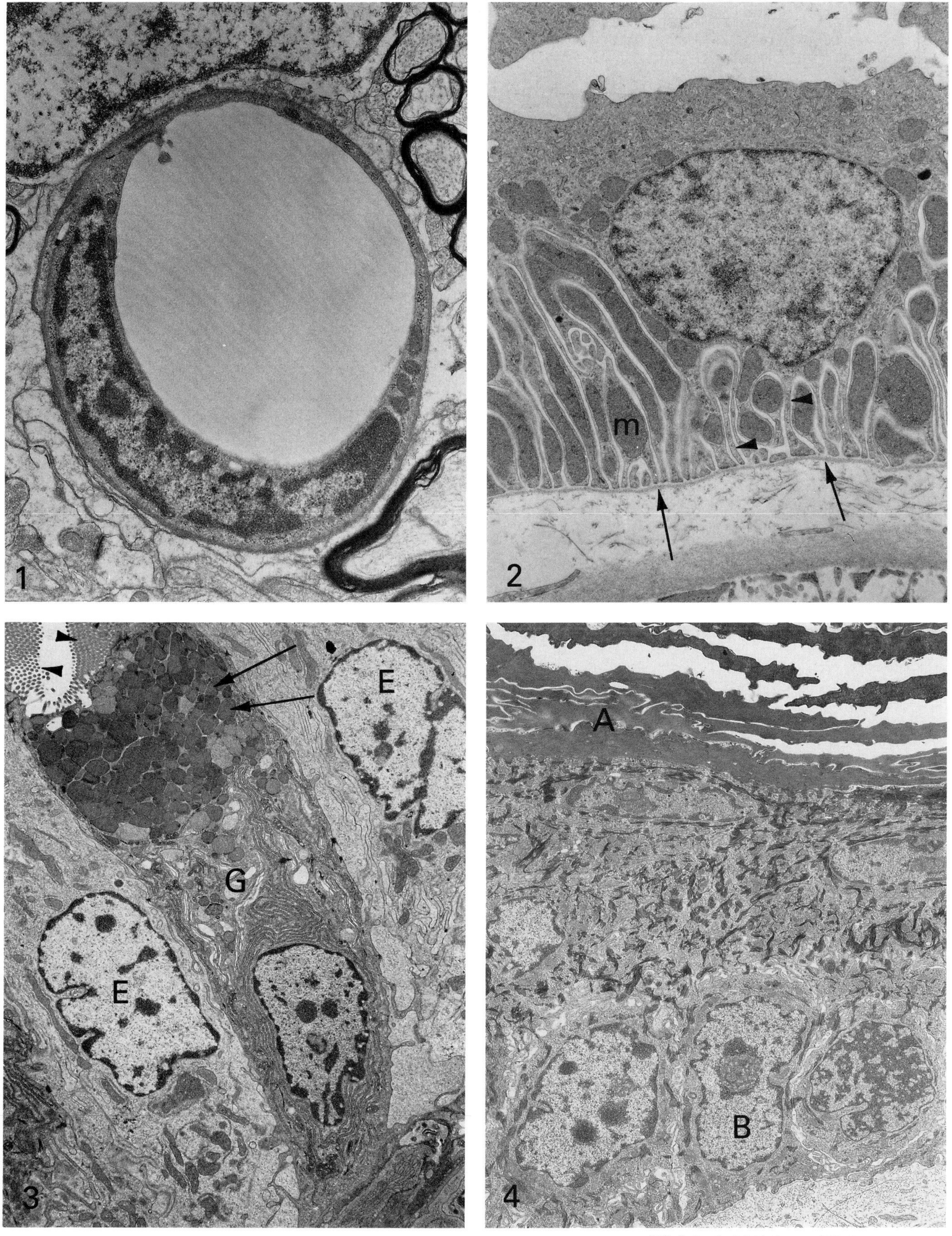

EM 1: 14,700× EM 2: 13,300× EM 3: 3,600× EM 4: 5,600×

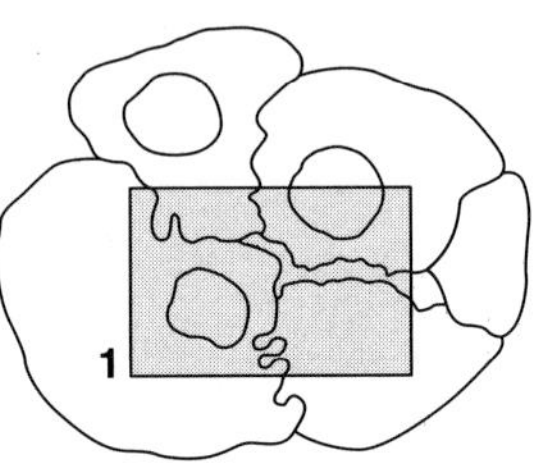

Kidney tubule

Epithelial cells that form sheets lining a lumen are typically joined together in their apical regions by **junctional complexes.** At low magnification, as in micrograph 1, these complexes appear as simple densities (arrows) at the boundary between adjacent cells. At the higher magnification shown in micrograph 2, it is obvious that a single complex is composed of individual junctions that have a distinct order with respect to the luminal surface: the zonula occludens, or tight junction (A), occupies the most apical position, then the zonula adherens (B), and next the macula adherens, or desmosome (C). Each junction has a unique role in coordinating the function of epithelial sheets.

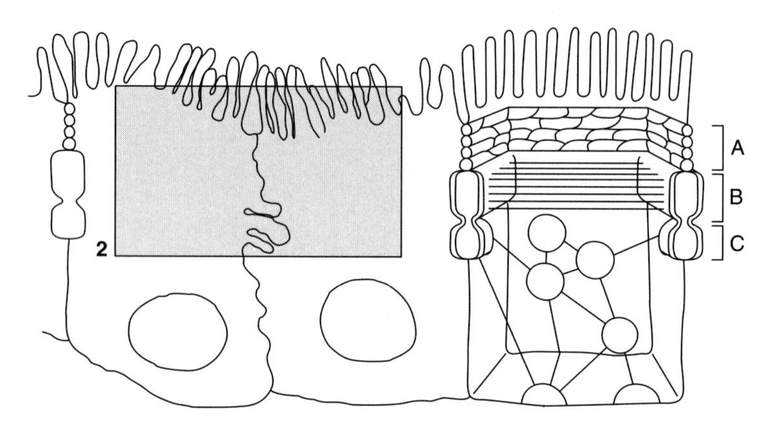

The **zonula occludens** (A) maintains polarity and controls the movement of substances across epithelial barriers. The membranes of adjacent cells come together at regular intervals to seal the two cells together. These areas of contact are actually anastomosing strips that continue around the entire circumference of the cell.

The **zonula adherens** (B), like the zonula occludens, is a beltlike junction that surrounds the entire cell. Actin filaments associated with this junction are part of an extensive network concentrated in the microvilli and apical cytoplasm. The actin–zonula adherens complex is important in contraction of the apical surface during morphogenesis and differentiation. Vinculin and α-actinin, found in focal adhesion plaques and the contractile unit of muscle, have been localized in the dense plaques of the zonula adherens.

The **macula adherens** (C) is a spotlike junction. Associated intermediate filaments extend from one spot to another on lateral and basal cell surfaces.

In contrast to the tight association of cell membranes in the zonula occludens, adjacent cell membranes of adherens junctions appear separated by a relatively wide extracellular space. In both the zonula adherens and macula adherens this space contains densities (particularly obvious in the desmosome in micrograph 2) composed of the glycosylated portions of membrane proteins of the cadherin family. The "interlocking" of these proteins provides cell-to-cell adherence. On the cytoplasmic side of each adherens junction other proteins form dense plaques into which the cytoskeletal elements insert, actin into the zonula adherens and intermediate-sized filaments into the macula adherens.

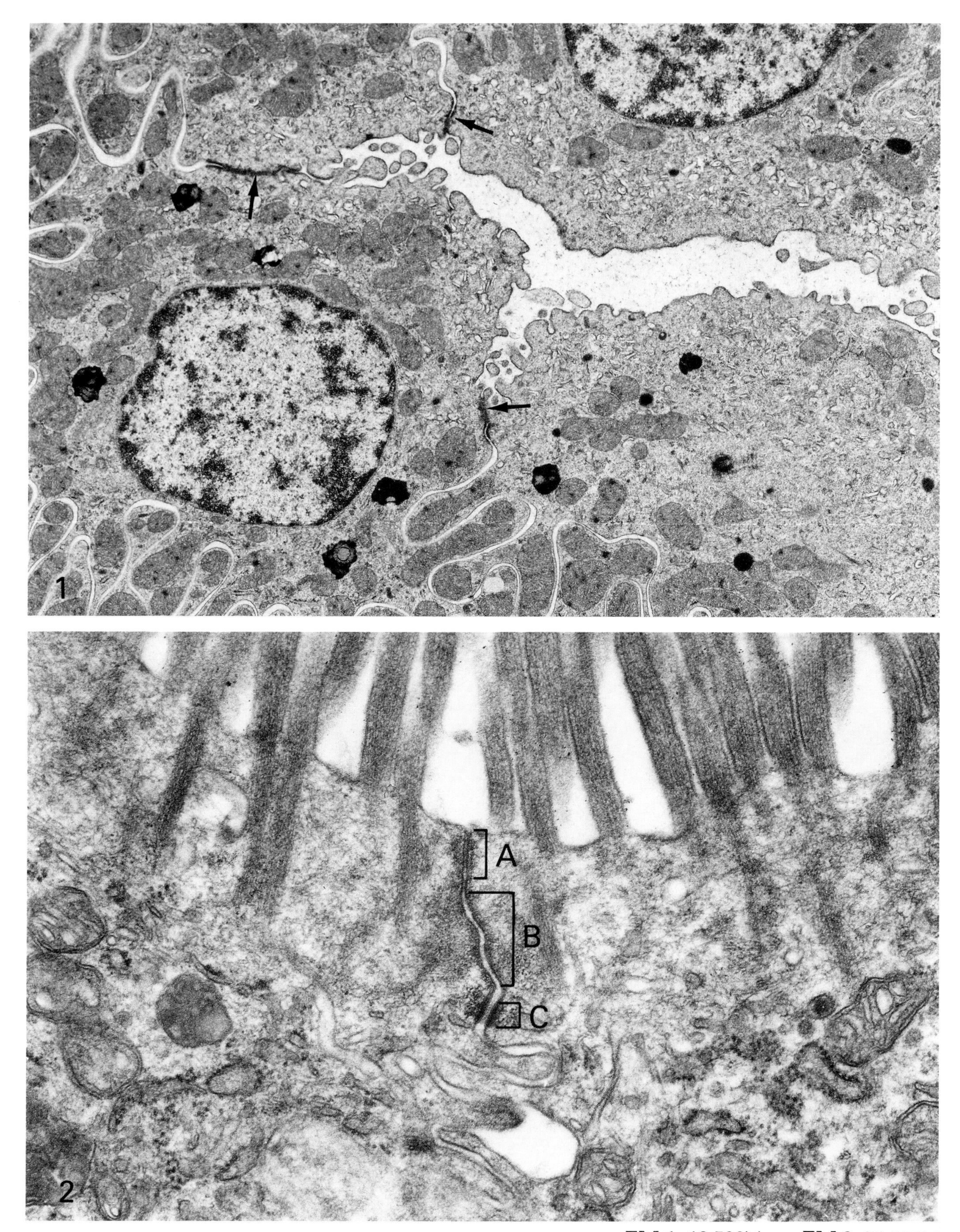

EM 1: 18,500× **EM 2: 51,000×**

Zonula Occludens (Tight Junction)

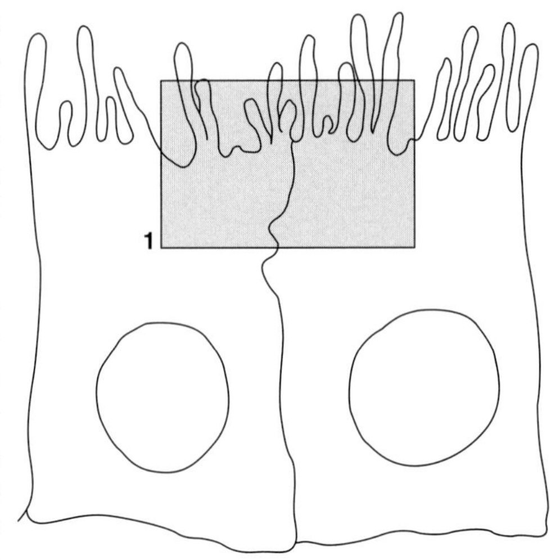

Two components of the junctional complex, the zonula occludens (ZO) and zonula adherens (ZA), are evident in micrograph 1. In the **zonula occludens,** the regions where the membranes of adjoining cells seem to fuse are visible. These areas of fusion regulate two types of movement between the lumen and underlying tissue, paracellular (between cells) and transcellular (through cells).

- **Paracellular movement:** Zonula occludens completely prevent the paracellular movement of macromolecules and polar molecules, and they restrict, to varying degrees, the movement of ions and small nonpolar molecules. Paracellular transport through tight junctions at different locations can vary from very leaky (electrical resistance as low as $5\ \Omega \cdot cm^2$) to very selective ($2000\ \Omega \cdot cm^2$). The degree of resistance has been correlated with the number of fusion areas that encircle the cell parallel to the surface.
- **Transcellular movement:** Certain membrane channels, enzymes, and receptors are localized in either the apical or basolateral cell membrane. Such restricted placement is essential to the localization of surface events adjacent to different environments (e.g., lumen vs. underlying connective tissue) and to the transport of molecules across cells. Tight junctions appear to contribute to the maintenance of this polarity by preventing (or restricting) the mixing of proteins and lipids between the apical and basolateral membranes.

In micrograph 2, the apical surfaces of three liver cells, joined via junctional complexes, define a bile canaliculus (bc). The formation of bile, the exocrine product of the liver, depends upon both transcellular and paracellular events. Bile acids, cholesterol derivatives important in the emulsification of fats, are transported vectorially from the blood to the bile canaliculi. This transcellular transport depends upon the segregation of different proteins to the apical (arrows) and basolateral domains (arrowheads: lateral cell surface). Once the bile acids are secreted into the canalicular lumen, an osmotic gradient is created that draws ions and water into bile through the tight junctions via the paracellular pathway. Most of the sodium content of bile is derived in this way.

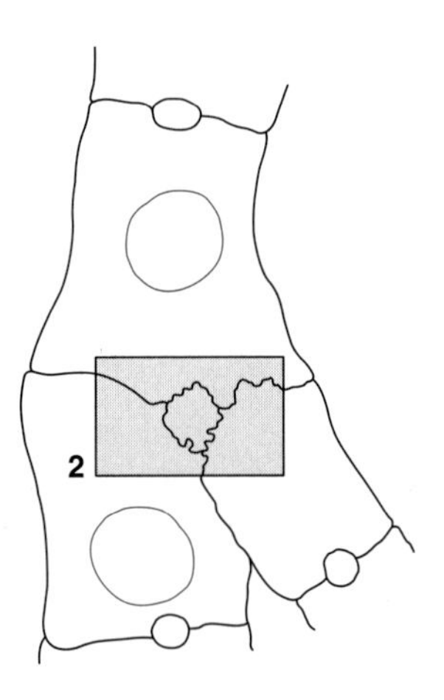

Tight-junction permeability is altered in response to diverse events. It has been suggested that actin filaments associate with proteins (ZO-1) unique to tight junctions and participate in adjusting permeability. Densities adjacent to these junctions may represent cytoskeletal elements.

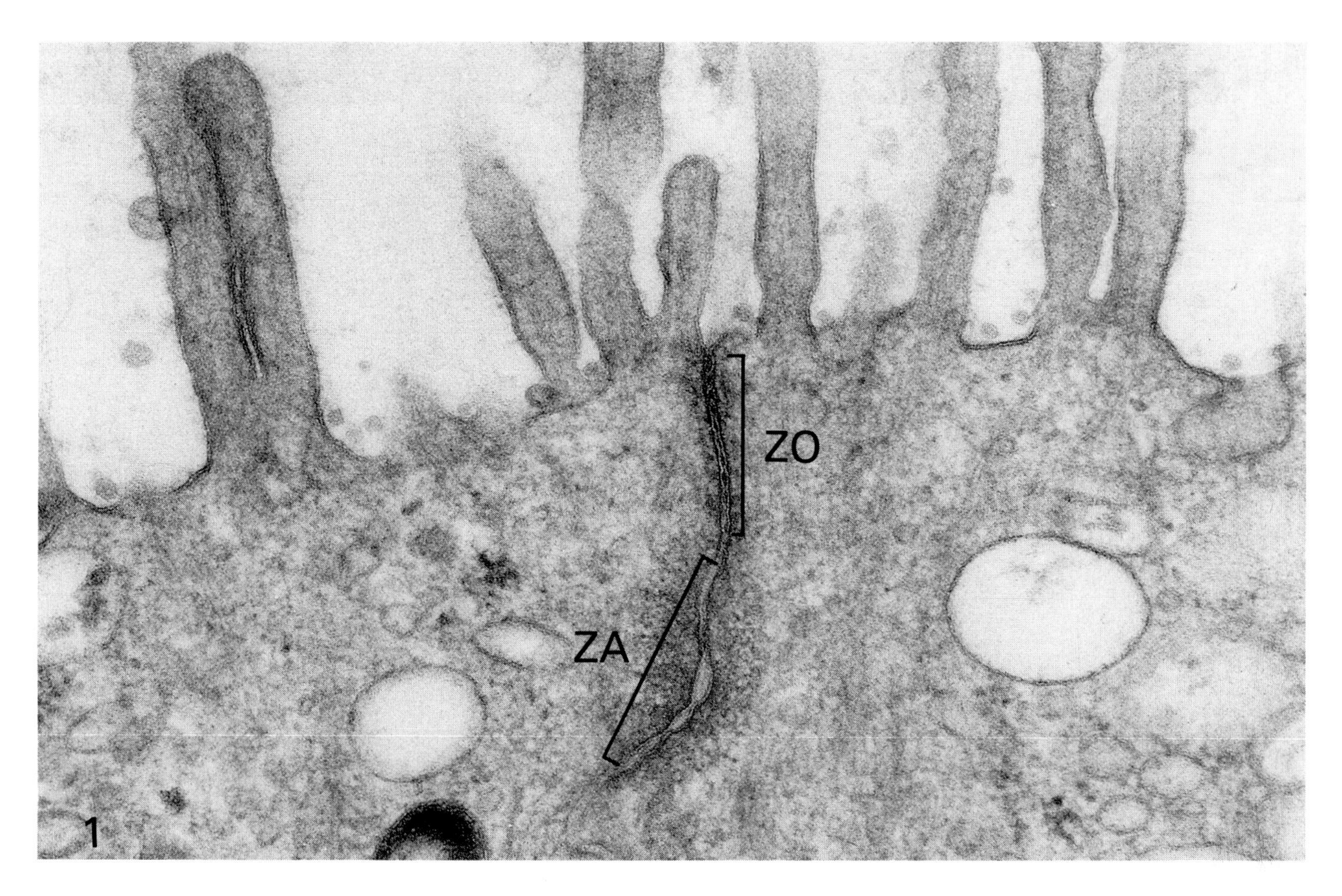

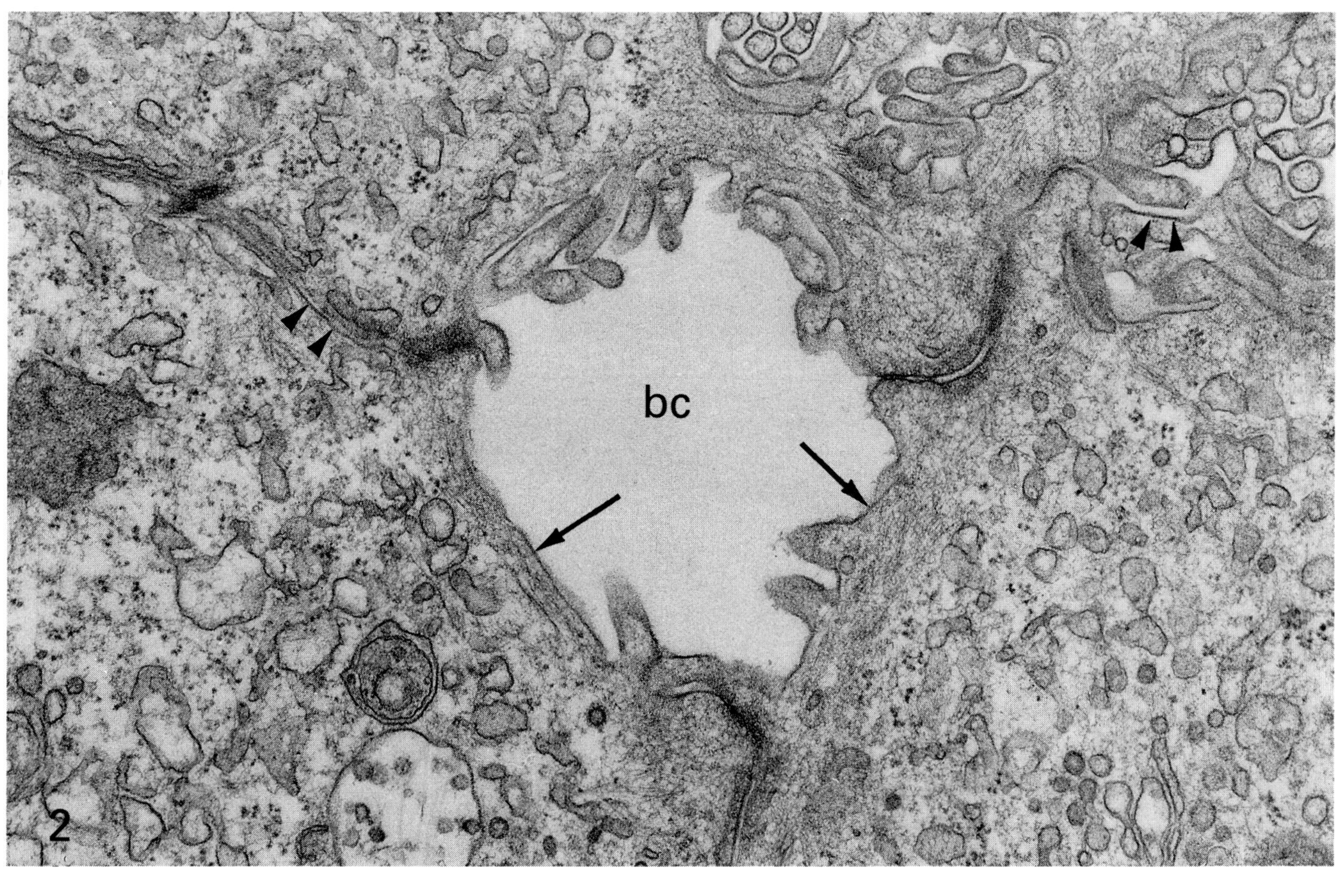

EM 1: 96,600× EM 2: 39,200×

Macula Adherens (Desmosome)

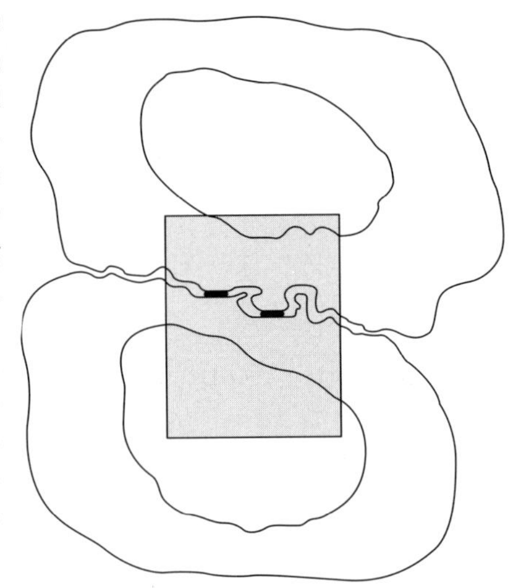

Desmosomes are common in skin (micrograph), where they are important in maintaining the integrity of an epithelium subjected to continual abrasion and distortion. In the encircled desmosome, characteristic components are particularly well illustrated: (1) the **dense line** bisecting the intercellular space where the glycosylated portions of integral membrane proteins meet; (2) the **dense plaque** (arrows) of concentrated proteins (including nonglycosylated membrane proteins) on the cytoplasmic side; and (3) **intermediate filaments** (arrowheads).

Desmosomes are not effective individually, but rather depend upon their association with other desmosomes in order to function. Stress applied to any one desmosome is rapidly distributed to others by the intermediate filaments (keratin tonofilaments in epithelium), which course within the cell from one desmosome to another.

Each of the desmosomal proteins plays an important part in the adhesive complex. Autoimmune disorders occur that affect desmosomal intercellular linking proteins (desmoglea) and plaque proteins (desmoplakins). In these diseases, bound antibodies interfere with cell-to-cell adherence, resulting in many cases in a group of diseases, pemphigus, characterized by blistering that is potentially lethal.

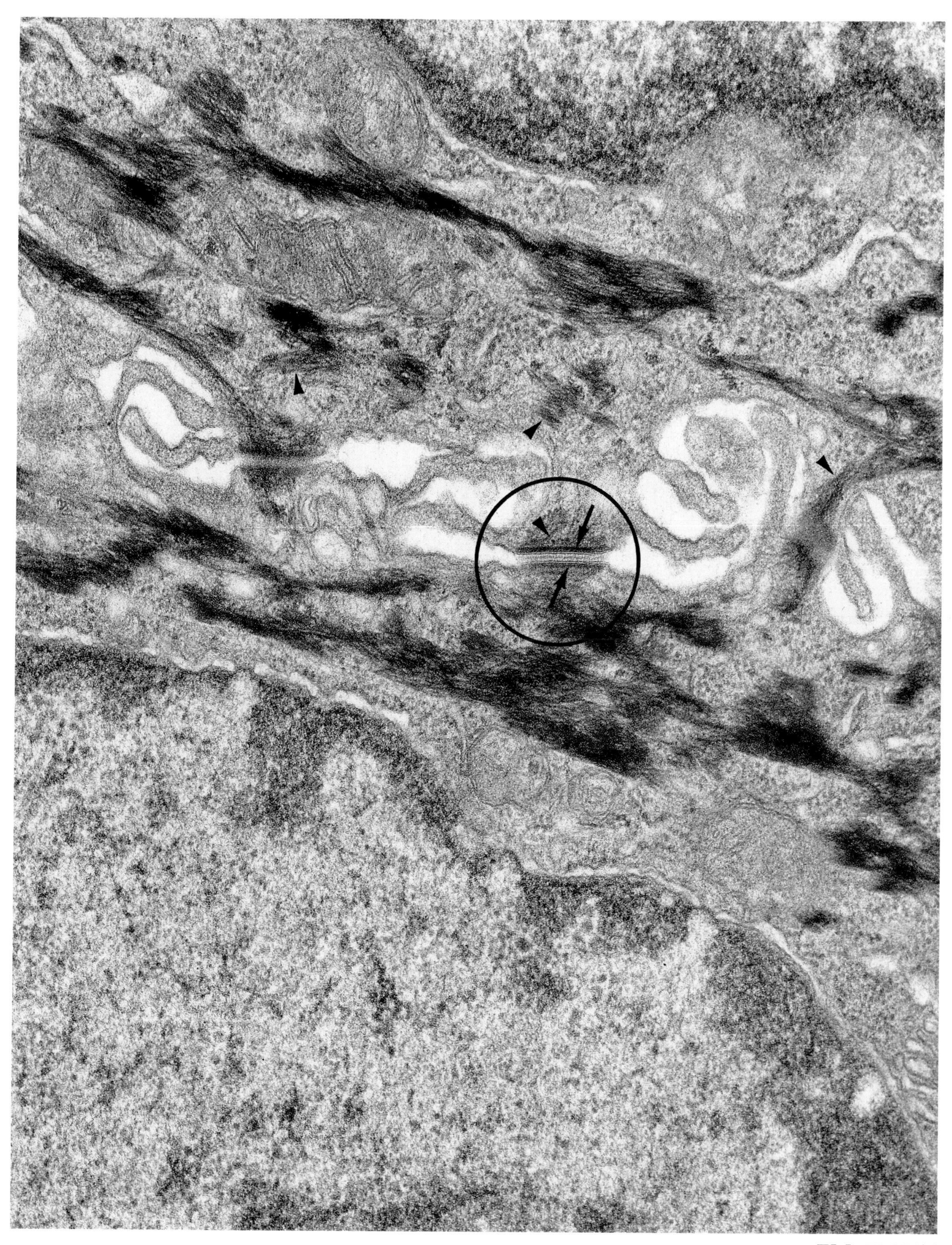

EM: 57,350×

Gap Junction

Cells in close contact, such as in epithelial sheets and glands, frequently communicate with one another via **gap junctions.** At low magnification, as in micrograph 1 of liver hepatocytes, gap junctions appear as electron-dense strips (arrows) between adjacent cells. The intercellular space seems occluded at this magnification, but in fact there is 2-nm gap between opposing cells. The increased electron density is an accumulation of integral membrane proteins specialized as channels between these two cells.

Gap junctions isolated from hepatocytes are shown in micrograph 2 (courtesy of Dr. E. Gogol). At this increased resolution, dark bands bridging the intercellular gap can be seen (areas at arrowheads). These bands are protein channels, **connexons,** that consist of a central pore of 1.5 nm, surrounded by six identical polypeptide subunits, connexins. A connexon spanning one membrane binds to another from the adjacent membrane to form a continuous channel.

Small molecules and ions (but not macromolecules) can readily pass from one cell to another through an open connexon pore. Increased concentration of calcium or hydrogen ions causes tilting and sliding of the connexon subunits to effectively close the pore and stop communication. The ability of cells to control communication may be important in a number of different ways, for example, (1) as a protective mechanism following the death of adjacent cells and (2) as an essential step during differentiation and the acquisition of unique properties.

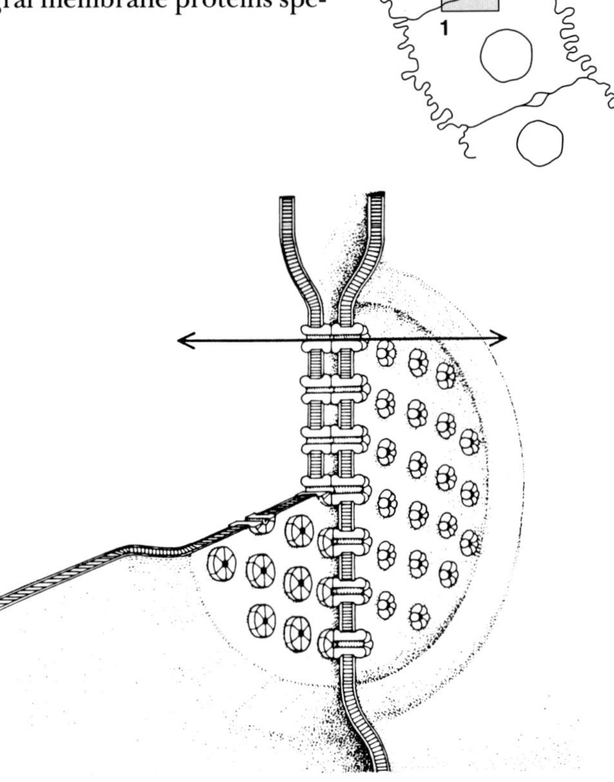

From L. A. Staehelin and B. E. Hull, *Sci. Amer. 238:* 150 (May 1978).

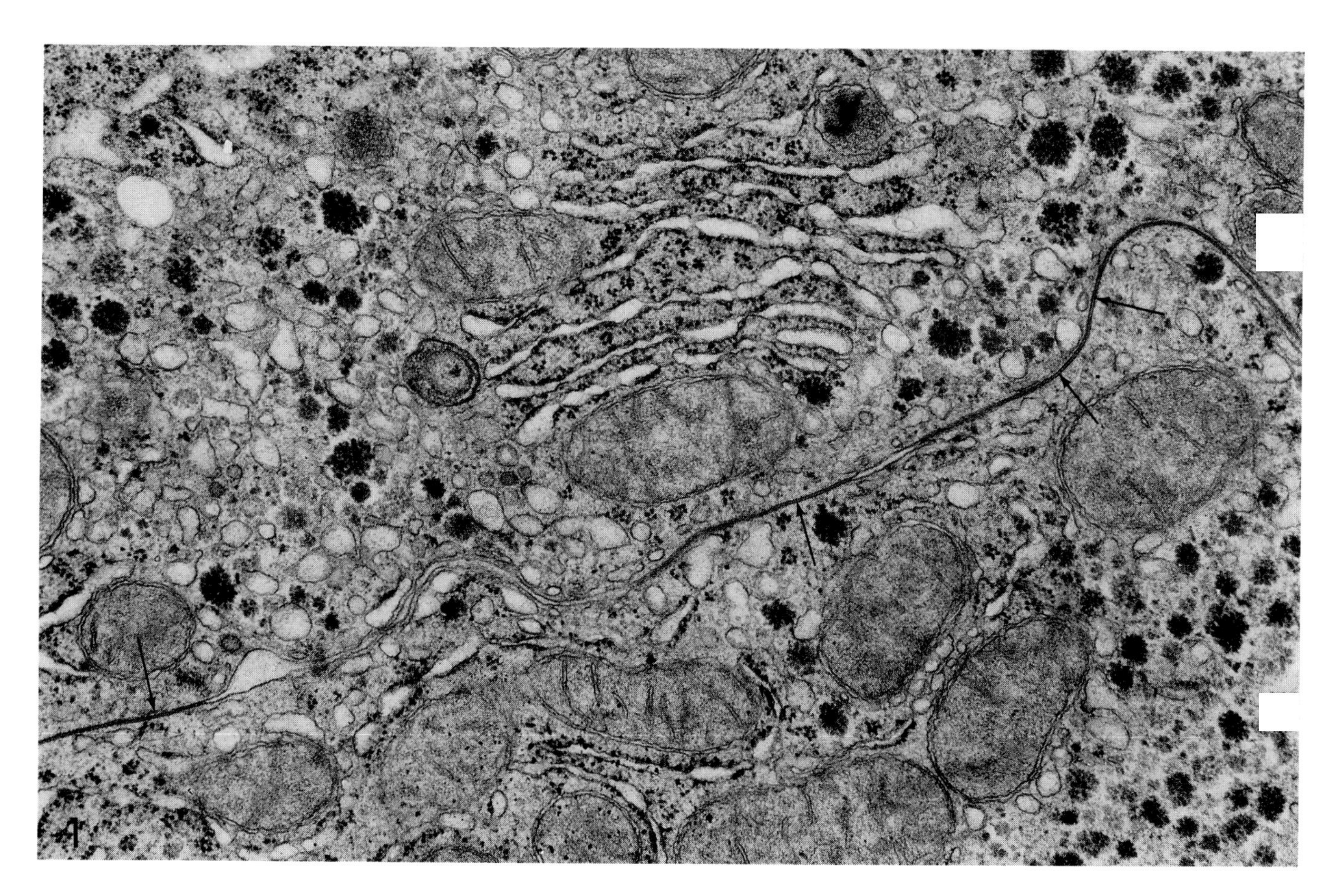

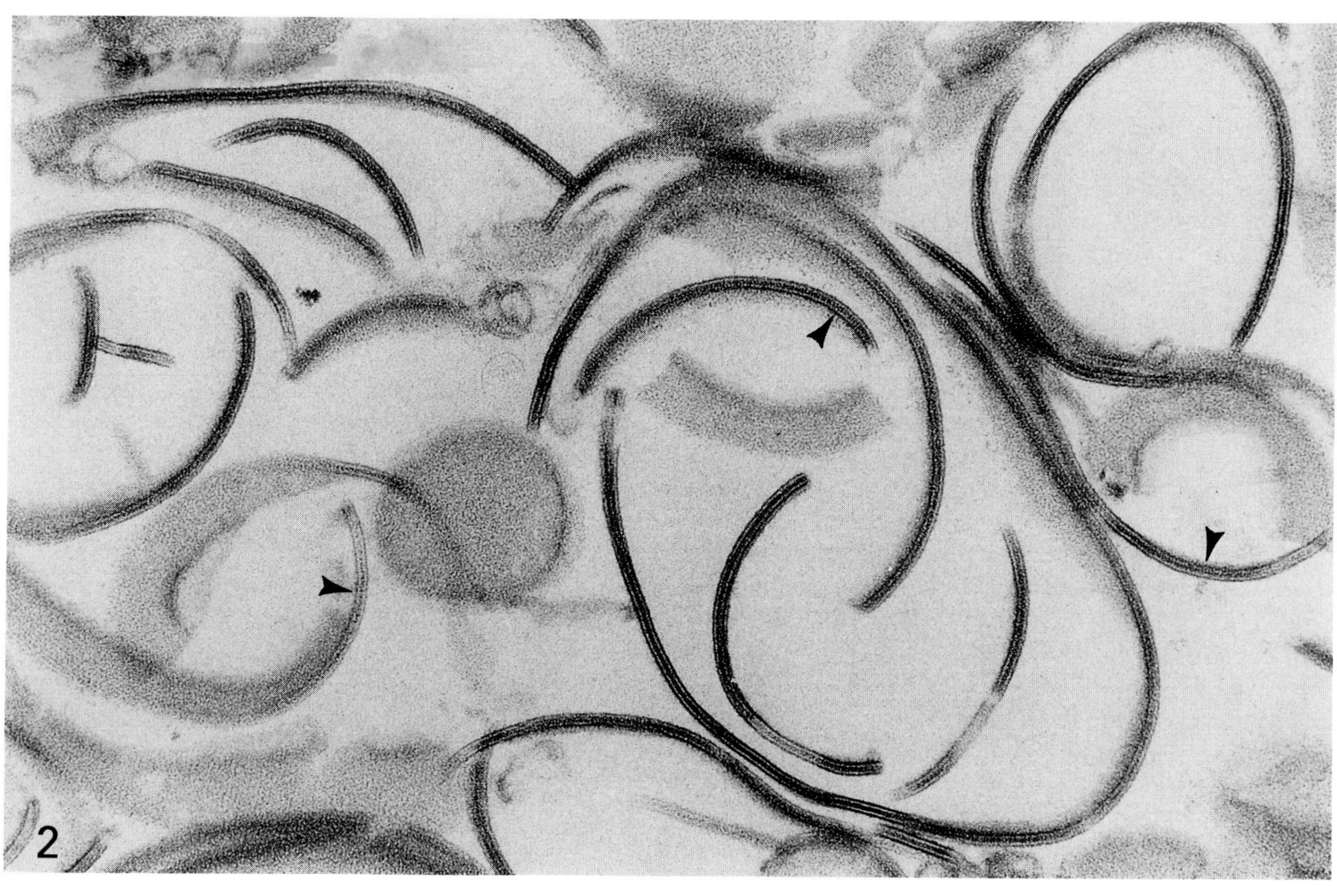

EM 1: 36,400× EM 2: 103,000×

APICAL SPECIALIZATIONS: Cilia

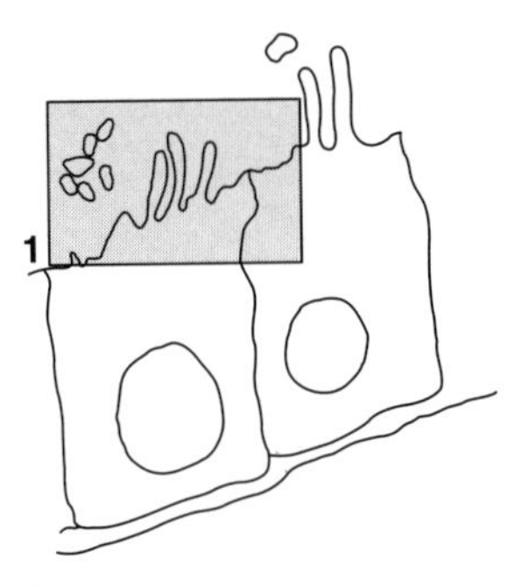

Cilia, 3 to 4 μm in length, project from the apical surface of certain epithelial cells and function in the coordinated movement of luminal contents. They are easily identified in longitudinal section (micrograph 1) by characteristic bands of microtubules (arrows) that extend from basal bodies (b) in the apical cytoplasm.

In cross section (micrograph 2) the core of each cilium is seen to contain precisely arranged microtubules, the **axoneme,** composed of one central unit containing two complete microtubules and nine peripherally distributed doublets. Each doublet consists of one complete tubule (subfiber A) attached to an incomplete tubule (subfiber B). Dynein side arms (arrows, micrograph 2) project from the A tubule of each doublet.

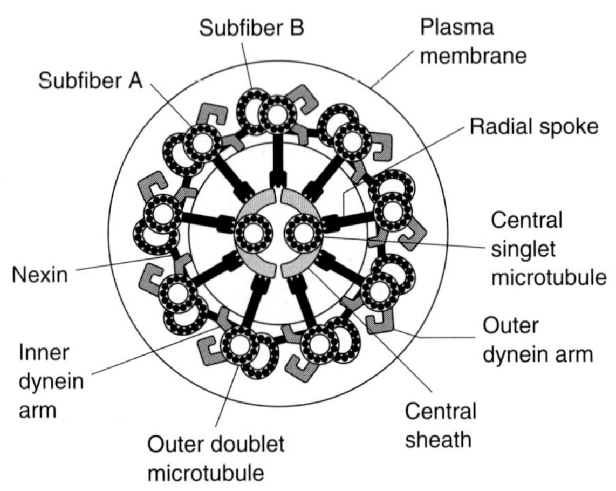

Modified from L. Stryer, *Biochemistry,* 3rd Ed., Freeman, New York, 1988.

Each cilium moves with a rapid forward effective stroke and a slower whiplike recovery stroke. Movement is initiated by chemical (e.g., neurotransmitters, hormones) or mechanical factors. It is propagated within sheets of cells via gap junctions. Thousands of cilia covering an epithelial sheet move in a given direction slightly out of phase with each other to create a wavelike pattern. This ciliary beating involves the controlled attachment and detachment of the **dynein side arms** to the facing B tubule such that each doublet "walks" along an adjacent doublet. Dynein, like the myosin head, is an ATPase, and the hydrolysis of ATP provides the force for the shearing between doublets. Evidence suggests that the outer dynein arm controls the final sliding velocity whereas the inner arm may initiate the sliding.

A sliding movement is converted to bending since all doublets are anchored to the basal body. Accessory proteins within the axoneme, such as the radial spokes that extend from the A tubules, nexin elastic filaments that attach adjacent doublets, and an inner sheath surrounding the central pair of microtubules, are important in stabilizing the complex to provide coordinated movement within each cilium.

Defects in cilia have profound effects during development and in the respiratory and reproductive systems in which their action is most vital to organ function. Congenital abnormalities (Kartagener's syndrome) and acquired abnormalities (particularly in the respiratory system exposed to environmental hazards) can lead to immotile cilia or altered beat frequency. Even though ultrastructure does not always reflect malfunction, deficiencies in cilia are frequently associated with absent or altered dynein side arms.

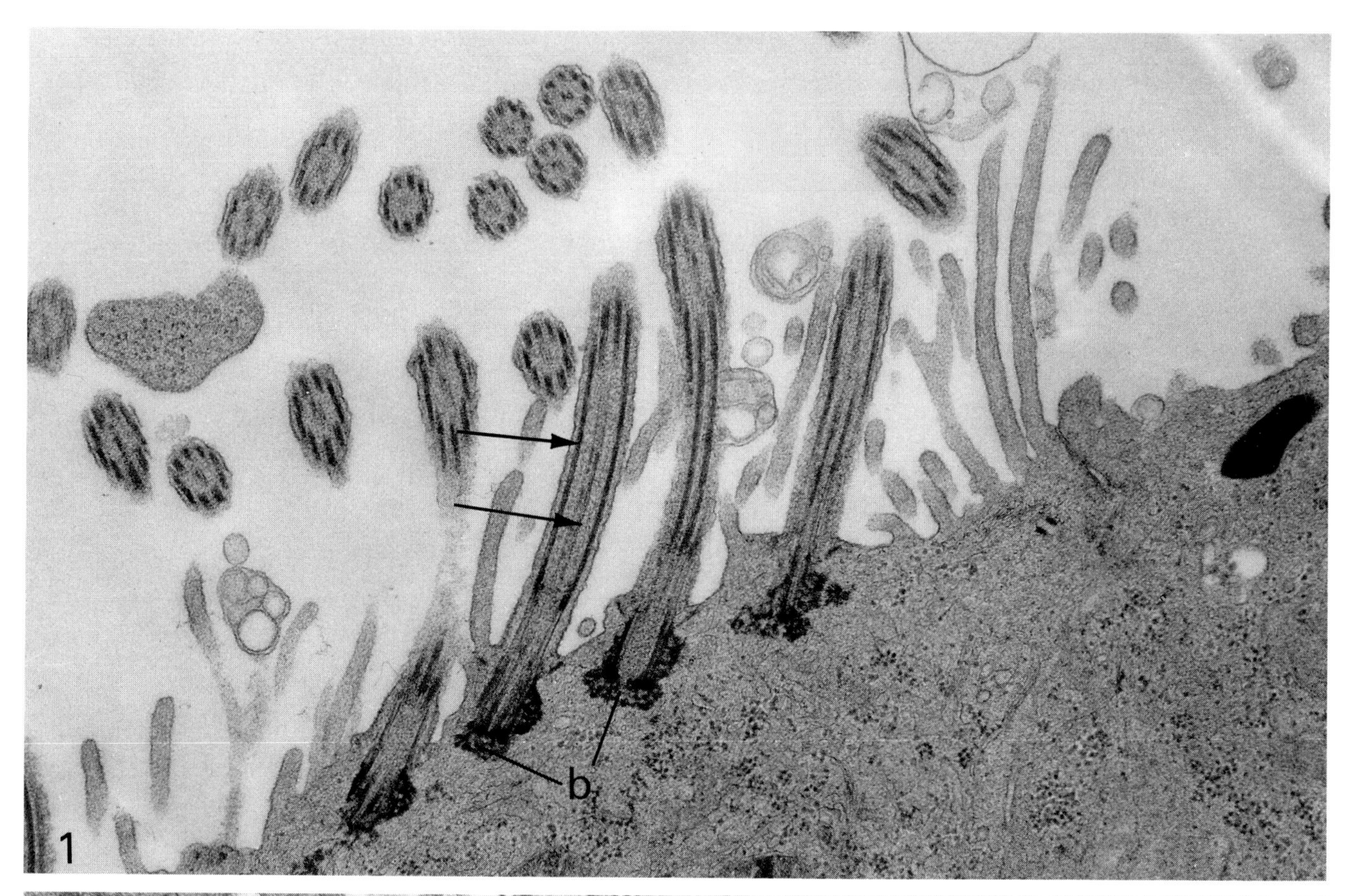

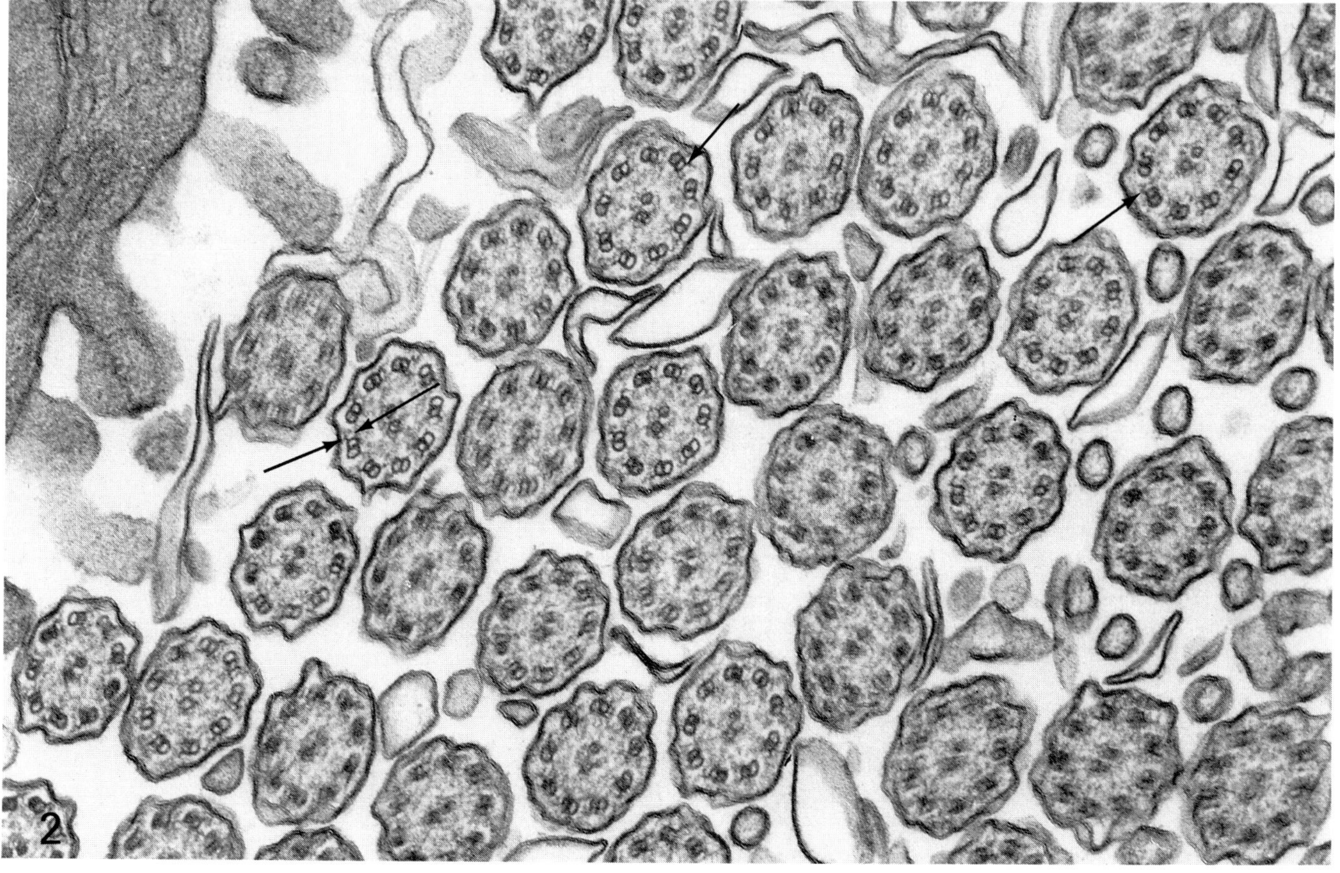

EM 1: 33,600× **EM 2: 70,000×**

APICAL SPECIALIZATIONS: Cilia, Basal Body

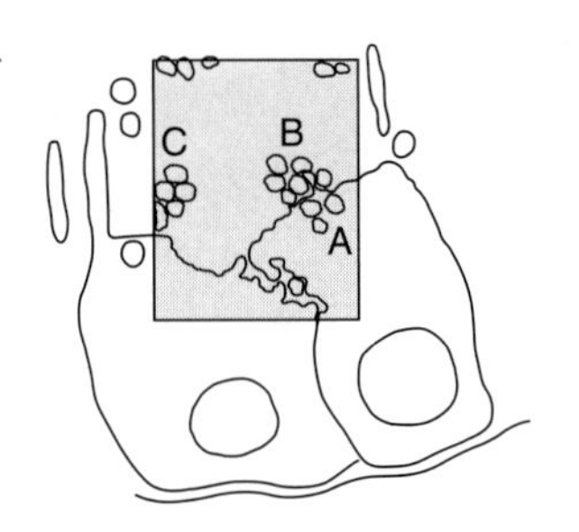

The micrograph depicts the apical surface of an epithelial cell and cross sections of cilia at various distances distal to the apical cytoplasm. Cilia develop from **basal bodies** (A, micrograph) in the apical cytoplasm. Basal bodies originate from and have a substructure similar to that of centrioles, with nine peripheral microtubule triplets. The two inner microtubules of each triplet in a basal body act as templates for the growth of the **outer doublets** in the cilium. The **central microtubules** arise distal to the basal body and are not present in cross sections of cilia near the apical surface (B, micrograph). In more distal cross sections the typical axoneme 9 + 2 arrangement is evident (C, micrograph). Radial spokes can be seen in the cilia indicated by curved arrows.

Basal bodies, like the centriole region of the mitotic spindle, are the site of nucleation of microtubules and are associated with their negative ends (see Cell, pages 34, 36). The tip of each cilium, like a kinetochore within the mitotic spindle, is the positive end region of the microtubules and the primary site of microtubule assembly and disassembly. A degree of homology has been demonstrated in certain species between proteins that "cap" the tips of cilia and proteins within the kinetochore.

Basal bodies develop from centrioles that divide and migrate to positions directly under the apical cell membrane. As the axoneme is forming at the apical region of the basal body adjacent to the cell membrane, striated filamentous rootlets (not seen on the micrograph) are extending from the opposite side into the cytoplasm to anchor each cilium. In at least one instance, the basal body is known to return to a centriole function. At fertilization the basal body of the sperm flagellum develops into the centrioles of the mitotic spindle of embryo cleavage.

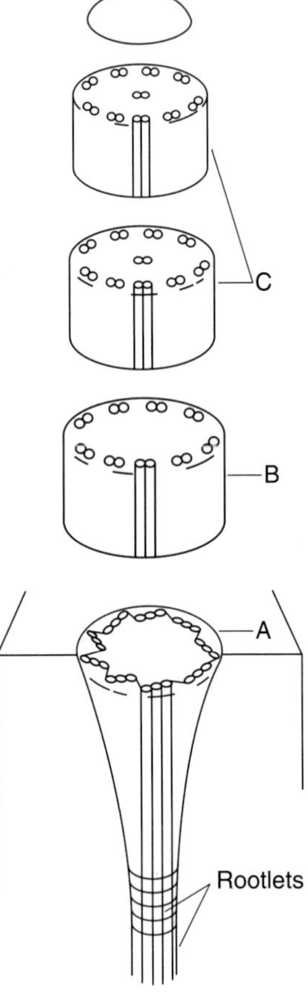

Modified from J. A. G. Rhodin, ***Histology, a Text and Atlas,*** Oxford University Press, New York, 1974.

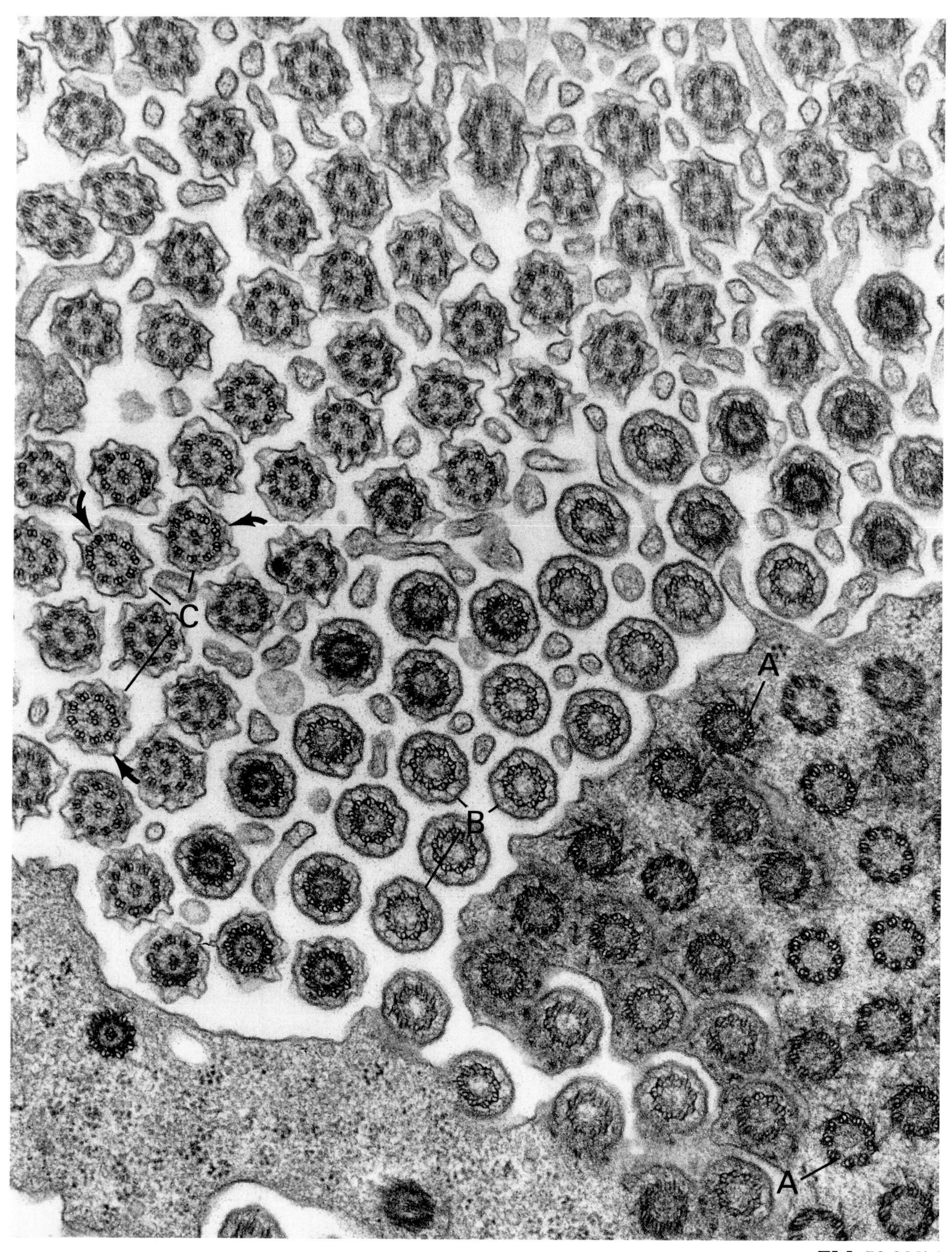

EM: 58,000×

Apical Specializations: Microvilli

Microvilli (seen in longitudinal section in micrograph 1 and cross section in micrographs 2 and 3) are relatively short (1 μm) surface projections that increase the apical surface area of the cell. Each microvillus (arrows, micrographs 1 and 2) has a core of actin filaments that extend into the apical cytoplasm to form a terminal web (t, micrograph 1). Actin filaments of the terminal web cross-link to form an extensive network throughout the apical region. This network, anchored at the lateral surfaces into the zonula adherens, stabilizes the microvilli and enables the apical surface to act as a unit.

Other microvillar proteins bind actin microfilaments together (villin, fimbrin) and to the cell membrane (myosin 1). This assembly of proteins provides additional strength and rigidity for microvilli. In addition, the myosin 1 has ATPase activity and binds to actin in an ATP-reversible manner. This activity may facilitate a type of membrane movement.

The section of microvilli in micrograph 3 illustrates the internal, regularly arranged actin filaments (seen as dots in cross section) and an external "fuzzy coat" (arrowheads), the **glycocalyx.** The glycocalyx consists of the carbohydrate side chains of membrane proteins and lipids that extend beyond the lipid bilayer. Even though present on all apical surfaces, the glycocalyx is most pronounced on cells lining the small intestine, where it can reach a thickness of 0.5 μm. This reflects a unique specialization of the intestinal surface in digestion. The final steps in the breakdown of proteins and carbohydrates are carried out directly on the cell surface by enzymes that are integral membrane proteins and a part of the glycocalyx.

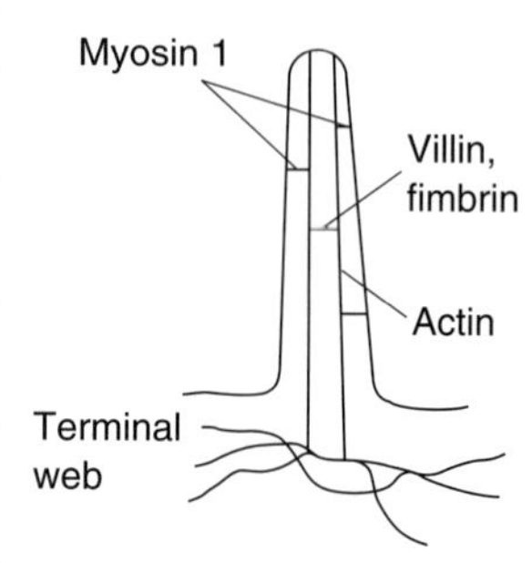

Modified from J. Darnell et al., *Molecular Cell Biology,* Scientific American Books, New York, 1986.

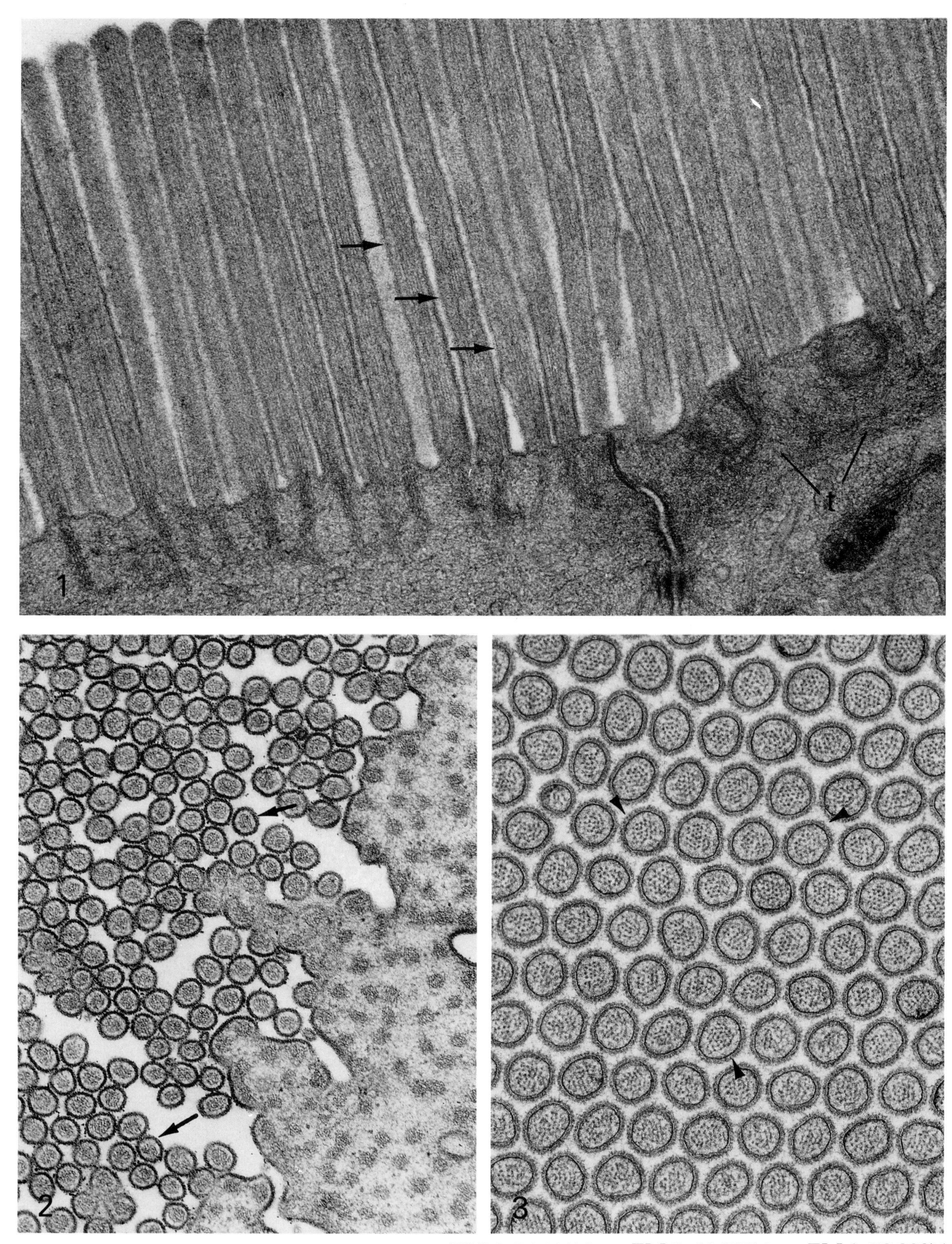

EM 1: 54,000× EM 2: 44,200× EM 3: 78,200×

Apical Specializations: Stereocilia

The apical surface of a few cell types (such as the epididymal epithelial cell shown here) contains extremely long microvilli. The supportive filamentous core of actin is less well developed in these projections than it is in the shorter, more typical microvilli of the intestinal lining cells. The reduced cytoskeletal support and their long length contribute to a greater flexibility, leading to a resemblance to cilia at the light-microscope level. These microvilli are referred to as **stereocilia,** even though their **actin core** structure is very different from the microtubule core of true cilia.

In the epididymis, fluid absorption by the epithelium creates a current that is essential to the transport of sperm. Both the increased surface area provided by the stereocilia and the vesicles (arrowheads) in the apical cytoplasm reflect this function. Epididymal cells also contribute unique proteins to the luminal fluid, some of which associate with sperm (arrow indicates a sperm head) and may be important in the development of the ability of sperm to fertilize ova. It is possible that the increased cell surface provided by the stereocilia is also significant in this secretory process.

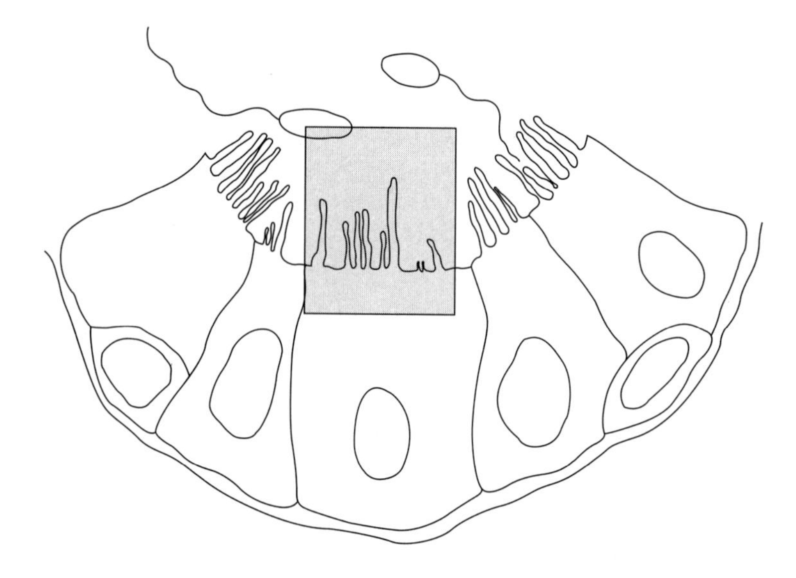

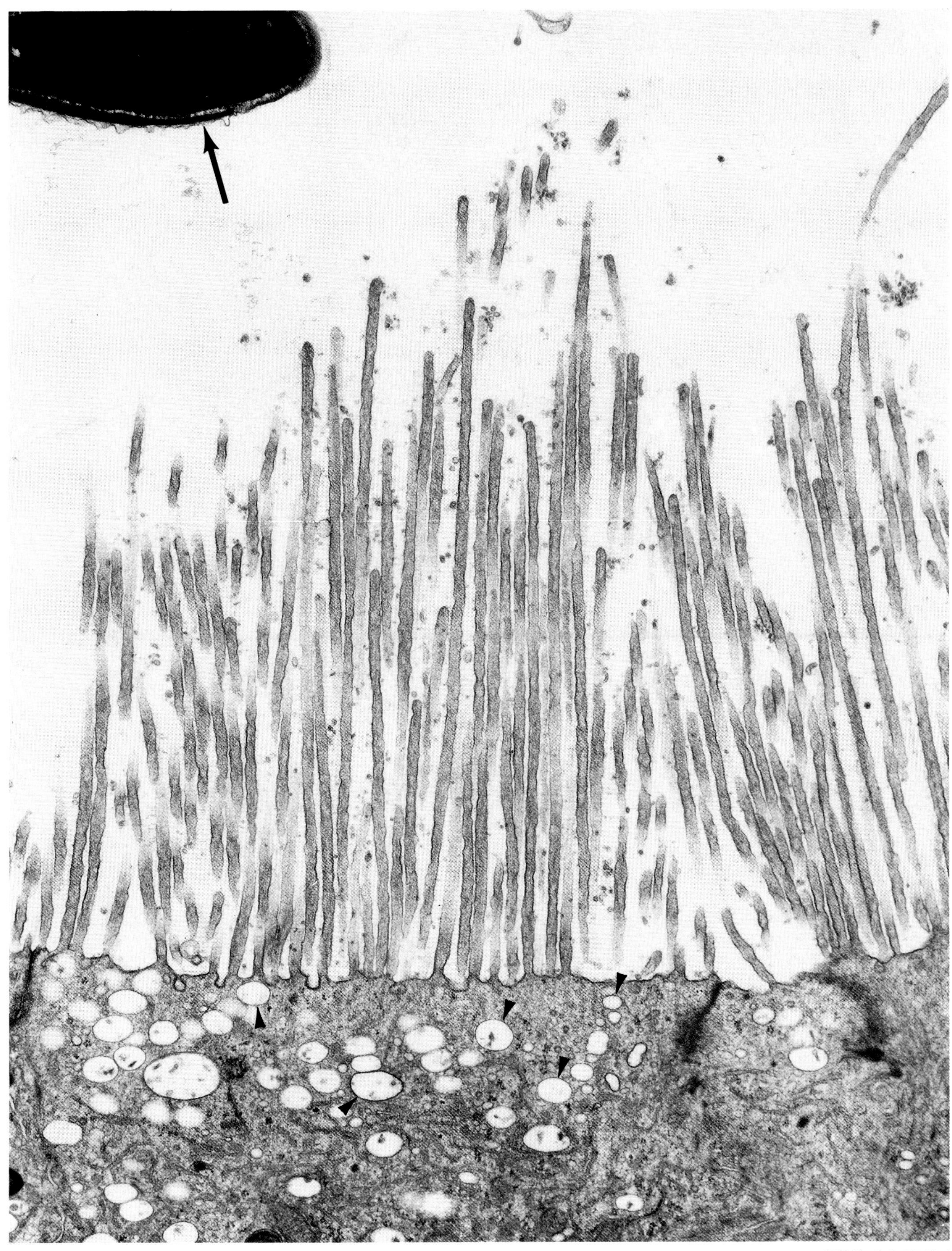

EM: 26,250×

APICAL SPECIALIZATIONS: Transitional Epithelium

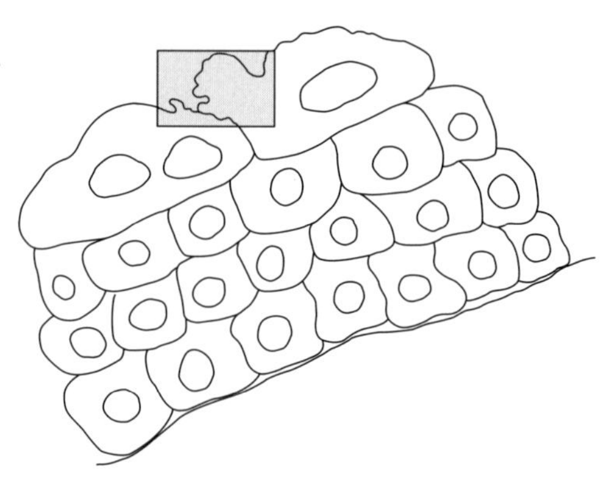

Transitional epithelium, found only in the urinary tract, is a stratified epithelium in which the apical cells change shape from cuboidal to squamous, depending on the degree of distention. In addition to a junctional complex typical of many epithelia, the surface cells of this epithelium possess a thick glycocalyx, abundant tonofilaments, characteristic membrane vesicles, and scallops along the luminal surface.

The ultrastructure of the apical cell layer of this stratified epithelium reflects its two major functions.

1. It is a **permeability barrier:** Tight junctions of the junctional complexes (JC, micrograph 1) between these cells prevent the paracellular movement of ions and water between the lumen and intercellular regions. This barrier maintains the

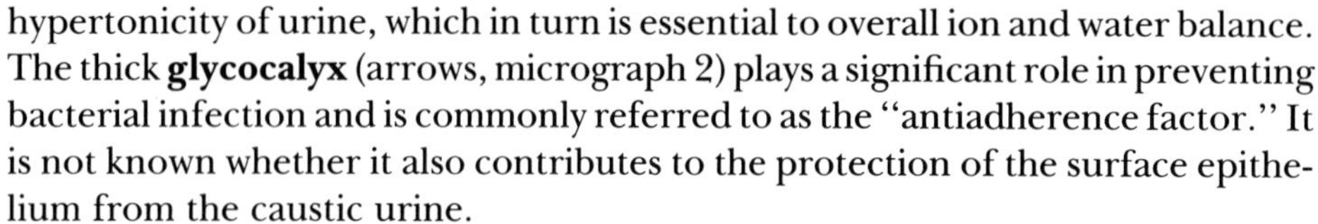

hypertonicity of urine, which in turn is essential to overall ion and water balance. The thick **glycocalyx** (arrows, micrograph 2) plays a significant role in preventing bacterial infection and is commonly referred to as the "antiadherence factor." It is not known whether it also contributes to the protection of the surface epithelium from the caustic urine.

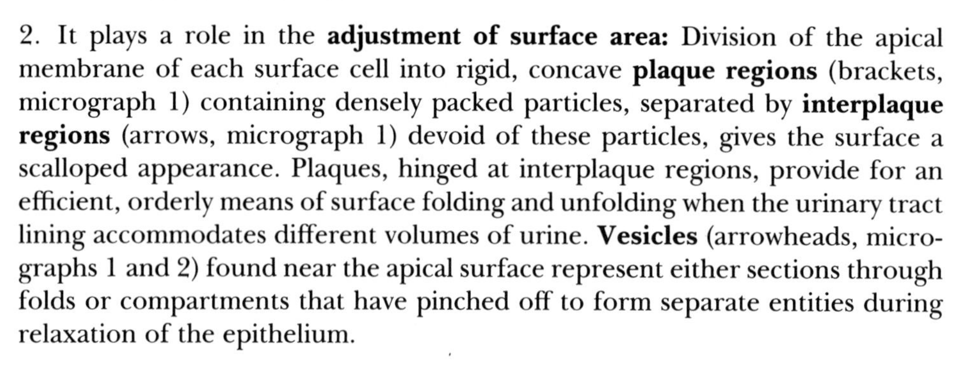

2. It plays a role in the **adjustment of surface area:** Division of the apical membrane of each surface cell into rigid, concave **plaque regions** (brackets, micrograph 1) containing densely packed particles, separated by **interplaque regions** (arrows, micrograph 1) devoid of these particles, gives the surface a scalloped appearance. Plaques, hinged at interplaque regions, provide for an efficient, orderly means of surface folding and unfolding when the urinary tract lining accommodates different volumes of urine. **Vesicles** (arrowheads, micrographs 1 and 2) found near the apical surface represent either sections through folds or compartments that have pinched off to form separate entities during relaxation of the epithelium.

Tonofilaments (curved arrows, micrograph 3) provide strength to maintain the integrity of the transitional epithelium during the distortion associated with stretching and relaxing. These filaments associate with both the plaque proteins and desmosomes.

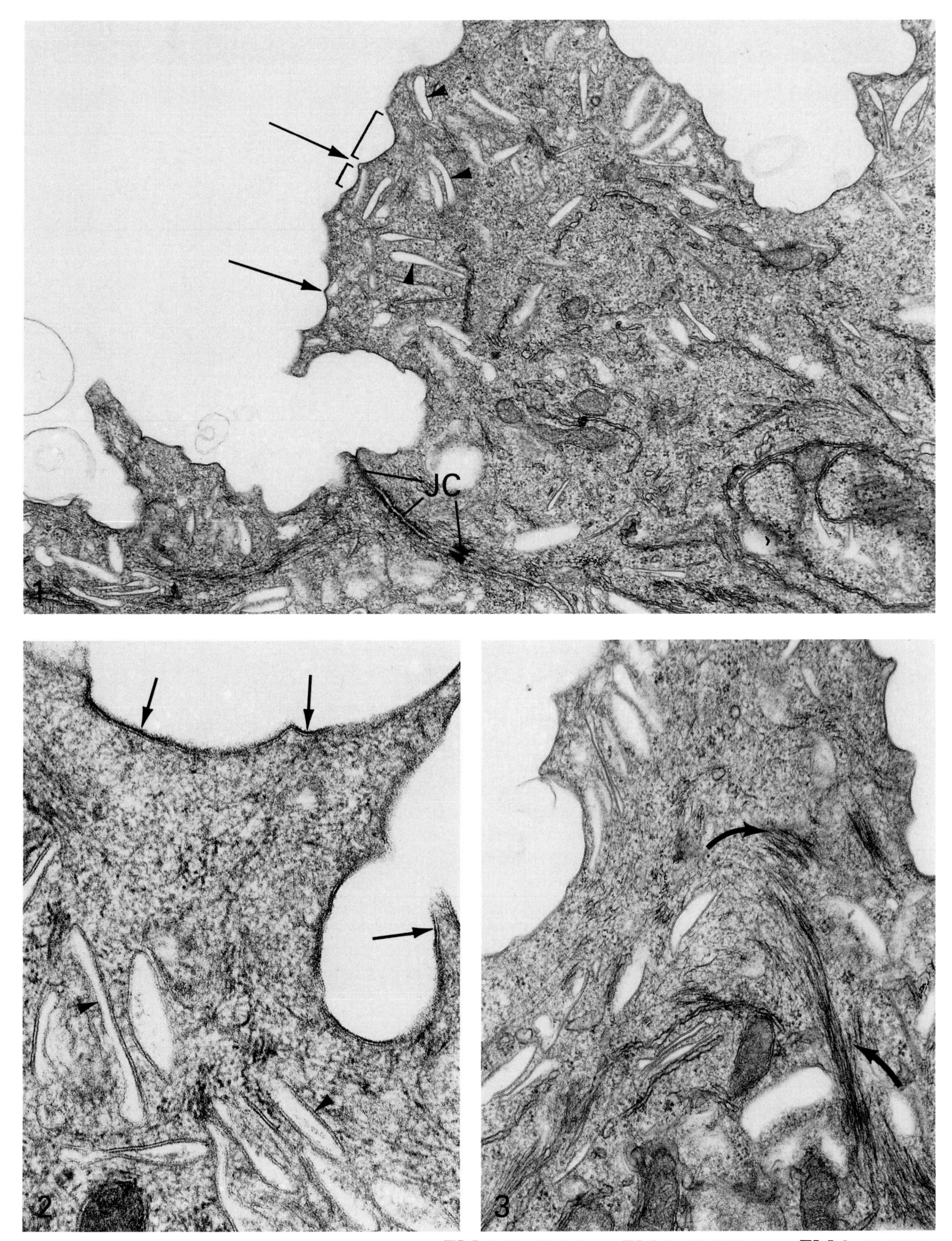

EM 1: 27,200× EM 2: 85,000× EM 3: 45,000×

Basal Specializations

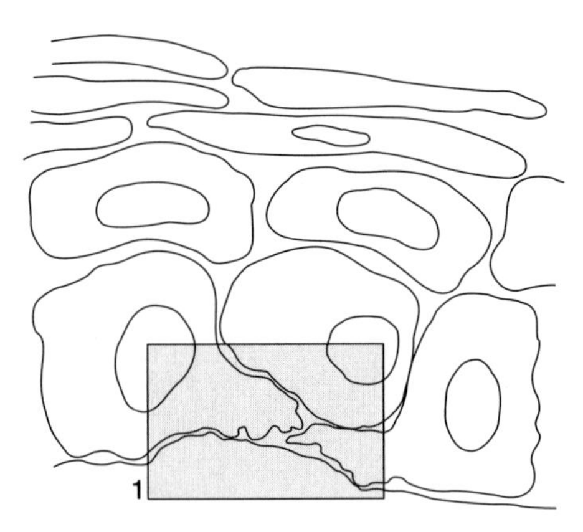

An electron-dense band, the **basal lamina,** or lamina densa (arrows, micrograph 1), follows the contours of the basal surface of epithelial sheets. The basal lamina is the site where epithelial cells attach to the extracellular matrix. Its functions are diverse, some a direct property of the structure itself (e.g., permeability barrier, neurotransmitter breakdown) and others expressed as effects on associated cells (e.g., induction of differentiation, including polarity). In many instances (as in micrograph 1) the basal lamina is associated with a connective tissue layer of fine reticular collagen fibers (arrowheads) that is referred to as the **reticular lamina.** The basal and reticular laminas are visible as a single structure, the **basement membrane,** at the light-microscope level.

The basal lamina is composed of over 100 different polypeptides. Type IV collagen, laminin, fibronectin and heparan (and heparin) sulfate proteoglycan are the most abundant. These proteins are also present in the space **(lamina rara)** between the cell and the basal lamina but are generally less abundant in this region. They bind to one another, to the epithelial cells, and to the underlying connective tissue to form complex patterns. Integrins, membrane proteins specialized for adhesion, mediate many of the signals from matrix to cell. Fibronectin extends from integrin receptors on the basal cell surface through the basal lamina to bind to reticular lamina collagen. Laminin and fibronectin have a particularly strong affinity for tumor cells and appear to facilitate the escape of cancerous cells across the basal lamina into the surrounding tissue.

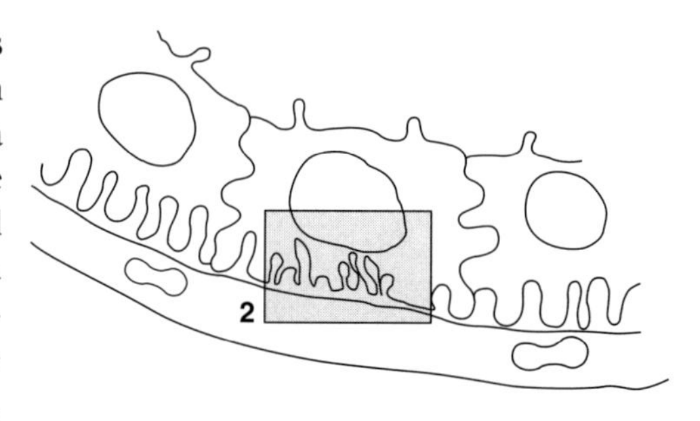

In regions where a major function of the epithelial cell is directed toward secreting into or absorbing substances from the blood, the epithelial basal lamina fuses with the basal lamina of the capillary lining. This brings the epithelial cell close to the blood supply to facilitate exchange. In micrograph 2 the basal laminas of a kidney proximal tubule epithelial cell (Ep) and a capillary endothelial cell (E) appear fused except at one location (arrow) where two separate laminas can be seen. In the kidney tubule shown here, the basolateral surface is specialized, with infoldings (arrowheads) and mitochondria (m) to accommodate an ion pump that retrieves 60 to 70% of the Na^+ lost in the initial filtration process.

Hemidesmosomes (white arrows, micrograph 1) are frequently found on the basal surface of epithelial cells. Their ultrastructure resembles half of a desmosome; however, their protein composition is unique. Tonofilaments insert on the cytoplasmic side. Other proteins project extracellularly, (circle), bridge the lamina rara, and appear to insert into the basal lamina.

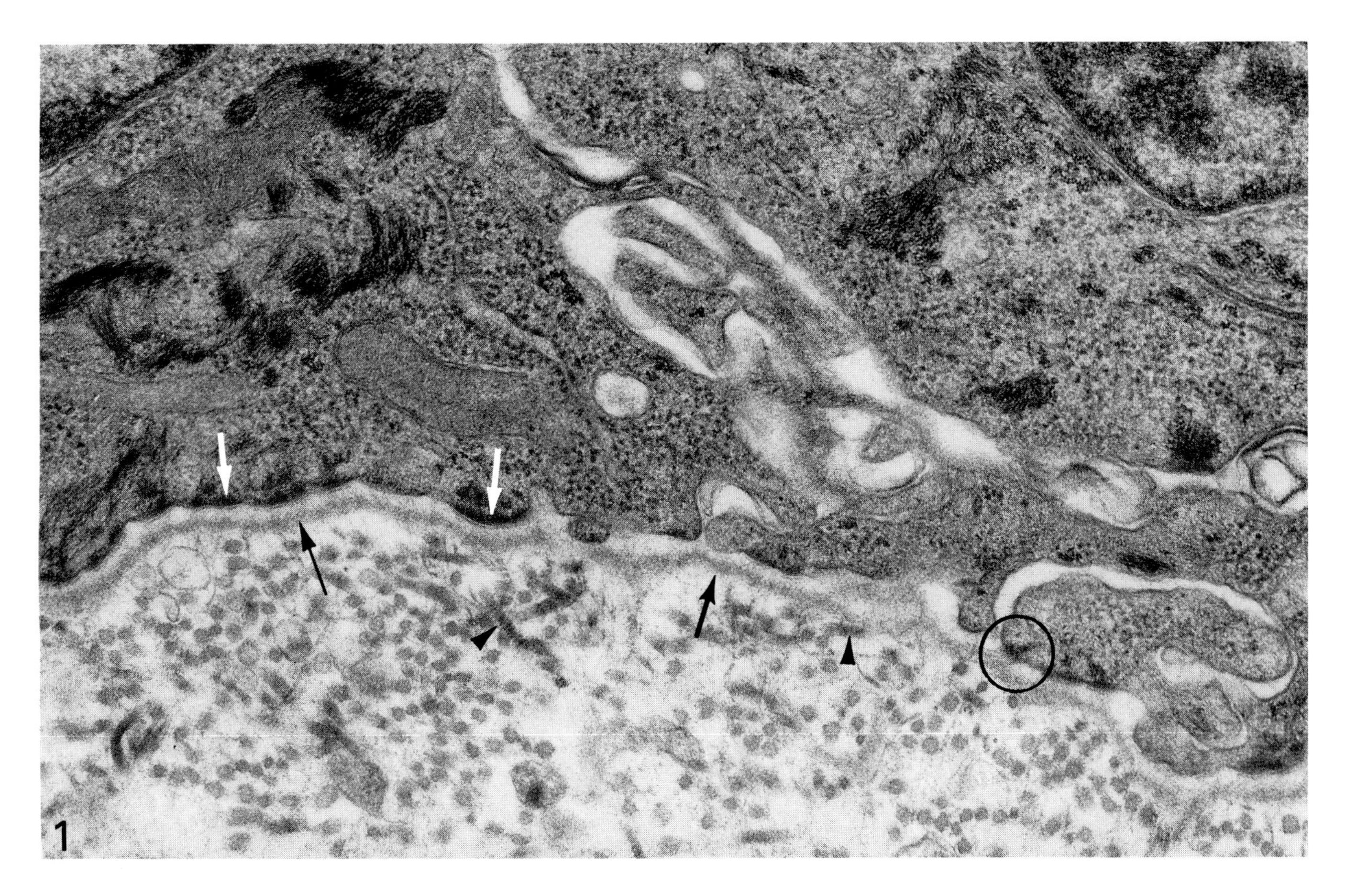

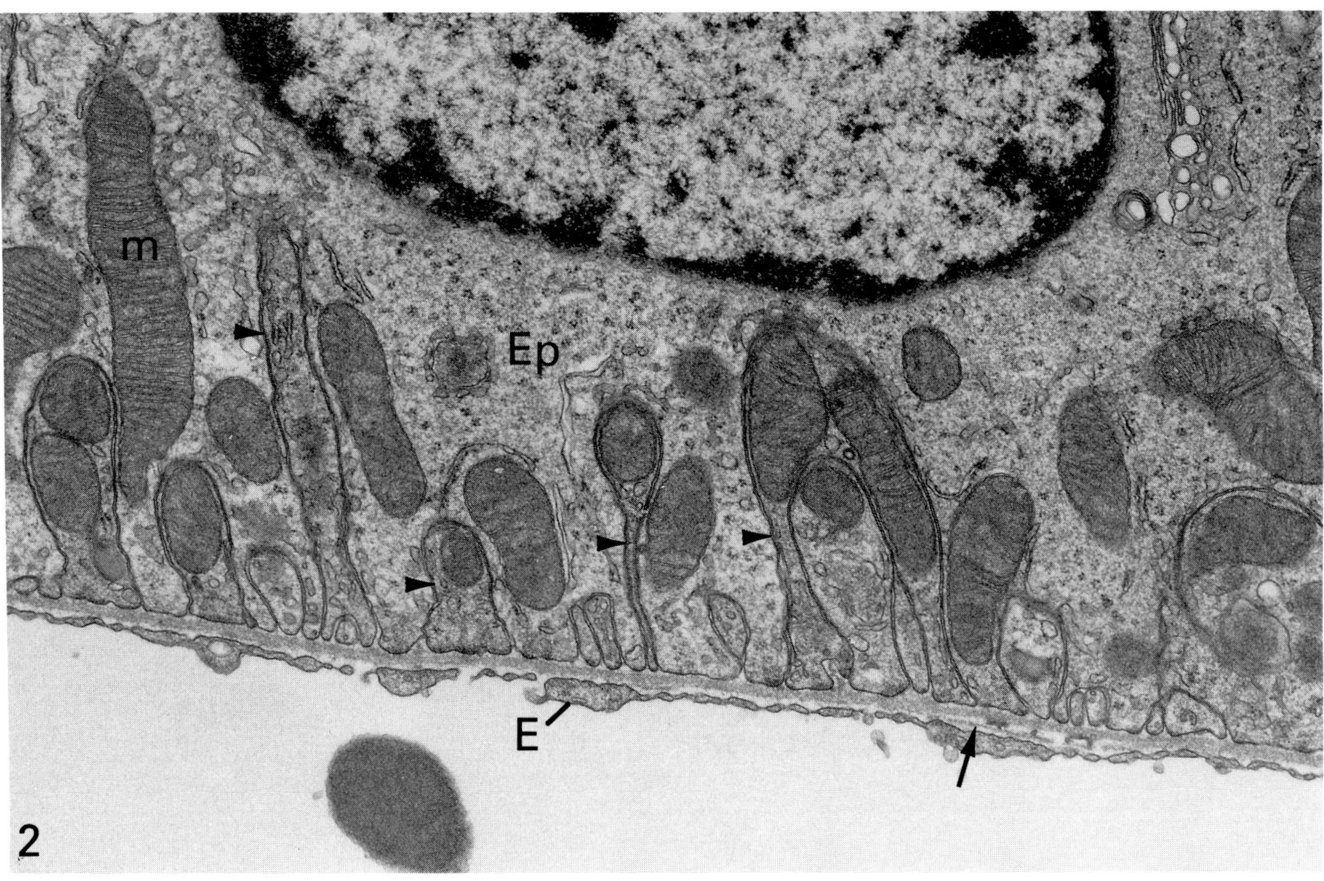

EM 1: 45,500×   EM 2: 18,900×

CONNECTIVE TISSUE

Fibroblast

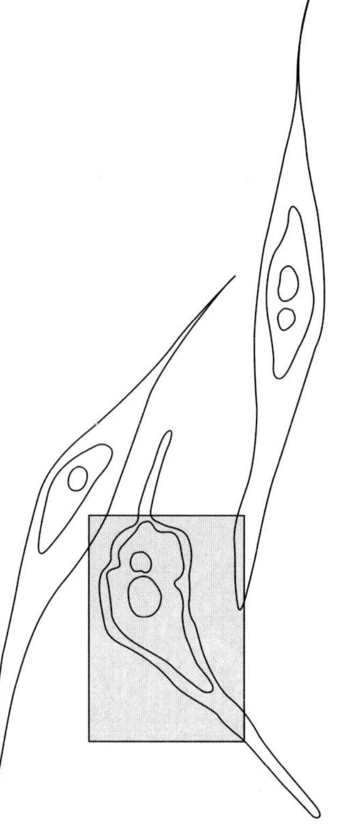

Connective tissue cells, unlike epithelial cells, are not tightly packed, but instead are surrounded by **extracellular fibers** and **ground substance.** The **fibroblast** (F, micrograph), the major cell of connective tissue proper, synthesizes **collagen** and **elastic** fibers and the **proteoglycan** component of the ground substance.

Collagen (c, micrograph) is the most abundant protein in the body and typically the predominant extracellular component of connective tissue. The process of collagen synthesis is complex and involves both intracellular and extracellular steps. Procollagen polypeptide chains are synthesized in the rough ER (r, micrograph), with hydroxylation of specific proline and lysine residues occurring on the growing chains. Sugars are added in both the rough ER and the Golgi complex (not shown).

Assembly of the **triple helical procollagen** molecule occurs in the Golgi and is facilitated by nonhelical peptides at both amino and carboxyl termini. These terminal peptides ensure solubility of the molecule within the cell and prevent the intracellular polymerization into fibrils. From the Golgi procollagen is packaged into secretory vesicles (arrows, micrograph). During or soon after exocytosis, the terminal nonhelical peptides are cleaved by procollagen peptidases and procollagen is converted to tropocollagen. Striated collagen fibrils (arrowheads) consist of precisely arranged tropocollagen molecules (see Connective Tissue, page 72, for details) assembled after cleavage.

Several different types of collagen have been identified. These types differ somewhat in amino acid content but retain such common characteristics as a high glycine content and the presence of two unusual amino acids, hydroxyproline and hydroxylysine. Hydroxyproline forms cross-links that provide intra- and intermolecular stability to collagen fibrils. A deficiency in vitamin C, an essential cofactor in the formation of hydroxyproline, leads to scurvy, a degenerative connective tissue disorder.

The most common types of collagen are Types I, II, III, and IV. Type I, shown in the micrograph, predominates in bone, tendon, and skin; Type II in cartilage; Type III (reticular fibers) in lymphoid organs, vessels, and muscle; and Type IV in basal lamina. Type IV does not organize as fibrils, but instead forms a network of individual tropocollagen molecules associated with other matrix proteins.

Proteoglycans are not observable in connective tissue proper on routine electron micrographs. They occupy the clear areas (*, micrograph) surrounding the cell and between collagen fibers and also maintain the spacing between individual collagen fibrils. Proteoglycans consist of a core protein covalently attached to a series of highly sulfated repeating disaccharide units called **glycosaminoglycans** (GAGs) (see Connective Tissue, page 84, for details). Proteoglycans associate with collagen and form a hydrated network that is important in the movement of cells and molecules through loose connective tissue.

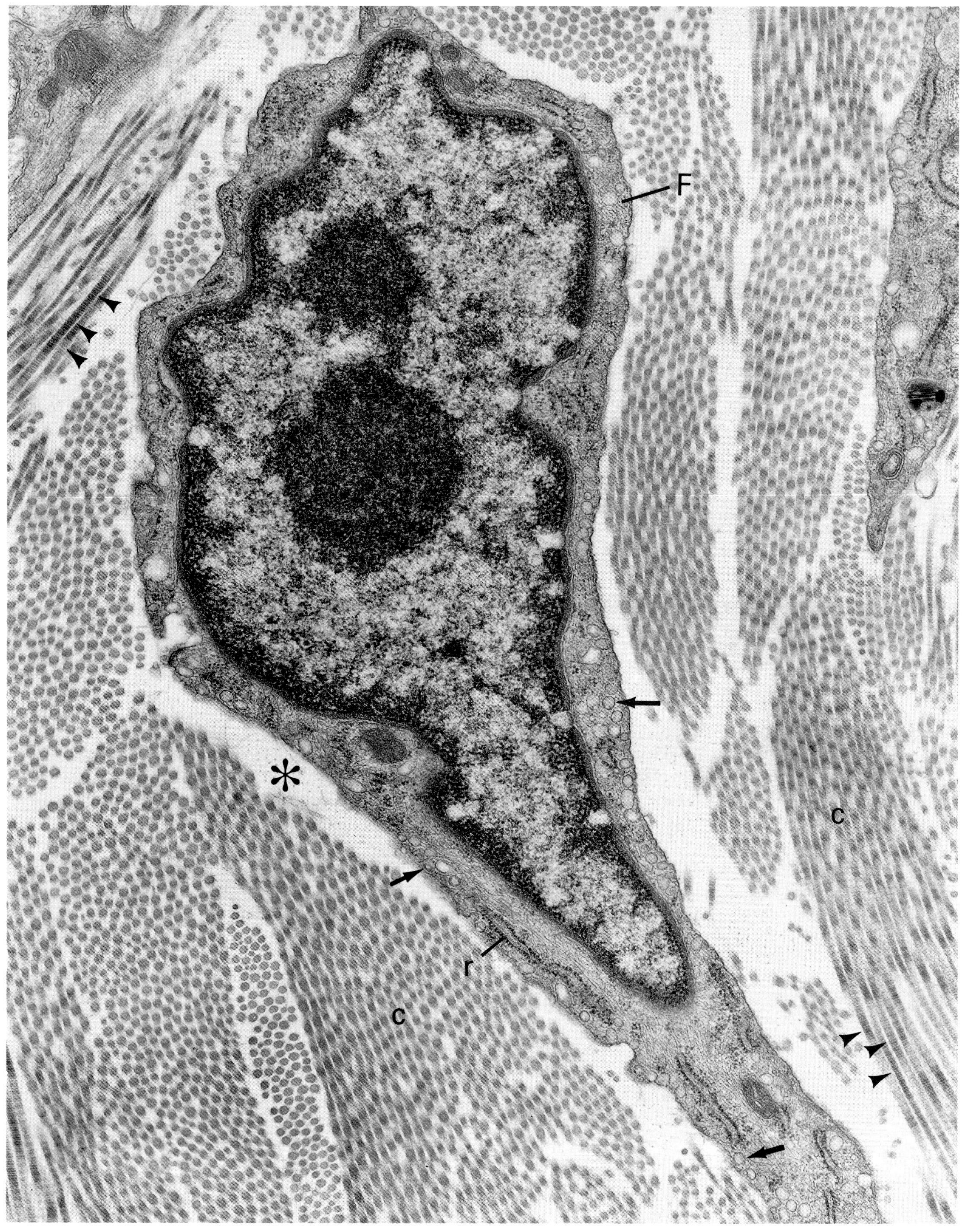

EM: 28,000×

Collagen and Elastic Fibers

Collagen, visible in the light microscope as fibers with a diameter of 0.5 to 20 μm, is made up of smaller units, **fibrils.** In electron micrographs of loose and dense connective tissue, collagen fibrils (c, micrograph 1), with a diameter of 20 to 100 nm, are visible in cross section. In longitudinal section striations with a periodicity of 67 nm are apparent. These are created by the staggered arrangement of the individual tropocollagen molecules separated end to end by a gap of 35 nm. Additional narrow bands within this pattern represent stain binding to polar residues of tropocollagen lined up in register.

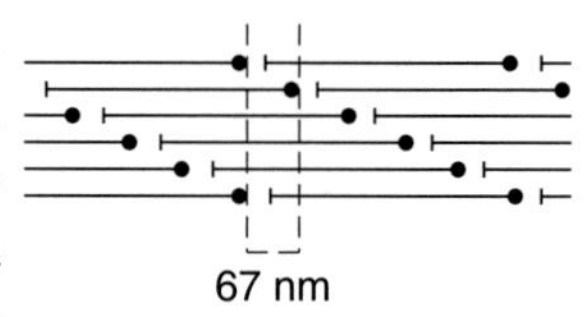

Modified from L. C. Junquiera et al. *Basic Histology,* 7th ed., Appleton and Lange Medical Publications, Norwalk, Conn., 1992.

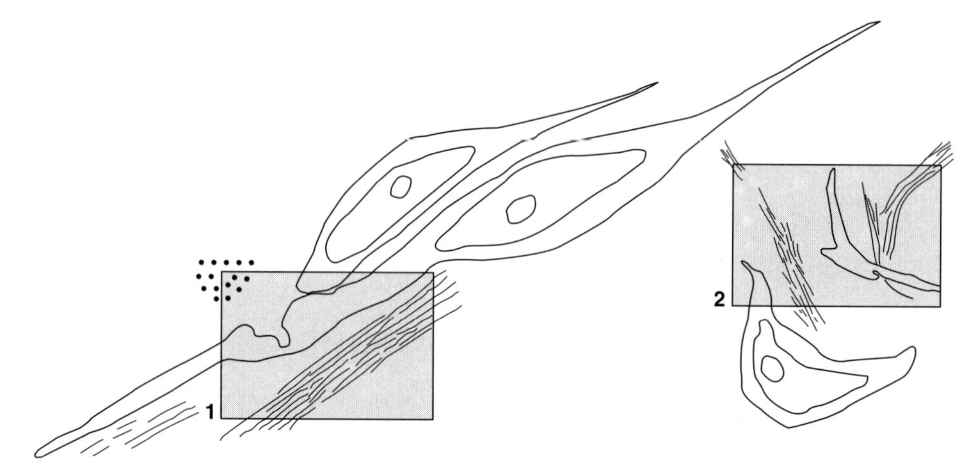

Like collagen, **elastic fibers** are synthesized by fibroblasts and are ubiquitous. In contrast to most collagen, these relatively thin (0.2–1 μm) branching fibers are not composed of smaller striated fibrils. One elastic fiber viewed under the light microscope is seen as a single structure in routine electron micrographs. The arrows in micrograph 2 identify an elastic fiber that appears to be branching. Each fiber consists of (1) **elastin** (e), a unique protein with primarily hydrophobic, nonpolar amino acids, which does not stain and appears as an amorphous strip, and (2) stained 10-nm **microfibrillar proteins** (arrowheads) containing hydrophilic residues that form a sheet around the elastin.

The resilient character of elastin, which is particularly important in the lung, aorta, and skin, is due in part to the intermolecular cross-links that form extracellularly between tropoelastin molecules. A defect in lysyloxidase, an enzyme necessary for both the cross-linking of tropocollagen and tropoelastin, results in hyperextensible skin and joints, one type of Ehlers–Danlos syndrome.

Extracellular fibers attach to cells either directly via membrane receptors (as with elastin) or indirectly via other molecules such as laminin (in the case of collagen Type IV) and fibronectin (with collagen Types I–III). This association can influence gene expression and functions in many diverse ways. The elastic fiber receptor holds recently synthesized tropoelastin in proper orientation for cross-linking and final fiber formation.

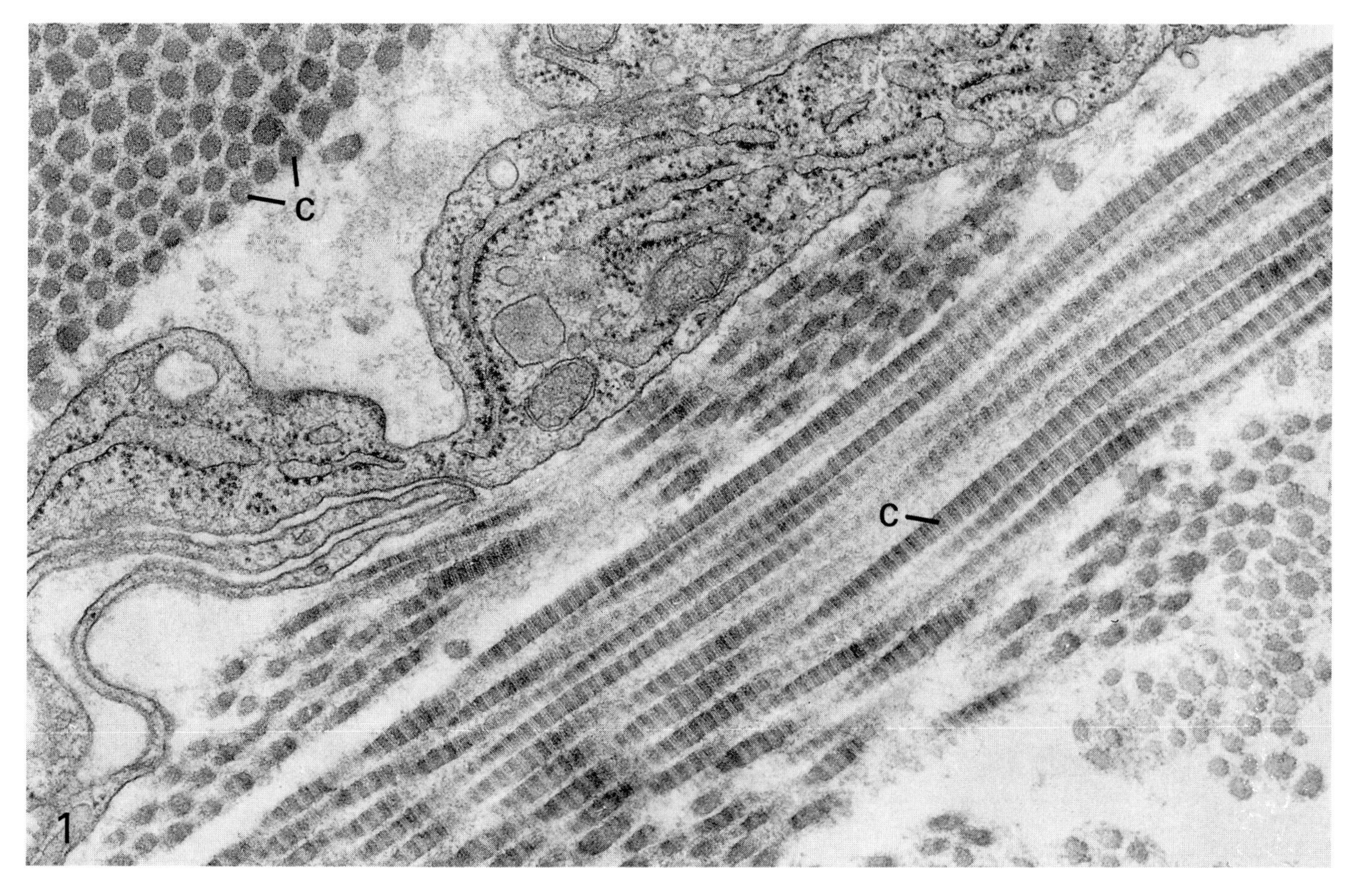

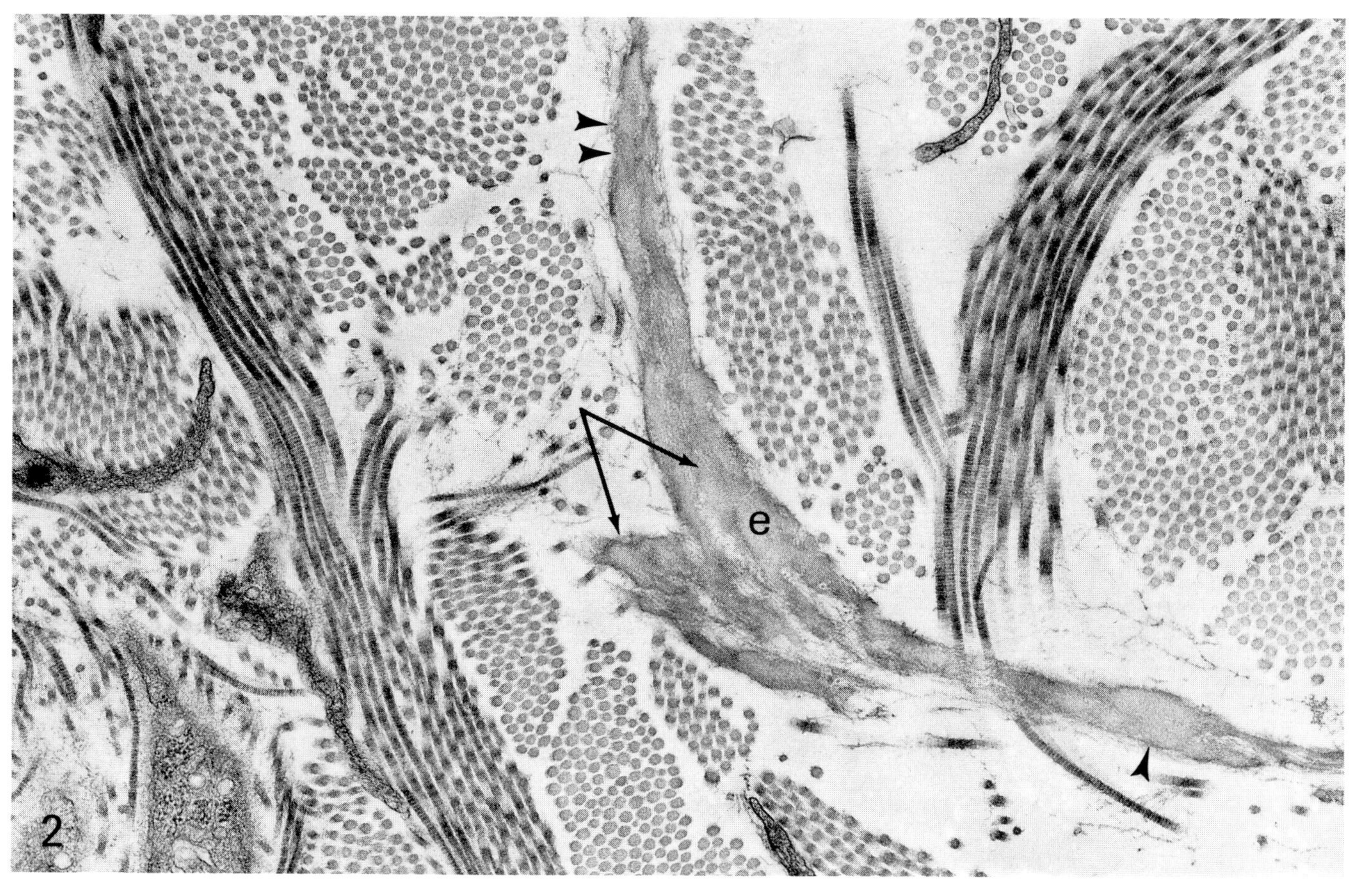

EM 1: 35,000× EM 2: 23,000×

MACROPHAGE

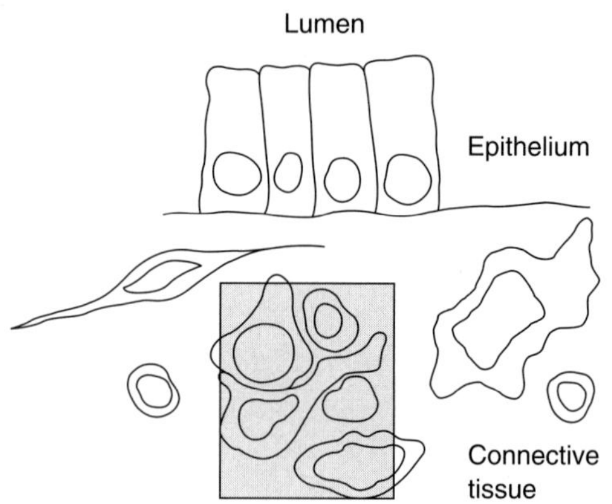

Macrophages (M, micrograph) are a diverse group of specialized cells that carry out a wide range of functions. The most common role of these cells is everyday scavenging as avid phagocytes. Other, more specialized functions include antigen presentation and the secretion of mediators of the immune response.

In section, macrophages frequently exhibit an irregular shape as they are caught in some stage of active movement. Characteristically they contain accumulations of heterogeneous bodies in **secondary lysosomes** (arrows, micrograph), which represent phagocytosed material in the process of digestion.

Phagocytosis by the macrophages in loose connective tissue provides defense against foreign antigens entering across epithelial barriers. More specifically, phagocytosis by macrophages (1) kills invading organisms (by using the respiratory burst; see Cell, page 20); (2) removes cellular debris, particularly following inflammation; and (3) is essential prior to presentation of antigens to lymphocytes in the immune responses.

Phagocytosis depends upon receptor-coupled activation, and in many cases the receptor is for the Fc portion of antibodies (see Immune system, page 198). Each antibody class (e.g., IgG, IgA, IgE) produced by plasma cells (P, micrograph) has a unique Fc region and is involved in a different type of immune response. Following secretion, many antibodies enter the vascular system (capillary, C, micrograph) and have effects at distant locations, while others act locally. Certain macrophages have receptors for the Fc portion of both IgG and IgA and can therefore act as scavengers for all antigen bound to both of these antibody classes.

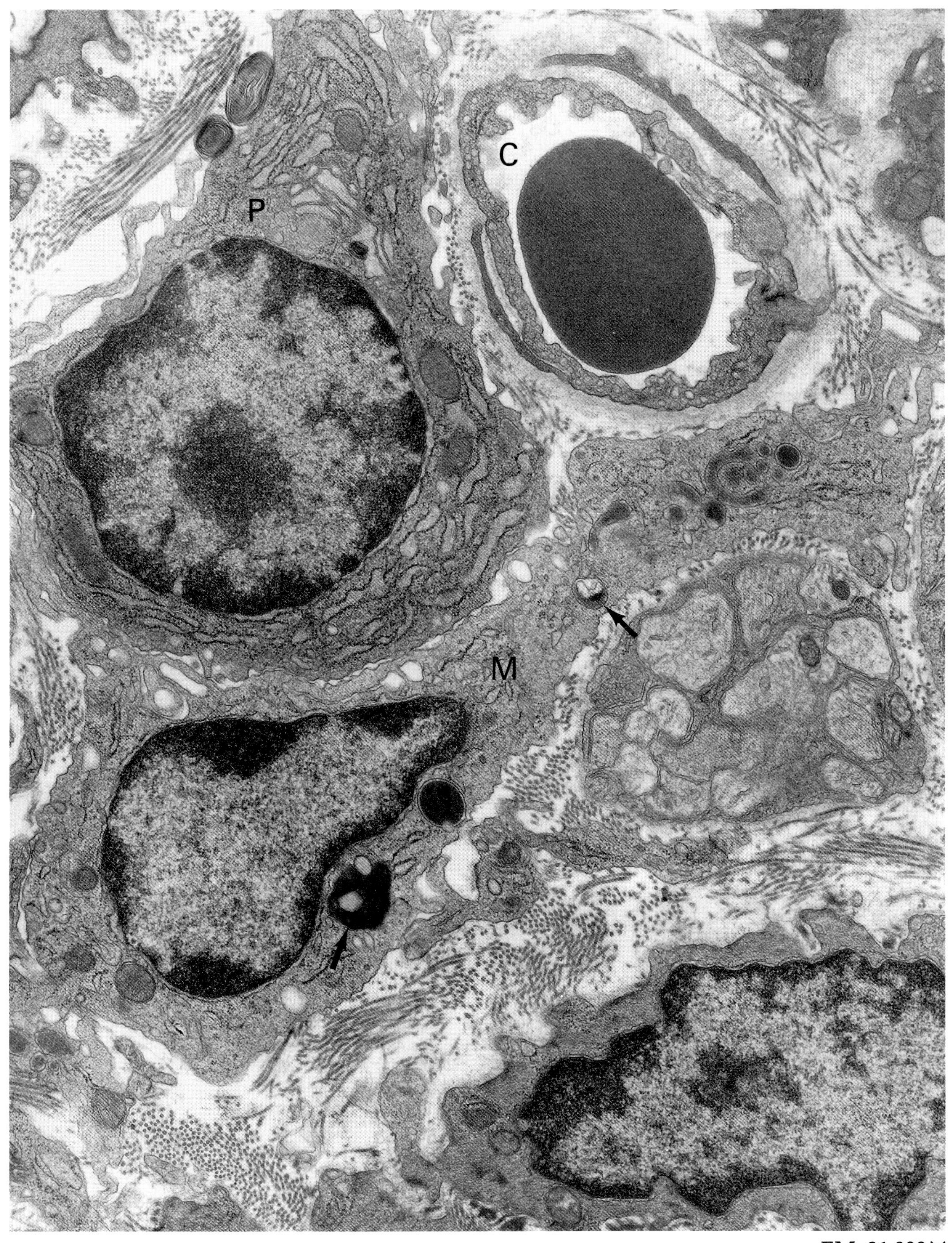

EM: 21,000×

MAST CELL

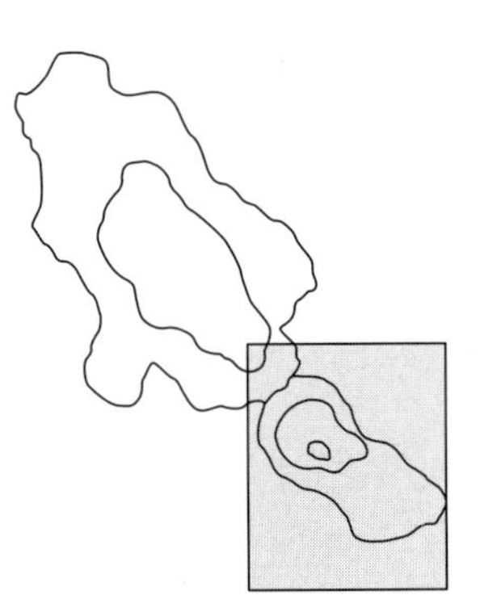

Mast cells (micrograph) are large cells, 20–30 μm, with a round nucleus (N), conspicuous granules (g), and folds of membrane (arrows) that project from the surface. They are found throughout the body in loose connective tissue, particularly under the epithelium of the respiratory and gastrointestinal tracts, and are frequently concentrated next to blood vessels. Mast cells are a heterogeneous population that can adapt to changes in the immediate environment by altering both type and quantity of synthetic activities.

In their role in maintaining health, mast cells facilitate the movement of defense molecules and cells into sites of foreign invasion. They increase vascular permeability allowing plasma exudates containing antibodies and other defense molecules to enter the "threatened" tissue. They also recruit leukocytes (particularly neutrophils and eosinophils) to the area and facilitate their movement across vessel walls. In certain cases the action of mast cells results in an early and late physiological response. The late phase reaction is associated with the presence of large numbers of leukocytes.

Even though cytokines and neuropeptides can directly activate mast cells, the most well-documented mechanism is via the antibody **IgE.** IgE, formed by plasma cells during initial exposure to certain antigens, binds (by the Fc portion) to receptors on the mast cell. Mast cells, packed with granules and coated with IgE specific to the antigen, migrate to regions under epithelia where they present concentrated antibody in an area where the same antigen may reappear. At the second exposure to the particular antigen, receptor aggregation resulting from the binding of a single antigen molecule to two or more IgE molecules activates the mast cell.

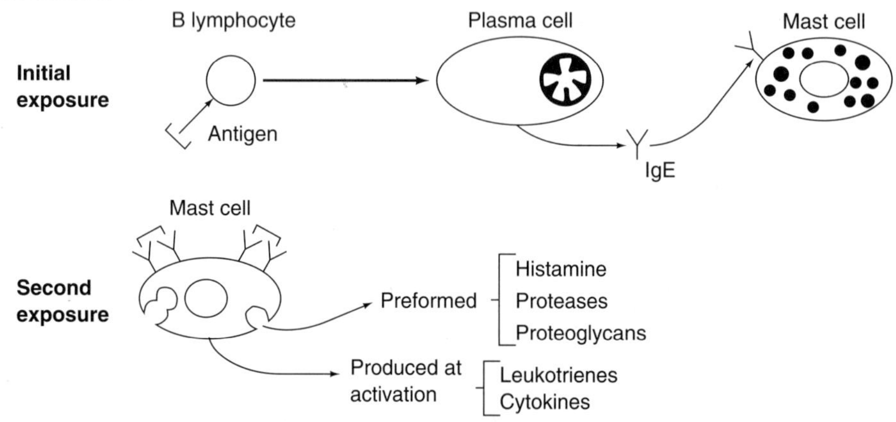

On activation mast cells release three classes of bioactive molecules that mediate their action: those preformed and stored in the granules, newly formed membrane-derived molecules (e.g., leukotrienes), and newly formed cytokines (e.g., tumor necrosis factor-α, or TNF-α). The most well-defined associations between mediator and defense actions are the increase in vascular permeability due to histamine release and the recruitment of leukocytes by leukotrienes and TNF-α. Proteoglycans function primarily to bind and concentrate the mediators within the granules. Secretory granules of other cell types possess proteoglycans that serve a similar packaging function.

The combined effects of mast cells and invited inflammatory cells on bronchiole constriction and airway mucus secretion can be life threatening, as in severe allergy and bronchial asthma.

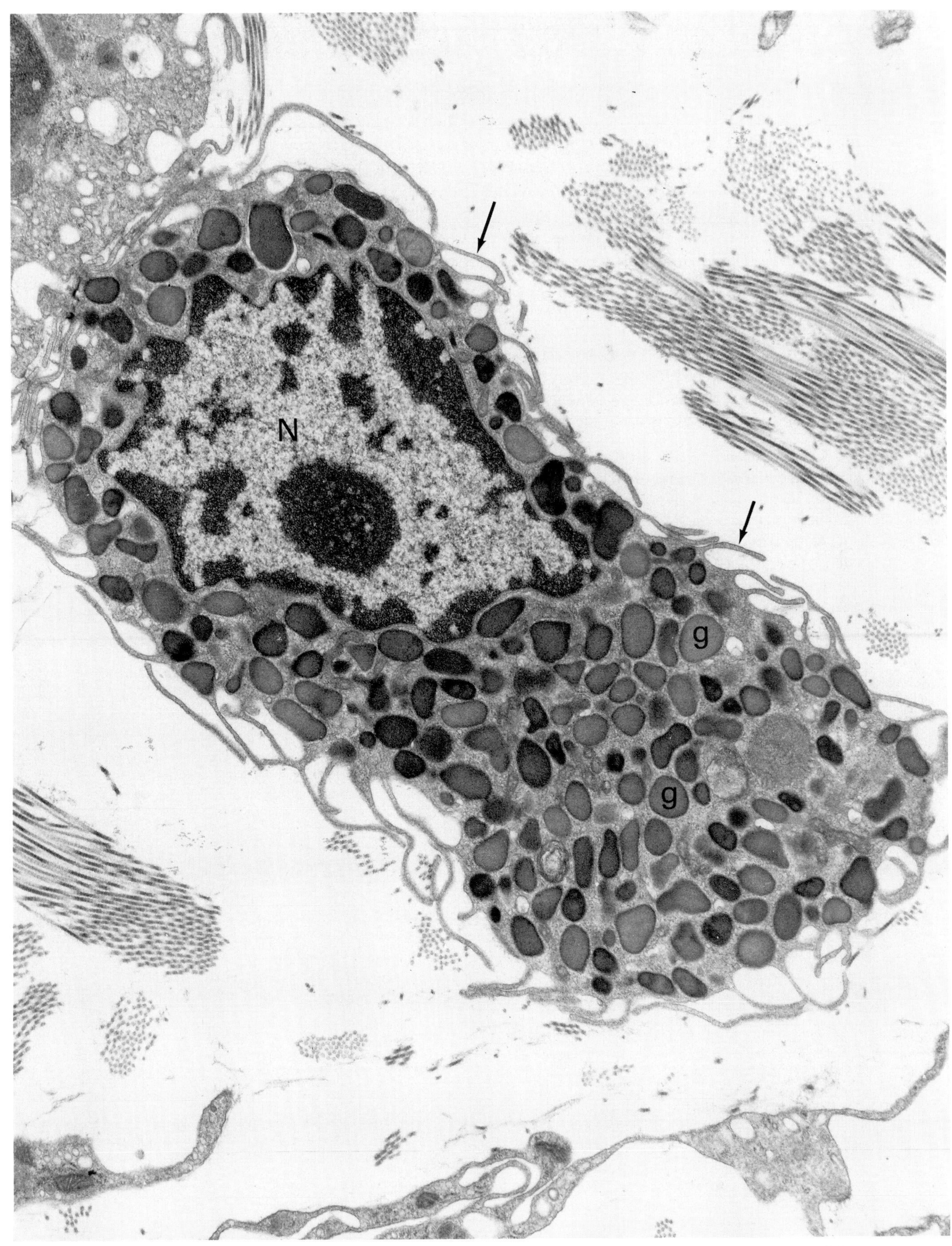

EM: 17,500×

Mast Cell–Eosinophil Interaction

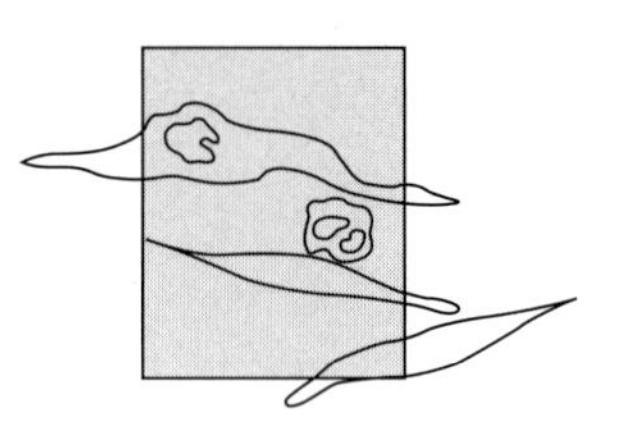

When **mast cells** are **activated,** the contents of their granules are released at once in a process known as compound exocytosis. During this event granules fuse with one another, forming channels to the cell surface for quick release. The two activated, degranulated mast cells (M) in the micrograph are easily recognized by the spaces (s) that represent sections through these channels after activation. Excess cell membrane, the result of fusion of the granule membrane with the cell membrane at exocytosis, is the source of the characteristic mast cell surface folds.

Eosinophils (E, micrograph) are found at the sites of mast cell activity, attracted from nearby blood vessels to the region by factors released during mast cell activation. Recruitment is an important means of bringing eosinophils to areas where they are most effective in defense, such as destruction of parasitic larval schistosomes (see Blood, page 176). Once at the site, eosinophils appear to modulate and neutralize the potentially deleterious effects of mast cell action.

Eosinophils inhibit the synthesis of certain mast cell mediators and also degrade mediators following their release. An eosinophil histaminase deaminates mast cell histamine, and a peroxidase converts certain mast cell leukotrienes to isomers that lack vasoactive bronchoconstrictive activity.

Aside from dampening mast cell activity, eosinophils seem to complicate defense reactions in certain instances. The release of basic proteins that make up the crystalline core (arrowheads, micrograph) of the eosinophil granule is the major factor responsible for certain allergic complications. The respiratory epithelial damage associated with asthma is characterized by the infiltration of both mast cells and eosinophils.

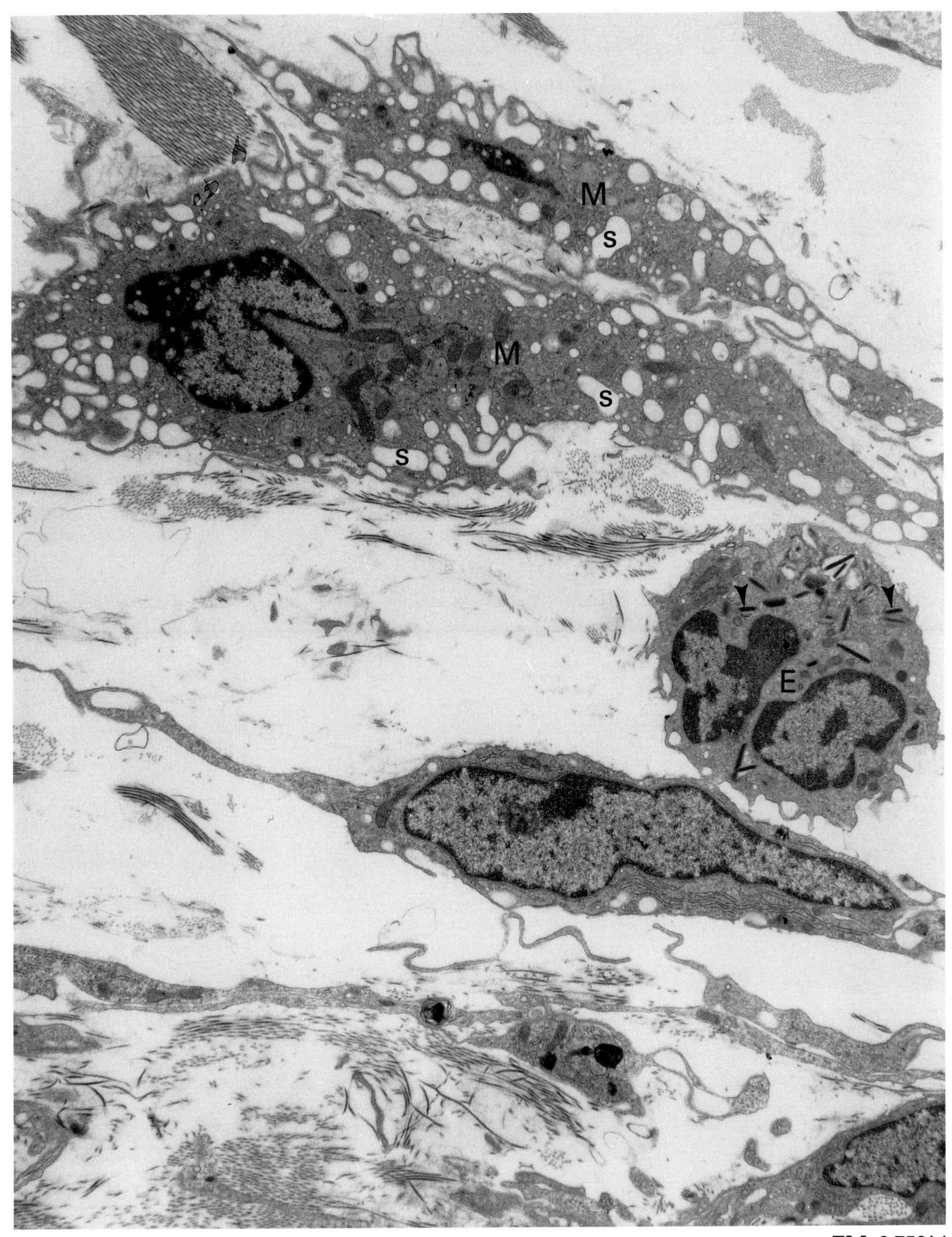

EM: 8,750×

Lymphocytes and Plasma Cells

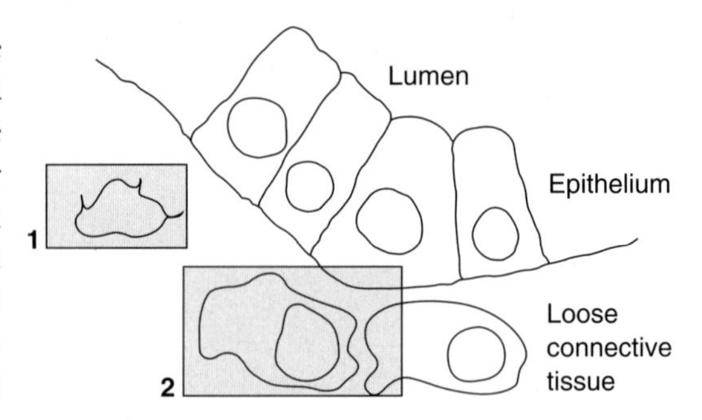

Lymphocytes are the principal cells of the immune system. In connective tissue, lymphocytes are organized into follicles or scattered between other connective tissue elements. Most lymphocytes (micrograph 1) have a similar ultrastructure; they are small cells (7 – 9 μm) with a thin rim of cytoplasm. In routine electron micrographs it is not possible to distinguish B from T types, or immature (virgin) from most products of activation (e.g., memory, T cytotoxic cells). One exception, however, is the **plasma cell** (micrograph 2), a product of B cell activation that synthesizes and secretes antibodies. This cell has an eccentric nucleus (N), a well-developed Golgi (G), and a cytoplasm packed with dilated rough endoplasmic reticulum (r) containing many copies of a single type of antibody.

Lymphocytes and plasma cells are commonly found in the loose connective tissue under secretory epithelia. An immune response in this region provides defense against foreign material entering the body by crossing an epithelial barrier. Local plasma cells specifically synthesize and release IgA, the antibody of secretions.

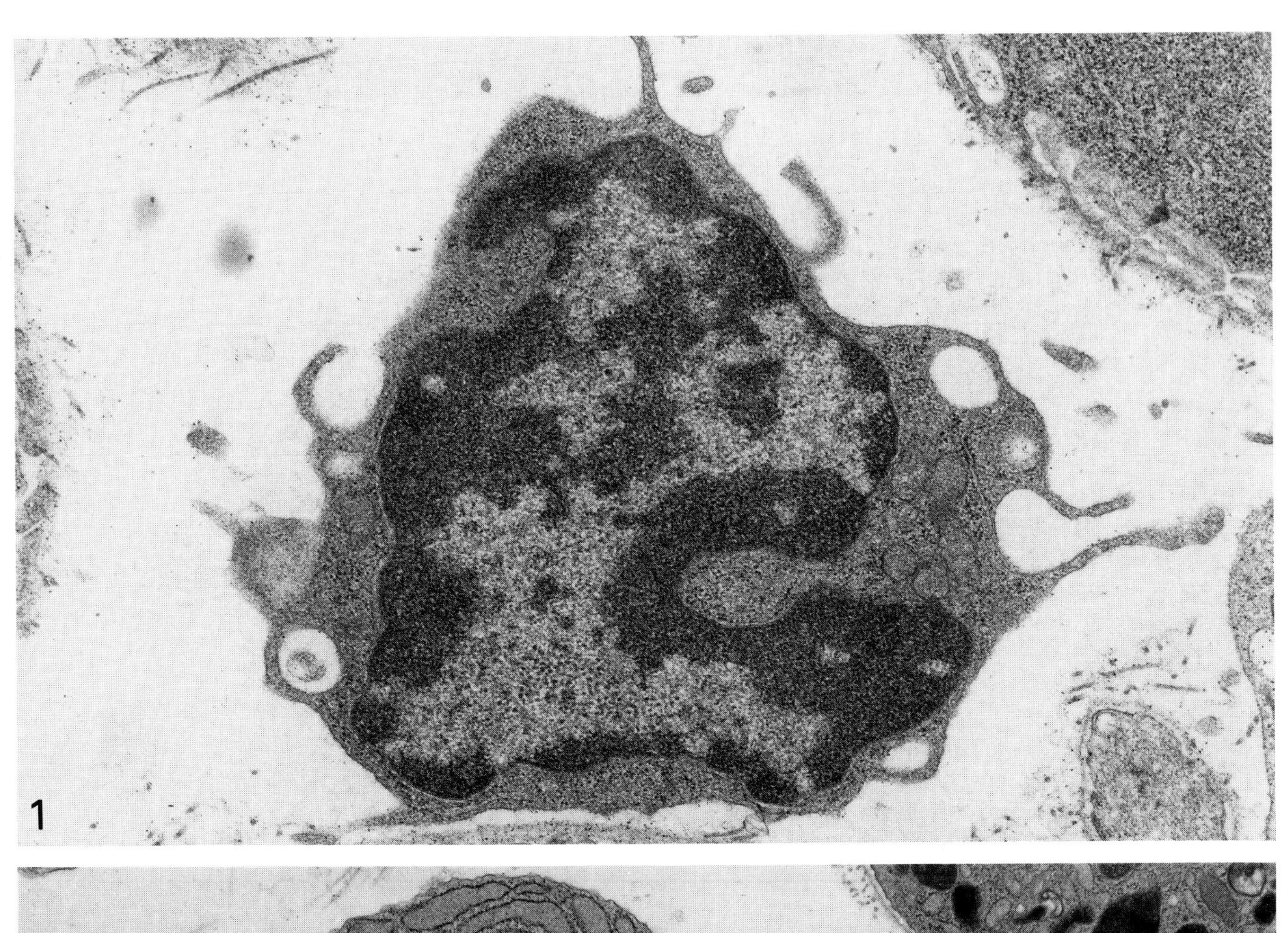

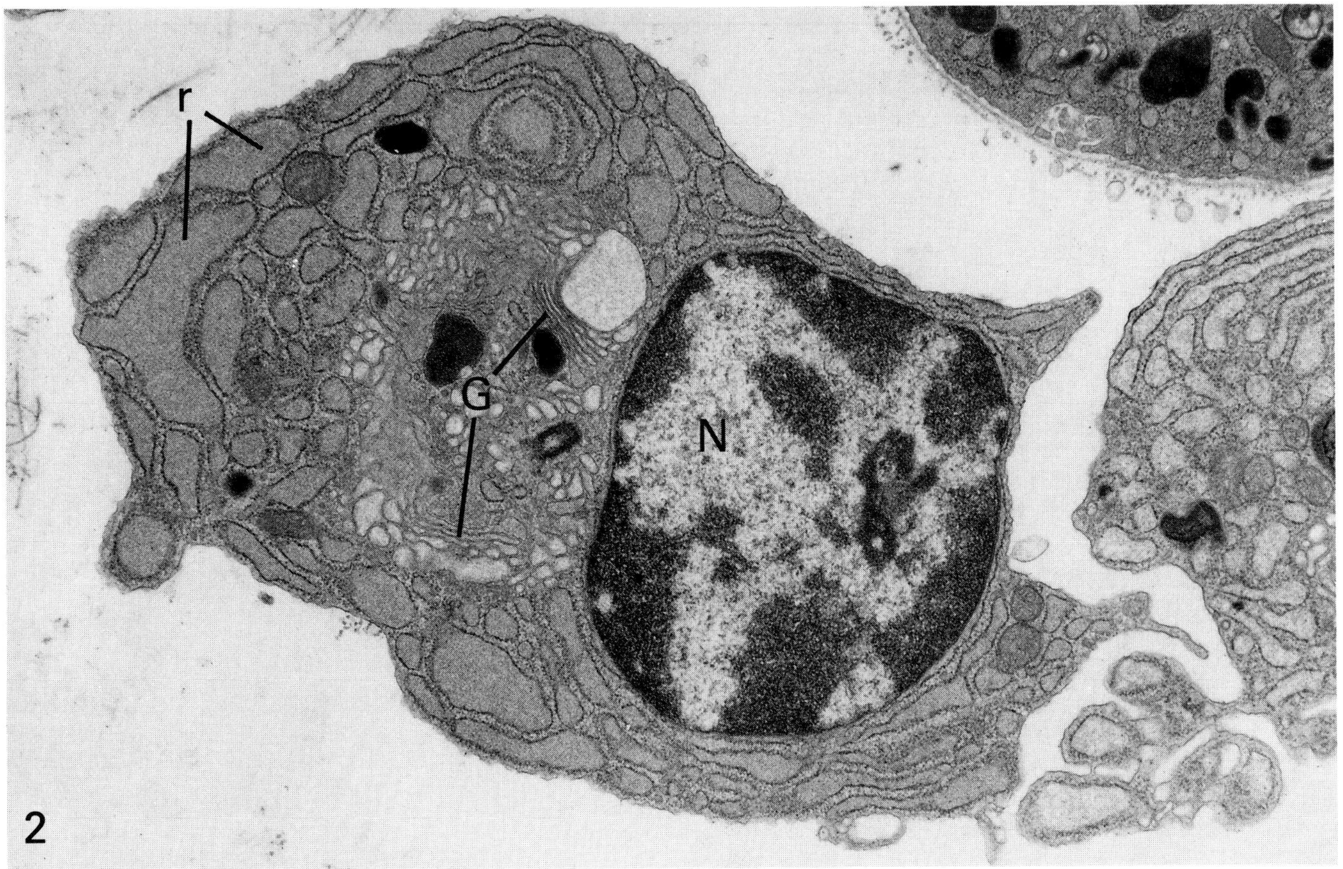

EM 1: 26,000× EM 2: 19,000×

ADIPOCYTE

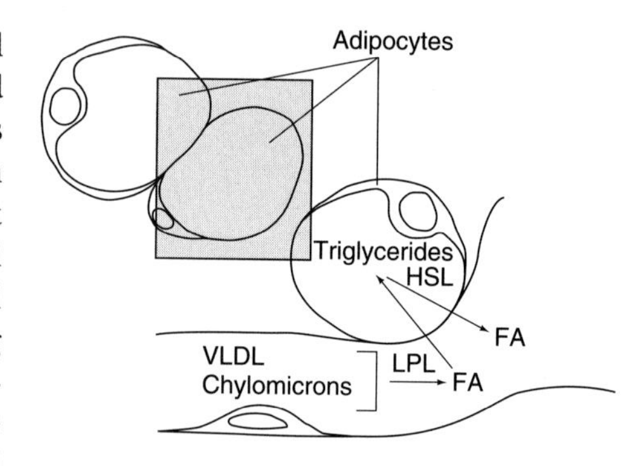

Adipocytes (A, micrograph), cells specialized for fuel storage, are found isolated in loose connective tissue or packed together in groups in certain locations. The fuel, **fatty acids,** is either stored in many small lipid droplets (multilocular, brown fat) or concentrated almost entirely into one large lipid droplet (unilocular, white fat) as in the micrograph. In adults, nearly all adipocytes are of the white type and function as the principal fatty acid energy store in the body. Two main distinguishing morphological characteristics of unilocular adipocytes are illustrated in the electron micrograph: (1) the large size of the cell (compare to the red blood cells, R, in the capillary, C) and (2) the massive lipid droplet that forces the nucleus (N) and cytoplasm to a peripheral location, creating a "signet ring" appearance.

Fatty acids are packaged in the lipid droplet as esters with glycerol, forming **triglycerides** (TGs). TGs are continually being synthesized and broken down as energy demands fluctuate. Some of the fatty acids are synthesized from glucose within the adipose cell itself. Many fatty acids, however, originate in the intestine and liver and are transported to adipocytes and other target organs packaged with proteins. These lipoprotein packages, chylomicrons from the intestines and very low density lipoproteins (VLDLs) from the liver, reach the adipocyte via capillaries. Each package that reaches the adipocyte contains thousands (in VLDLs) to millions (in chylomicrons) of TG molecules.

Fatty acids and monoglycerides are released from chylomicrons and VLDLs by an enzyme, **lipoprotein lipase** (LPL), synthesized by adipocytes and adsorbed (probably via a heparan sulfate) to the surface of the endothelium (arrowheads, micrograph) of adjacent capillaries. The freed fatty acids and monoglycerides are transported to the adipose cell, where TGs are resynthesized within the smooth endoplasmic reticulum and added to the lipid droplet.

Lipoprotein lipase is found associated with endothelial cells in many different tissues and functions in a similar way in the supply of fatty acids to local cells for storage or oxidation. The activity of this enzyme changes in response to foreign invasion. Cachectin (tumor necrosis factor), a protein synthesized by macrophages in response to pathogens and endotoxins, inhibits lipoprotein lipase, causing the weight loss characteristic of wasting in chronic disease.

The breakdown, or lipolysis, of triglycerides in cells is regulated by a **hormone-sensitive lipase** (HSL) within the adipocyte that is controlled both by hormones (e.g., epinephrine and glucagon) and neurotransmitters (e.g., norepinephrine). Each adipocyte is covered with many different receptors for these mediators.

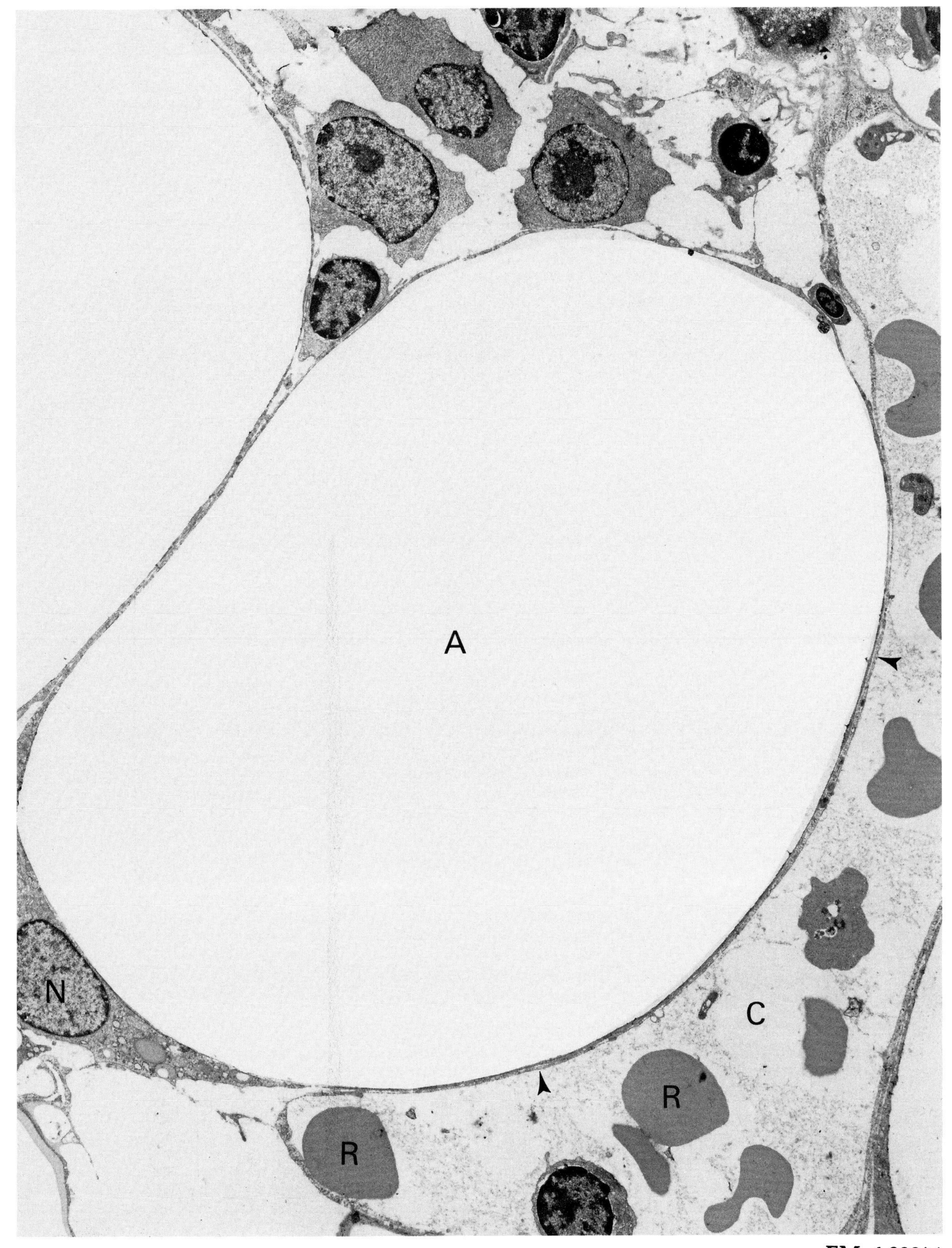

EM: 4,800×

Hyaline Cartilage

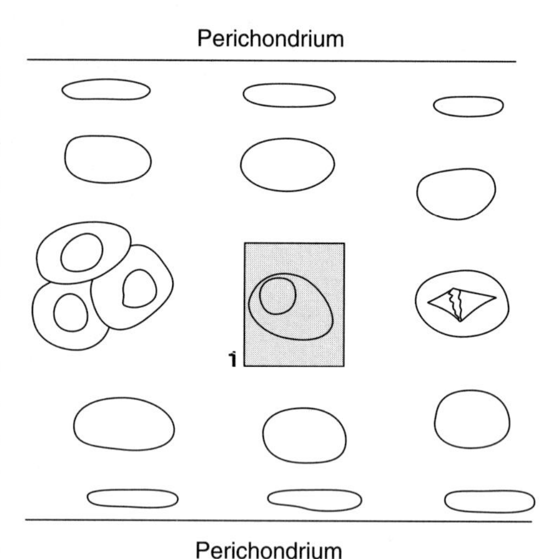

Cartilage is found in regions requiring support in conjunction with tensile strength (fibrocartilage), elasticity (elastic cartilage), and relative rigidity (hyaline cartilage). The functional characteristics of cartilage are carried out primarily by the fibers and proteoglycans of the extracellular matrix. The cells that synthesize these matrix components, **chondrocytes** (C, micrograph 1), have well-developed Golgi (G, micrograph 1) and rough endoplasmic reticulum (r, micrograph 1) characteristic of protein-secreting cells.

In both hyaline and elastic cartilage, chondrocytes develop from progenitor cells located in a surrounding connective tissue perichondrium (appositional growth). Even as differentiated cells are surrounded by matrix, they retain their ability to undergo division (interstitial growth).

In routine electron micrographs, collagen (arrows, micrograph 1) is the only obvious component of the extracellular matrix. **Type II collagen,** characteristic of hyaline and elastic cartilage, is composed of thin (20–30 nm), faintly striated fibrils. The fibrils are not grouped together into fibers and therefore not typically observed in the light microscope.

Proteoglycans, an important component of all types of cartilage, are particularly prominent in hyaline cartilage (10% of the wet weight of the tissue). The **glycosaminoglycans** (GAGs) chondroitin sulfate and keratan sulfate are covalently linked to a **protein core,** forming the basic proteoglycan unit. In cartilage, large numbers of these units form **aggregates with hyaluronic acid** (an extremely large GAG) to form molecular structures 1200 nm long. The proteoglycans, like collagen, are synthesized via rough ER and Golgi. Hyaluronic acid, however, is synthesized at the plasma membrane, with the growing chain passing to the cell exterior. The final assembly into the aggregates of proteoglycans and hyaluronic acid occurs outside of the cell. Each hyaluronic acid molecule binds up to 100 proteoglycan molecules. Following isolation, hyaluronic acid aggregates appear as shown in the diagram; however, in routine electron micrographs they form 70-nm particles (arrows, inset).

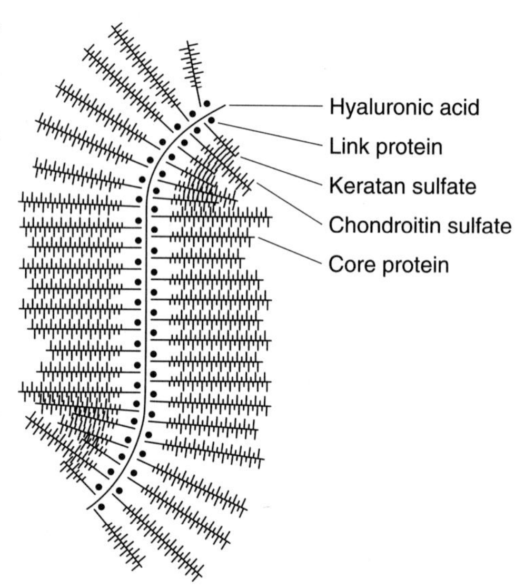

Modified from L. Stryer, ***Biochemistry,*** **Freeman,** New York, 1988.

The rigidity of cartilage is a result of swelling due to the high concentration of water associated with the anionic sites on the GAGs. The hydrated GAGs can be displaced, resulting in some deformability, but the large proteoglycan aggregates, "bound" between collagen fibrils, return to their original location, drawing water and providing resiliency. Bound water functions also as the medium for the transport of nutrients and wastes in this avascular tissue.

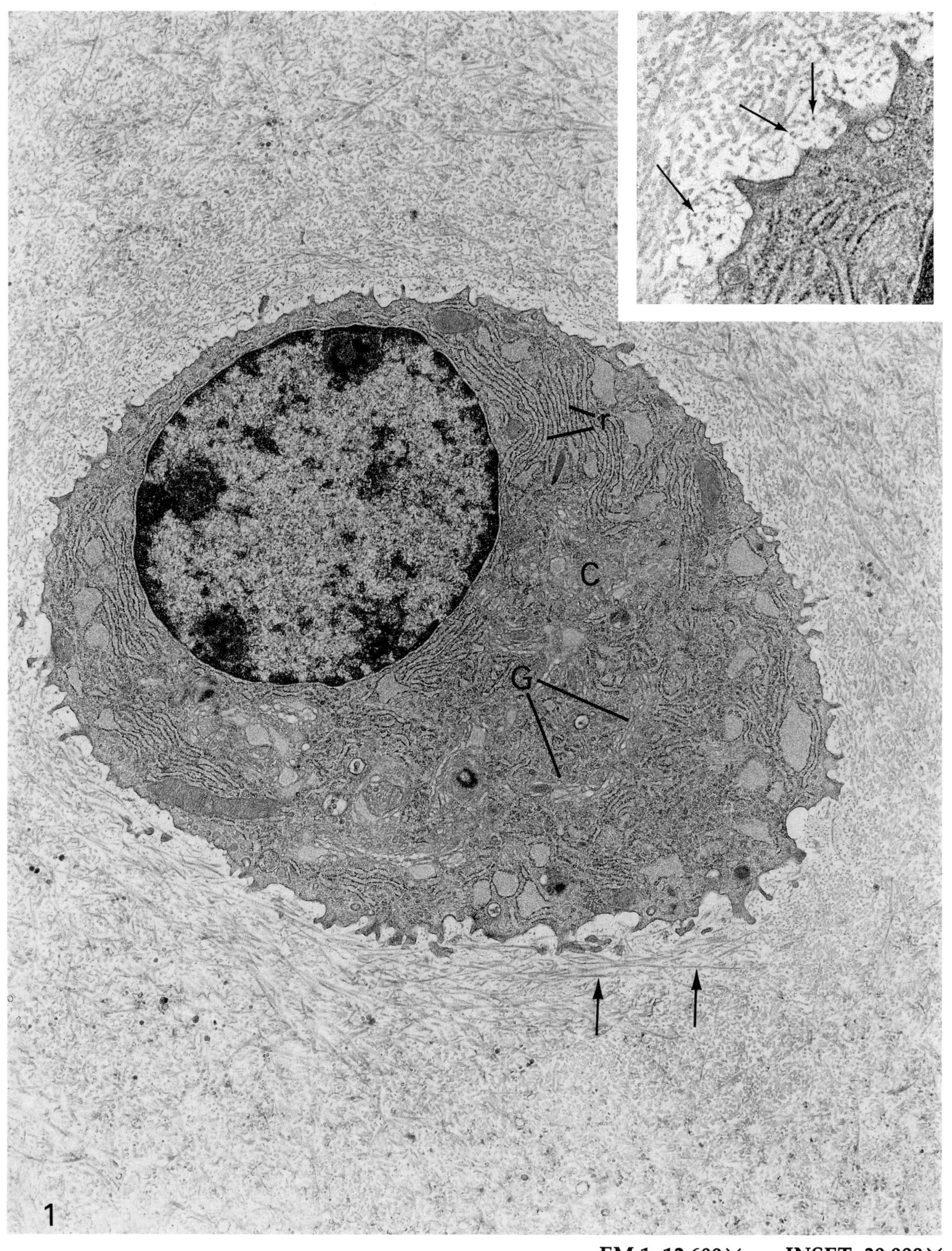

EM 1: 12,600× INSET: 30,000×

BONE: Osteoblasts and Osteocytes

Bone, the skeletal support system of the body, is a dynamic tissue that is continually being formed by osteoblasts, removed by osteoclasts, and maintained by osteocytes.

Osteoblasts (micrograph 1) synthesize the organic matrix of bone and regulate its mineralization. They are highly polarized cells with eccentric nuclei and a cytoplasm bulging with organelles synthesizing and sorting proteins. The collagen and proteoglycan matrix secreted by these cells is not mineralized initially and is referred to as **osteoid** (o, micrograph 1). Areas of mineralization can be seen as black regions in the lower left corner of micrograph 1.

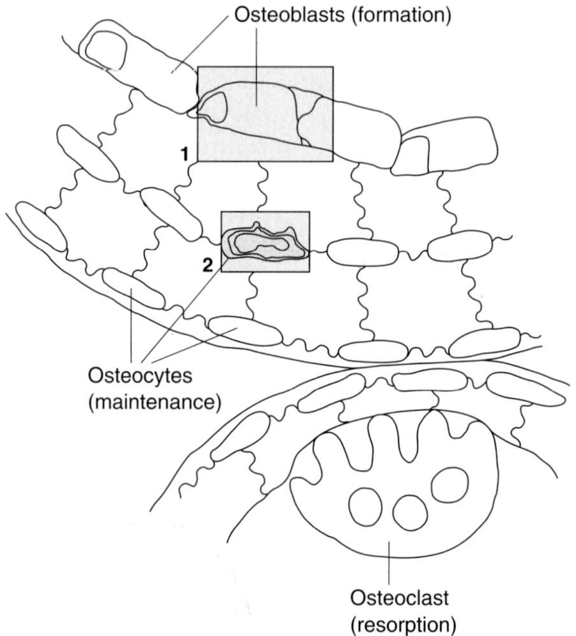

Osteoblasts originate from sheetlike mesenchymal cells lining the outer (periosteal) and inner (endosteal) bone surfaces. They form layers adjacent to the bone, attached to each other by gap junctions. At some point osteoblasts lose their polarity, secrete matrix around their circumference, and become trapped in spaces called lacunae (L, micrograph 2). Once enclosed by bone, these cells, now known as **osteocytes** (micrograph 2), become smaller and less active. Osteocytes are not responsible for a net increase in bone matrix; however, they are essential to the maintenance and routine turnover of the matrix.

Cellular processes of osteocytes (arrowheads, micrograph 2) remain attached to each other and to osteoblasts through channels in the bone referred to as **canaliculi.** Nutrients from blood vessels outside the bone matrix diffuse to the osteocytes both through the canaliculi surrounding the cell processes and through the cells themselves via gap junctions. Since the effective diffusion distance is limited, osteocytes cannot survive more than 0.2 mm away from a blood vessel. This limitation defines the size of spongy bone and the basic structural unit of compact bone, the Haversian system.

Defining osteoblasts as "bone forming" is a correct but simplistic view of these cells. Bone formation and bone resorption are intimately tied together, so much so that one normally does not occur without the other. Considerable evidence suggests that osteoblasts are essential for both. In culture, osteoclasts fail to resorb bone in the absence of osteoblasts. In addition, receptors for parathyroid hormone, which increases bone resorption and osteoclast activity in vivo, are found on osteoblasts but not osteoclasts.

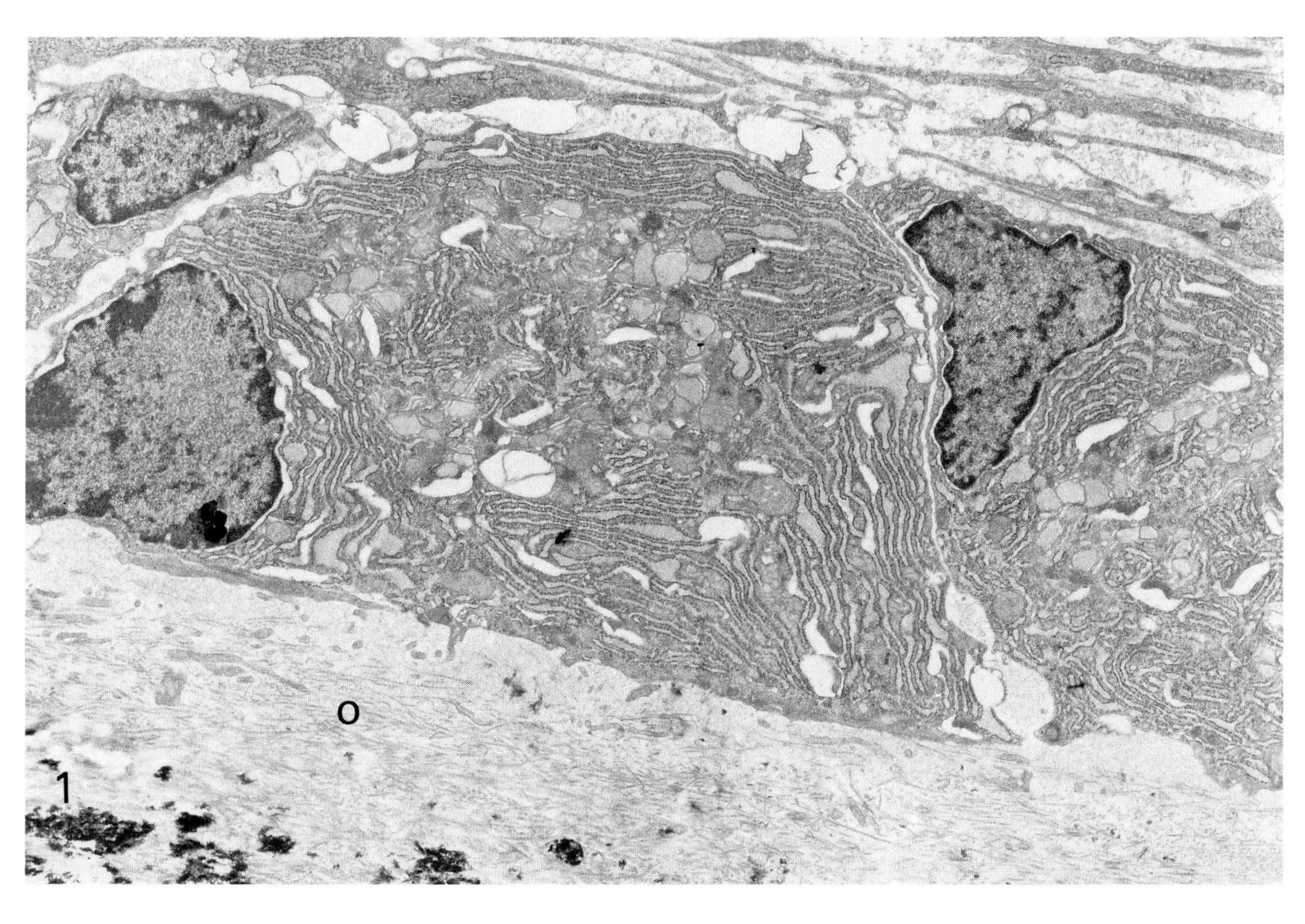

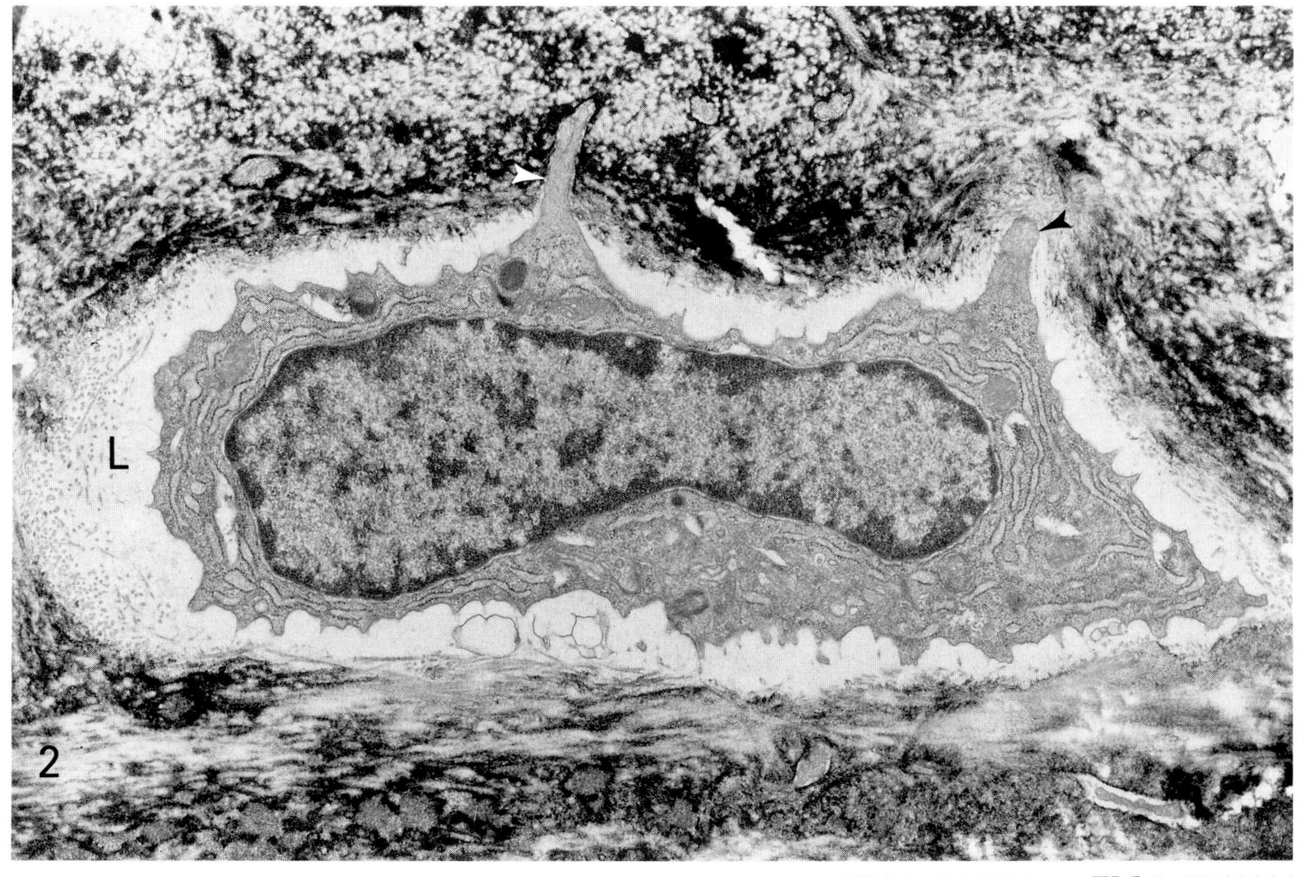

EM 1: 8,000× **EM 2: 12,000×**

Extracellular matrix accounts for 90% of the total weight of compact bone. The inorganic component of extracellular matrix, **microcrystalline hydroxyapatite,** contributes 60% of total bone weight, and the organic component, primarily proteins, contributes 30%.

Nearly all of the organic matrix of bone is **Type I collagen,** and much of the activity observed in **osteoblasts** (O, micrographs 1 and 2) is directed toward its synthesis. The fine threads (straight arrows, micrograph 1) within Golgi saccules are the rigid triple helices of procollagen that form prior to secretion.

In this preparation, the nonmineralized osteoid (o, micrographs 1 and 2) is clearly differentiated from the region where mineralization and hydroxyapatite crystal deposition have begun (curved arrows, micrographs 1 and 2). In the area of mineralization, matrix vesicles (v, micrograph 2), believed to originate from osteoblasts, accumulate calcium, and contain enzymes (e.g., alkaline phosphatase) that could bring about its precipitation. The first crystals of hydroxyapatite are formed within the 35-nm gaps between tropocollagen molecules.

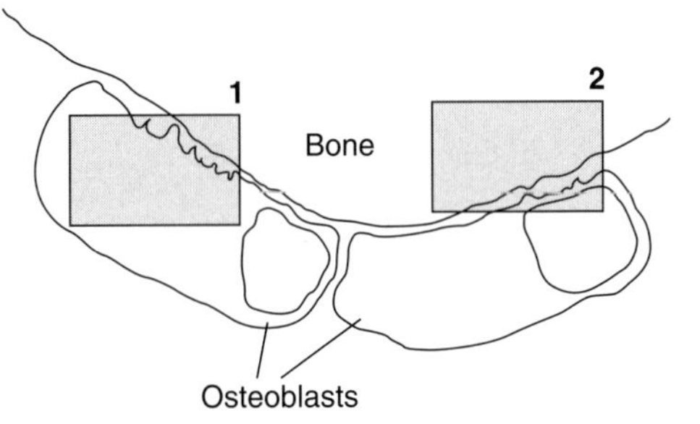

Osteoblasts synthesize and secrete several noncollagenous proteins, including osteopontin and osteocalcin, that appear to have central roles in matrix organization and calcium homeostasis. Osteopontin, a phosphorylated glycoprotein, is concentrated in areas where cartilage is transforming into bone, in particular where osteoclasts attach to the bone surface. It has been suggested that osteopontin binds to osteoclasts and facilitates their association with calcified matrix. Osteocalcin, a small protein that contains the modified amino acid γ-carboxyglutamic acid, has also been implicated in osteoclast activity. This protein, bound to bone mineral crystals, attracts and activates osteoclasts and is important in bone turnover. The severe bone disorders that resulted as a side effect of the clinical use of the anticoagulant warfarin were found to result from interference in the carboxylation of glutamic acid residues. This led, as expected, to interference with the synthesis of certain clotting proteins but also had the detrimental effect of interfering with the synthesis of osteocalcin.

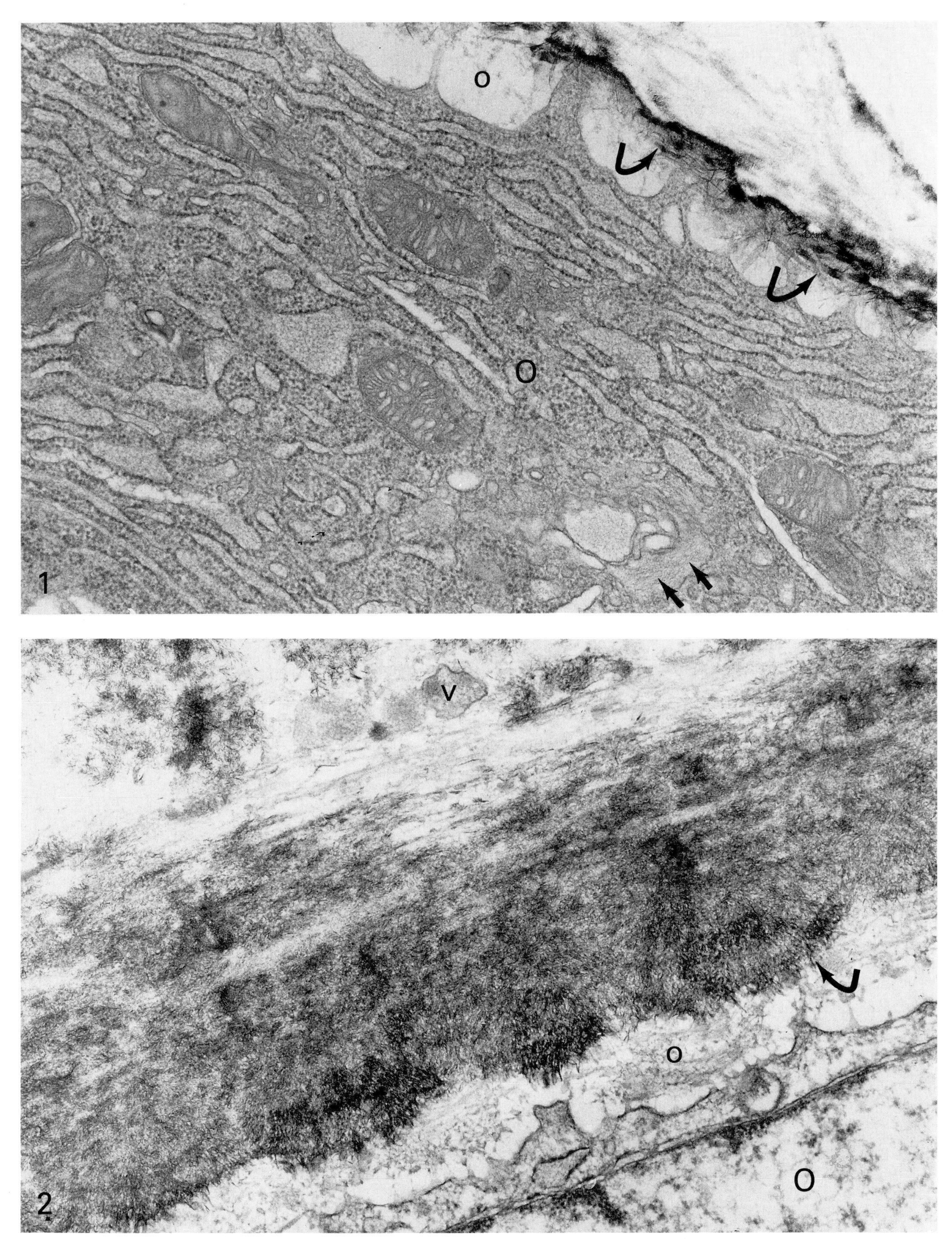

EM 1: 43,500× EM 2: 43,500×

BONE: Long Bone Growth

All long bones are initially composed of **hyaline cartilage** (black regions in the diagram). Beginning in the fourth week of fetal development, this cartilage is gradually replaced by bone. Even though bone replaces cartilage and strengthens the skeletal framework, cartilage remains in defined regions **(epiphyseal plates)** at the ends of growing long bones as a source of dividing cells. The division and development of cartilage in this region is responsible for the growth in length of long bones. Epiphyseal cartilage is replaced by bone as growth terminates.

The progression of events in the epiphyseal plate is highly ordered, with a **proliferative, hypertrophic,** and **degenerative** sequence in the programmed cell death of chondrocytes. The chondrocytes in micrograph 1 are recent products of division of a single cell in the proliferative stage. They are characteristically arranged in a row and still close enough together to be in a common matrix region (territorial matrix) containing a high concentration of proteoglycans.

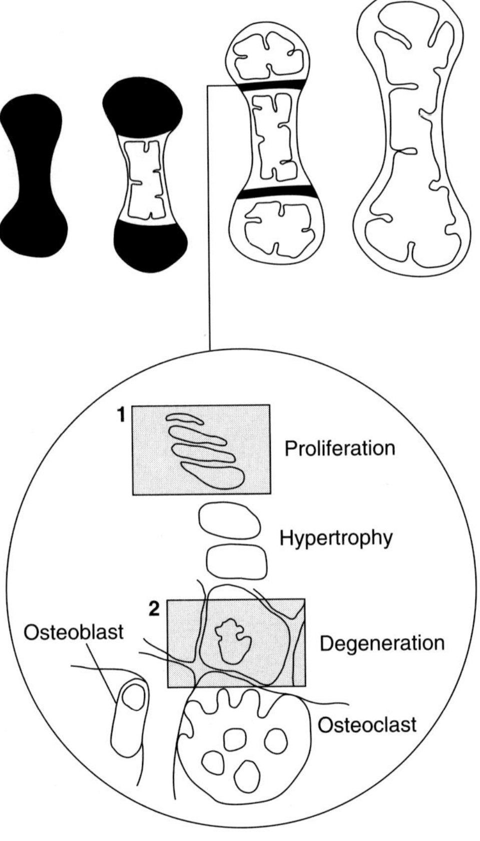

Each chondrocyte becomes separated from its neighbor in the developing row as it grows in size (hypertrophy) and forms more matrix. At this stage of development the matrix in the longitudinal septa (arrows, micrograph 2) between rows of cells calcifies. The resulting isolation of these cells from their nutrient supply facilitates chondrocyte death, leaving remnants of cells (N, nucleus of a degenerating chondrocyte, micrograph 2) within lacunae. The spaces once occupied by chondrocytes are remodeled in the metaphysis region by the removal of matrix and cells by osteoclasts and the addition of bone by osteoblasts.

The proliferation of chondrocytes in the epiphyseal plate is regulated by growth hormone from the anterior pituitary. Evidence indicates that this hormone acts directly on chondrocytes, which in turn synthesize somatomedins, peptide growth factors that promote clonal expansion in an autocrine manner.

Vitamin D is essential for the normal mineralization of bone. In children with rickets, a disease caused by vitamin D deficiency, the cartilage in the epiphyseal regions fail to calcify, resulting in the prolonged life of chondrocytes and thickening of the epiphyseal plate. This increased amount of cartilage, along with an increase in the proportion of nonmineralized osteoid, leads to skeletal weakness and deformity.

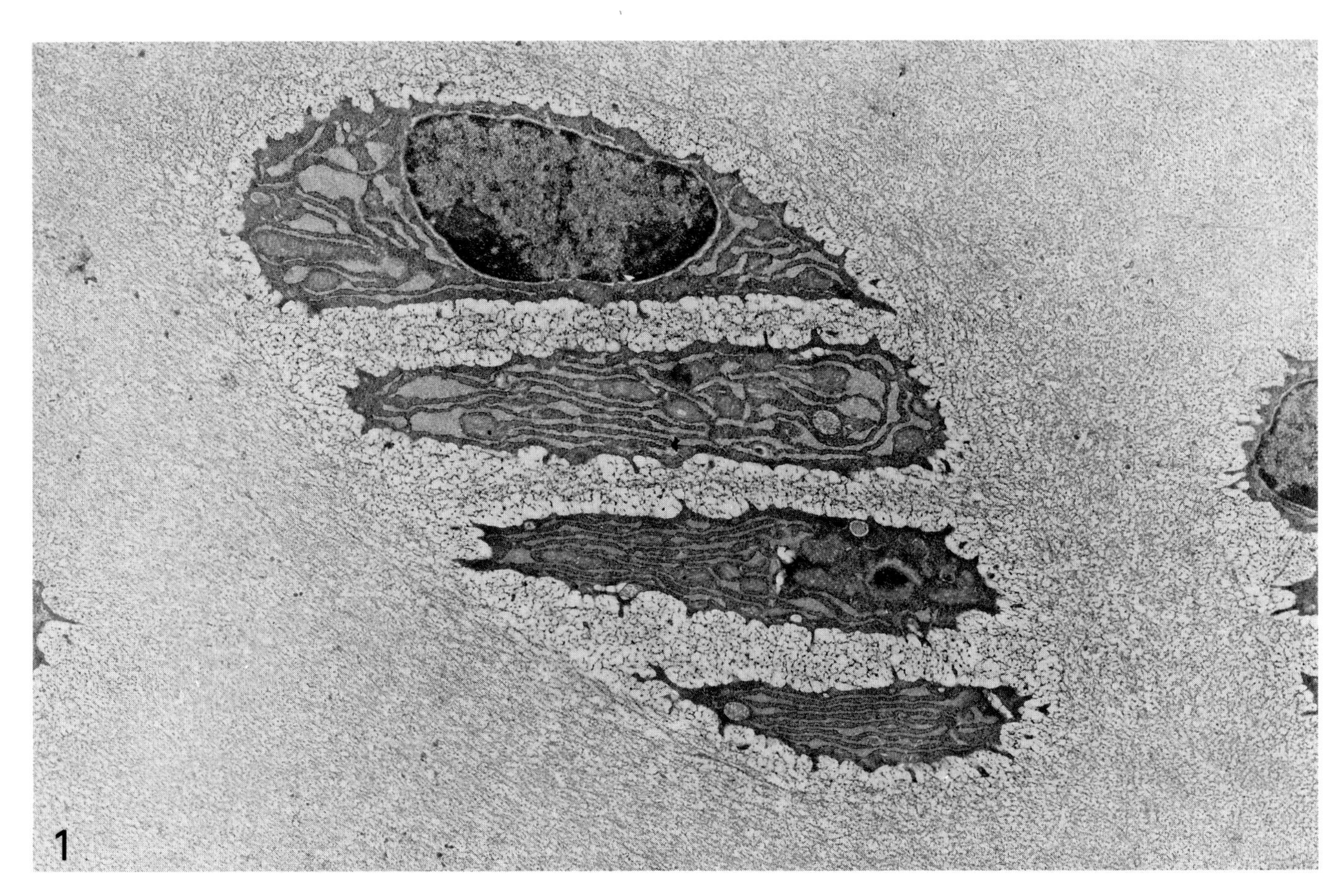

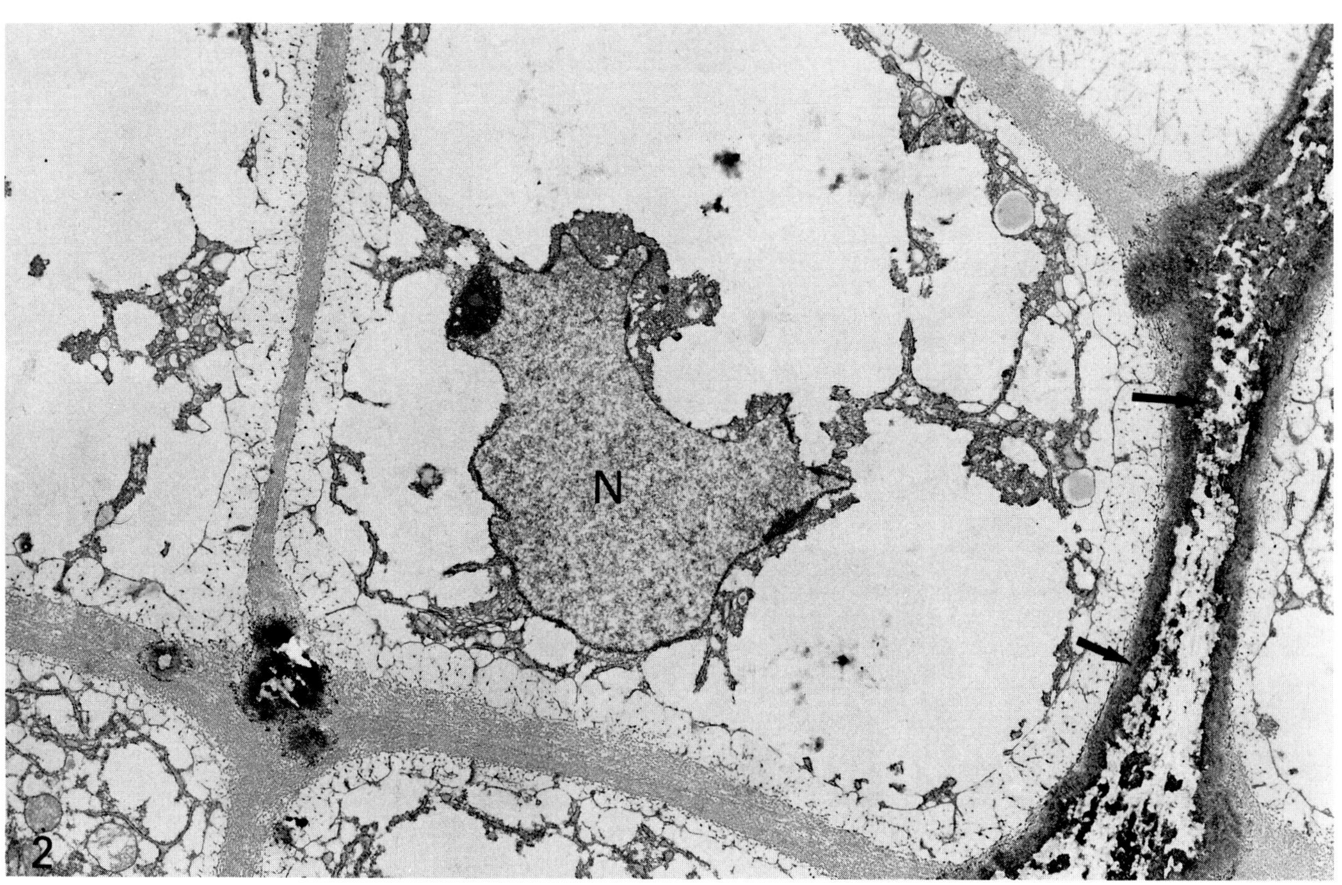

EM 1: 7,800× EM 2: 5,400×

BONE: Osteoclasts

Osteoclasts (micrograph 1) are giant (100-μm) **multinucleate** cells that originate from the fusion of several monocytes. These cells resorb bone by attaching to the bony surface (upper right, micrographs 1 and 2), sliding back and forth, and dissolving the matrix components in a pit underneath the cell. The osteoclast membrane adjacent to the bone forms a **ruffled border** of deeply infolded membranes (arrows, micrographs 1 and 2).

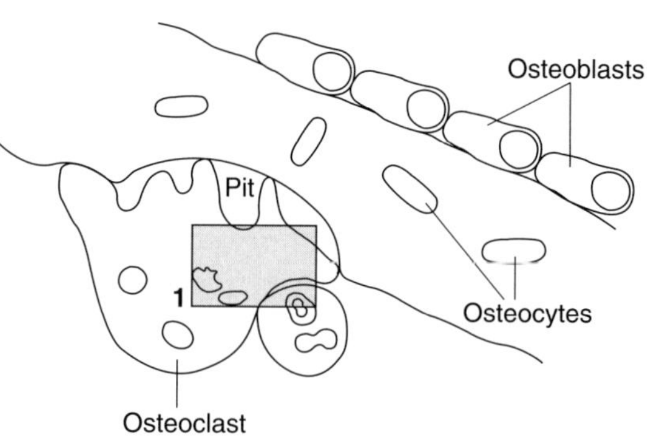

The **extracellular pit** between the osteoclast and the bone is the site of resorption. A proton pump localized in the ruffled border in this region maintains high concentrations of hydrogen ion in the pit. The resulting low pH dissolves the inorganic component, hydroxyapatite. Lysosomal enzymes released into the pit via exocytosis are activated by the low pH. These enzymes degrade the organic components such as collagen and proteoglycans. The extra surface area provided by the ruffled border and the large number of associated mitochondria (particularly evident in micrograph 1) are typically found in regions of cells involved in ion pump activity.

The extracellular pit is analogous to a secondary lysosome in other cell types. In both, digestion occurs in a low pH maintained by a membrane pump. What is unique in the action of the osteoclast is the exocytosis of lysosomal enzymes and their action outside instead of inside the cell. The large amounts of calcium released during osteoclast digestive activity would be incompatible with intracellular functioning. Inorganic and organic products of osteoclast activity enter capillaries (c, micrograph 1) and are recycled to other locations.

During the growth and continual remodeling of bone, osteoclasts and osteoblasts work together in the balance of resorption and formation. The placement of osteoclasts and osteoblasts determines the shape of bones as they grow. In the developing skull, resorptive activities of the osteoclast on the inside surface are balanced by bone formation by osteoblasts on the outside surface. Such an interaction maintains the thickness of the skull as it is expanding to accommodate the developing brain.

From age 20 on, the balance between these two events shifts, such that resorption by osteoclasts is not completely repaired by osteoblasts. The resulting reduction in bone mass increases susceptibility to fracture. In women, accelerated bone loss appears to follow the reduction of estrogen levels at menopause. Estrogen replacement at this time increases bone mass. One effect of estrogen is to inhibit the osteoblast synthesis of interleukin-6, a cytokine that stimulates the development of osteoclasts, thus maintaining bone mass by balancing bone synthesis and resorption.

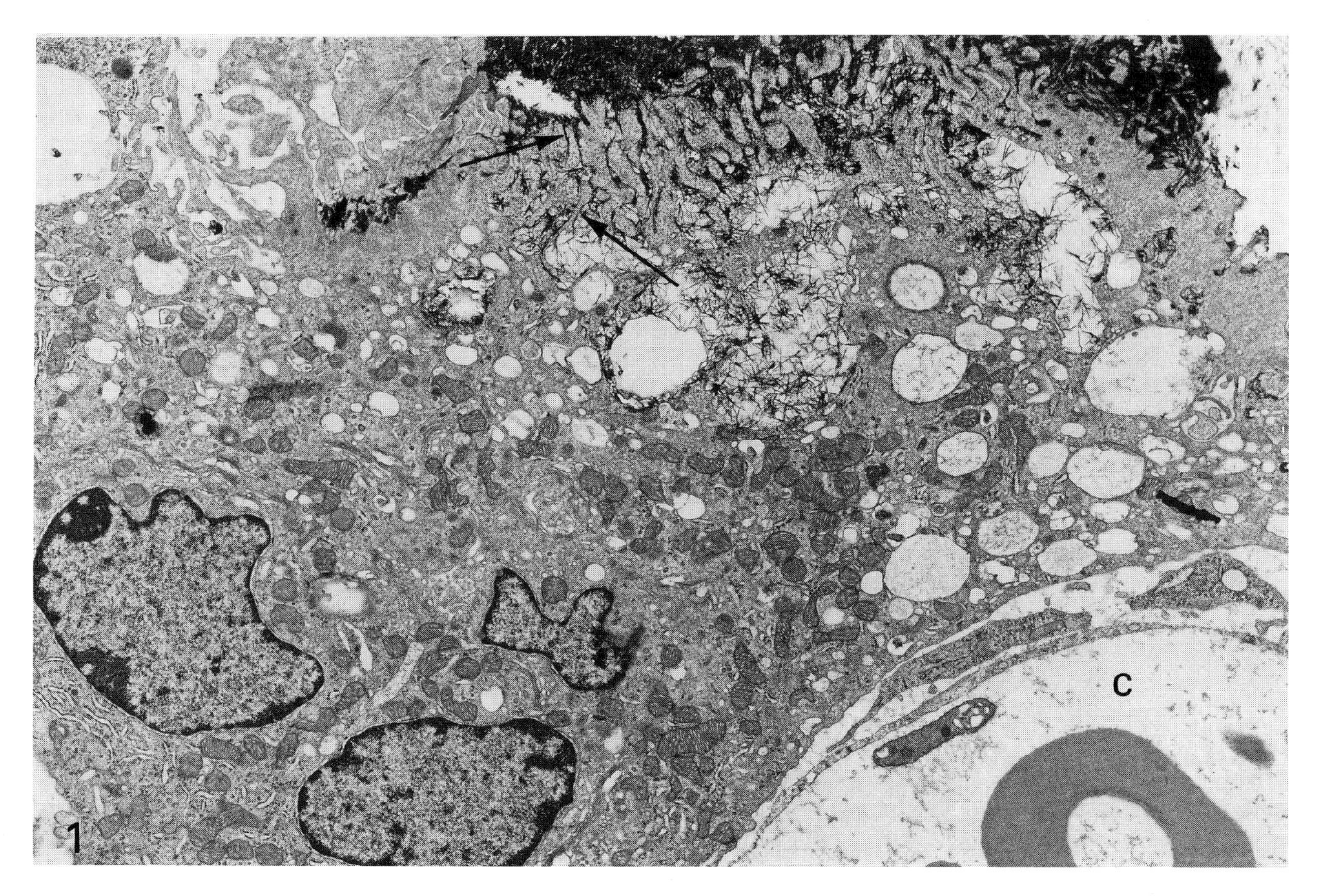

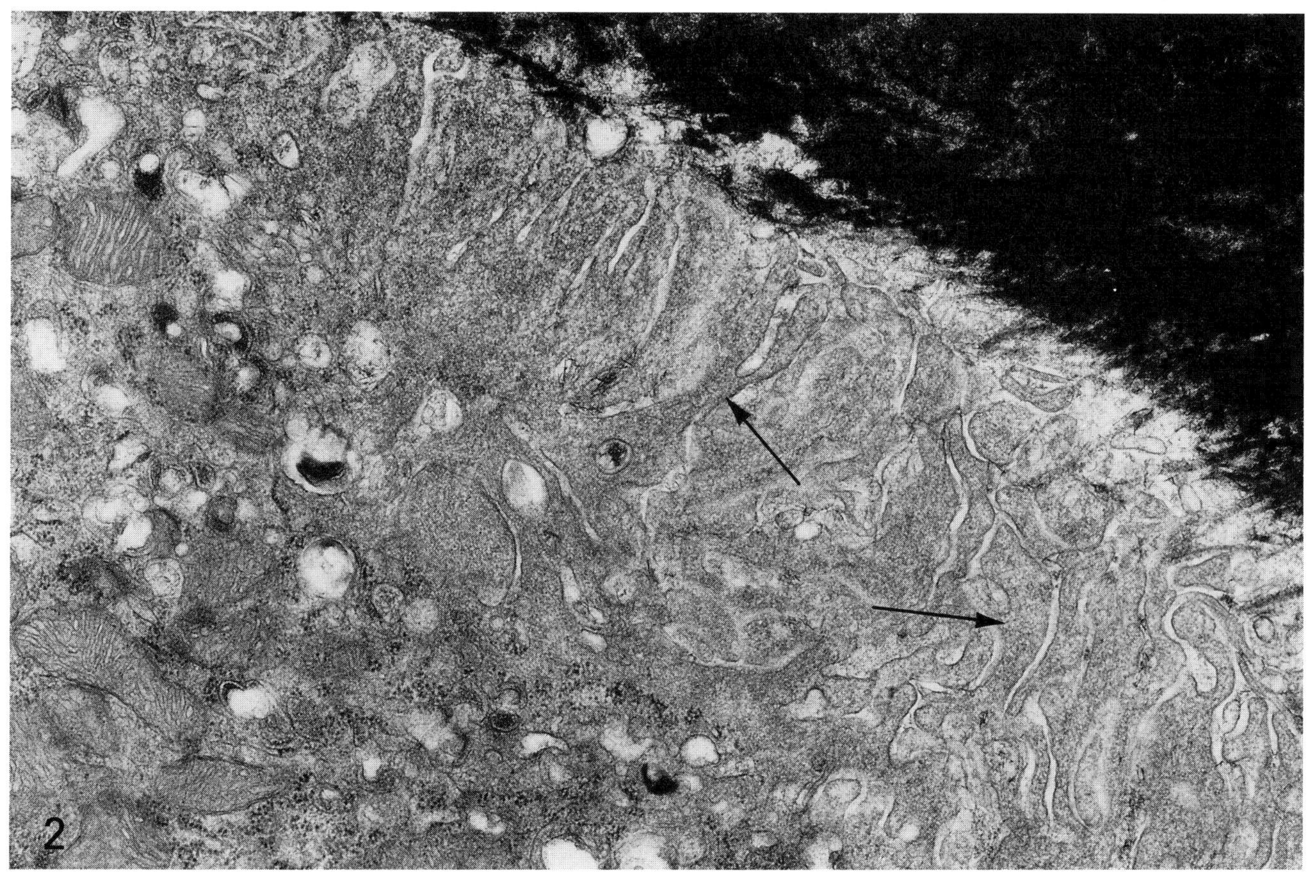

EM 1: 8,000× **EM 2: 27,000×**

MUSCLE

OVERVIEW

Muscle fibers are specialized contractile cells. By working together they are able to control the movement of hollow organs and vessels (smooth muscle), the skeletal system (skeletal muscle), and the heart (cardiac muscle). Each type of muscle cell has a different morphology that is suited to its location and type of contraction.

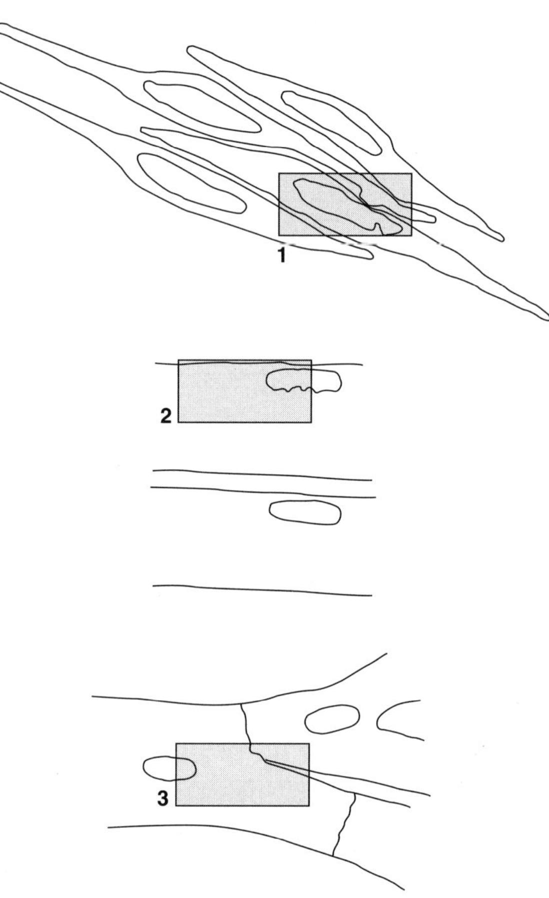

Smooth muscle fibers (micrograph 1) are small, with a central nucleus (N) and tapered ends. They group together in a staggered fashion in the walls of many organs. The fusiform shape facilitates the twisting that takes place during contraction. Even though well-defined contractile units are not obvious, a certain level of organization exists. Actin inserts into dense bodies (arrows) within the cytoplasm and at the cell margins.

Skeletal muscle fibers (micrograph 2) are large multinucleated syncytia that take a direct path from the point of origin to insertion. Contraction is linear and results from the shortening of discrete units, sarcomeres (S), arranged in series. Nuclei (N) sit in the peripheral cytoplasm, where they direct the synthesis of proteins important in contraction and synaptic activity.

Cardiac muscle fibers (micrograph 3) are intermediate in size between smooth and skeletal muscle fibers. Typically each fiber has several branches and one centrally placed nucleus (N). Individual fibers are bound together by intercalated discs (arrows), elaborate junctional complexes that occupy the entire ends of the cells where they contact each other. The tightly associated branching fibers form a network that acts as a functional syncytium, carrying contraction on an elaborate path through the atria and ventricles.

Whereas voluntary skeletal muscle fibers are individually innervated and only contract with neural input, involuntary smooth and cardiac muscle fibers can contract spontaneously. The autonomic innervation to involuntary muscle alters the speed and force of contraction, and the electrical signal is transmitted by gap junctions between cells. All three muscle types are surrounded by a network of reticular fibers that facilitates the coordination of contraction.

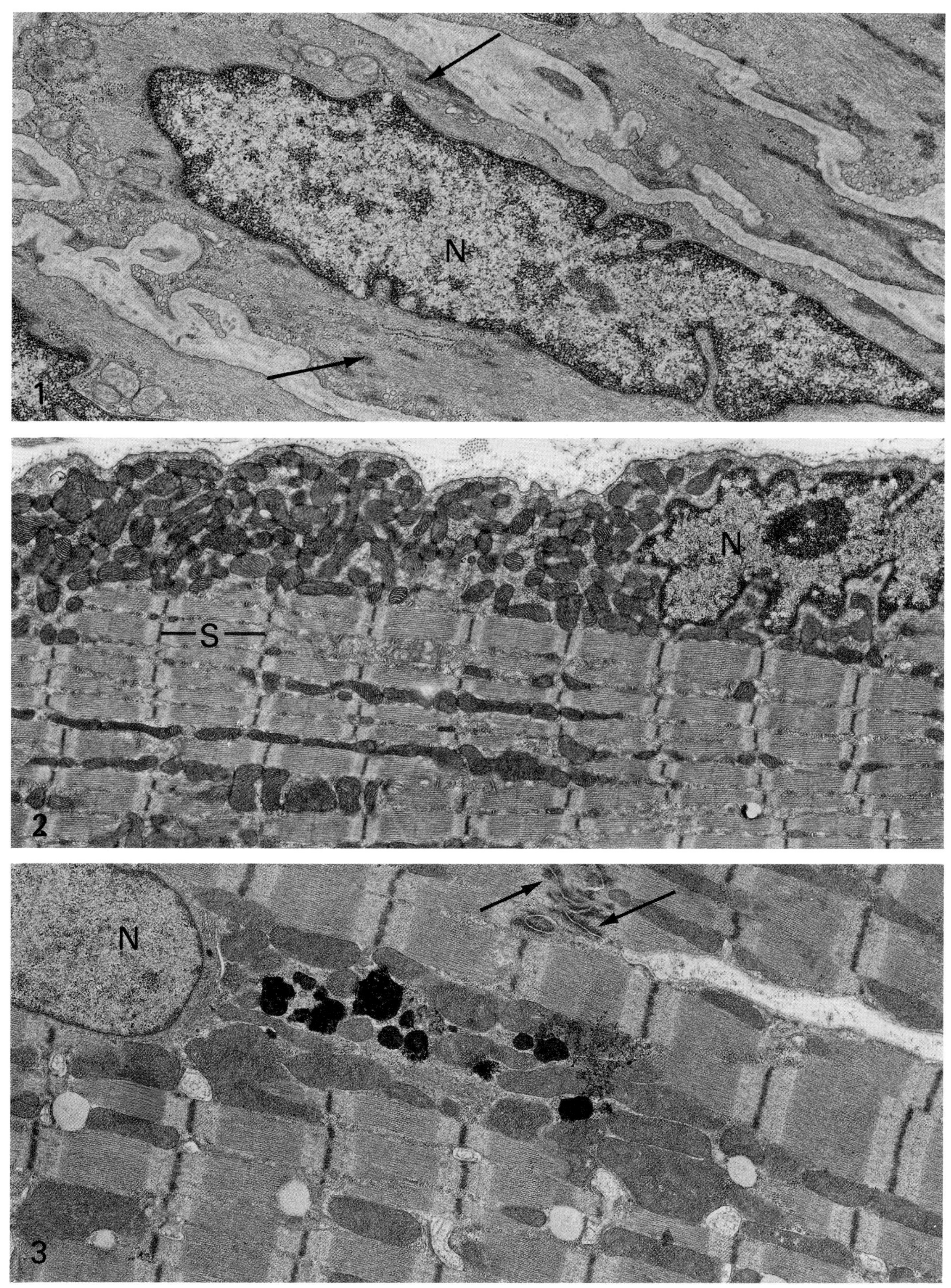

EM1: 17,500× EM2: 9,000× EM3: 18,000×

Skeletal Muscle: Fibers and Myofibrils

Skeletal muscle fibers are easily distinguished by their large size, peripheral nuclei (N, micrograph), and cytoplasm packed with well-defined units, **myofibrils** (circles, micrograph). Organelles such as mitochondria (m, micrograph) and smooth endoplasmic reticulum (sarcoplasmic reticulum) occupy the cytoplasm (sarcoplasm) between myofibrils. The particular muscle in this micrograph undergoes a slow, continuous contraction with little fatigue. The extensive blood supply (note the two capillaries, c, on either side of the fiber) and many mitochondria reflect the high oxygen demand associated with its aerobic metabolism.

Skeletal muscle fibers are grouped into **fascicles,** and groups of fascicles are bound together to form a muscle. Individual fibers are surrounded by an **endomysium** (e, micrograph) of loose connective tissue, fascicles are surrounded by **perimysium,** a thin cellular sheath, and groups of fascicles are bound together by an **epimysium** of dense connective tissue.

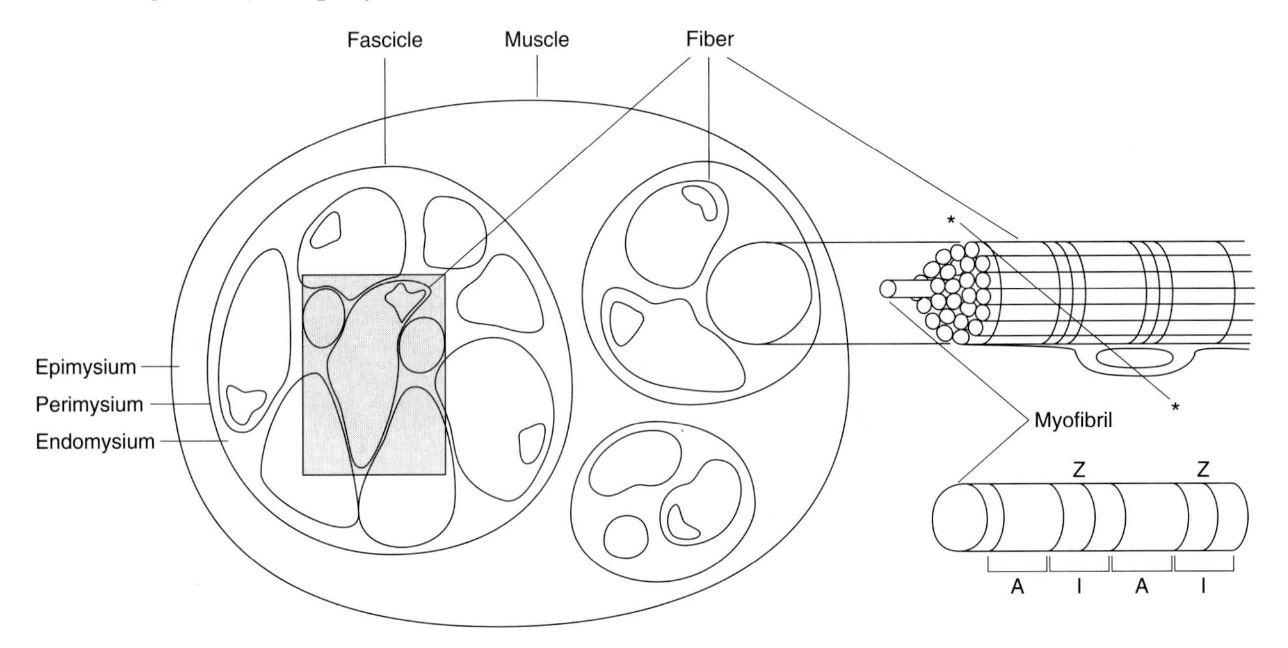

In longitudinal section, each myofibril exhibits a series of light **I-bands** and dark **A-bands.** Thin dark **Z-lines** bisect the I-bands. This arrangement reflects a precise organization of the contractile proteins actin and myosin within individual myofibrils (see Muscle, page 100, for details). The micrograph was taken from an oblique section (*, diagram) that cut through the nucleus and different bands in a single cell. Consequently, sequential A, I, Z, I, A areas are evident in the micrograph.

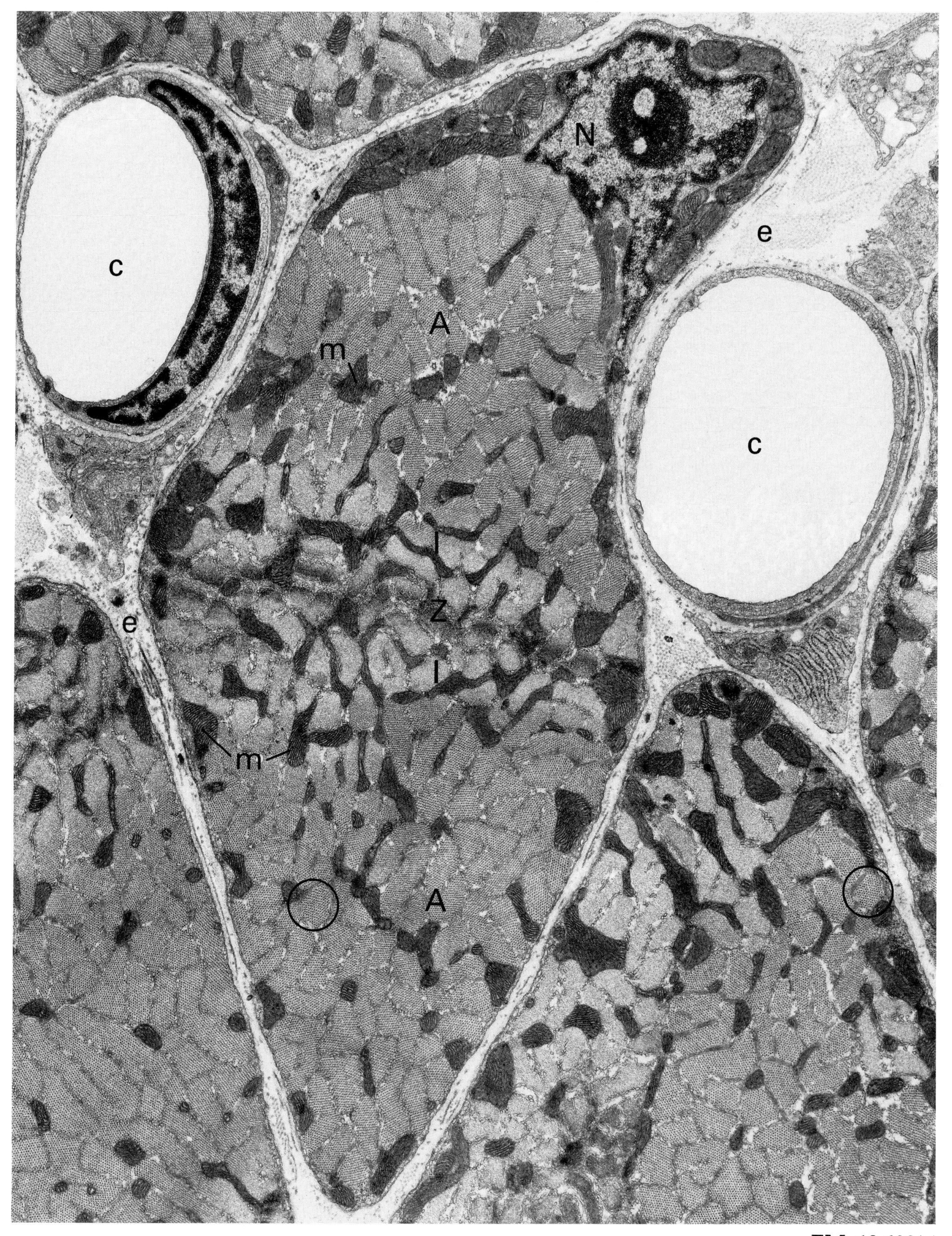

EM: 13,600×

The contractile unit of skeletal muscle, the **sarcomere,** consists of highly ordered arrays of thin and thick filaments. One sarcomere (S, micrograph) is defined at each end by a Z-line (Z, micrograph). **Thin filaments,** anchored in the α-actinin-rich Z-line, project part way into the myosin-containing A-bands (A, bracket, micrograph) of adjacent sarcomeres. They are polarized in opposite directions on each side of the Z-line (arrows, diagram). **Thick filaments** are anchored in the center of the A-band at the denser M-line (M, micrograph). Crossbridges form between the thick and thin filaments in the region of overlap. The movement of the myosin heads within the thick filaments pulls the thin filaments toward the M-line, shortening the sarcomere length and thus the muscle fiber. With sarcomere shortening, both the I-band (I, bracket, micrograph), which contains only thin filaments, and the H-band (H, bracket, micrograph), which contains only thick filaments, become narrower.

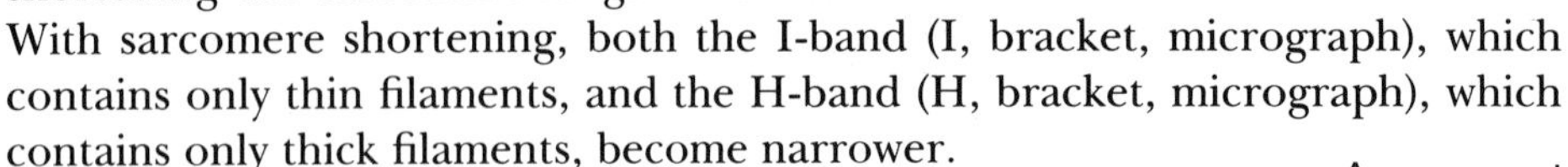

Each thin filament is composed of **F-actin** and the associated proteins **tropomyosin** and **troponin.** The filamentous tropomyosin wraps around the F-actin while troponin, consisting of **C** (calcium-binding), **T** (tropomyosin-binding), and **I** (actin-binding) subunits, binds to the actin–tropomyosin complex at regular intervals.

Each thick filament is composed of many **myosin** molecules, each consisting of two heavy chains and associated light chains. The two heavy chains (200 kd each) are wound around each other to form an alpha helical coiled-coil tail with two globular heads. The tail region functions in the assembly of the thick filament while the heads contain the ATPase activity and the actin binding site. Two light chains (20 kd and 17 kd) associate with each heavy chain head. Myosin molecules are grouped in the thick filament in a staggered bipolar arrangement with a central region devoid of heads.

Modified from D. W. Fawcett, *A Textbook of Histology*, Saunders, Philadelphia, 1986, and from L. C. Junquiera et. al., *Basic Histology,* 7th ed., Appleton & Lange Medical Publications, Norwalk, CN, 1992.

An action potential generated at the neuromuscular junction is propagated quickly to all regions of the cell via invaginations of the cell membrane (sarcolemma) called transverse tubules **(T-tubules)** (arrows, micrograph). In skeletal muscle, T-tubules occur at the A–I junctions. In response to the action potential, calcium is released from the sarcoplasmic reticulum and initiates contraction. Calcium binds to troponin C, altering the conformation of both troponin and tropomyosin, thus exposing the myosin binding site on the actin. Crossbridges form and contraction initiates, powered, in this "fast" muscle, by ATP generated from glycogen (arrowheads, micrograph) breakdown.

EM: 50,500×

In the cross section in the micrograph, **myofibrils** (My) are clearly defined by surrounding sarcoplasmic reticulum (arrows). This section cut through myofibrils within a region of the **A-band** (A, micrograph), where thick and thin **myofilaments** overlap, and through the **I-band** (I, micrograph), which contains thin, but not thick, filaments. In the A-band, thin filaments are arranged in a hexagonal pattern around the thick filaments, with each thin actin filament associated with three thick myosin filaments (circles, micrograph).

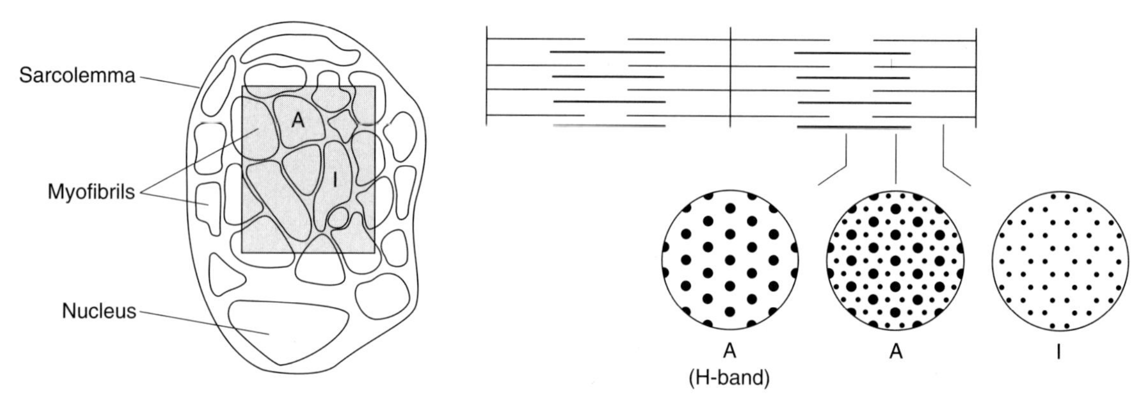

Modified from D. W. Fawcett, *A Textbook of Histology,* Saunders, Philadelphia, 1988.

A proposed sequence for **force generation** between thick and thin filaments in the A-band is shown below. As illustrated, when myosin binds to actin, P_i and then ADP are released, resulting in a change in the orientation of the myosin head, the power stroke that moves the thin filament. ATP binds, myosin is released from actin, and ATP is hydrolyzed to ADP and P_i as the myosin head is cocked for another cycle.

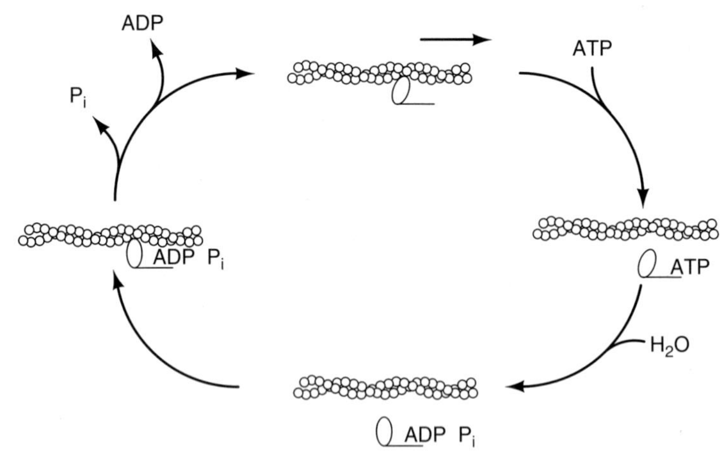

Courtesy of Dr. J. Spudich.

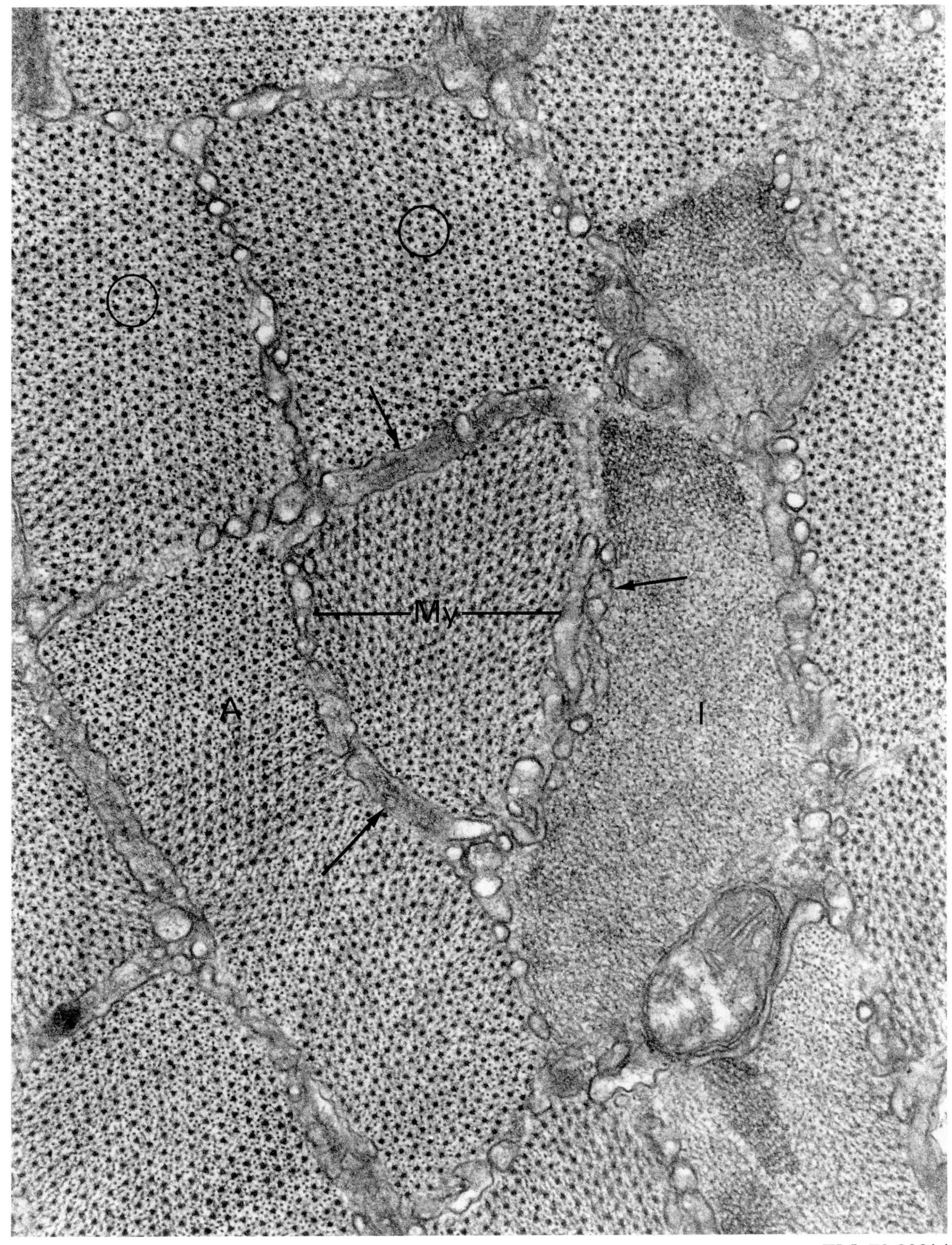

EM: 72,000×

SKELETAL MUSCLE: Triad

In skeletal muscle, the concentration of calcium within the sarcoplasm is regulated by the **sarcoplasmic reticulum** (s, micrograph). This organelle consists of membranous cisternae that encase each myofibril in a distinct arrangement. The sarcoplasmic reticulum comes into close contact with invaginations of the cell membrane, the **T-tubules,** at the **A–I junctions** and forms **triads** (circles, micrograph). Each triad consists of a central T-tubule (arrowheads, micrograph) flanked by two terminal cisternae (short arrows, micrograph) of the sarcoplasmic reticulum. Electron-dense "feet" (long arrows, micrograph) project from the sarcoplasmic reticulum across a 12-nm gap and attach at regular intervals to the T-tubule. Events in the triad area, where the terminal cisternae and T-tubules are intimately associated, are thought to couple the depolarization of the T-tubule to the release of calcium from the sarcoplasmic reticulum.

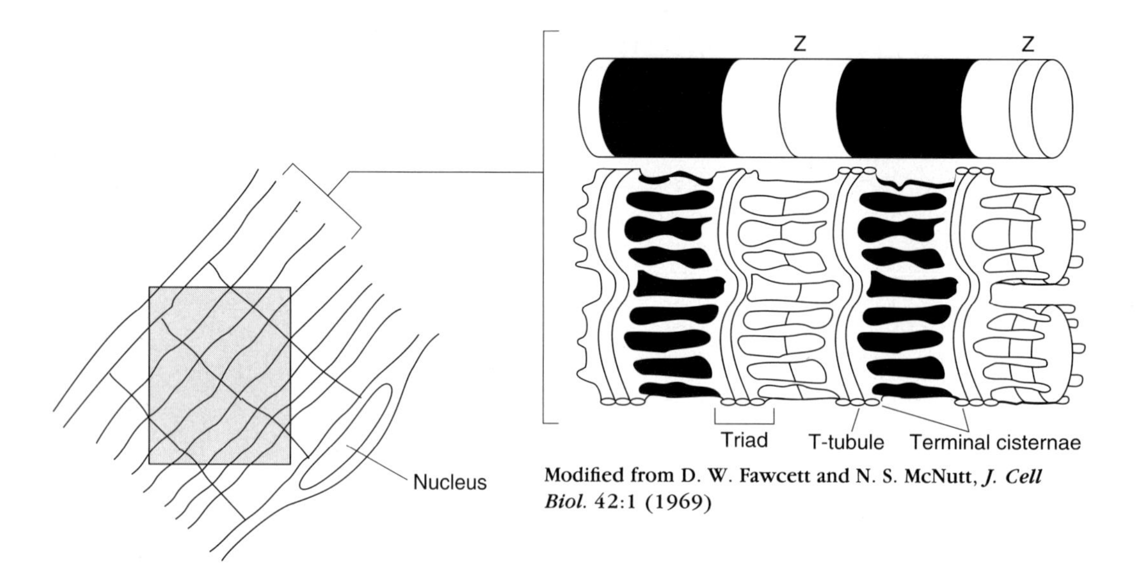

Modified from D. W. Fawcett and N. S. McNutt, *J. Cell Biol.* 42:1 (1969)

Most calcium is concentrated in the sarcoplasmic reticulum of the triads, bound to the acidic protein **calsequestrin** (note electron density in terminal cisternae). Evidence suggests that calcium is not only stored in the terminal cisternae, but is also preferentially released in this area. **Calcium-release channels** seem to be the same structures as the junctional feet seen on electron micrographs. Small alterations in the membrane charge of the T-tubules may lead to conformation changes in the junctional feet, thus opening the calcium-release channels.

During relaxation, calcium is pumped back into the sarcoplasmic reticulum throughout the nontriad regions by specific **calcium–ATPase complexes** that make up 90% of the membrane protein content of this region.

EM: 78,200×

SKELETAL MUSCLE: Fiber Type

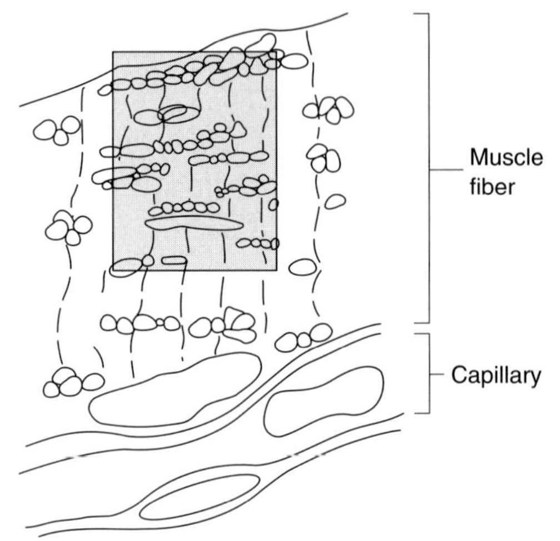

Skeletal muscle fibers vary in their manner of contraction, biochemistry, and ultrastructure. Some undergo a **slow, continuous contraction** (e.g., postural muscles) and are fatigue resistant, while others undergo a **fast contraction for short periods** of time (e.g., extraocular muscles of the eye) and fatigue quickly. Slow muscle fibers (micrograph) have many mitochondria (m) but sparse sarcoplasmic reticulum (SR) separating relatively small myofibrils.

Slow muscles use aerobic metabolism almost exclusively for their energy demands. The numerous mitochondria are packed with cristae that contain high concentrations of enzymes involved in oxidative metabolism. Triacylglycerols stored as lipid droplets (L, micrograph) provide one source of energy. Oxygen is readily available from (1) myoglobin, an oxygen-binding heme protein concentrated in the cytoplasm, and (2)

hemoglobin in the numerous capillaries associated with each fiber.

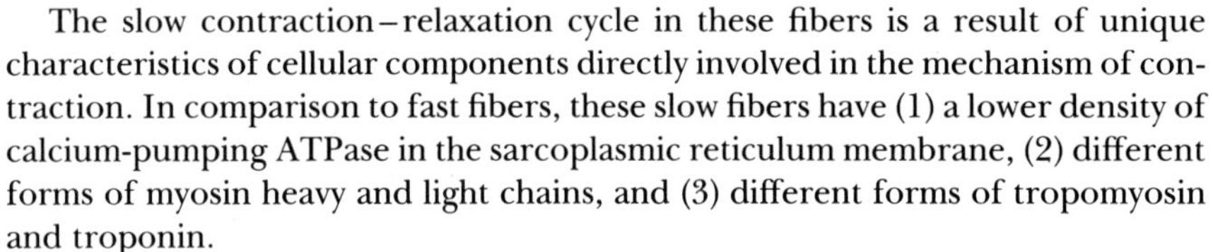
The slow contraction–relaxation cycle in these fibers is a result of unique characteristics of cellular components directly involved in the mechanism of contraction. In comparison to fast fibers, these slow fibers have (1) a lower density of calcium-pumping ATPase in the sarcoplasmic reticulum membrane, (2) different forms of myosin heavy and light chains, and (3) different forms of tropomyosin and troponin.

All the characteristics of a single muscle type, including myosin type, metabolic activity, fiber size, and density of capillaries, can be reversed by altering the **neural input.** When the innervation to fast and slow fibers is reversed, the muscle fibers change to the type corresponding to the new innervation. It is also known that exercise frequently results in a shift in muscle type. The relative plasticity of these phenotypes represents alterations in gene transcription and translation.

EM: 27,200×

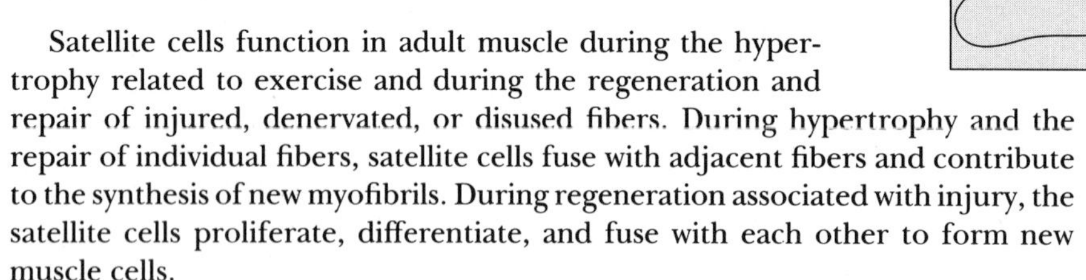

During development skeletal muscle fibers form by the fusion of myoblasts. At the completion of development, some myoblasts remain undifferentiated on the periphery of the multinucleated mature muscle fibers. These undifferentiated cells are the **satellite cells** (SC, micrograph) found within the basal lamina (arrows, micrograph) of the muscle fiber and tightly associated with the sarcolemma. There is approximately one satellite cell (with a single heterochromatic nucleus) for every 100 muscle fiber nuclei.

Satellite cells function in adult muscle during the hypertrophy related to exercise and during the regeneration and repair of injured, denervated, or disused fibers. During hypertrophy and the repair of individual fibers, satellite cells fuse with adjacent fibers and contribute to the synthesis of new myofibrils. During regeneration associated with injury, the satellite cells proliferate, differentiate, and fuse with each other to form new muscle cells.

Satellite cell proliferation is tightly regulated in normal adult muscle. In vitro studies suggest that activation that occurs following injury may be a response to (1) a factor released from crushed muscle cells themselves and (2) laminin, the major glycoprotein of the basal lamina. An intact basal lamina is essential for regeneration and provides more that just a physical scaffold.

The number of satellite cells decreases with age and is reduced in certain disease states. Duchenne muscular dystrophy (DMD), a fatal X-linked disorder affecting 1 in 3500 males, is associated with the wasting of skeletal muscle. The gene affected in DMD has been characterized, and its product, dystrophin, is absent in DMD muscle. In normal muscle cells, dystrophin, which shares structural homology with α-actinin and spectrin, is found associated with the protoplasmic side of the cell membrane. A possible role for dystrophin as a link between the surface membrane and cytoskeleton has been suggested, and its absence in DMD muscle may compromise cellular integrity. The number of satellite cells in DMD is only 2% of normal, and the degenerating muscle cells are not replaced.

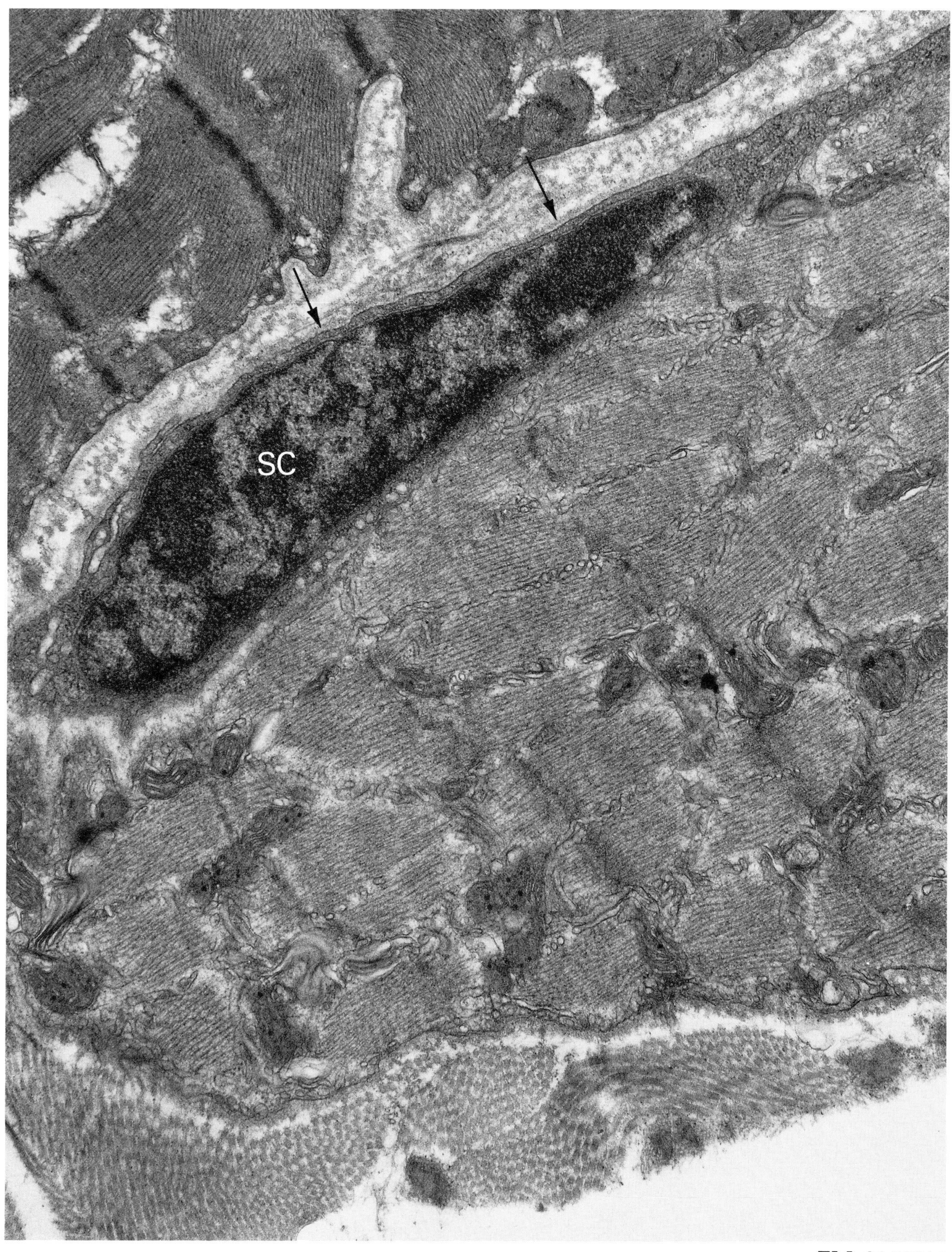

EM: 31,500×

SKELETAL MUSCLE: Motor End Plate

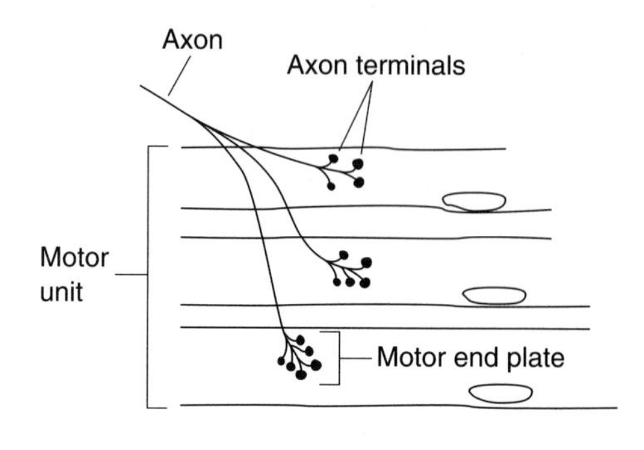

Muscle fibers are arranged in **motor units** composed of from 3 to over 150 fibers. All of the fibers in each motor unit are innervated by branches of a single axon and respond together and to the same degree to the axon impulse supplying the unit. While a single neuron may innervate many muscle fibers, each muscle fiber receives innervation from only one neuron. At the **neuromuscular junction,** the axon branches to form several **terminal swellings** that together constitute the **motor end plate.**

Two terminal swellings are present in the micrograph. Each swelling sits in a depression in the muscle fiber called a **primary synaptic cleft** (arrows, micrograph). The axon terminal contains many mitochondria (m, micrograph) and small synaptic vesicles (v, micrograph) holding the neurotransmitter acetylcholine. The region of the muscle cell adjacent to the nerve terminal forms deep surface junctional folds giving rise to **secondary synaptic clefts** (arrowheads, micrograph). In the synaptic region the distance between nerve terminal and sarcolemma is 50 nm.

At the neuromuscular junction, the myelin sheath covering the axon is lost and Schwann cell cytoplasm (S, micrograph) comes into direct contact with the upper region of the nerve terminal. Schwann cells in this region exhibit ionic activity in response to synaptic events. In addition, if the axon is severed, the Schwann cell phagocytoses the axon terminals, clearing and remodeling the synapse region in preparation for reinnervation.

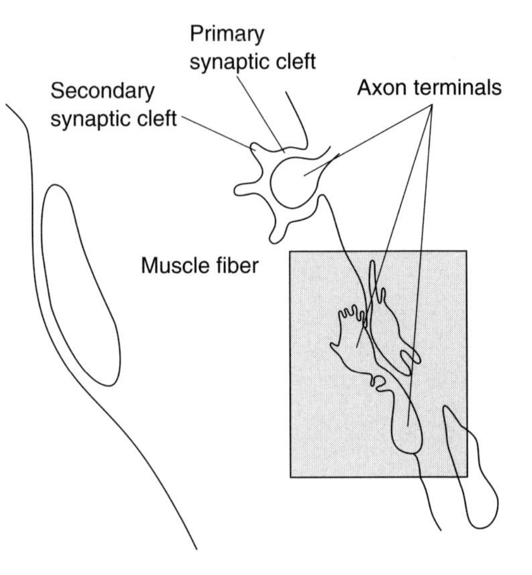

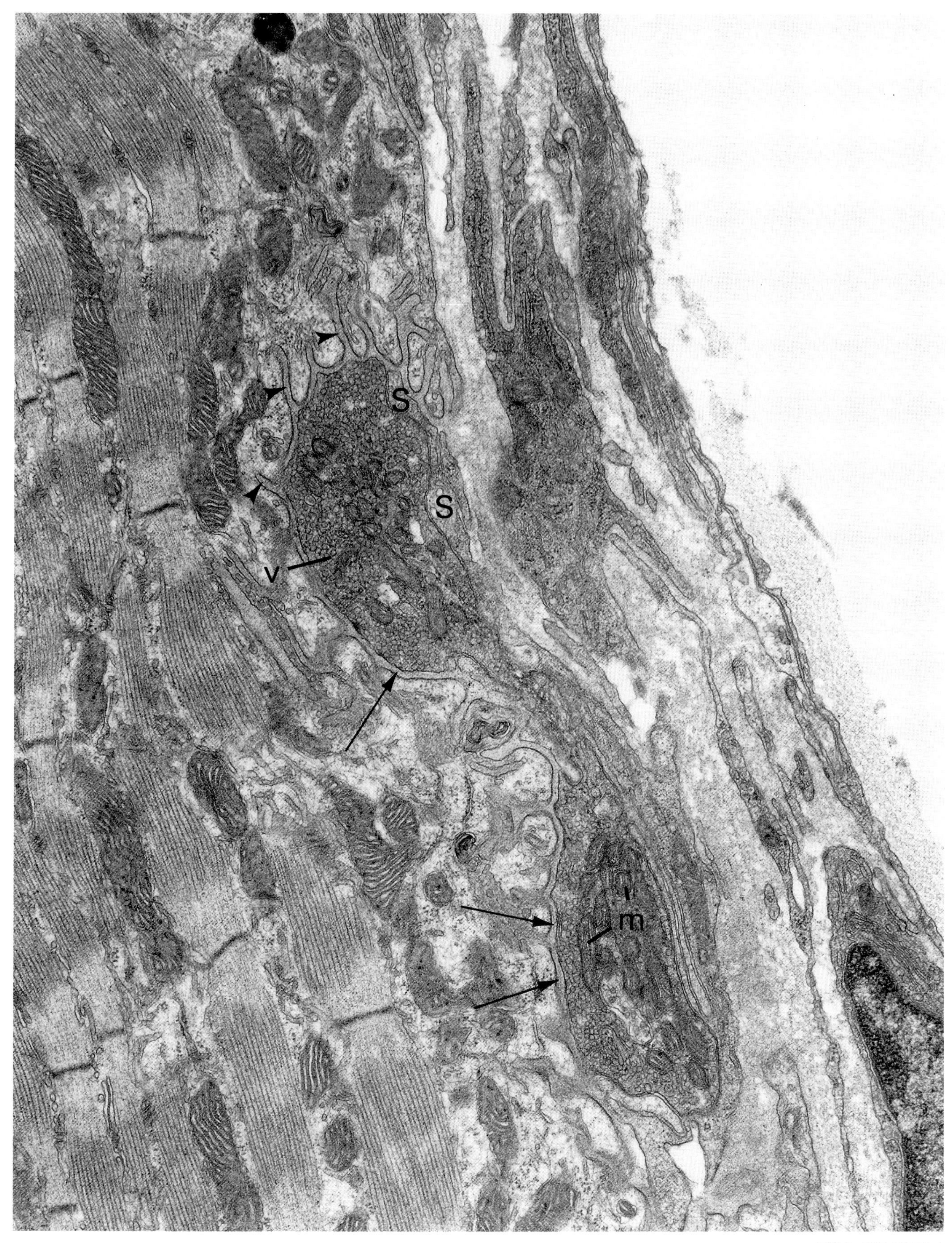

EM: 27,200×

SKELETAL MUSCLE: Synaptic Region

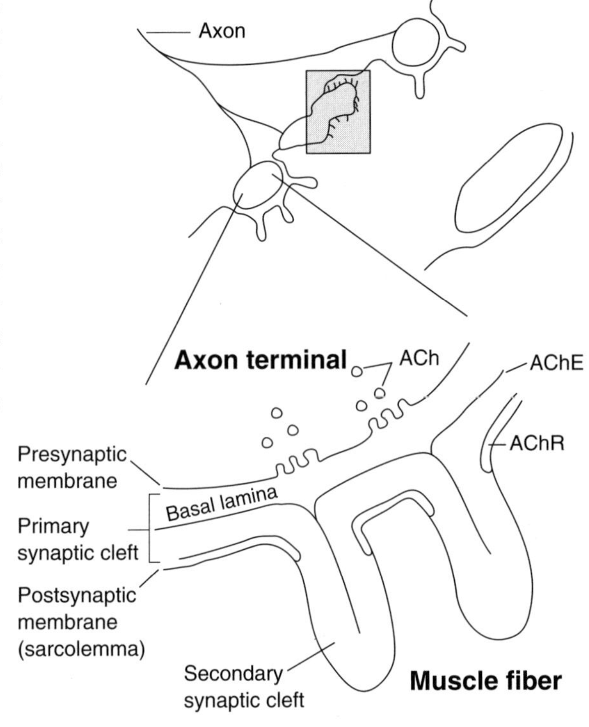

In skeletal muscle each nerve terminal contains tens of thousands of 50-nm **synaptic vesicles** (straight arrows, micrograph), each containing a defined amount (~ 10,000 molecules) of the neurotransmitter **acetylcholine** (ACh). When the nerve action potential reaches the terminal, calcium channels in the nerve presynaptic membrane open and calcium enters the cell. This triggers the fusion of a few hundred synaptic vesicles with the presynaptic membrane and the release of large quantities of ACh into the extracellular synaptic cleft. Synaptic vesicles fuse with the presynaptic membrane at **active zones** (curved arrows, micrograph), electron-dense areas directly across from the junctional folds.

Acetylcholine diffuses across the basal lamina (arrowheads, micrograph) and binds to **receptors** (AChR) in the sarcolemma. These receptors are concentrated on the crests and uppermost regions of the junctional folds. Each AChR is also an ion channel. When ACh binds to the receptor, the entry of sodium and other ions results in a depolarization of the muscle sarcolemma that is propagated as an action potential. ACh is rapidly hydrolyzed by **acetylcholinesterase** (AChE), which is bound to the **basal lamina** by a collagenlike tail. The rapid removal of ACh by AChE is necessary for the electrical recovery of the postsynaptic sarcolemma during normal muscle activity.

In addition to housing the AChE, the basal lamina in the synaptic region contains agrin, a protein that regulates the clustering of AChE and AChR. Components of the basal lamina are capable of inducing the differentiation of all postsynaptic specializations. When muscle fibers are destroyed, they regenerate within the basal lamina sheath of the original fibers and form junctional folds and concentrated receptors in the original synapse region of the sheaths, even in the absence of nerve terminals.

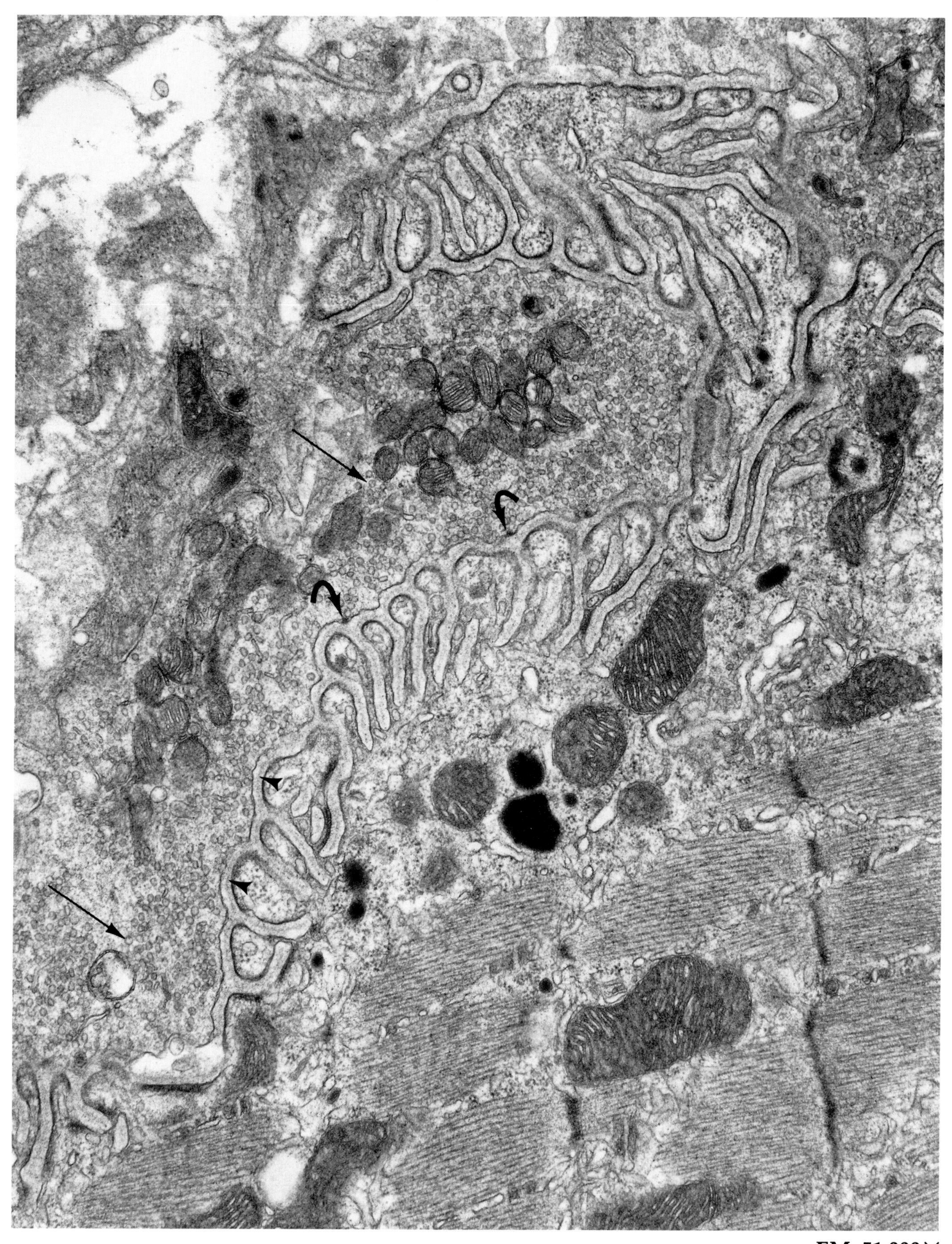

EM: 51,000×

SMOOTH MUSCLE: Overview

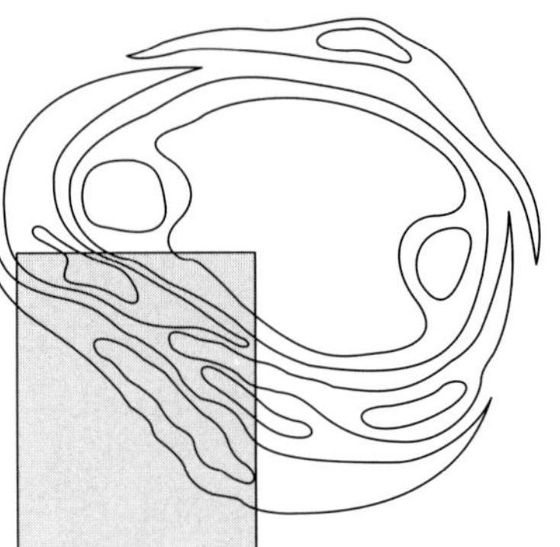

The **smooth muscle** fiber extending diagonally across micrograph 1 has an elongated, centrally placed nucleus (N), **dense bodies** (arrows), fibrillar material in the cytoplasm, and relatively few mitochondria (m). This fiber is in the wall of an arteriole (L: lumen; E: endothelial cell), where its contraction plays an important role in controlling peripheral resistance and thus blood pressure.

The fibrillar material of smooth muscle cells is shown in higher magnification in the inset. The largest structures are the **microtubules** (curved arrows), 25 nm in diameter, and the smallest are the **thin** (actin) **filaments** (arrowheads), 7 nm in diameter. Filaments between these two sizes represent primarily 10-nm **intermediate filaments. Myosin,** though present in smooth muscle and essential for its contraction, is not easily preserved in these cells. The assembly of smooth muscle myosin filaments is sensitive to the level of phosphorylation, and when these filaments do assemble, they do so in a wide range of sizes not easily distinguished from those of other smooth muscle filaments.

Contraction in smooth muscle depends upon actin–myosin crossbridges and ATP force generation as in other muscle types, but the mechanism differs and a highly organized arrangement of actin and myosin filaments is absent.

Even though smooth muscle does not have the regularly arranged sarcomeres of striated muscle, the following evidence suggests that a contractile unit exists.

1. α-actinin is localized in dense bodies (as it is in the Z-line).

2. Actin inserts into dense bodies, with polarity directed away from the dense body (as true for actin insertion into the Z-line).

3. Myosin lines up with actin at a certain distance from dense bodies (suggesting some form of I- and A-banding arrangement).

Of the three muscle types, smooth muscle is unique in its ability to divide, to synthesize a wide variety of factors (e.g., prostacyclin, elastic fibers), and to respond to many different physiological mediators (e.g., hormones, mechanical stretch, and growth factors).

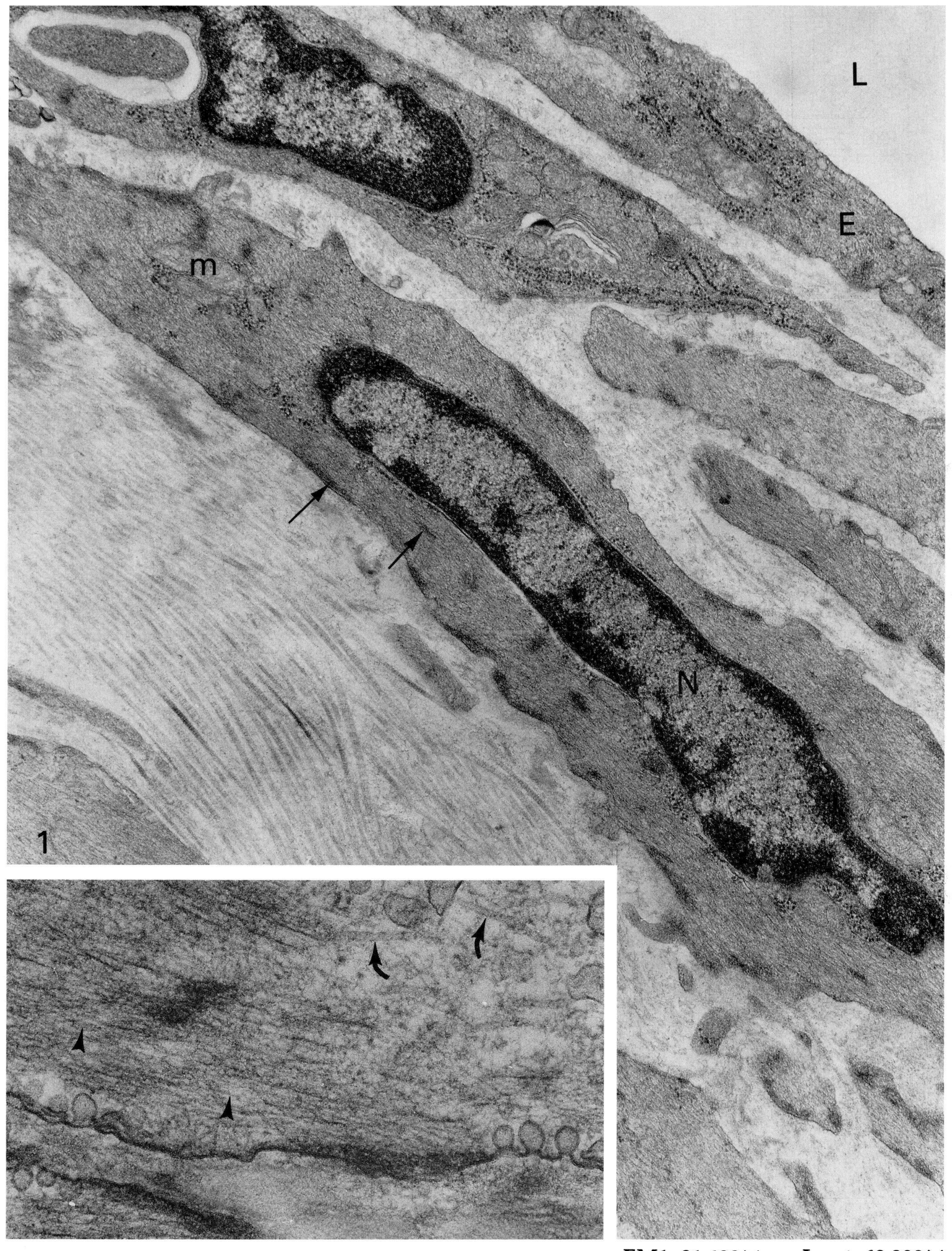

EM1: 31,400× Inset: 69,300×

SMOOTH MUSCLE: Contraction

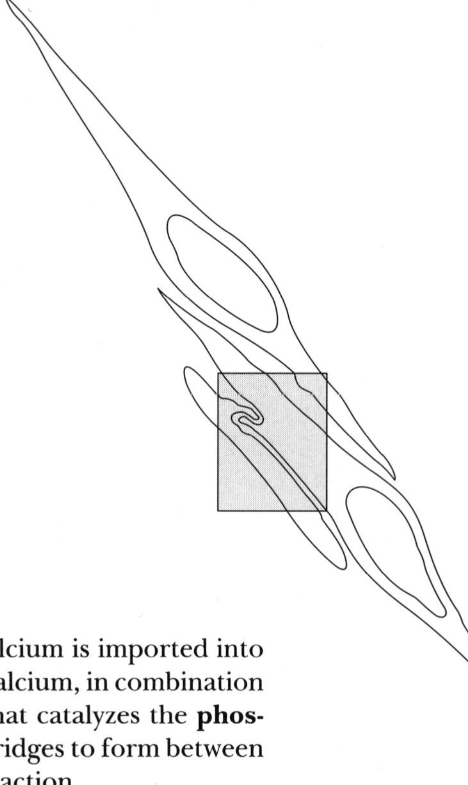

In this micrograph showing cross and longitudinal sections of smooth muscle cells, the **myosin filaments** (arrows) are well preserved and can be seen adjacent to **actin filaments** (arrowheads).

Smooth muscle thick and thin filaments differ from those in striated muscle in ratio (less myosin in smooth) and in structure. The thin filaments contain actin with a unique composition and associated tropomyosin, but not troponin. Thick filaments are variable in size and contain less organized, not necessarily bipolar, arrays of myosin.

The surface area of smooth muscle cells is increased by as much as 70% by flasklike invaginations, **caveolae** (c, micrograph), and it is often suggested that these structures are important in calcium regulation, perhaps by facilitating exchange with the extracellular fluid. As in all muscle, **calcium** regulates contraction, but the mechanism of regulation in smooth muscle involves myosin rather than the thin filament complex. Following receptor-mediated activation, extracellular calcium is imported into the cell and intracellular stores of calcium are released. Calcium, in combination with a soluble protein, **calmodulin,** activates a kinase that catalyzes the **phosphorylation of a myosin light chain.** This enables crossbridges to form between actin and myosin, with ATP hydrolysis resulting in contraction.

Cardiac and smooth muscle display an inherent electrical activity that is (1) modified by the action of the autonomic nervous system and (2) propagated to other cells via gap junctions.

Studies using digital video microscopy to track marker beads fixed to the outer surface of isolated smooth muscle cells show that twisting occurs along with contraction. These studies fit in nicely with the classical observation of corkscrew-shaped nuclei in light microscope preparations of smooth muscle. The arrangement of contractile units oblique to the long axis of the fiber and in parallel (versus in series as in striated muscle) permits large force generation and facilitates the unique capacity of smooth muscle to shorten to 20% of its length.

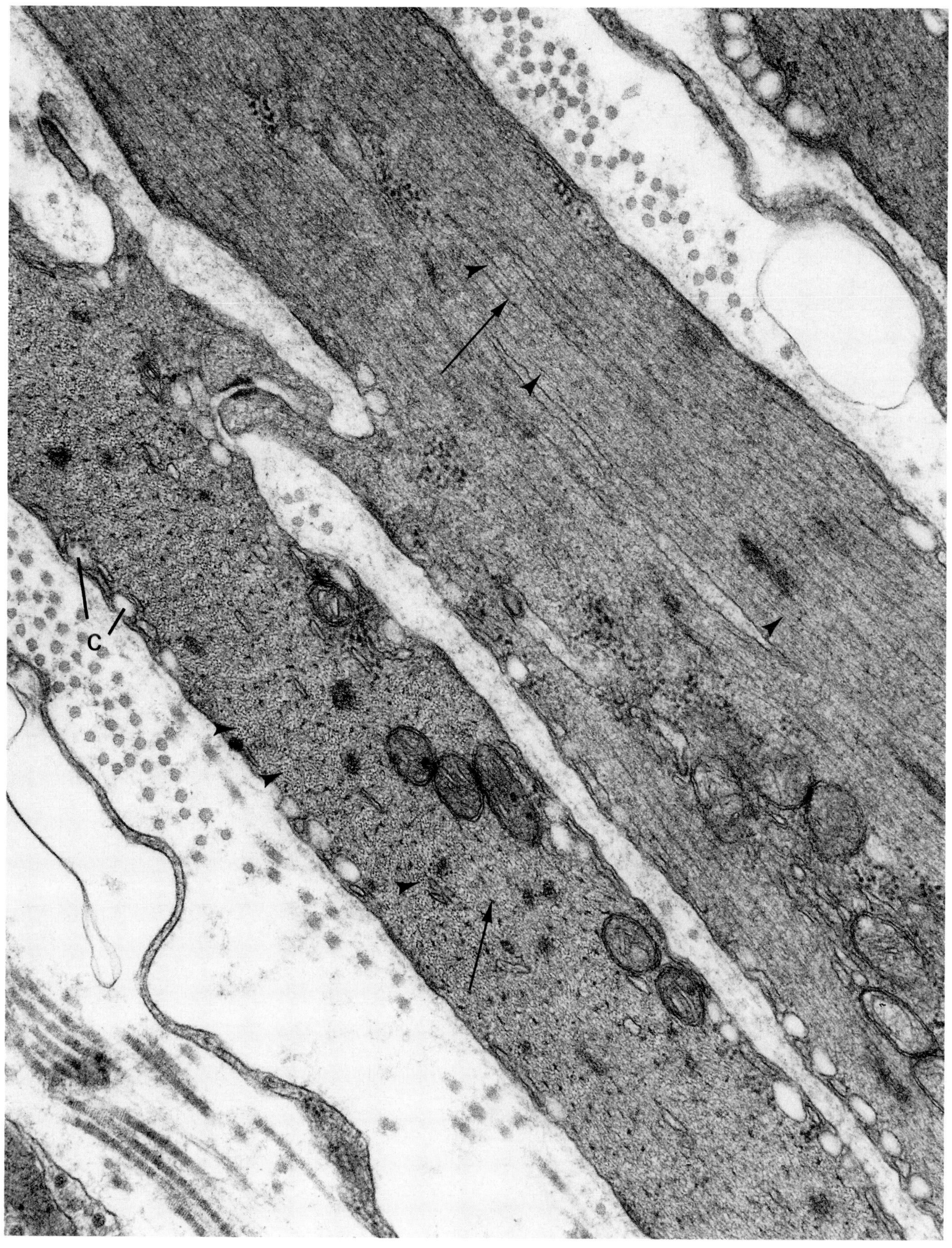

EM: 57,600×

CARDIAC MUSCLE: Overview

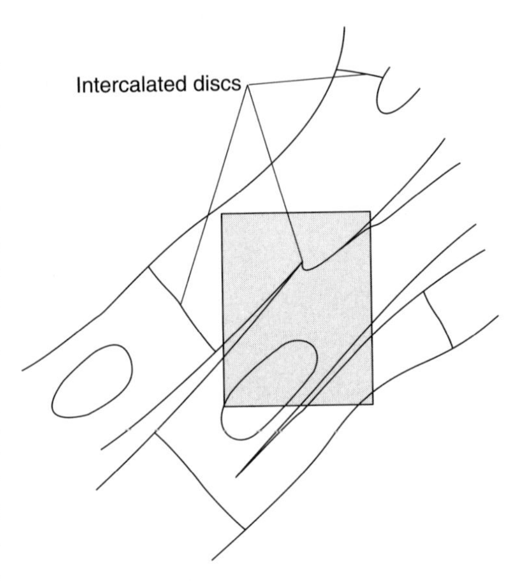

Cardiac muscle fibers form a **branching network** (M1 branches bind to M2 in the upper right of the micrograph) of cells tightly apposed at specific junctional regions called **intercalated discs** (straight arrows, micrograph). The nucleus lies in the center of the fibers, surrounded by sarcomeres.

The large number of mitochondria (m) and lipid droplets (L) in the cardiac fibers shown in this micrograph reflects the slow oxidative contraction pattern of these **ventricular cells.** Other regions of the heart contain fibers specialized for fast contraction, excitation conduction (e.g., Purkinje fibers), nerve input (e.g., nodes), and hormone production.

Excitation in cardiac muscle is spread along the cell surface and membrane invaginations **(T-tubules)** as in skeletal muscle. In cardiac muscle there is only one T-tubule per sarcomere that invaginates at the Z-line. T-tubules in section are identified by the presence of a basal lamina (arrowheads, micrograph), confirming that the space is extracellular. The action potential propagated along the cell surface and invaginations results in the flow of calcium into the cell from the extracellular fluid. This initial entry of calcium may cause a subsequent release of calcium from the sarcoplasmic reticulum. The sparsity of sarcoplasmic reticulum in the micrograph is characteristic of cardiac muscle in general, and the role of this organelle in the heart is not as well defined as it is in skeletal muscle. Calcium plays the same role in actin–myosin interaction and the resulting force generations in both cardiac and skeletal muscle.

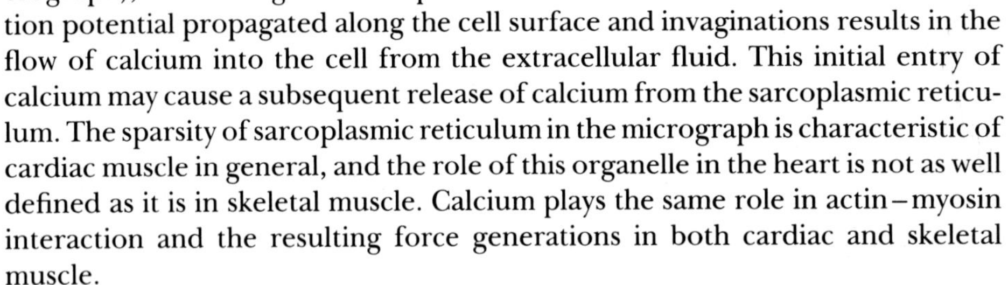

Adult cardiac muscle cells do not normally divide and are not replaced. During their long life they accumulate **residual bodies** (curved arrow, micrograph) that contain aged cellular components removed in the process of routine turnover. The heart responds to increased demands for cardiac output resulting from disorders or normal increased activity (e.g., exercise) with an increase in mass; however, this reflects a hypertrophy of individual cells, not hyperplasia. With prolonged or extensive demands or injury, scar tissue replaces cardiac cells.

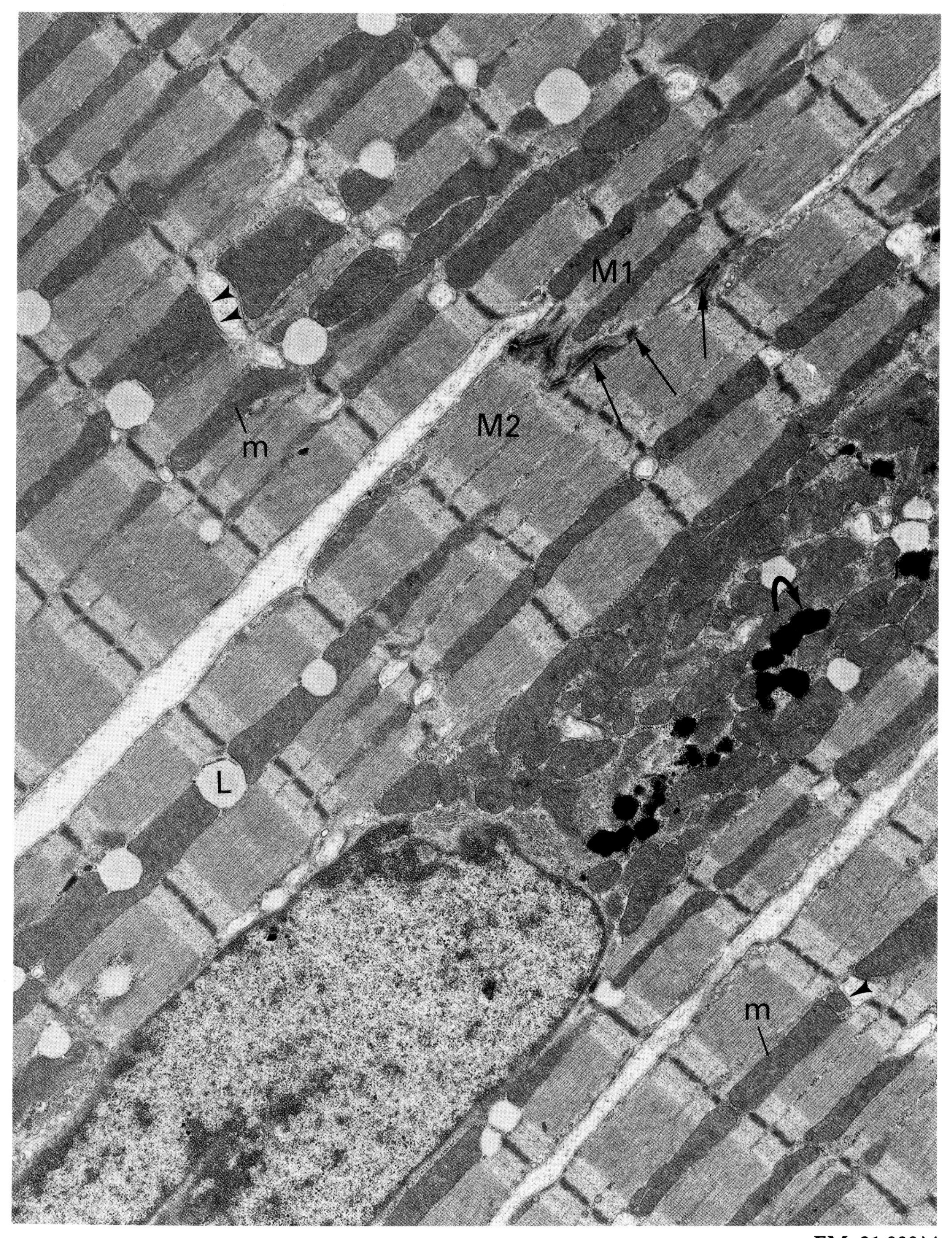

EM: 21,000×

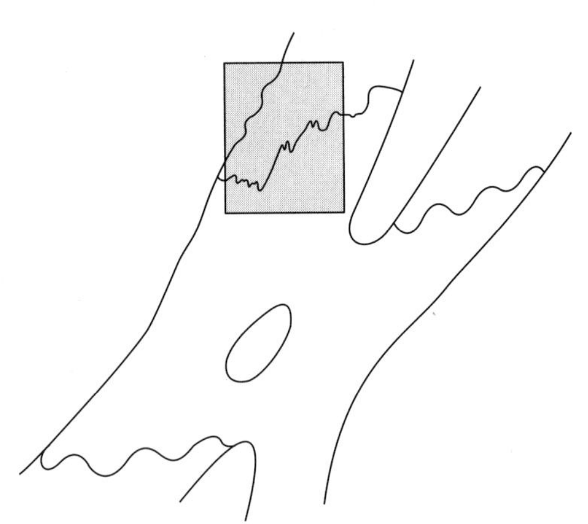

The function of cardiac muscle depends upon a tight association and selective communication between individual cells. In micrograph 1 two cardiac muscle cells (M1 and M2) are seen attached end to end by an **intercalated disc.** Each disc consists of a stepwise arrangement of transverse and longitudinal components. The transverse part of the step, localized at the Z-line, contains primarily adherens junctions and is easily recognized by the cytoplasmic densities of associated proteins. Both **macula** (desmosome) and **fascia adherens** are found in this region but are not easily distinguished. The fascia adherens is the attachment site for actin filaments and contains both α-actinin and vinculin, two proteins also characteristic of other regions of actin–membrane association. The desmosome is the attachment site for **desmin,** a type of intermediate filament. Desmin functions like other intermediate filaments associated with desmosomes, to distribute stress within individual cells.

Longitudinal parts of intercalated discs are composed primarily of **gap junctions** (arrows, micrograph 1 and inset). The cross striations bridging the intercellular space (inset) are connexons, the protein channels through which ions and small signal molecules move.

Gap junctions allow passage of the action potential through an extensive, intricate journey from its site of origin. The action potential is generated in the sinoatrial (S-A) node (the heart's most rapid pacemaker) by spontaneous depolarization of specialized cardiac muscle cells. It is transmitted in a defined sequence through the atria to a second specialized region, the atrioventricular (A-V) node, then to the bundle of His, Purkinje fibers, and ventricular fibers. Both the S-A and A-V nodes are innervated by sympathetic and parasympathetic neurons that control the speed of conduction and the rate and force of contraction.

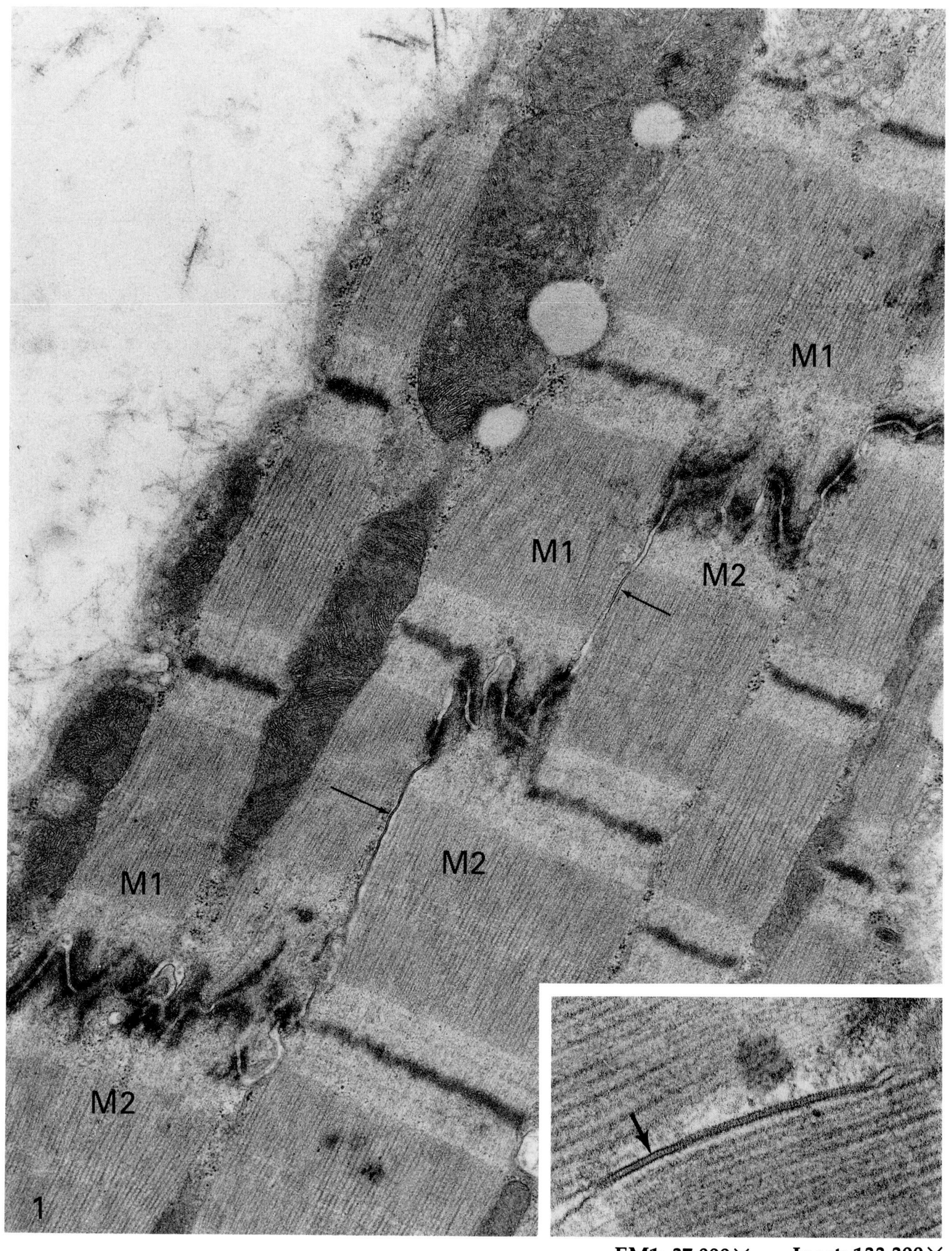

EM1: 37,000× Inset: 133,300×

NERVE

Overview

Most neurons consist of (1) a **cell body** with a large euchromatic nucleus and prominent nucleolus, (2) **dendrites,** thick, relatively short processes that carry signals to the cell body, and (3) the **axon,** a single long, thin process that carries signals away from the cell body. The size and shape of neurons vary considerably; however, within this diverse group most nerve cells retain this basic morphology.

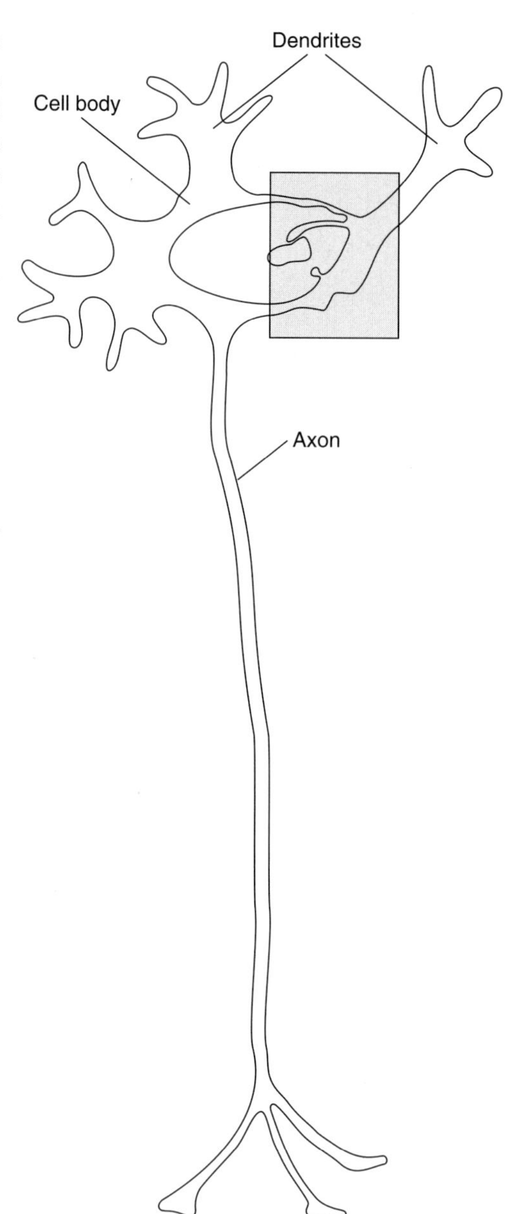

The cell body of a neuron from the central nervous system (CNS) occupies the center of the micrograph. A single large nucleolus (n) with its associated heterochromatin (h) lies in the indented nucleus (N). The cytoplasm surrounding the nucleus contains mitochondria (m), free (arrowheads) and attached (arrows) ribosomes, and several Golgi (G). In this micrograph only one process (*) can be observed extending from the cell body.

Most of the area outside the cell body consists of neuron processes (p, micrograph, in cross and oblique section) packed directly adjacent to one another. These processes represent over 90% of nerve cell volume; in humans one axon can have a volume 10,000 times that of a liver cell. Components within the processes are maintained in large part by synthetic activity within the cell body.

Neuron processes associate with one another at **synapses** (circle, micrograph) to form a cellular network that extends into every region of the body. Information is carried long distances along these cellular pathways as self-propagated electrical signals, **action potentials.** These signals are involved in coordinating most activities in the body and are essential for events ranging from the stimulation of acid secretion in the stomach to complex thought processes in the brain.

Glial cells, nonneuronal cells that perform critical functions unique to nervous tissue, outnumber neurons 10 to 1 and carry out many functions essential to neuron survival. Glial cells in the CNS include astrocytes, oligodendrocytes, microglia, and ependymal cells; in the peripheral nervous system (PNS) they include Schwann cells and satellite cells. Some axons are surrounded by a myelin sheath (curved arrow, micrograph), a dense lipid encasement synthesized by oligodendrocytes in the CNS, and by Schwann cells in the PNS.

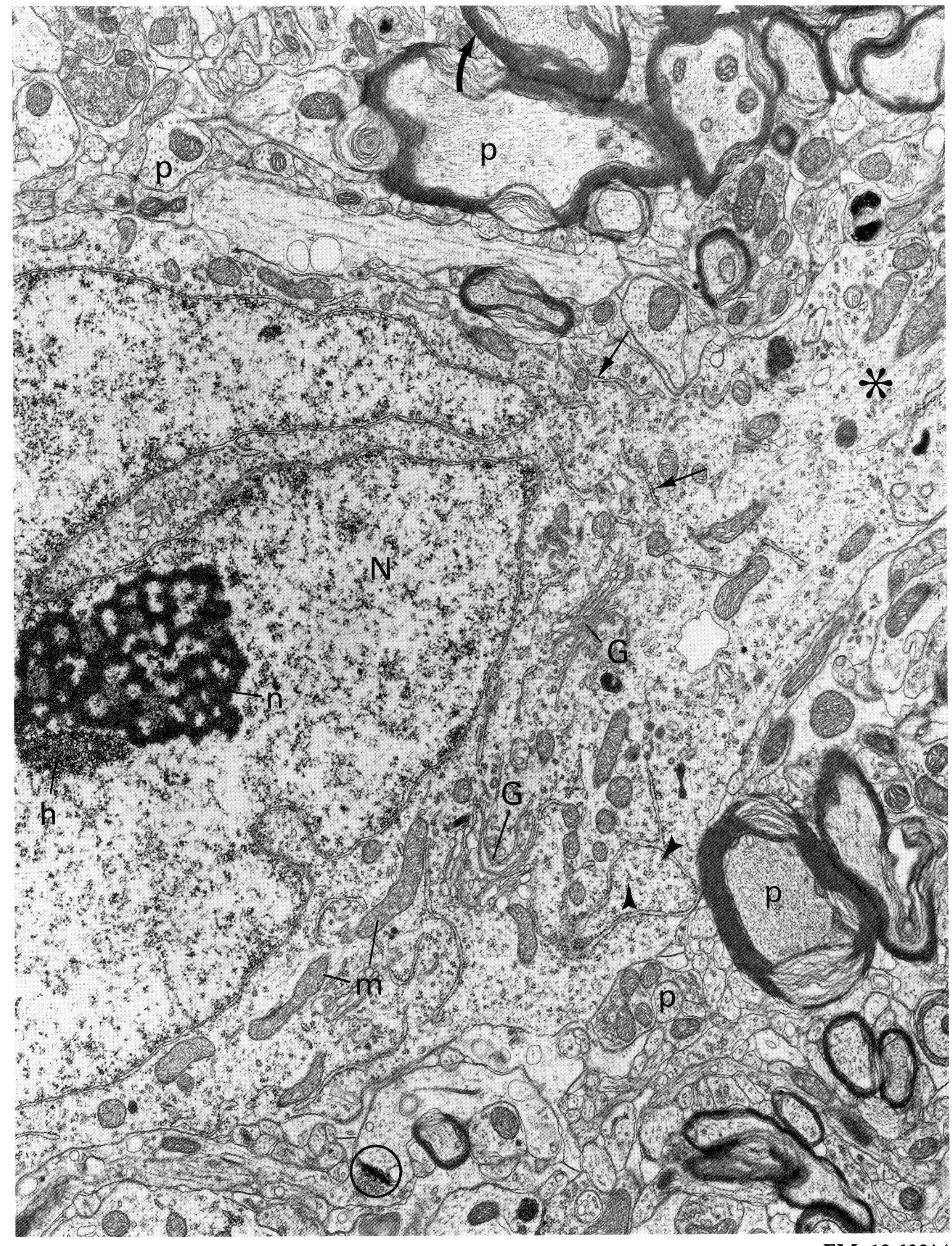

EM: 13,600×

NEURON: Nissl Bodies

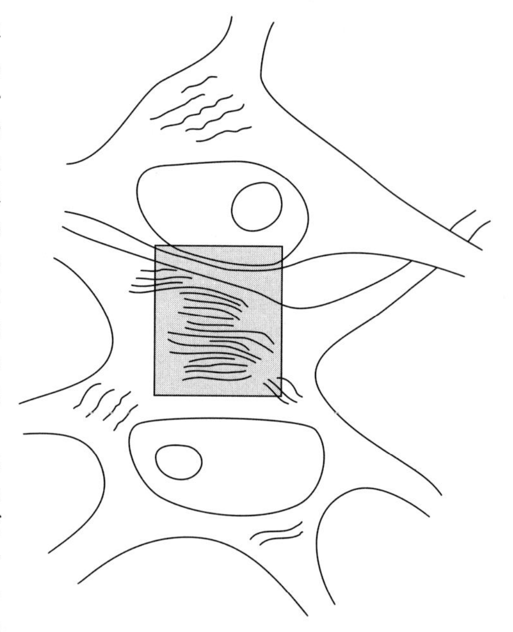

The **ribosomes** in many nerve cell bodies are arranged in a distinctive pattern. Flattened cisternae of rough ER (arrows, micrograph) alternate with groups of free polysomes (arrowheads, micrograph) to form **Nissl bodies.** The entire ribosome assembly in the micrograph represents a single Nissl body, seen in light microscope preparations as a blue dot when stained with a basic dye such as toluidine blue. Nerve cells express more of their DNA than any other cell type. The euchromatic nucleus (N, micrograph) and large number of polyribosomes within nerve cell bodies reflect active transcription and translation. In the micrograph, the Nissl body and the euchromatic nucleus are actually within two different neurons; nerve processes (p) can be seen between the two cell bodies.

Proteins synthesized on the Nissl bodies are critical to a variety of essential neuron activities. Those produced on free polyribosomes include (1) enzymes used in the synthesis of neurotransmitters and (2) cytoskeletal elements needed for support and transport within nerve processes. Proteins synthesized on polyribosomes attached to ER include (1) membrane proteins that comprise ion channels and receptors and (2) synaptic vesicles and neuropeptide neurotransmitters. These components pass from the rough ER to the Golgi (G, micrograph), where they are sorted and directed to specific locations.

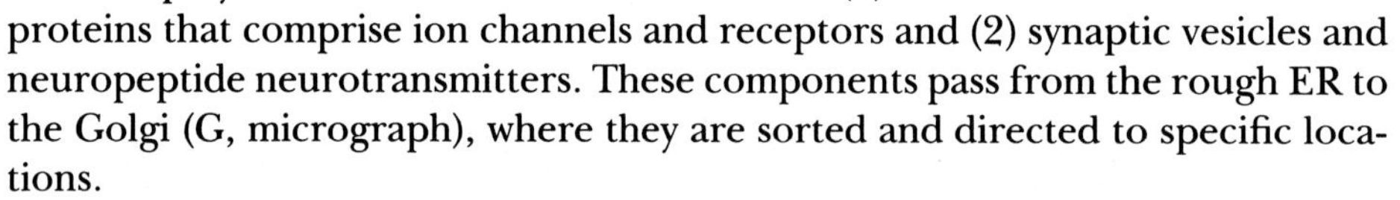

Ribosomes are found in the cell body and at the base of dendrites, but are not present in axons. Axonal proteins are synthesized on both free and attached ribosomes of the Nissl bodies and transported a considerable distance along the axon to their site of functioning. Certain neurotransmitters and synaptic vesicles are carried all the way to the end of the axon, in some cases a distance of 1 meter. The movement of these secretory vesicles to the site of exocytosis is an interesting example of cellular polarity.

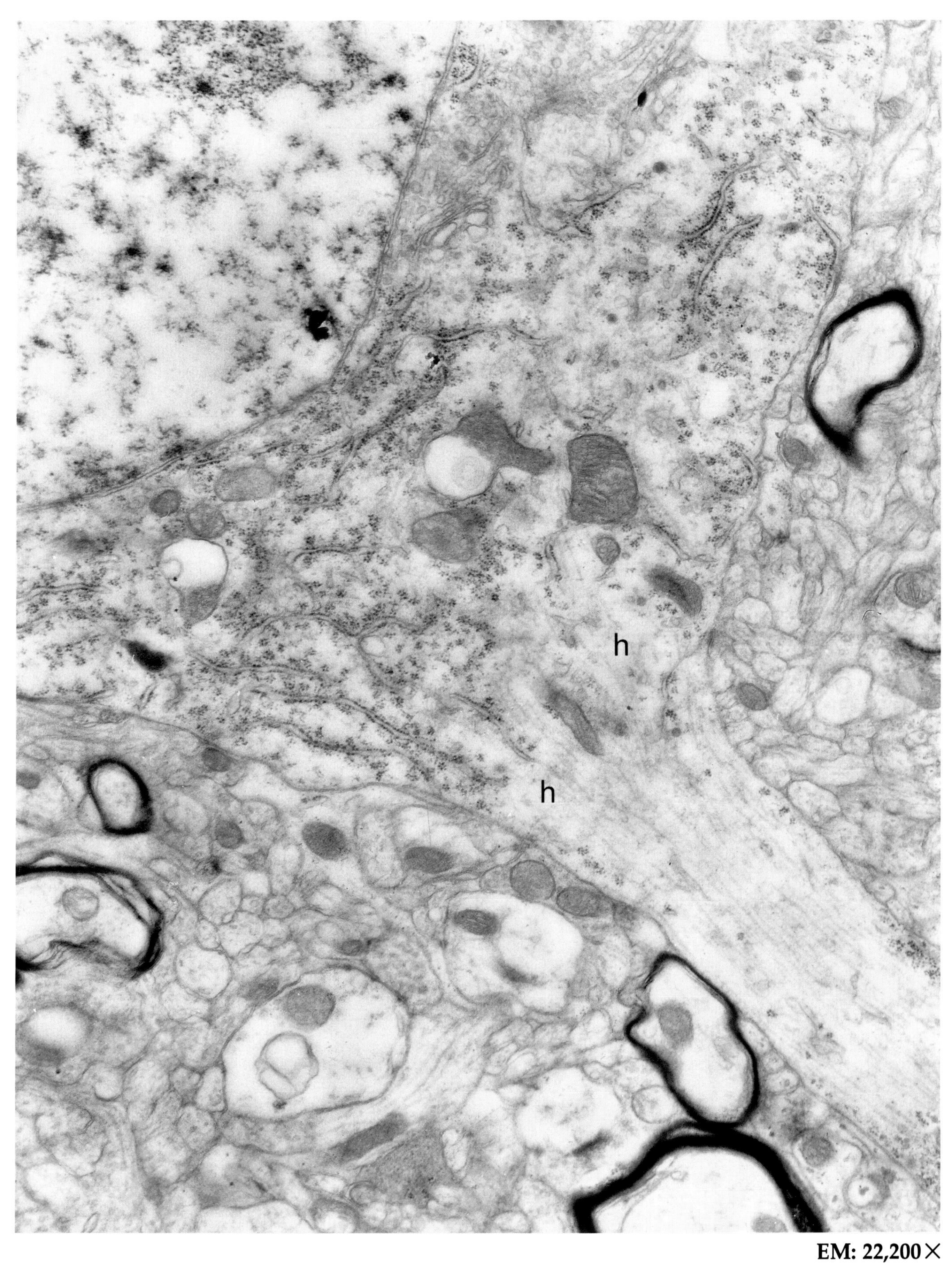

EM: 22,200×

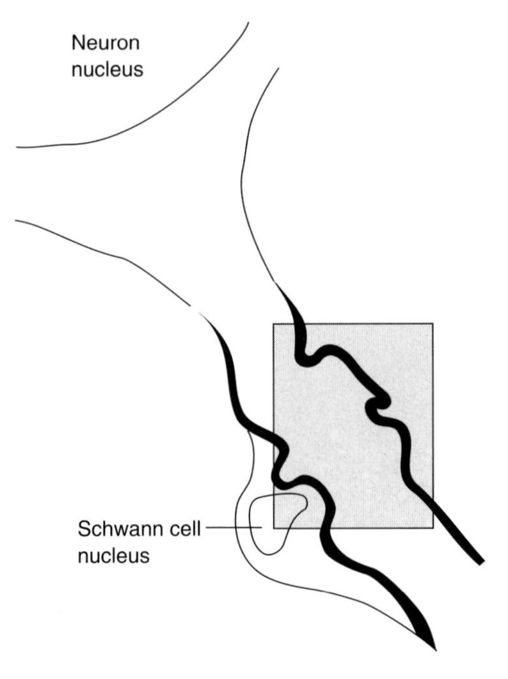

All axons have an abundance of **cytoskeletal elements** running parallel to their long axis. These filamentous proteins are the most prevalent proteins in axons. **Microtubules** (curved arrows) and **neurofilaments** (arrowheads) can be distinguished in the axoplasm of the myelinated axon in the micrograph. These structures, along with the third cytoskeletal element, **microfilaments** (not seen in the micrograph), are arranged within axons in a lattice that provides compartments and organization for the other cytoplasmic structures. The lattice is maintained by extensive crossbridges (circles, micrograph) both within and between cytoskeletal classes. This protein scaffolding is maintained by precursors that move down the axon via slow transport (0.1 to 3 mm per day). Slow transport is also used to carry soluble enzymes that are needed a considerable distance from the nerve cell body where they are synthesized.

Membranous organelles are carried by a fast transport system that is capable of rates of 100–400 mm per day. Organelles attach to microtubules and, in an energy-dependent fashion, are moved in both anterograde and retrograde directions (Cell, page 40). Vesicles carrying molecules specific to synaptic functioning, such as norepinephrine, move to the axon terminal via fast axon transport. Mitochondria (m, micrograph) supply the energy for this type of movement and are themselves moved on microtubules. Retrograde microtubule transport carries a varied assortment of organelles and molecules back to the cell body. Within the cell body, some (e.g., synaptic vesicles) are degraded in lysosomes, while others can have life-promoting (e.g., nerve growth factor) or life-threatening (e.g., viruses and toxins) effects.

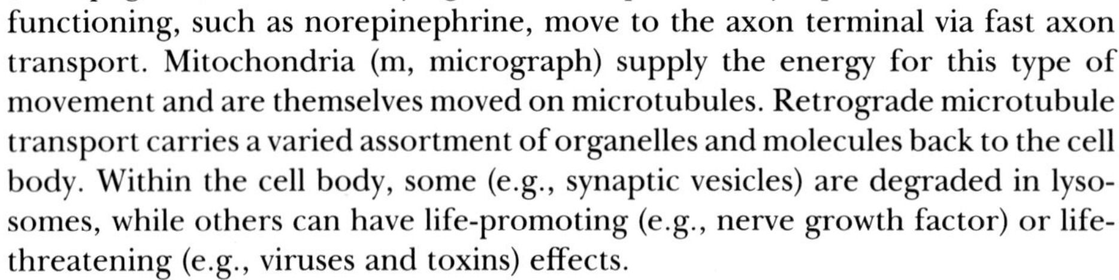

The axon in this micrograph is myelinated. The myelin (M), synthesized by Schwann cells (S), is interrupted at regular intervals by **Schmidt–Lanterman clefts** (straight arrow). In these areas some Schwann cell cytoplasm remains between the tightly packed cell membranes that comprise the myelin. It has been suggested that the Schmidt–Lanterman clefts provide a route for the exchange of nutrients and metabolites between the axoplasm, Schwann cell cytoplasm, and interstitial fluid.

Axons induce the Schwann cell synthesis of important myelin proteins and are thus essential for myelination. In turn, myelination is essential to the normal conduction in these axons. If a myelinated axon is demyelinated either experimentally or during disease (e.g., multiple sclerosis), impulse conduction is slower and sporadic.

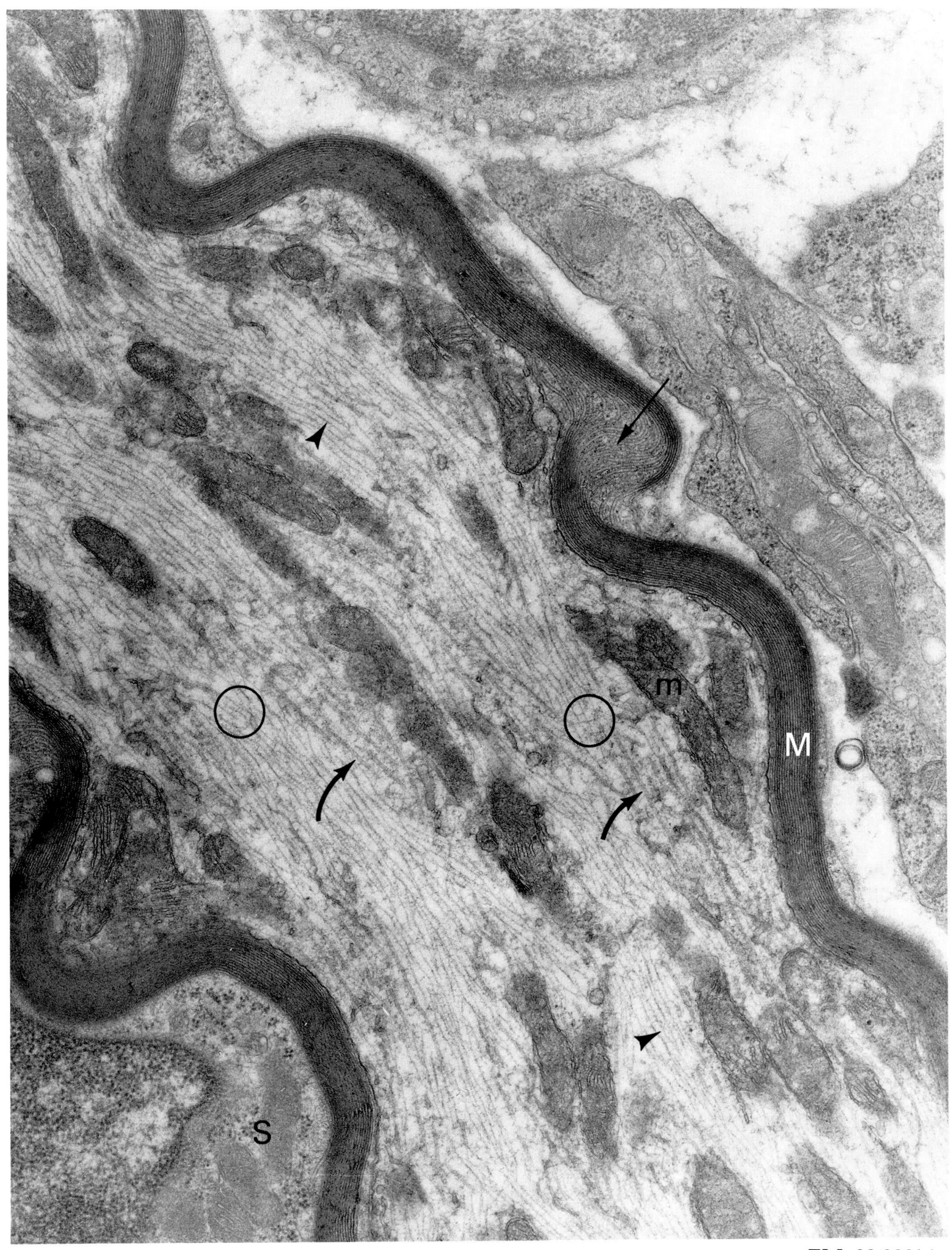

EM: 33,000×

Information is transmitted from neurons to other cells in specialized regions known as **synapses.** Synapses occur either between neurons and the effector cells they innervate, such as muscles and glands, or between two neurons, as in the central nervous system (shown here) and peripheral autonomic ganglia.

In micrograph 1, nerve-to-nerve synaptic regions are circled. At this relatively low magnification, the synaptic region is recognized by **electron densities** associated with the synaptic membranes and by the accumulation of small **synaptic vesicles** in the presynaptic terminal. When an action potential reaches the presynaptic nerve terminal, calcium channels in the cell membrane open and calcium rushes into the terminal. Synaptic vesicles then fuse with the cell membrane, releasing **neurotransmitter.** Neurotransmitter crosses the 20- to 30-nm synaptic cleft and binds to membrane receptors within the postsynaptic membrane, opening ion channels.

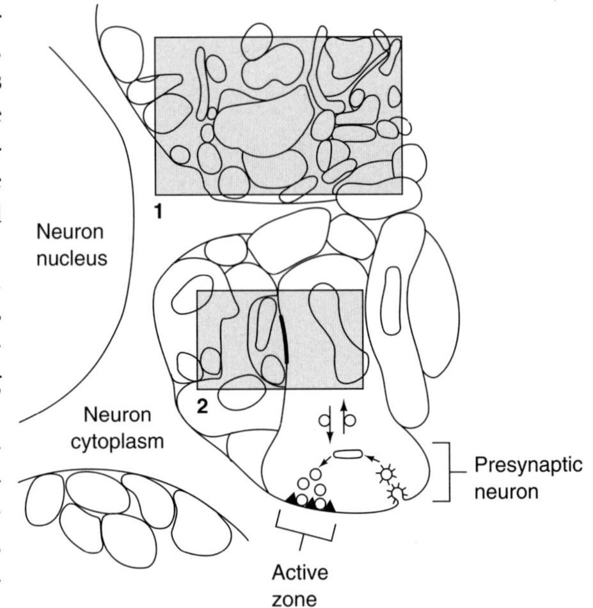

Synaptic vesicle exocytosis occurs between conical projections of electron-dense material (arrows, micrograph 2) in an **"active zone"** associated with the presynaptic membrane. Following exocytosis, excess membrane is recovered outside the active zone by pinocytosis in coated pits. Much of this membrane is locally recycled within the nerve terminal. However, some **vesicle turnover** occurs over considerable distance and involves retrograde transport of vesicles to the cell body, where they are digested by lysosomes. New vesicles are formed in the rough ER, packaged in the Golgi, and transported to the nerve terminal. Microtubules (arrowheads, micrographs 1 and 2), the predominant component of axons, carry vesicles to and from the cell body. The endocytosis and recycling of cell membrane during neurotransmitter release requires energy that is supplied by the mitochondria (m, micrograph 2), which are a typical component of presynaptic terminals.

Synapses in the brain can undergo long-lasting changes that increase their efficiency of operation (i.e., long-term potentiation). This process, a form of memory, involves, in part, information transfer from the postsynaptic neuron to the presynaptic neuron (a direction opposite to the classic synapse flow) and the release of increased amounts of neurotransmitter by the presynaptic neuron. Nitric oxide appears to be an important retrograde synaptic messenger needed for long-term potentiation.

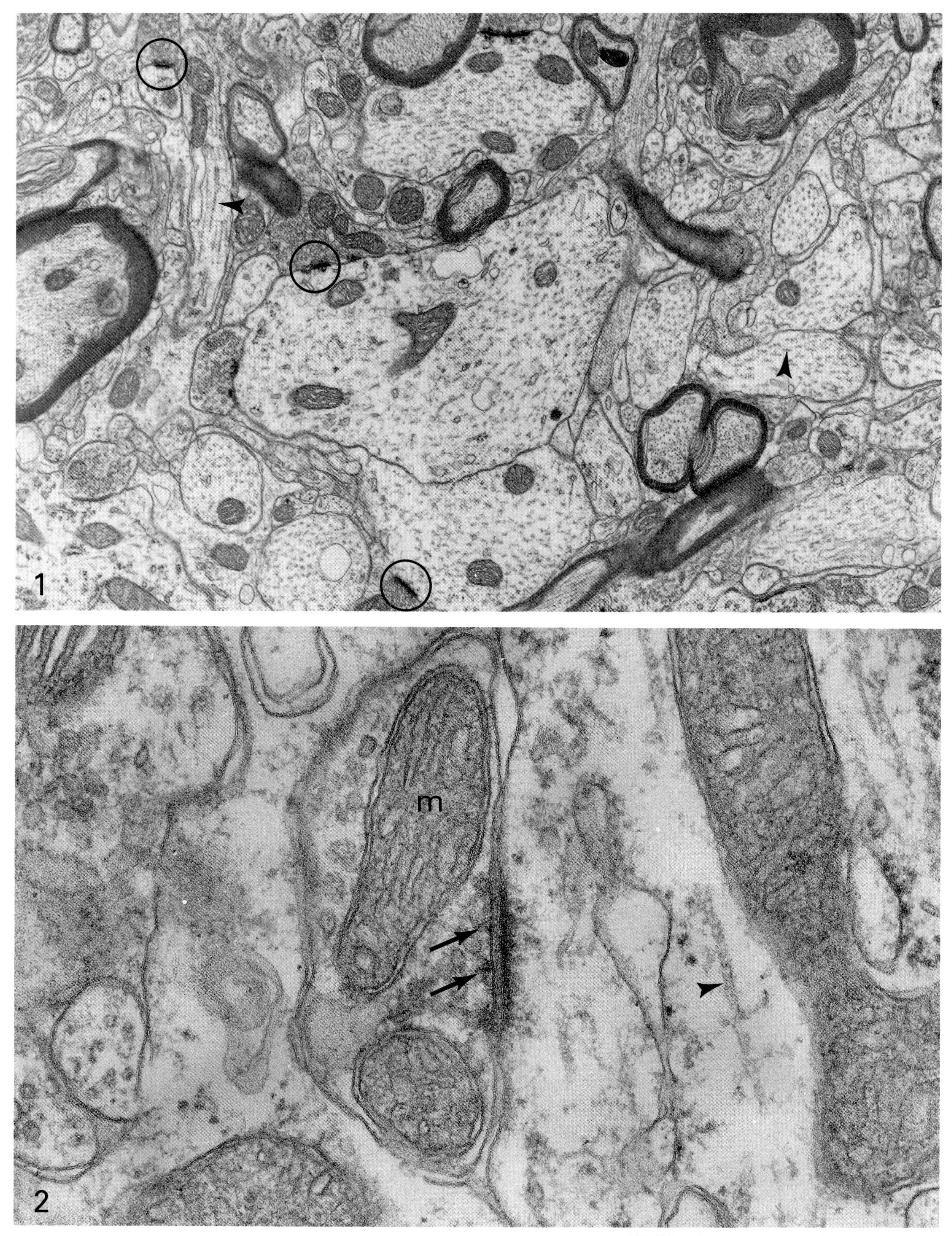

EM1: 9,000× **EM2: 88,000×**

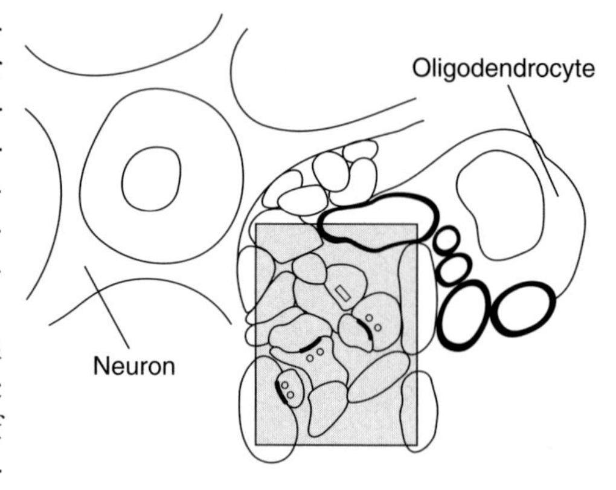

Chemical synapses between neurons are often classified into two groups based on observed differences in their ultrastructure. **Type I** has the appearance of synapse A (micrograph 1), with round synaptic vesicles and a prominent postsynaptic density. **Type II** has the appearance of synapse B (micrograph 1), with some flattened synaptic vesicles and symmetrical presynaptic and postsynaptic densities. Flattened synaptic vesicles have been associated with inhibitory action.

There are many types of **neurotransmitters.** Even though some neurotransmitters are associated with a specific neuronal response (e.g., GABA with inhibition), the action of other neurotransmitters, such as acetylcholine, can be inhibitory or stimulatory, depending upon postsynaptic neurotransmitter receptors. Neurotransmitters may have additional roles such as hormonal coordination of gastrointestinal activity (e.g., cholecystokinin) and building blocks of proteins (e.g., glycine).

Small-Molecule Neurotransmitters

- Acetylcholine
- Modified amino acids (catecholamines, dopamine, epinephrine, norepinephrine)
- Unaltered amino acids (γ-aminobutyric acid, or GABA, aspartic acid, glutamic acid, glycine)

Large-Molecule Neurotransmitters

- Neuropeptides (cholecystokinin, β-endorphin, gastrin, secretin)

Synaptic efficiency depends upon the quick removal of neurotransmitter following its action on the postsynaptic cell. At the neuromuscular junction, acetylcholinesterase within the basal lamina carries out this function, however, at most nerve-to-nerve synapses excess neurotransmitter is taken up and degraded by the presynaptic neuron. Certain drugs (e.g., some antidepressants) act by inhibiting this reuptake and thus increasing the concentration of neurotransmitter remaining within the synaptic cleft.

Synapses are not fixed structures; they are continuously moving and reestablishing contacts. Changes in synaptic size, shape, and location occur in response to such diverse events as learning and exposure to anesthetics. Electron densities associated with postsynaptic cell membranes are composed of globular proteins resting in a filamentous cytoskeleton. One of the major globular proteins, a calcium-activated neutral protease, may regulate the turnover of the postsynaptic cytoskeleton and thus contribute to synaptic plasticity.

In addition to the coupling of neurons by neurotransmitters, some neurons are coupled electrically via gap junctions (g, inset). In this type of "synapse," nerve processes, separated by an extremely small intercellular space (2-nm), are associated via connexons (see Epithelium, page 54) that carry current from one neuron to another. Gap junctions have been localized in regions of chemical synapse. Certain neurotransmitters (e.g., GABA) may control the opening and closing of gap junctions, and thus control the pattern of nerve firing.

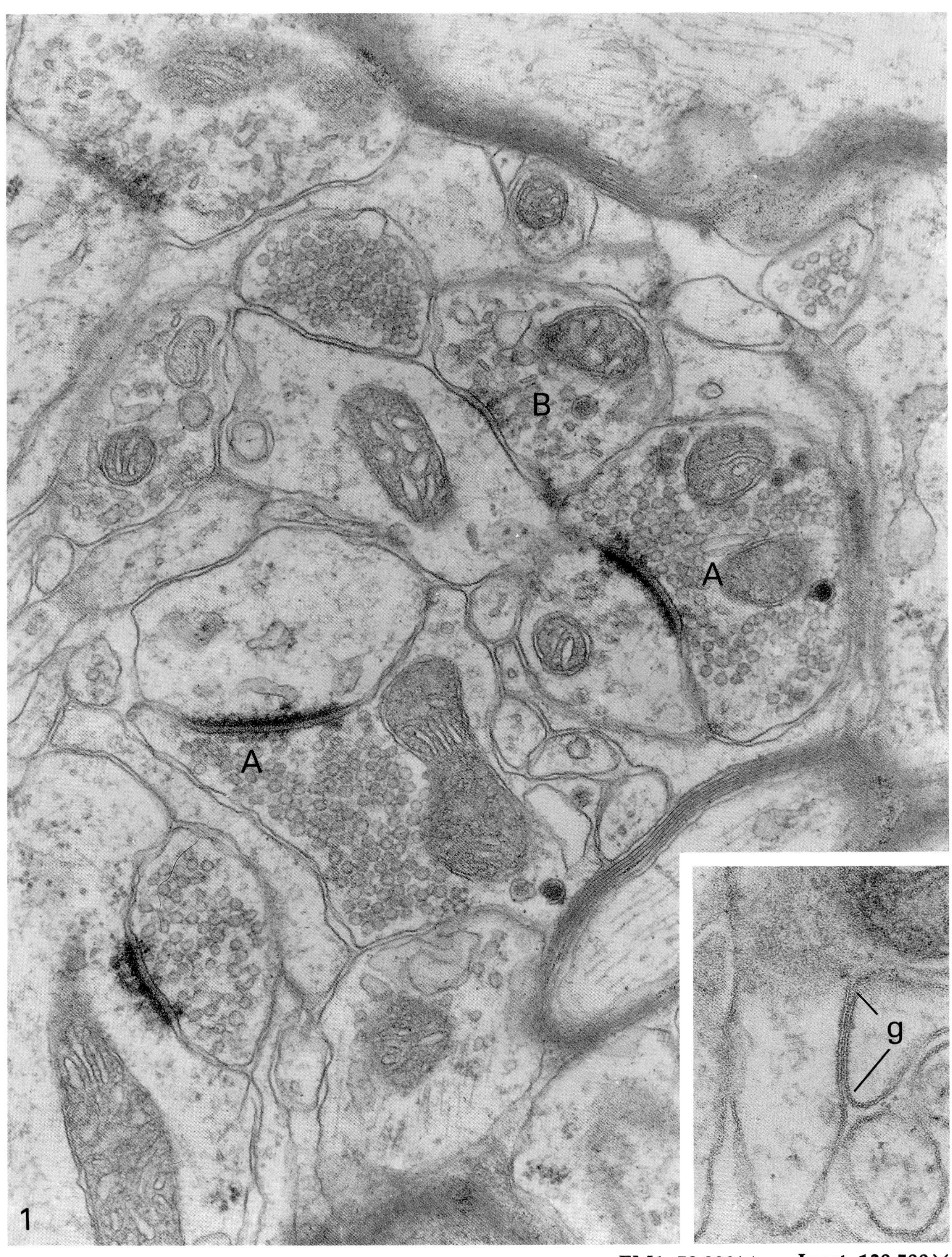

EM1: 58,000× Inset: 130,500×

GLIAL CELLS: Unmyelinated and Myelinated Axons

Schwann cells surround all axons in the peripheral nervous system (PNS). The association between Schwann cells and axons is intimate but it is different in complexity in **unmyelinated** and **myelinated** axons. Unmyelinated axons (a, micrograph 1) simply indent into the cytoplasm of a Schwann cell (S, micrograph 1). In this type of association, each Schwann cell can encase many axons. The connecting channel between each axon and the external surface of the surrounding Schwann cell is called the **mesaxon** (arrows, micrograph 1).

In the PNS, myelinated axons (a, micrograph 2) are each wrapped with many layers of Schwann cell membrane, which form a **myelin sheath** (m, micrograph 2). Myelin is formed by the growth, elongation, and spiral wrapping of the mesaxon (solid arrows, micrograph 2) and associated Schwann cell cytoplasm (S, micrograph 2) around the axon. A basal lamina (b, micrographs 1 and 2) defines the outer boundary of each Schwann cell.

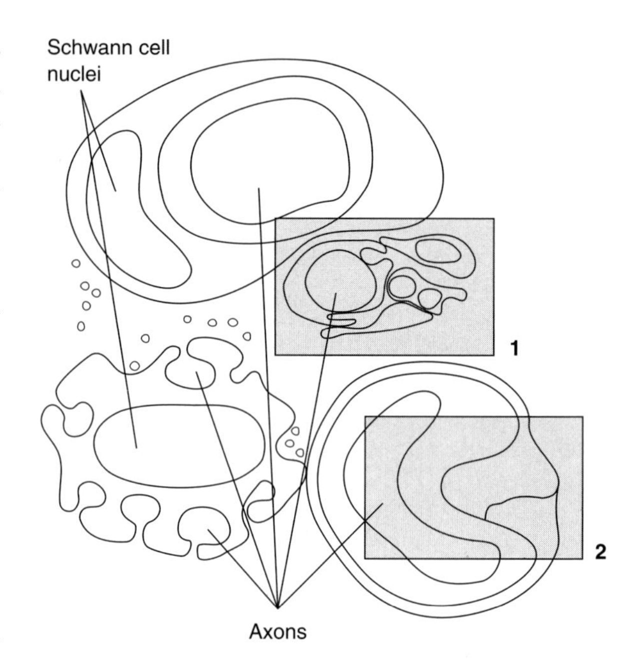

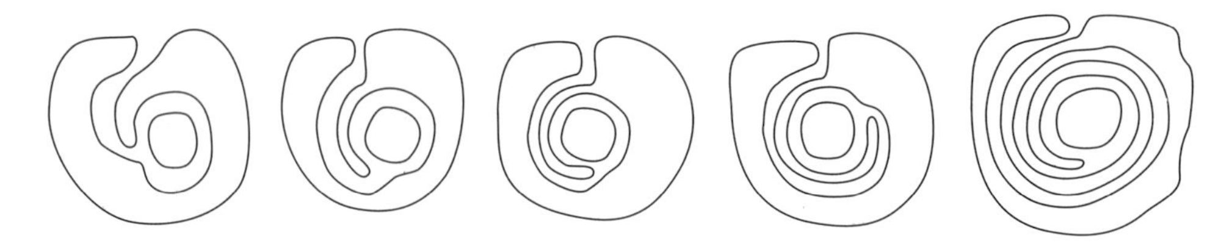

In the process of myelination, Schwann cell cytoplasm is squeezed out so that the inner leaflets of the cell membrane pack together to form the **major dense line** (arrowheads, micrograph 2). As each successive wrapping overlaps the previous one, a periodicity of 12 nm is established between the major dense lines. This light-staining 12-nm area includes the outer membrane leaflets as well as the intercellular space between the outer membrane leaflets. The diameter of this intercellular space is determined by the size of the membrane components extending from the cell surface. At the point where the myelin wrapping begins (open arrow, micrograph 2), new myelin proteins with smaller extracellular domains are inserted into the outer surface of the membrane, which results in a reduced intercellular space. In the higher magnification inset, an interperiod line that represents these extracellular domains can be resolved between the major dense lines.

The composition of myelin is different from the composition of most cell membranes in several ways: myelin (1) has a high proportion of lipid to protein, (2) is deficient in several standard membrane proteins such as ion channels, and (3) contains unique proteins, such as myelin basic protein, that seem to be involved in the tight compaction of adjacent membranes.

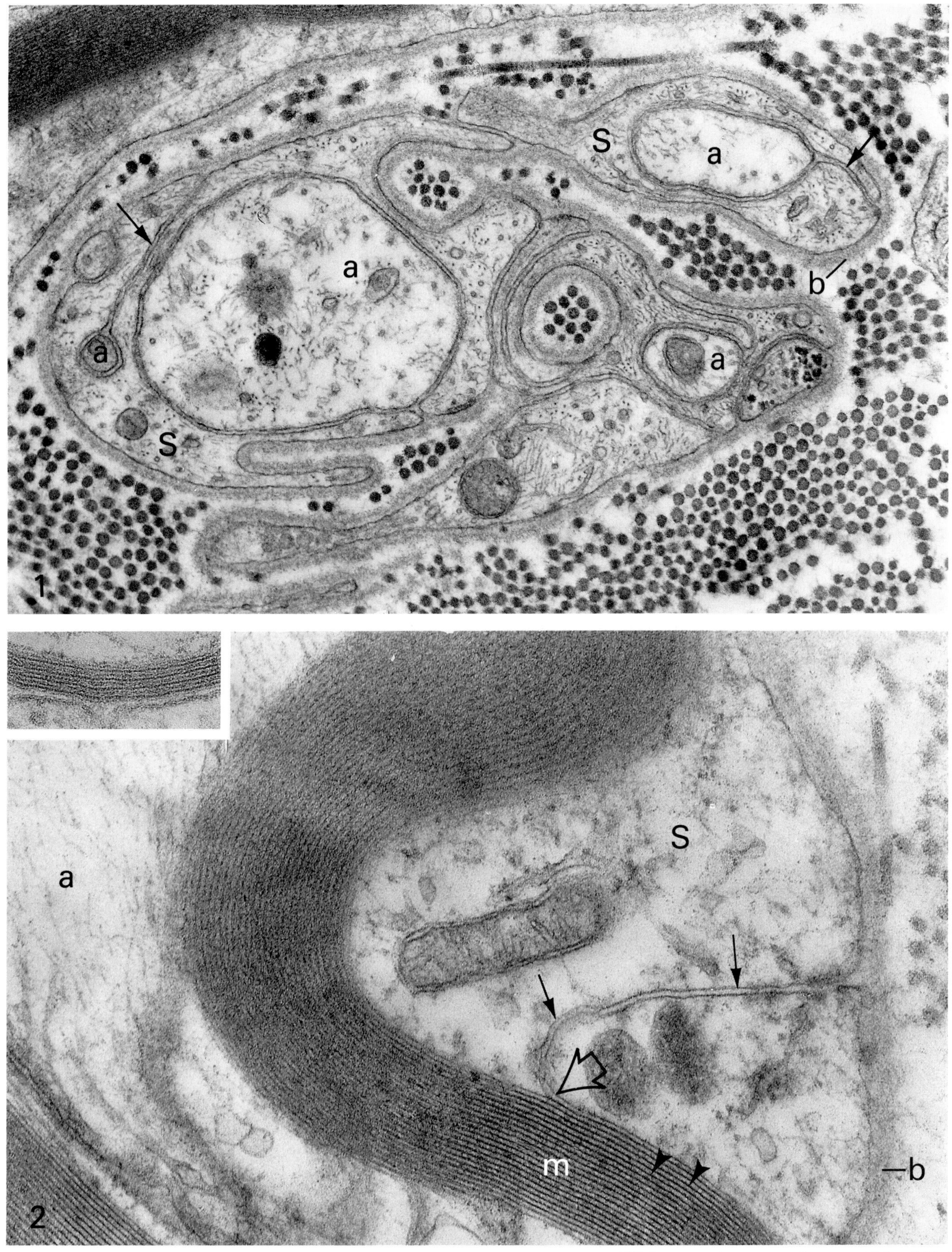

EM1: 49,300× EM2: 76,000× Inset: 130,500×

GLIAL CELLS: Oligodendrocytes and Schwann Cells

In the central nervous system, myelination is carried out by **oligodendrocytes** (O, micrograph 1). A single oligodendrocyte extends several processes that each myelinate a section (internode) of a different axon or of the same axon. The oligodendrocyte in micrograph 1 is seen in association with five axons (a) that it is myelinating. Some of the myelinated axons that appear to be separate (arrows) from the oligodendrocyte cell body may be connected by thin cytoplasmic processes that are not seen in this section.

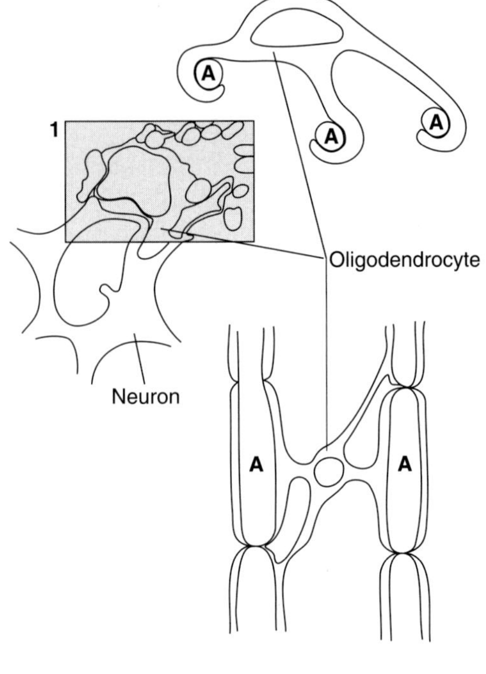

One oligodendrocyte can myelinate up to 50 internodes, which involves the formation of myelin membrane 600 times the amount of membrane covering its cell body. The synthetic demands on this cell are reflected in its ultrastructure. The polysomes packed in the cytoplasm direct the synthesis of (1) enzymes needed to form the vast amounts of cell membrane that comprise myelin and (2) structural proteins within myelin itself. A comparison between the ultrastructure of the neuron cell body (Ne) and the oligodendrocyte (O) in micrograph 1 highlights the characteristic electron density of this glial cell.

The amount of protein synthesized by an oligodendrocyte that is myelinating several internodes is considerably greater than that of a **Schwann cell,** which myelinates only one internode of one axon in the PNS. In micrograph 2 the section passes through the nucleus (N), cytoplasm (C), and myelin (m) of a Schwann cell encased around a single axon (a).

In contrast to the CNS, myelinated axons in the PNS are typically separated from one another by collagen (c, micrograph 2). Like muscle, peripheral nerves are packaged by a hierarchy of connective tissue sheaths. Individual neurons bound together by an endoneurium are grouped into fascicles surrounded by a cellular sheath, the perineurium (p, micrograph 2). Fascicles, in turn, are surrounded by the epineurium, a dense connective tissue sheath that defines the nerve.

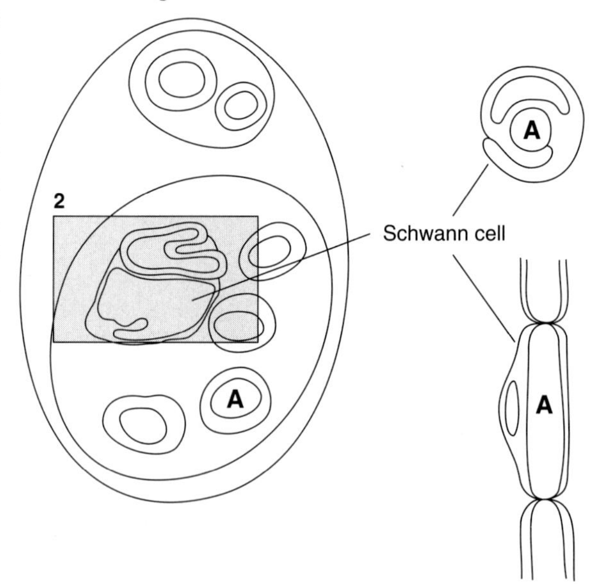

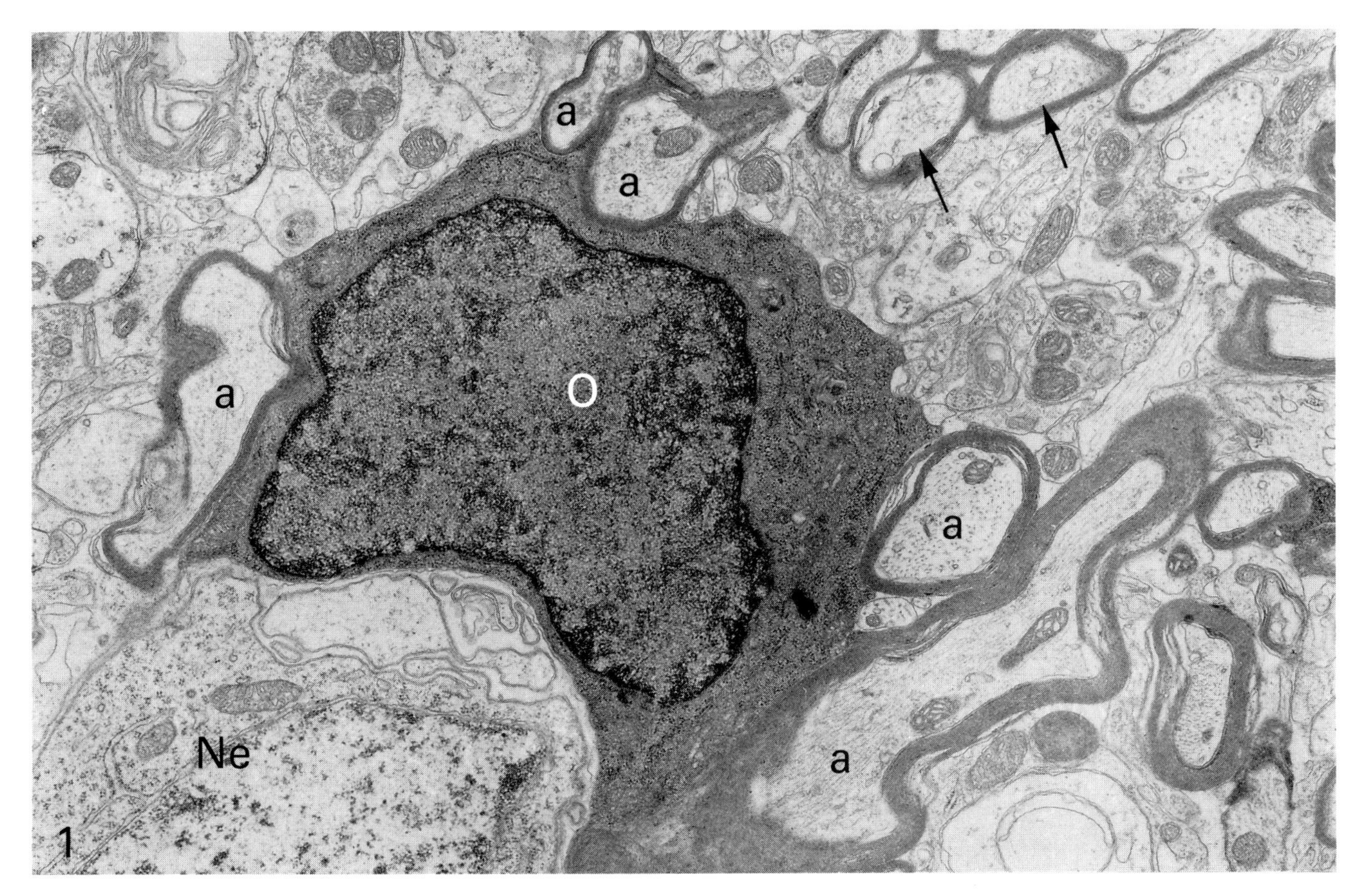

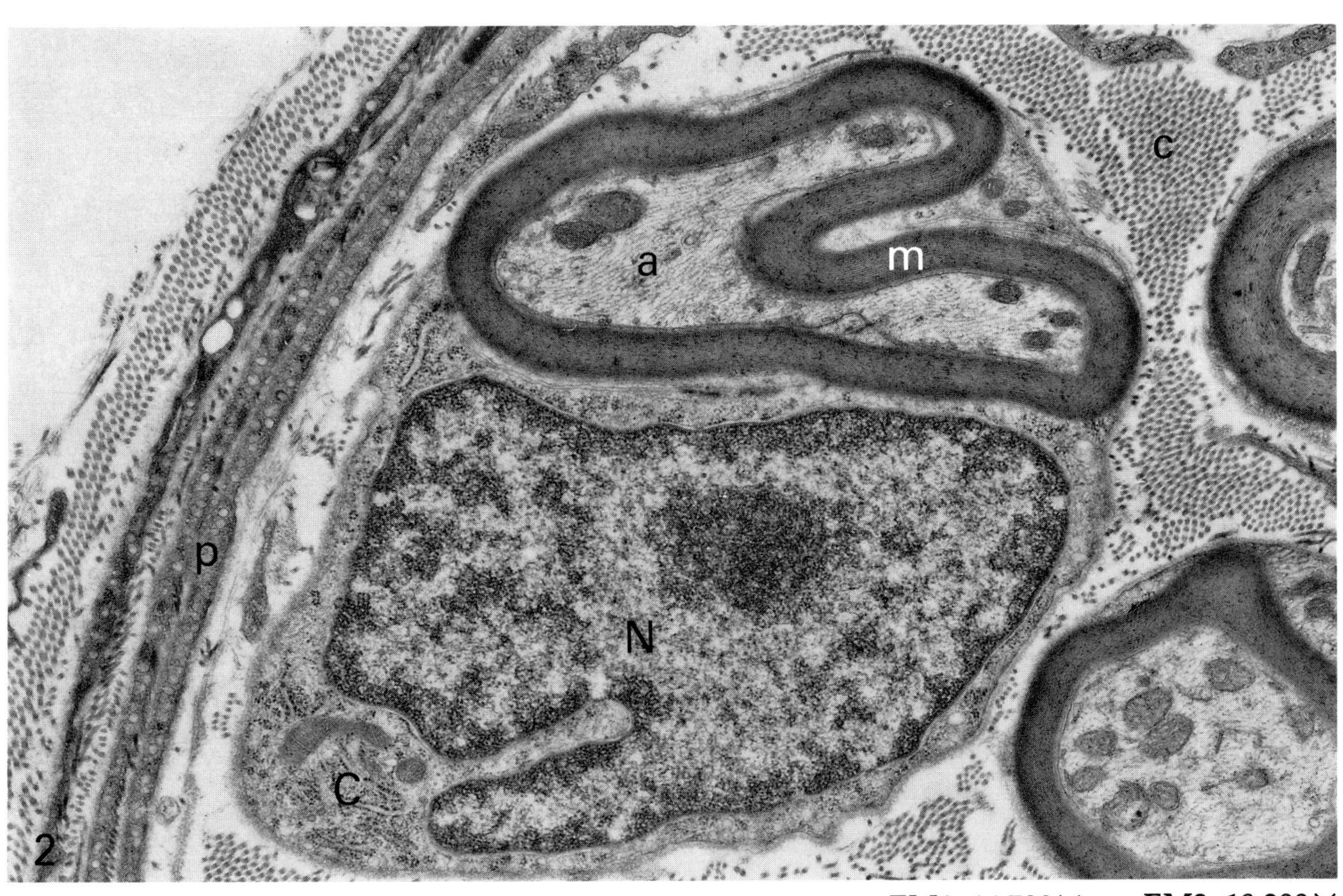

EM1: 14,700× EM2: 18,200×

GLIAL CELLS: Node of Ranvier

Action potentials initiated at the axon hillock are propagated down axons at different velocities, ranging from 1 to 100 meters/second (220 miles/hour). The **speed of impulse conduction** is related to axon diameter and the extent of myelination. As the diameter of an axon increases, internal resistance to ion flow is reduced and, consequently, impulse velocity is greater. Variation in axon (a) diameter is evident in micrograph 1 (courtesy of Dr. Larry Mathers), where many nerve processes are seen in cross section.

Most axons with a diameter of one micron and greater undergo myelination. The myelin coat increases the velocity of impulse conduction by increasing the membrane resistance and thus insulation, reducing membrane capacitance and ion loss during current flow. Ideally, the myelin insulation would be continuous from the axon hillock to the terminal bouton, much like the continuous insulation around a wire. In neurons, however, the self-propagated action potential has a tendency to fade and thus needs to be reinforced at regular intervals. Reinforcement of the action potential occurs in regions devoid of the myelin sheath, known as **nodes of Ranvier** (n, micrographs 1 and 2).

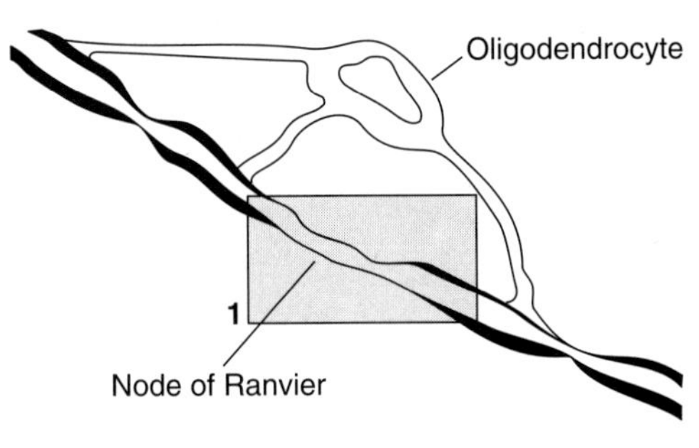

In the node regions, current passing longitudinally within the axon activates Na^+ channels. When these voltage-gated channels open, Na^+ rushes in, bringing positive charges to the inside of the axon to renew the action potential. This potential change is propagated rapidly to the next node. The distance between each node (1 – 2 mm) represents the maximum separation that still allows for undiminished axon current.

At the node of Ranvier in both CNS (micrograph 1) and PNS (micrograph 2), oligodendrocytes and Schwann cells terminate in a characteristic manner; several "fingers" of glial cytoplasm (arrowheads, micrographs 1 and 2) make intimate contact with the axon. These specializations are formed as the glial cell cytoplasm wraps around the axon. Junctions that form between glial cell and axon at the borders of the node maintain the different ion channel composition characteristic of node and internode regions. Sodium channels are concentrated in the node, where the axon is exposed to the extracellular fluid and most ion exchange occurs, and are sparse in the myelinated internode regions where very little ion exchange occurs.

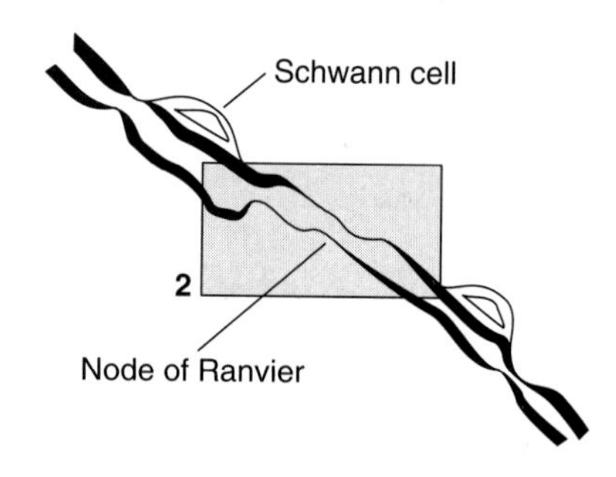

Nodes in the PNS are surrounded by a basal lamina (arrows, micrograph 2) that is continuous with that of the Schwann cells. The basal lamina covering is important to the regeneration of PNS fibers. When axons are severed and the distal portion resorbed, regeneration will occur if the basal lamina and associated collagen remain to form an "endoneurial" tube. The absence of a defined basal lamina and the inhibitory effects of oligodendrocyte myelin are two factors that may prevent axon regeneration in the CNS.

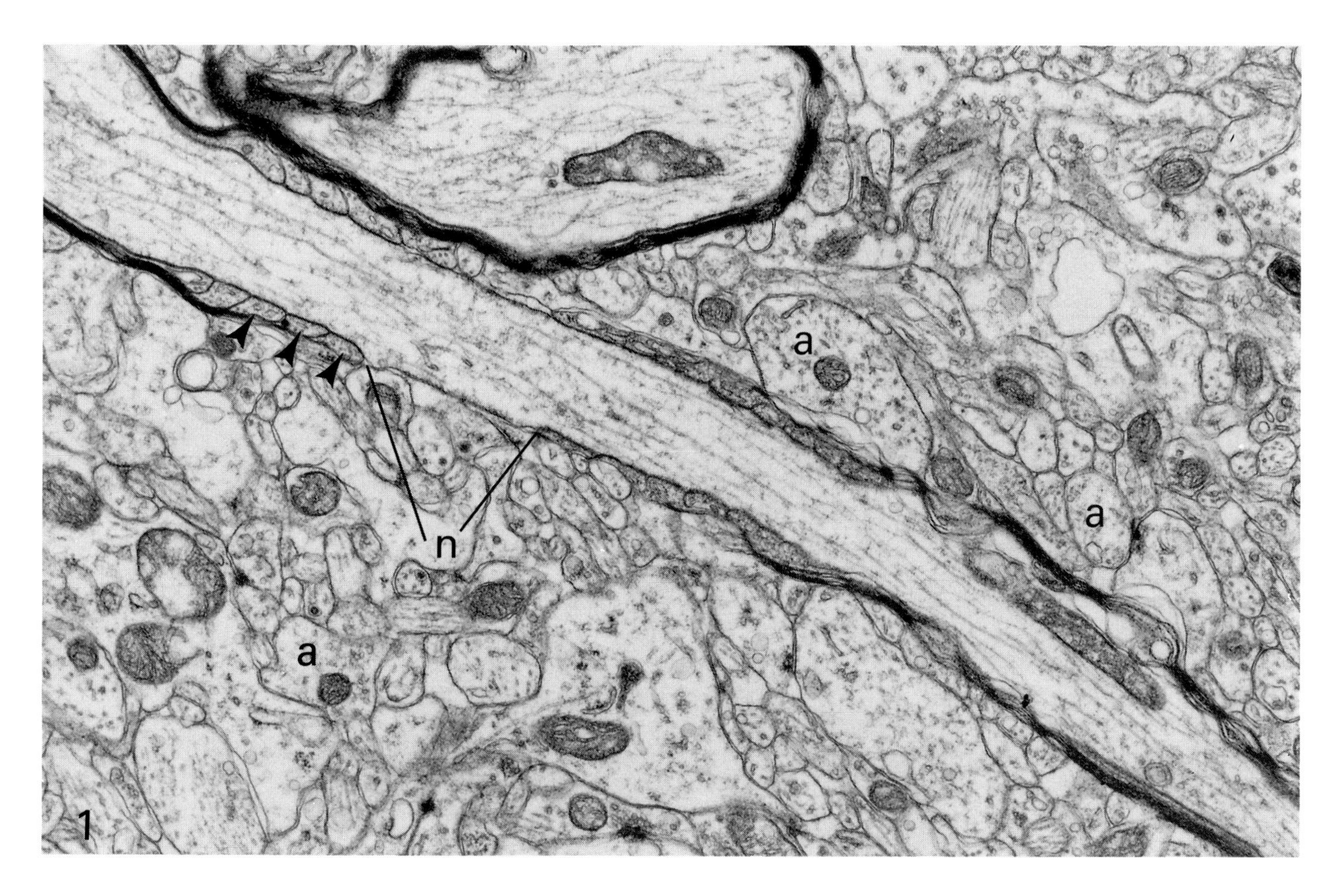

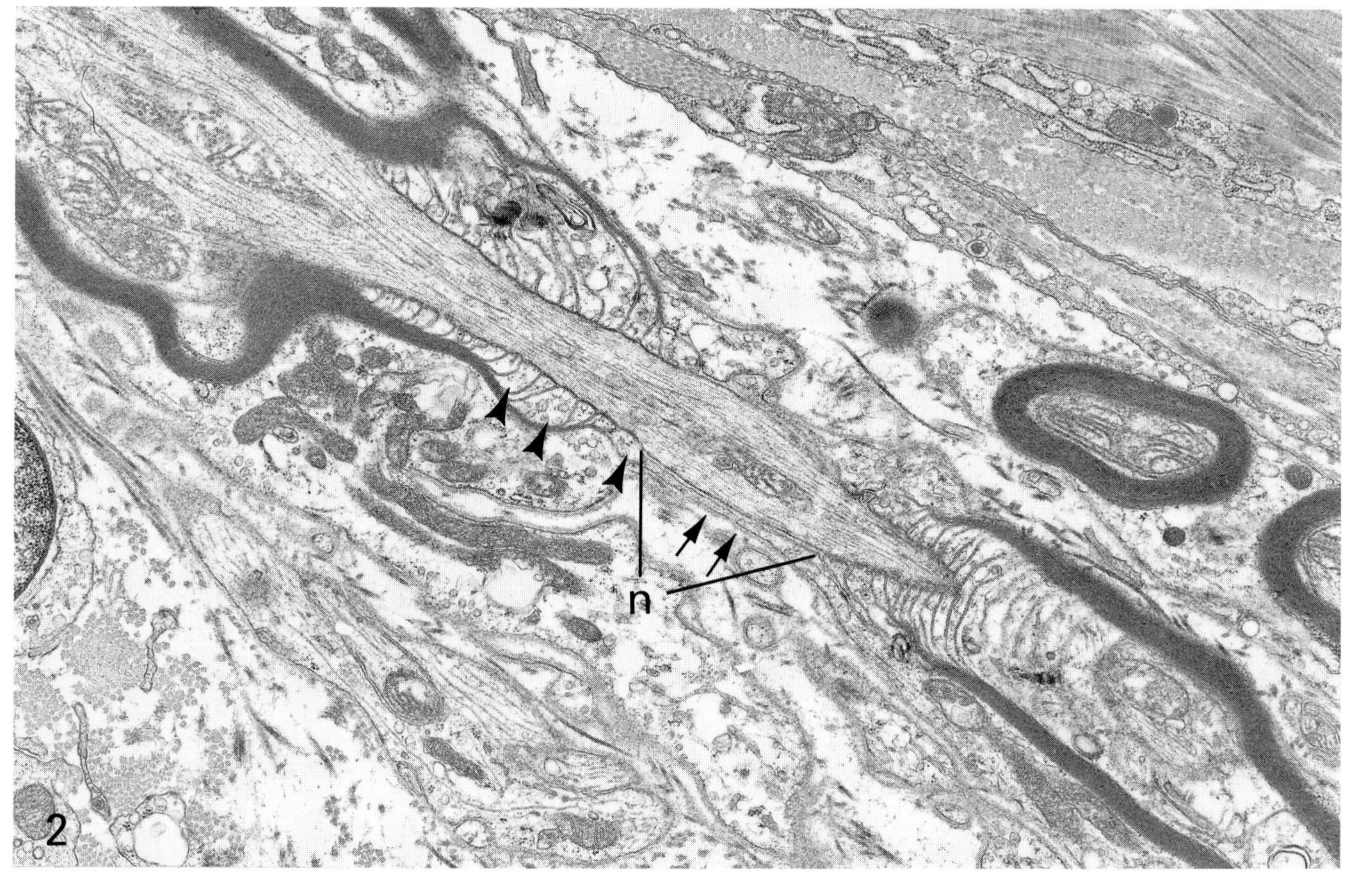

EM1: 21,000× EM2: 14,500×

GLIAL CELLS: Astrocytes

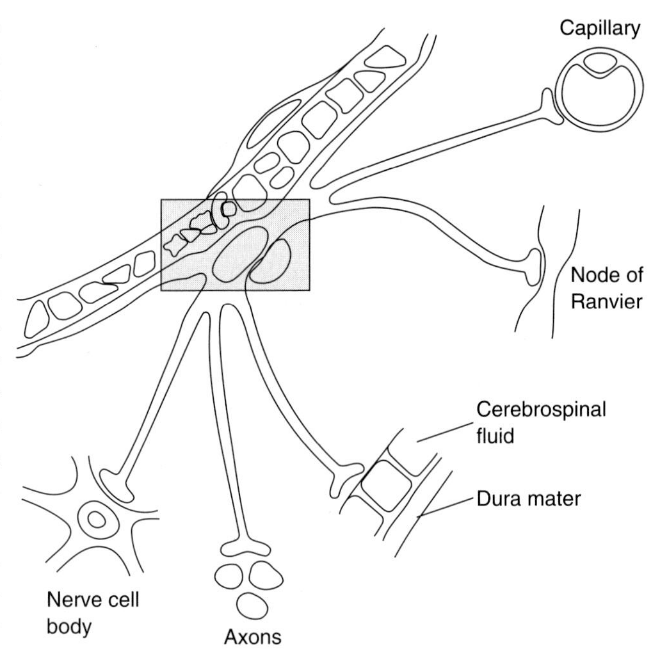

Astrocytes (A, micrographs 1 and 2, micrograph 2 courtesy of Dr. Larry Mathers), the most common glial cell in the central nervous system, are shaped like neurons, with many long processes extending from a central cell body. Each elongated process terminates in an **"end-foot"** that rests on nerve cells bodies, nerve processes, blood vessels, other astrocytes, or the inner surface of the meninges covering the brain. Astrocyte cytoplasm contains a characteristic **intermediate-sized filament** composed of glial fibrillary acidic protein. These filaments (f, micrographs 1 and 2) are obvious in routine electron micrographs, both next to the nucleus and within the processes that pass between axons and dendrites.

The euchromatic nucleus (N, micrographs 1 and 2) of astrocytes reflects, in part, the synthesis of the cytoskeleton. However, most of the nuclear activity is needed to maintain the extensive volume of cytoplasmic processes. Oligodendrocytes (O, micrographs 1 and 2), even though smaller cells, are also extremely active. Their activity, however, is more specific and directed toward the synthesis of myelin. Rough ER (arrows, micrograph 2), where myelin protein synthesis occurs, dominates their cytoplasm.

Astrocytes have unique functions relating to their intimate association with blood vessels and neurons. In capillaries, they induce and maintain the blood–brain barrier (see Blood Vessels, page 150). Even in larger vessels, such as the arteriole in micrograph 1, astrocytes form a layer between the vessel wall (W) and neuron processes (p).

A network of astrocytes may form a significant nonneural communication system. Potassium, which has a pronounced effect on neuron functioning, accumulates in astrocyte end-feet associated with nerve processes. This ion may be moved from the microenvironment surrounding neurons, through an astrocyte network, to the blood or cerebrospinal fluid. Gap junctions between astrocytes could account for the transportation of potassium for considerable distances. A separate indication of sophisticated ion transport in astrocytes is their ability to respond to the neurotransmitter glutamate with a calcium wave that is transmitted from cell to cell. These waves travel over long distances without diminishing, like the neuron axon potential.

Following nerve degeneration, astrocytes proliferate and accumulate in areas of injury. This repair activity may interfere with neuron regeneration in the CNS.

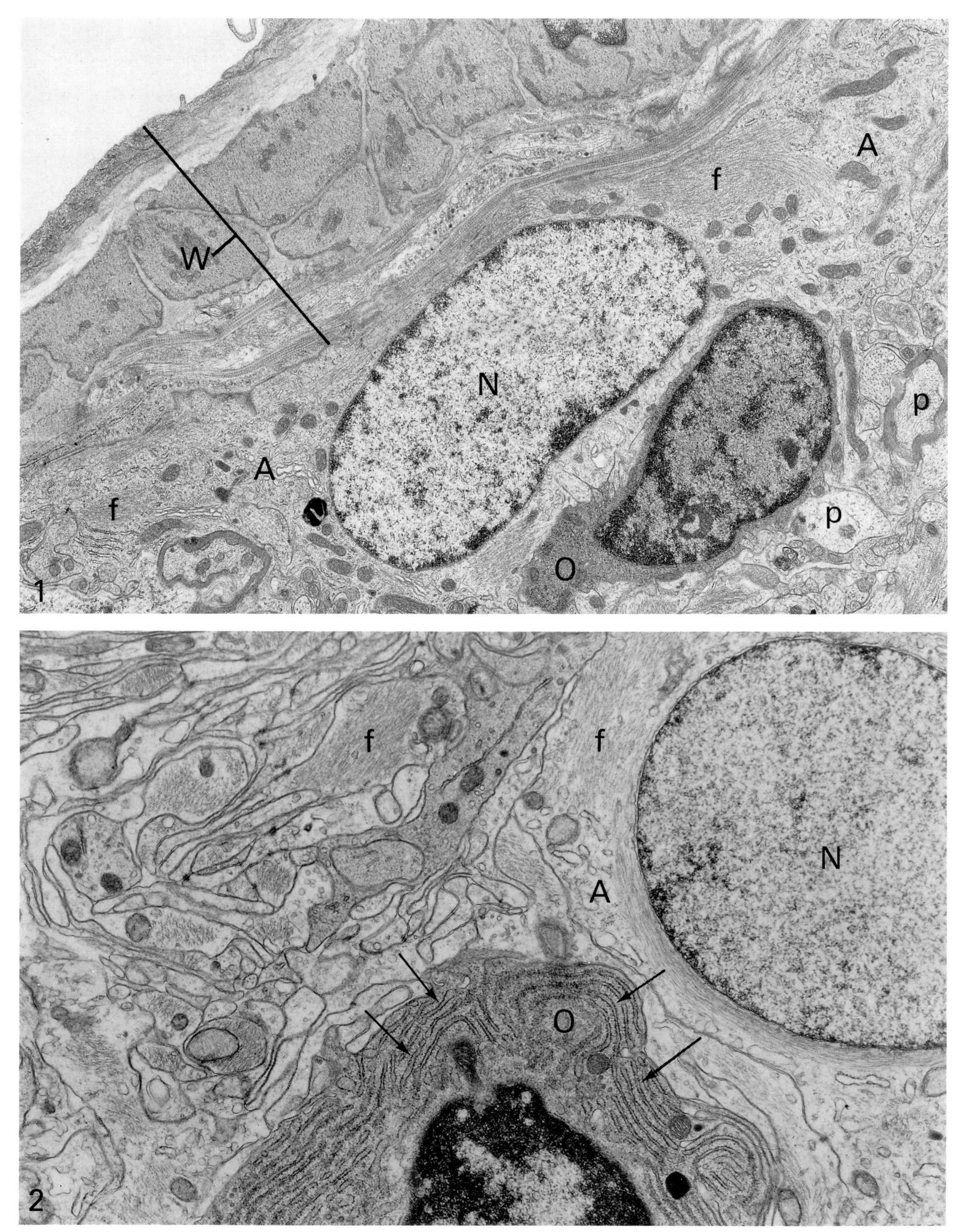

EM1: 9,000× EM2: 15,500×

GANGLIA

In the peripheral nervous system, neuron cell bodies are localized within **ganglia.** In both **dorsal root (sensory)** and **autonomic** (sympathetic and parasympathetic) ganglia, the cell bodies are covered with specialized glial cells called **satellite cells.** In the dorsal root ganglia (micrograph 1) the nuclei of three satellite cells (S) can be observed tightly apposed to the nerve cell body. In the autonomic ganglia (micrograph 2), even though nuclei of satellite cells are not seen in the section, a thin rim of satellite cytoplasm (cy) wraps the nerve cell body. In both types of ganglia, collagen (c, micrographs 1 and 2) provides the supportive framework.

Sensory neurons of the dorsal root ganglia are **pseudounipolar,** with only one process extending from the cell body. A short distance from the cell body the nerve process divides, with one branch leading to the periphery and one to the CNS. Information from the region of sensory input (e.g., pain receptors in skin) is carried directly to the CNS, bypassing the cell body. No signal information is processed in the cell body, which is solely "nutritive"; i.e., it maintains the turnover of cellular components. This, in itself, requires extensive synthetic activity as evidenced by the large cell body, euchromatic nucleus (N, micrograph 1), and prominent nucleolus (n, micrograph 1).

Unlike sensory neurons, autonomic neurons have a typical **multipolar** arrangement. In autonomic ganglia, postsynaptic dendrites and cell bodies receive and process information from synaptic contact with presynaptic neurons. The numerous nerve fibers (p, micrograph 2) that surround and separate each cell body in autonomic ganglia reflect the role of these ganglia in processing and communication. In contrast, in dorsal root ganglia, which are not regions of synaptic information transfer, neuron cell bodies are tightly grouped together, with few intervening nerve processes. In micrograph 1, note the presence of three other neuron cell bodies (arrows) directly adjacent to the central one.

All of the nerve processes observed in micrograph 2 are unmyelinated and grouped together by the Schwann cell cytoplasm that encases them. In one group of nerve fibers, the section cuts through the nucleus of a Schwann cell (Sc, micrograph 2).

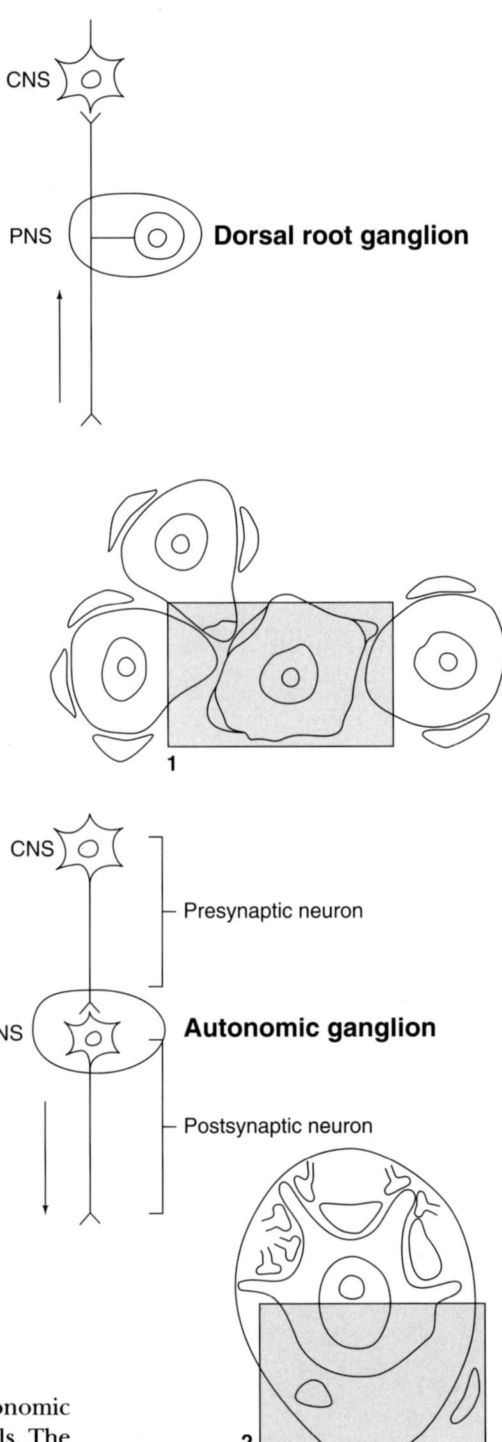

Synapses continually rearrange on the surface of the autonomic ganglia nerve cell bodies. As synapses move, so do satellite cells. The neuron cell bodies and surrounding satellite cells are connected by gap junctions and may communicate in ways important in this synaptic adjustment.

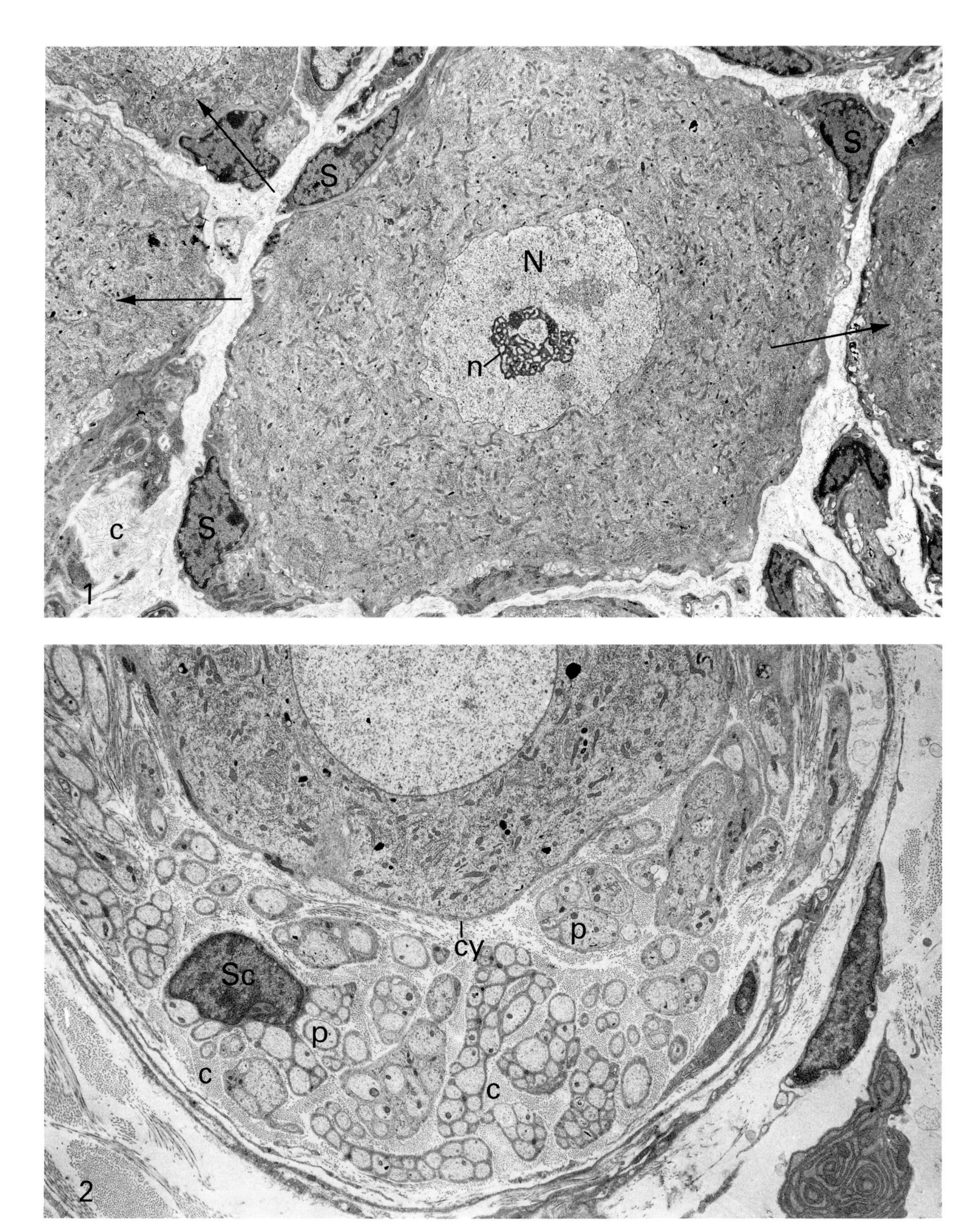

EM1: 2,600× **EM2: 5,000×**

Chapter Six

Blood Vessels

CONTINUOUS CAPILLARIES: Overview

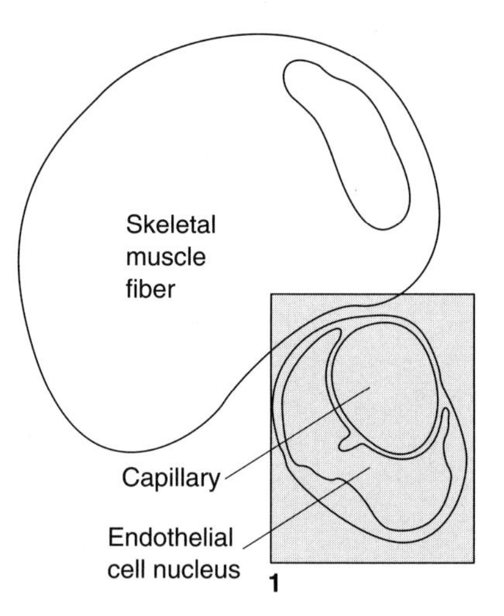

The inner lining of all blood vessels consists of a single layer of squamous **endothelial cells** (E, micrograph 1). These cells contain a full complement of organelles. Proteins synthesized on the rough endoplasmic reticulum and processed in the Golgi (G, micrograph 1) are important not only in facilitating exchange across the endothelial lining, but also in regulating overall vascular function. Endothelial cells secrete factors that control blood pressure (e.g., angiotensin II), prevent clotting (e.g., prostacyclin), and provide structural integrity (e.g., basal lamina proteins).

In **capillaries** (micrograph 1), endothelial cells and an underlying basal lamina (arrowheads) are the principal components of the vessel wall. **Continuous capillaries,** such as the capillary adjacent to the skeletal muscle fiber (M) in micrograph 1, are the most common type. These are characterized by a nonfenestrated lining of endothelial cells bound by junctions (circle, micrograph 1). Continuous capillaries frequently contain numerous 70-nm vesicles (arrows, micrograph 1 and inset), many of which transport fluid and molecules across the cell in a process referred to as **transcytosis.** In contrast to most endocytotic vesicles, transcytotic vesicles move directly across the cell without fusing with lysosomes. Vesicles move in both directions and have been observed to fuse with one another to create a continuous channel across the endothelium.

Transcytotic vesicles may represent the small and large pores postulated by physiologists studying the rate of movement of different-sized water soluble molecules across endothelial cells. Insulin and both monomers and aggregates (>14 nm) of albumin are transported selectively across endothelia within these vesicles.

Endothelial cells are capable of division and locomotion. The process of new vessel formation, **angiogenesis,** occurs following tissue injury, during developmental processes such as bone formation, and in disease states. The control of angiogenesis is an important clinical issue; solid tumors survive only when new vessels grow into the tissue mass. A method of inhibiting this vessel invasion might be used to control tumor growth. In other instances accelerated proliferation of vessels directly interferes with normal functioning of an organ. For example, when the vessels that cover the inner surface of the retina develop beyond their programmed pattern, blindness results.

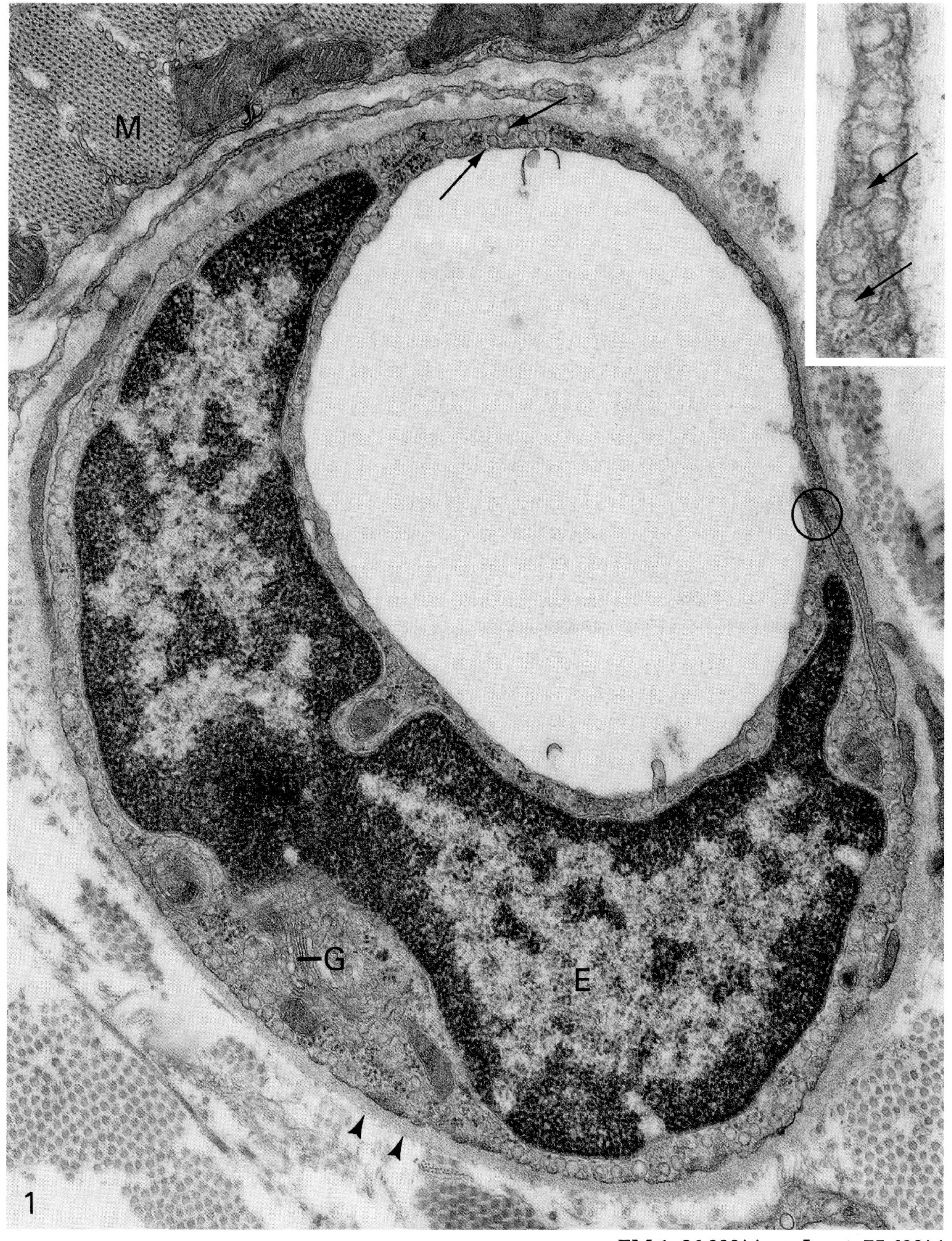

EM 1: 36,000× Inset: 75,600×

Continuous Capillaries: Blood-Brain Barrier

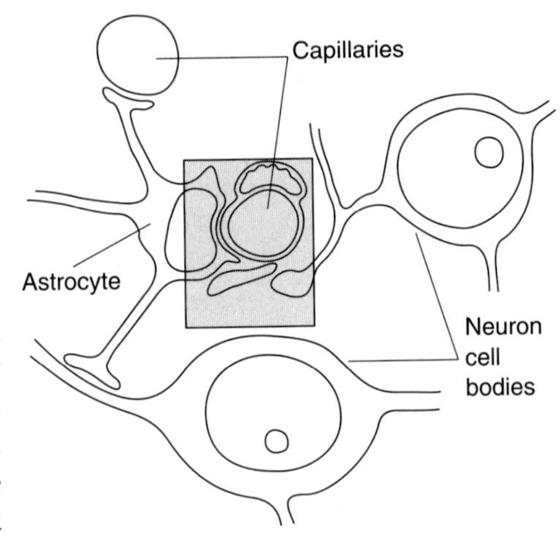

Precise control of transport between blood and interstitial fluid is particularly important in the brain, where many neurons (N, micrograph) are exposed and vulnerable to fluctuations in metabolite levels. The environment of CNS neurons is regulated by a capillary endothelium (E, micrograph) that is one of the least permeable and most specialized in the body, comprising the **blood-brain barrier** (BBB). The relative impermeability of the capillary in the micrograph is suggested by the continuous endothelial lining, pronounced junctional complex (arrow), and absence of large transport vesicles commonly found in other continuous capillaries. Whereas glucose and large neutral amino acids move relatively freely back and forth across the blood-brain barrier, potassium ions and glycine move only from brain to blood and require energy-coupled transport. Both potassium ions and glycine have important effects on the transmission of nerve impulses, and thus the levels need to be carefully regulated. Potassium ions affect the threshold of firing, and glycine is a CNS inhibitory neurotransmitter.

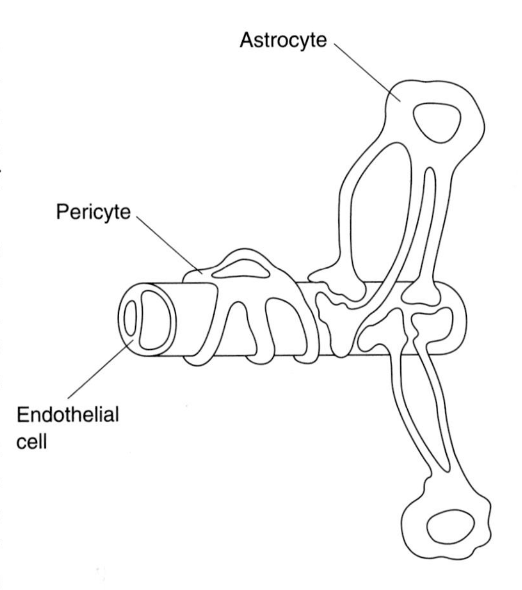

In the micrograph, the end-foot (f) of an **astrocyte** (A) rests on the capillary basal lamina. In addition to regulating the ionic environment of neurons (see Nerve, page 142), current evidence suggests that astrocytes are responsible for the induction of the BBB. The BBB is not an intrinsic characteristic of brain endothelial cells, but rather is induced by neural tissue. When avascular coelomic cavity tissue of chick embryos is transplanted into brain, the vessels that grow into the transplanted tissue are leaky like coelomic vessels, whereas, when avascular brain tissue is transplanted into coelomic tissue, the new vessels that grow into the transplanted brain tissue are characteristic of the BBB.

Pericytes (P, micrograph) are found associated with capillaries and venules in many tissues. These cells possess several long processes that wrap the vessels and may undergo contraction to regulate blood flow. During angiogenesis they seem to have the potential to develop into vascular smooth muscle.

The permeability of the BBB within pathological areas, such as tumors and regions associated with Alzheimer's disease, is altered, with a probable effect on disease progression.

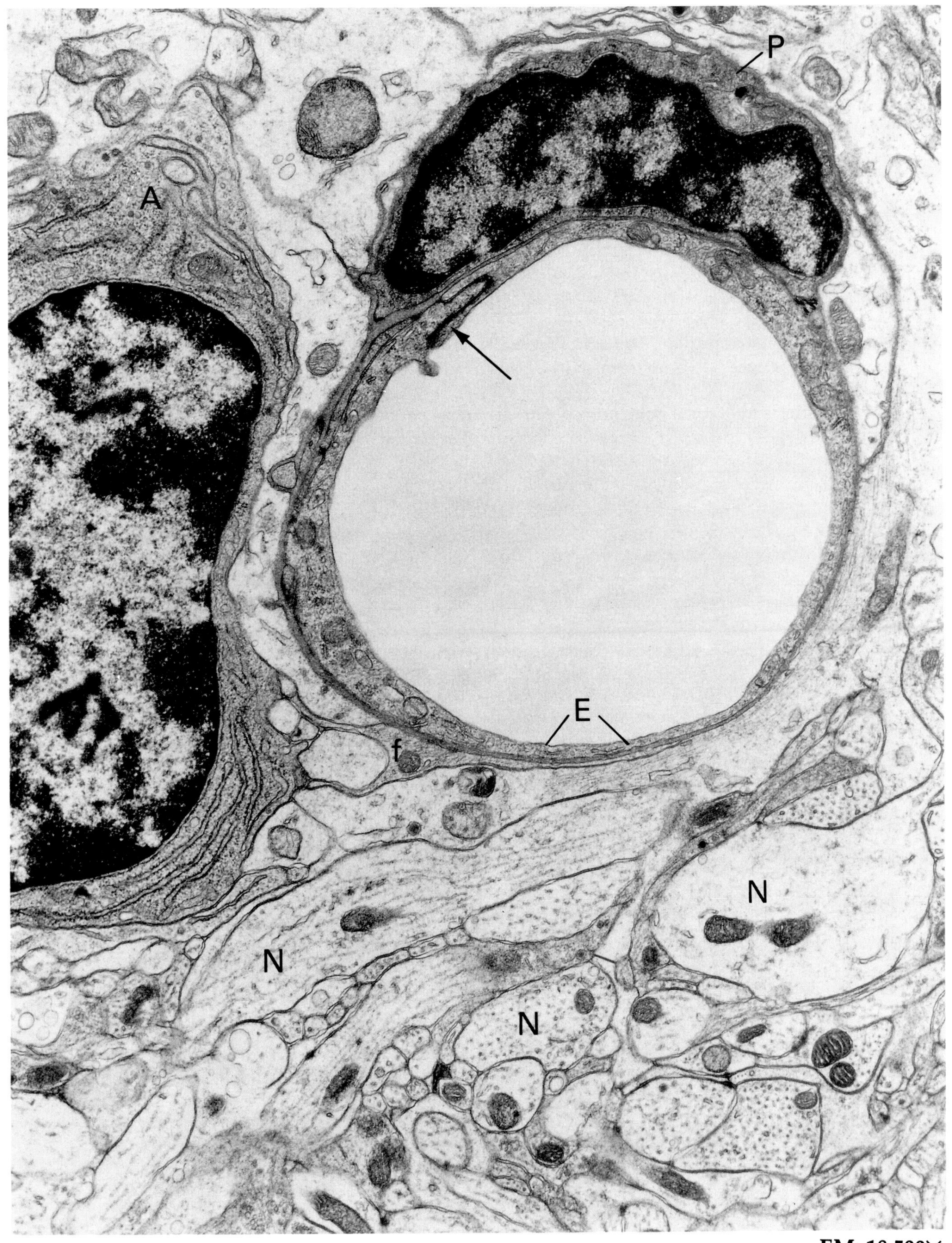

EM: 18,500×

FENESTRATED CAPILLARIES: No Diaphragm

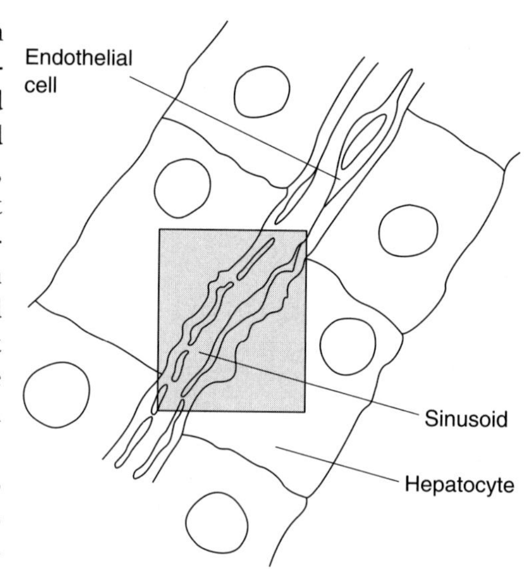

Liver hepatocytes (H, micrograph) are bathed with blood carried in **sinusoids** (S, micrograph), capillaries specialized to accommodate extensive exchange between plasma and underlying tissue. Open **fenestrations,** or "holes" (curved arrow, micrograph), in the endothelial lining (straight arrows, micrograph) provide most components of plasma with direct access to the hepatocyte surface. Hepatocyte microvilli (arrowheads, micrograph) facilitate an intimate association with blood, enabling the liver to efficiently monitor and filter blood coming from the intestine via the hepatic portal system before it enters the general circulation. Hepatocytes regulate both the types and quantities of carbohydrate, lipid, and protein in blood plasma.

Particles larger than 100 nm do not generally pass through the fenestrations. This size barrier prevents blood cells and chylomicron particles (triglyceride packages from the intestine) from reaching the hepatocyte surface. Chylomicron

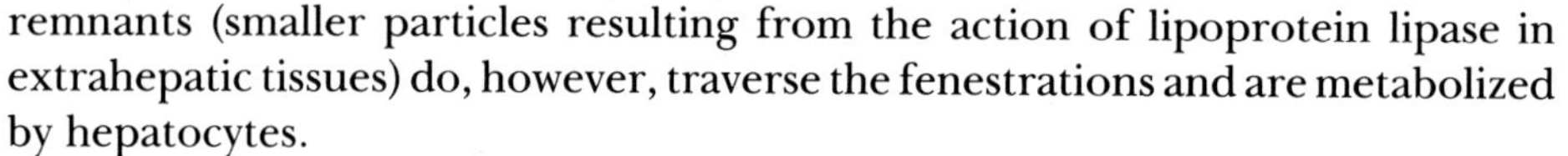

remnants (smaller particles resulting from the action of lipoprotein lipase in extrahepatic tissues) do, however, traverse the fenestrations and are metabolized by hepatocytes.

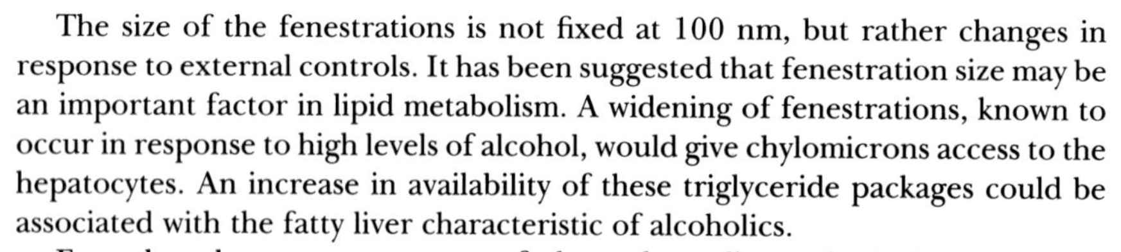

The size of the fenestrations is not fixed at 100 nm, but rather changes in response to external controls. It has been suggested that fenestration size may be an important factor in lipid metabolism. A widening of fenestrations, known to occur in response to high levels of alcohol, would give chylomicrons access to the hepatocytes. An increase in availability of these triglyceride packages could be associated with the fatty liver characteristic of alcoholics.

Even though most components of plasma have direct physical access to the hepatocyte surface via these fenestrations, some molecules preferentially bind to and cross endothelial cells prior to uptake by hepatocytes. For example, a significant proportion of iron bound to its transport glycoprotein, transferrin, is carried in transcytotic vesicles across the endothelial lining rather than passing through the fenestrations. Endothelial cells, therefore, may partially control the availability of iron to the hepatocyte, the major cell associated with iron metabolism.

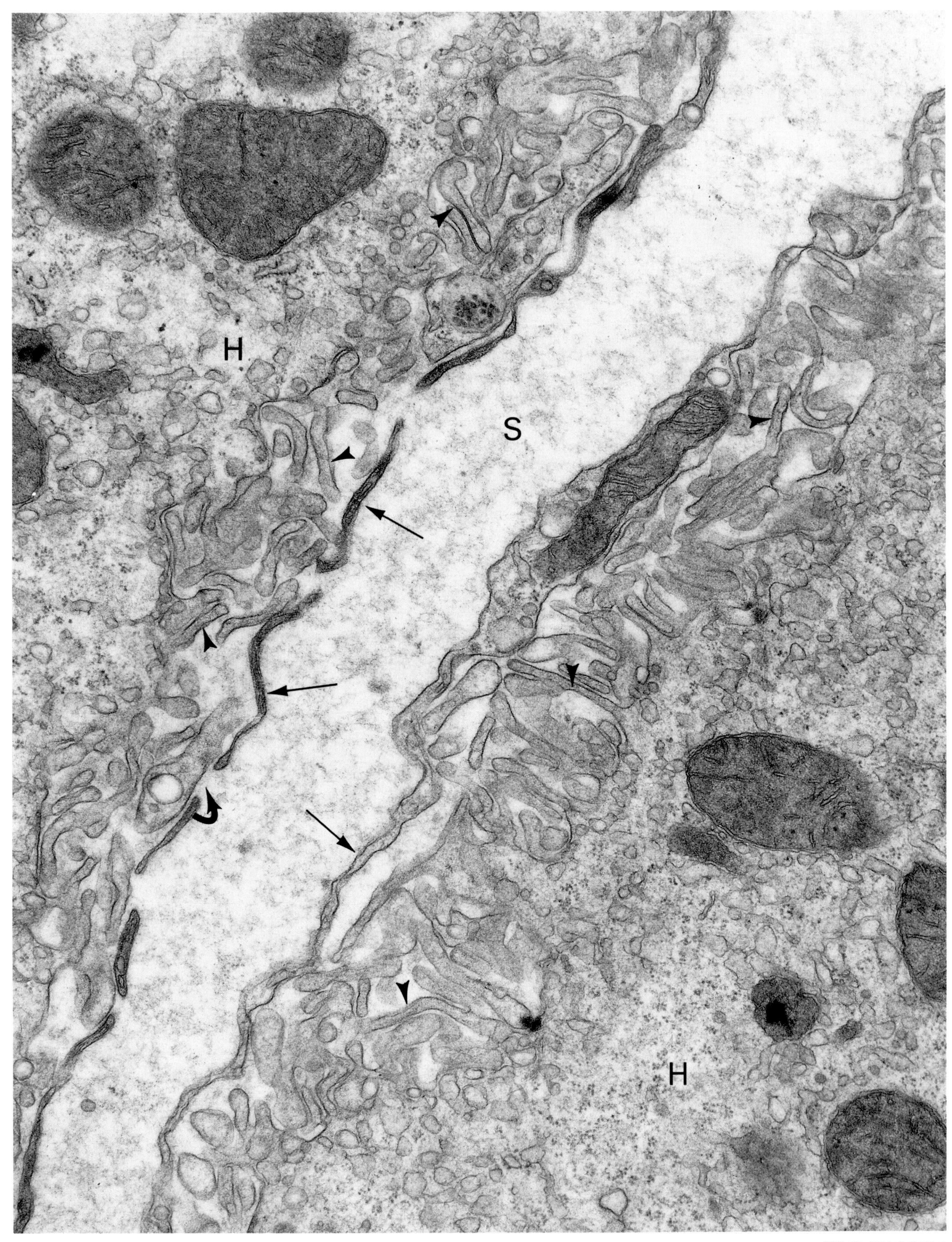

EM: 35,000×

FENESTRATED CAPILLARIES: Diaphragm

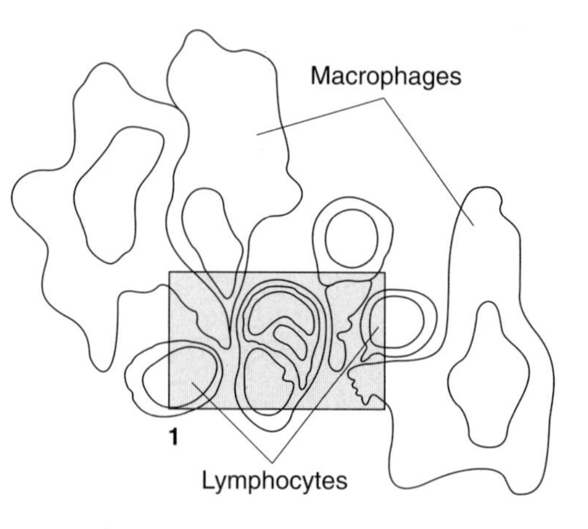

In certain locations (e.g., gastrointestinal tract, endocrine and exocrine glands) endothelial cells have **fenestrations** partially closed by **diaphragms.** A fenestrated capillary surrounded by lymphocytes (L) and macrophages (M) is shown in micrograph 1. Two distinguishing characteristics of capillaries in general are well illustrated in this micrograph: (1) the lumen is approximately one red blood cell (7 μm) in diameter, and (2) the rim of cytoplasm separating the blood and interstitial fluid is extremely thin.

The diaphragms of fenestrated capillaries are seen in cross section (straight arrows, micrographs 1 and 2) as thin ($\sim$ 7 nm) densities stretching between parts on an endothelial cell. The basal lamina (curved arrow, micrographs 1 and 2) is continuous across the fenestrated regions. *En face,* the diaphragms are seen to consist of radially oriented fibrils converging on a central meshwork, creating wedgelike channels. The fibrils are highly negatively charged and appear to consist of heparan sulfate-containing proteoglycans.

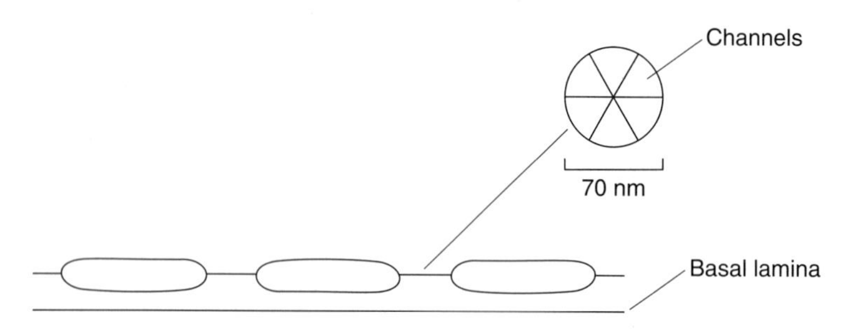

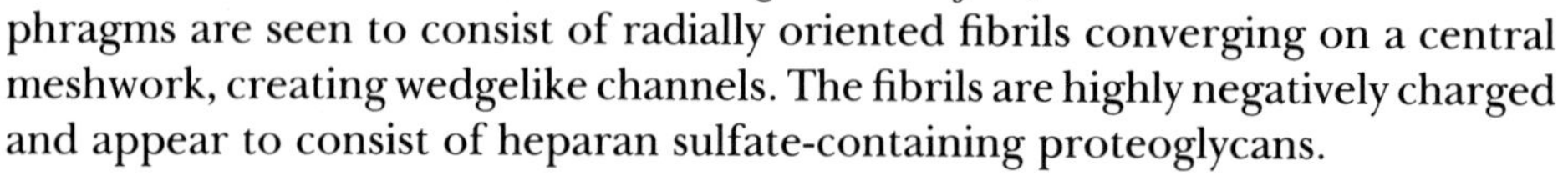

Fenestrated capillaries with diaphragms are found in locations in which large and small molecules regularly move across the endothelial lining. Even though it is known that these capillaries are more permeable than continuous capillaries, it is still unclear how (particularly for large molecules) transport is facilitated and what role the diaphragm plays.

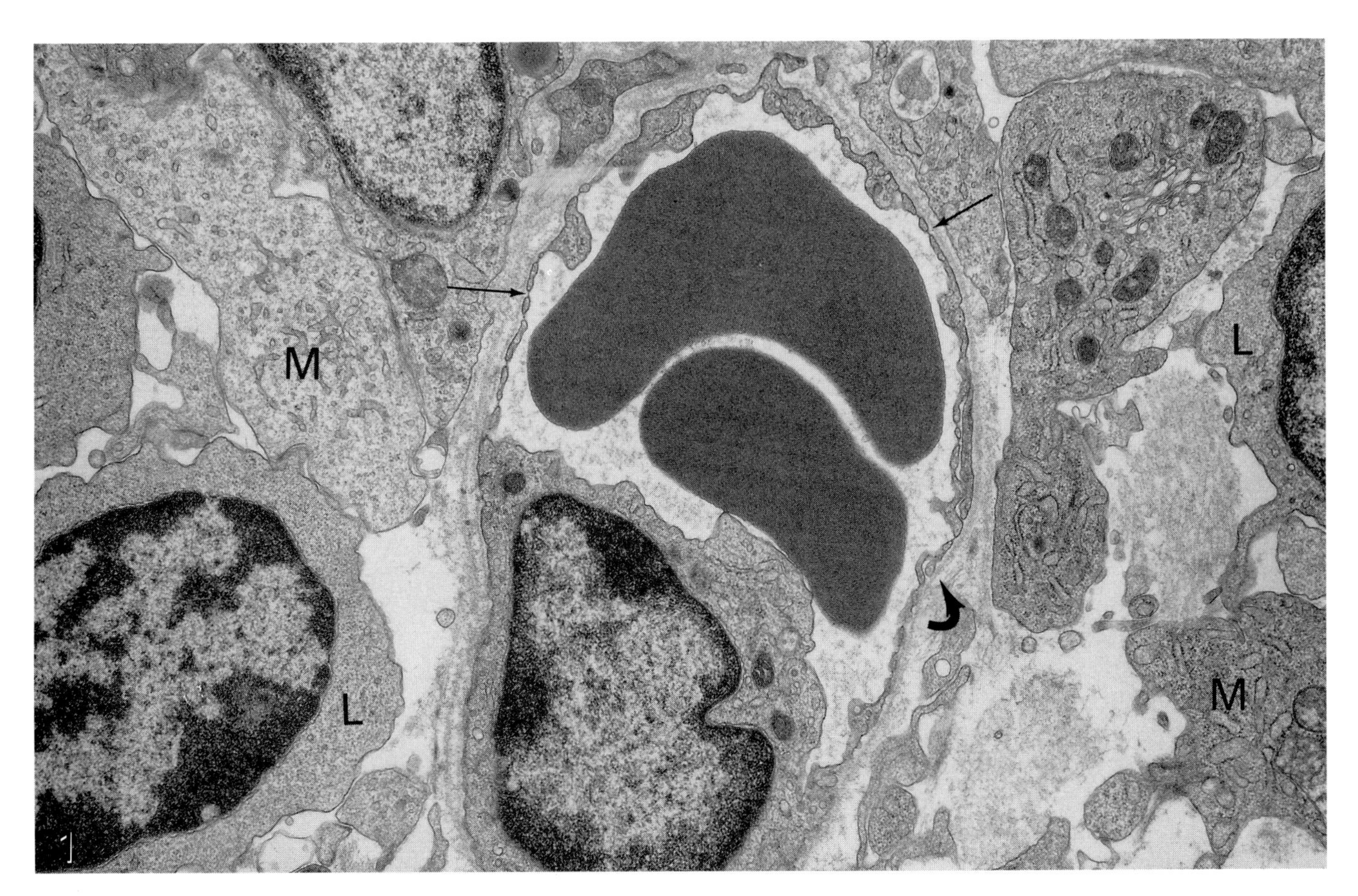

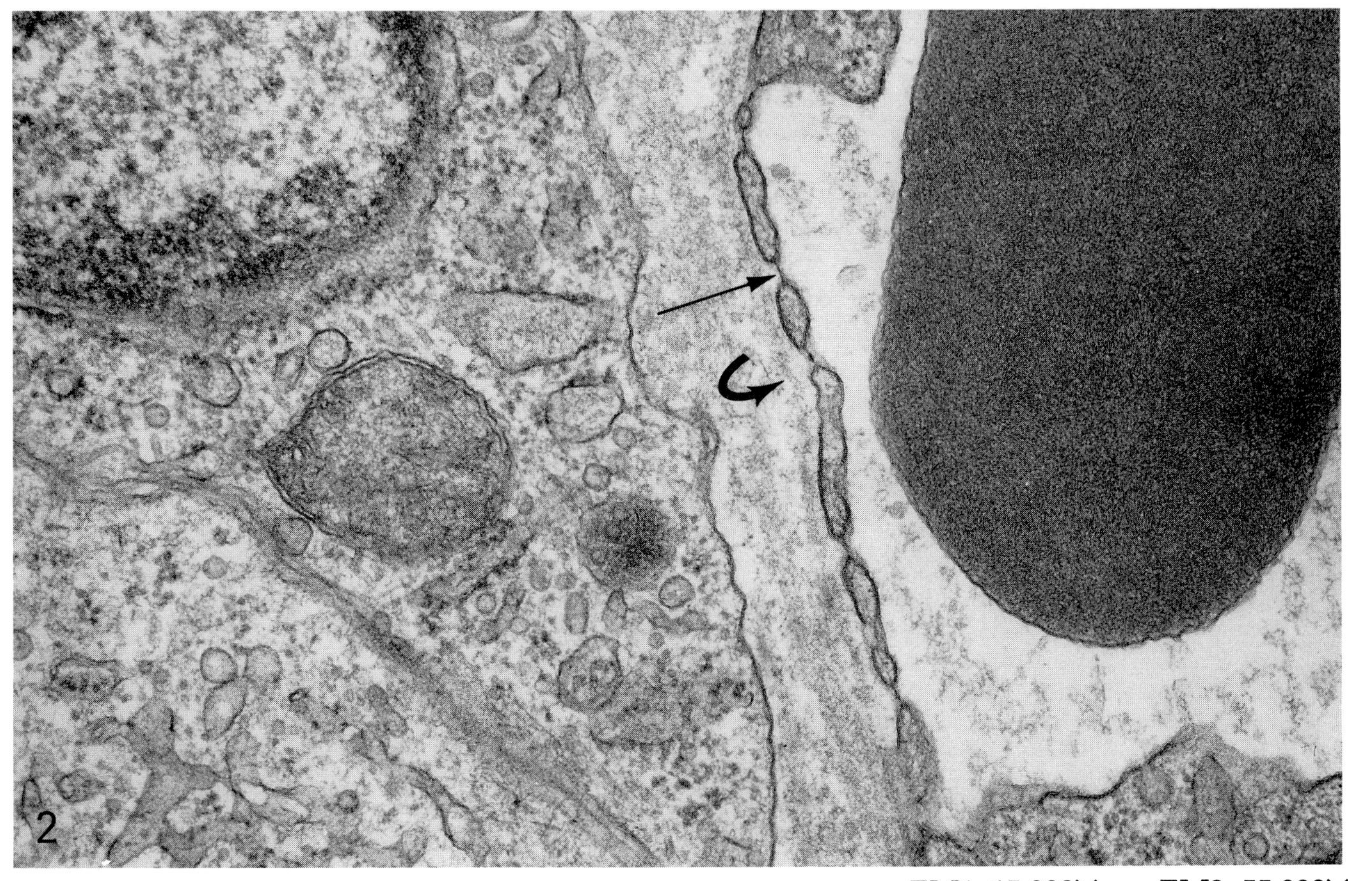

EM1: 15,000× **EM2: 55,000×**

Arteriole, Venule

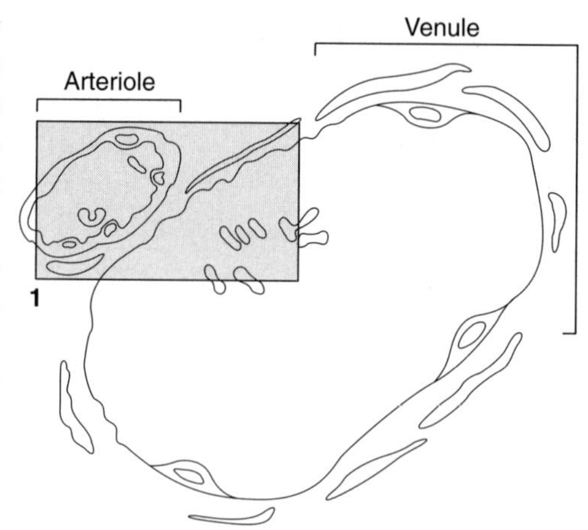

Arterioles (A, micrograph 1) deliver blood to capillary beds, and **venules** (V, micrograph 1) drain capillary beds. The arteriole, with its thicker muscular wall and smaller lumen, is easily distinguished from the venule. Typically, as in micrograph 1, arterioles and venules (and arteries and veins) that service a particular area travel side by side.

The wall of arterioles and venules (and larger vessels) is divided into three regions, each characterized by a certain cell type (micrograph 1): the **intima** lining the lumen contains endothelial cells (E), the **media,** contractile cells (S: smooth muscle in arteriole; P: pericyte in venule), and the **adventitia,** fibroblasts (F). All layers contain both elastic and collagen connective tissue fibers. In the electron micrographs, collagen fibers (c) in cross and longitudinal section are the most obvious extracellular components. In micrograph 1, the adventitia of the arteriole blends in with that of its companion venule.

The difference in vessel wall structure between arterioles and venules clearly reflects their special functions. The continuous smooth muscle lining of arterioles contracts and relaxes to adjust lumen size, thus contributing to the control of blood pressure and the distribution of blood. Smooth muscle in the media of arterioles maintains a continuous partial contraction (tone) that is important in the **peripheral resistance** to blood flow. Peripheral resistance maintains the diastolic (ventricular relaxation) blood pressure, whereas cardiac output accounts for systolic (ventricular contraction) pressure. Contraction of the smooth muscle layer during tissue preparation typically causes the endothelial cells to fold and project into the lumen (arrows, micrograph 1).

The relatively thin lining of the venule is suited for exchange between blood and the underlying connective tissue. The endothelial cells of venules contain specific receptors involved in the transport of cells and molecules across the vessel wall. Some receptors bind to cells and mediate diapedesis (crossing the vessel wall), while other receptors bind to substances released by cells activated during inflammation. Histamine released by mast cells binds specifically to venule endothelial cell receptors localized in junctional areas (curved arrows, micrograph 2). Changes in the amount and form of actin occur and gaps form between these cells. Plasma proteins and fluid leak into the underlying tissue, bringing defense molecules into the area. Edema can result and, in some cases, a reduction in blood pressure leads to anaphylactic shock.

Even though the lumen size does not change to the same degree in venules as in arterioles, there are contractile cells in the media of many venules. The thin cytoplasmic processes and dense cytoplasm suggests that these cells may be **pericytes** (P, micrographs 1 and 2) rather than smooth muscle. In micrograph 2, the basal laminas (arrowheads) of the pericytes and endothelial cells are particularly obvious. The repair of damaged vessels is initiated when platelets (pl, micrographs 1 and 2) adhere to exposed protein components of the endothelial basal lamina.

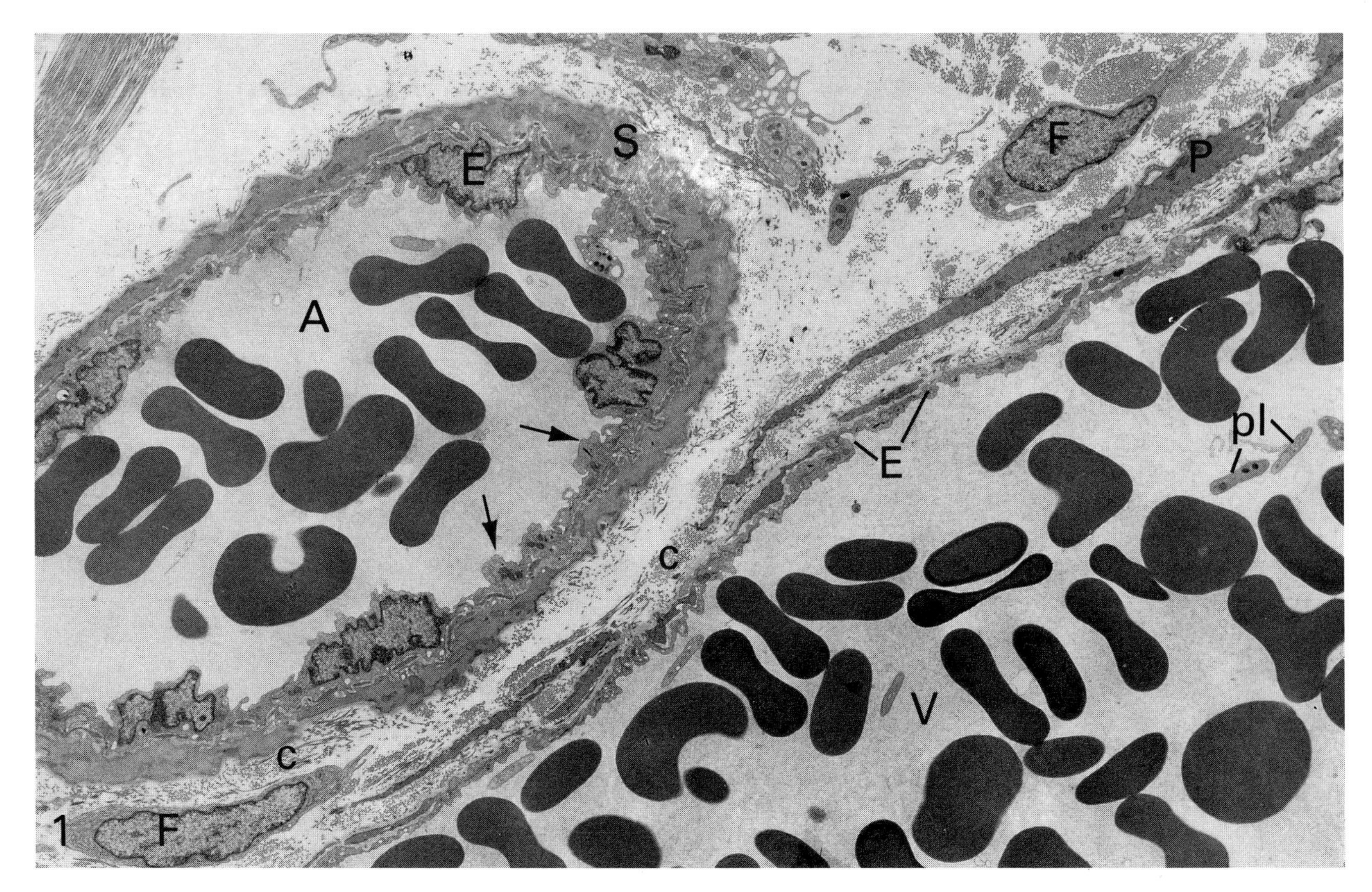

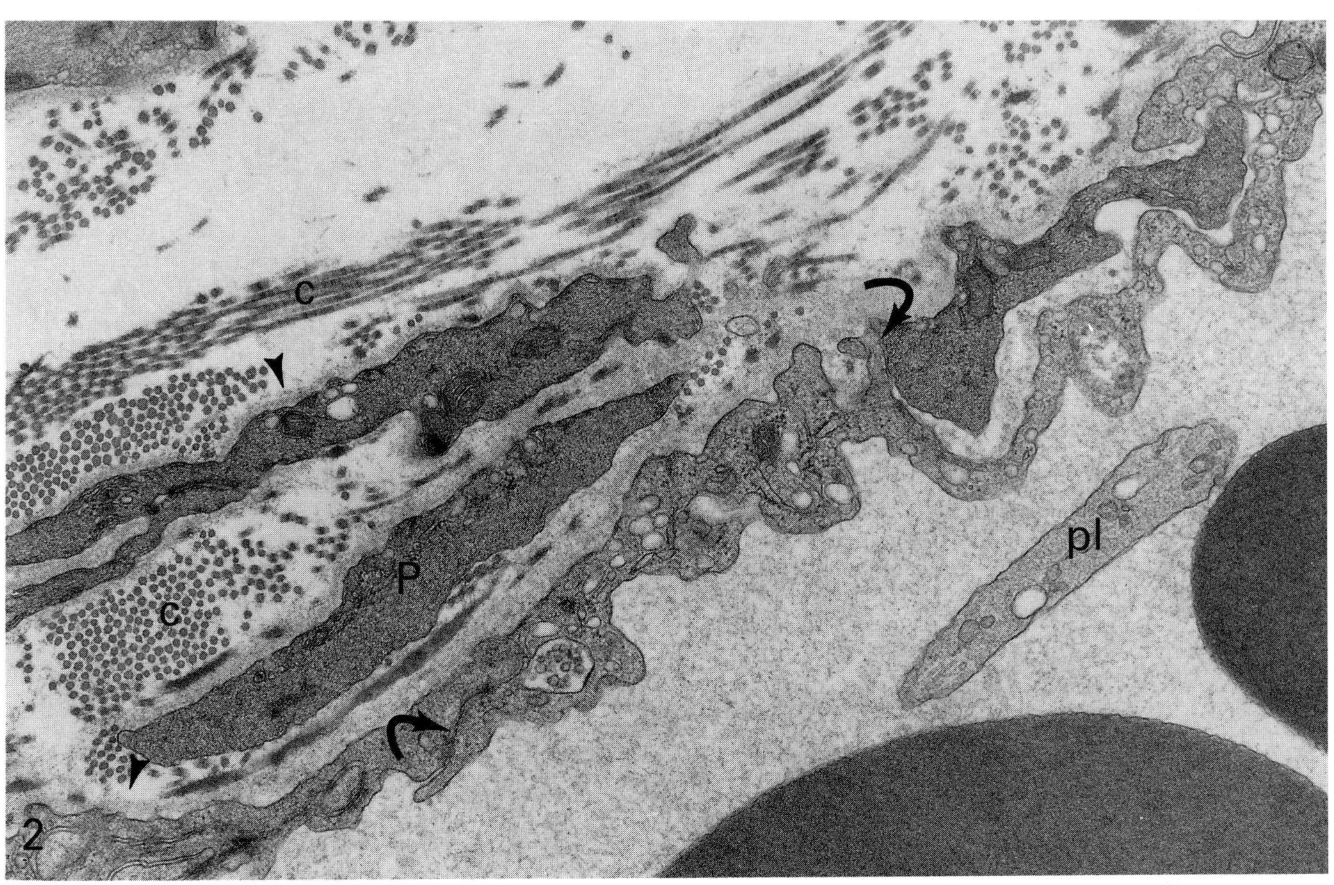

EM1: 4,050× EM2: 27,000×

ARTERIOLE

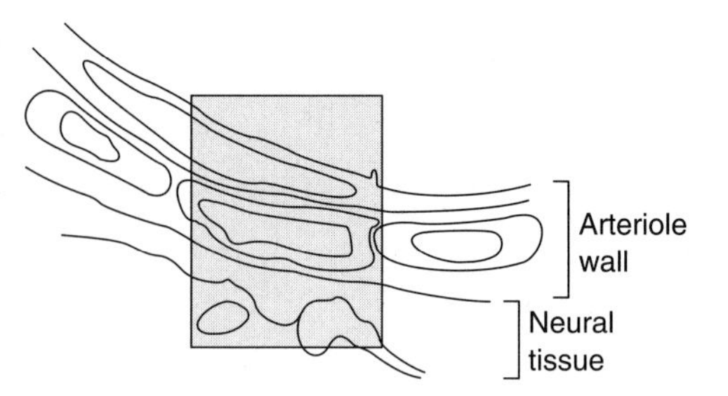

The micrograph illustrates a small arteriole in the cerebral cortex. Between the lumen (L) and the underlying nervous tissue (glial cell, G, and axons, A), the arteriole wall is seen to contain the cell types characteristic of the intima (endothelial cell, E), media (smooth muscle cell, S), and adventitia (fibroblast process, F).

Even though all cells contain a complete cytoskeletal complement, it is not unusual for one particular structure to predominate in specialized cells. In the micrograph, **microtubules** (arrows) are most obvious in the axons, **intermediate filaments** (arrowheads) in the glial cells, and **actin filaments,** visible as fine fibrils packed throughout the cytoplasm, in the smooth muscle cell. Actin filaments are also evident in the endothelial cell, particularly in the encircled area adjacent to the cell membrane. Endothelial cells are capable of contraction; however, most filamentous actin in endothelial cells forms stress fibers that function in cellular stability.

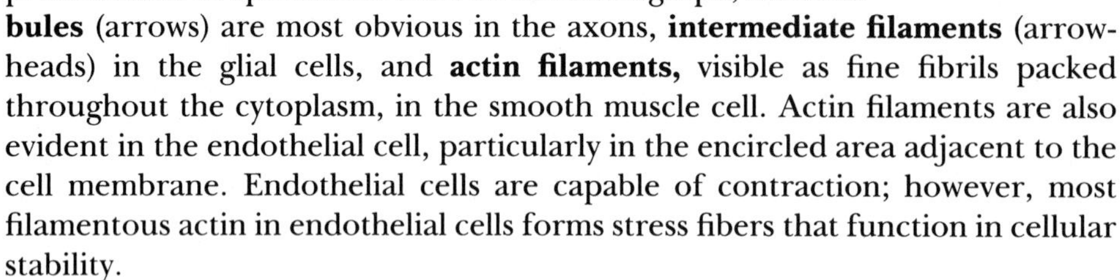

Stress fibers in endothelial cells, as in cultured cells, seem to play a structural role in adhesion and the distribution of forces. In arterioles (and arteries) this role is particularly important since blood flow rate is faster and more variable than it is within the venous system.

Considerable evidence suggests that hemodynamics is a significant factor in certain disease processes (e.g., hypertension). One disease-promoting effect of hemodynamic stress may involve alterations in the appearance and location of stress fibers, with subsequent compromise of endothelial integrity. A correlation between stress fiber number and blood velocity has been found.

Vascular smooth muscle contracts in response to factors as diverse as mechanical stretch, local mediators (e.g., angiotensin II and serotonin), and catecholamines from the adrenal gland and local sympathetic nerve endings. Events within the vessel lumen are frequently transmitted to the muscle by way of the endothelial cells.

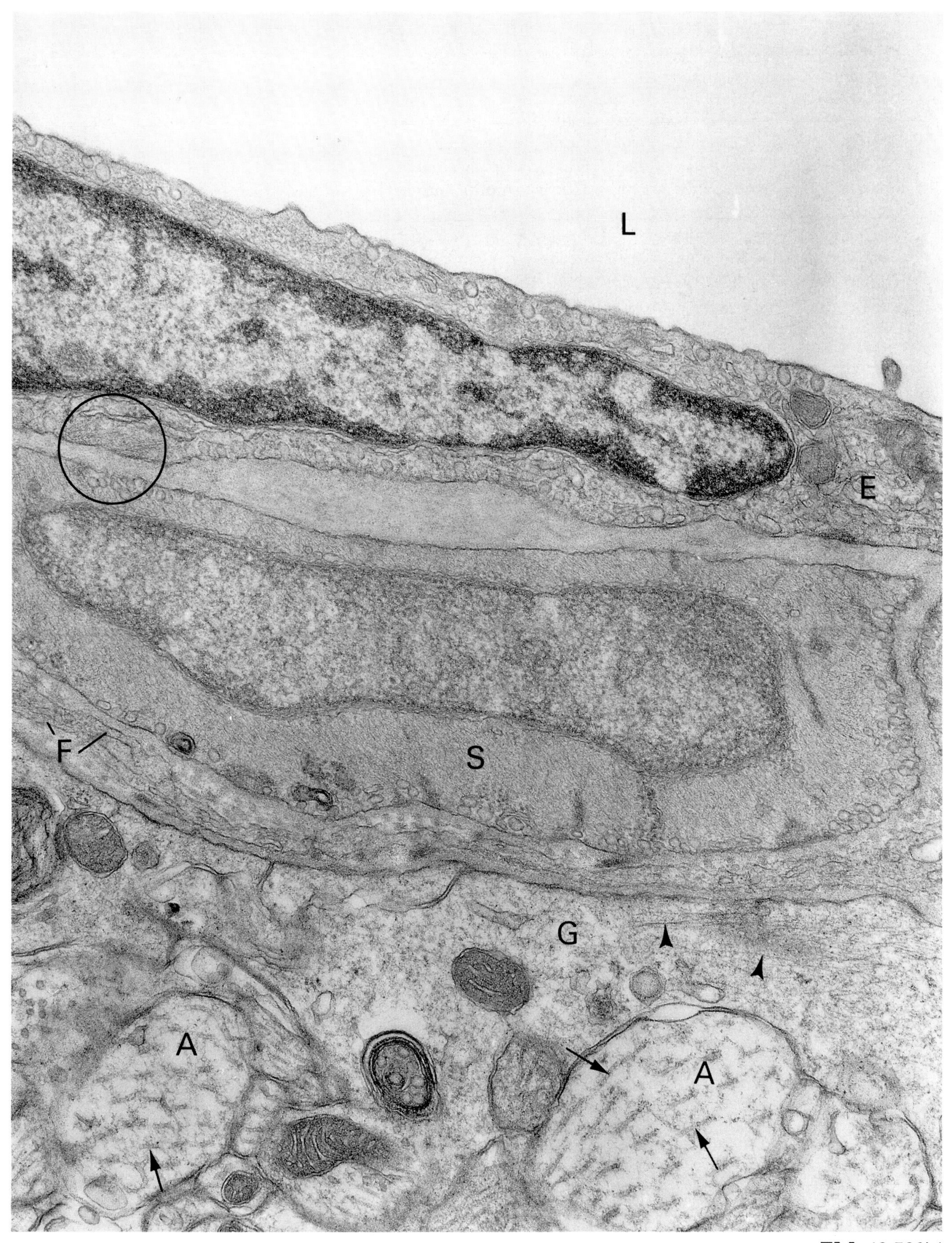

EM: 40,500×

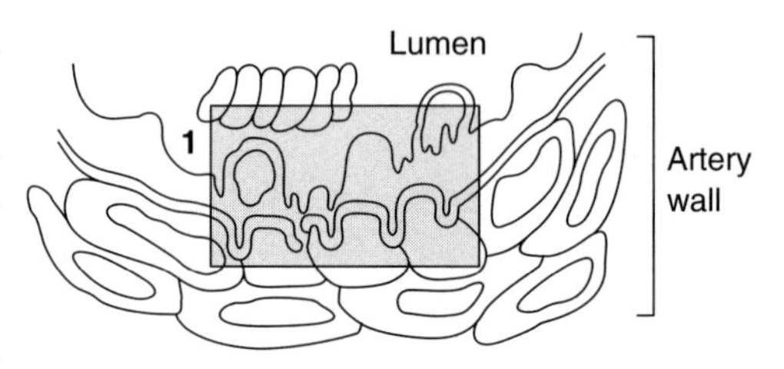

All vessels contain **elastic** fibers in the intima, media, and adventitia. In certain vessels in which the role of elastin is particularly critical, elastin forms sheets referred to as **lamina** (L, micrograph 1). These sheets, synthesized by smooth muscle cells (S, micrographs 1 and 2) of the media, are made up of the same components as elastic fibers. At high magnification (micrograph 2), the amorphous elastin (e) can be distinguished from the microfibrillar protein (arrowheads). Organized collagen fibrils (curved arrows, micrograph 2) stand out in comparison to the irregular elastic network.

In muscular arteries, a prominent **internal elastic lamina** (L, micrograph 1) forms the basal layer of the intima, and a less prominent external elastic lamina (not shown) lies between the media and the adventitia. In elastic arteries, which are closer to the heart, the elastic laminas extend throughout the media. These lamina allow the wall of the elastic artery to stretch to accommodate increased blood volume during ventricle contraction (systole) and to recoil during ventricle relaxation (diastole), and thus maintain a luminal pressure that dampens the pulsatile cycle of the heart.

Fenestrations in the elastic lamina are regions of communication between endothelial cells and underlying smooth muscle cells. At the arrow in micrograph 1 an extension of an endothelial cell can be seen to penetrate the elastic lamina and make contact with underlying cells. Gap junctions have been observed between endothelial and smooth muscle cells. The intimate association between these two diverse cell types is important in transmitting signals from the lumen to the vessel wall. Events in the lumen stimulate endothelial cells to produce relaxing factors (e.g., nitric oxide) or contracting factors (e.g., endothelin) that act on underlying smooth muscle.

The communication between endothelial and smooth muscle cells is also significant in disease. In damaged vessels smooth muscle cells respond to endothelial signals by migrating through the elastic lamina fenestrations, dividing and accumulating in the subendothelial regions. Platelet-derived growth factor (PDGF), a cationic protein released from platelets that stimulates smooth muscle division and migration, is synthesized by endothelial cells. PDGF is considered a critical factor in the development of vessel thickening associated with atherosclerosis.

Endothelial integrity is compromised during vascular surgical procedures and even lost during prosthetic vessel and valve replacement. As more is known about the complex functioning of this layer, more emphasis is being directed towards minimizing damage and encouraging reendothelialization.

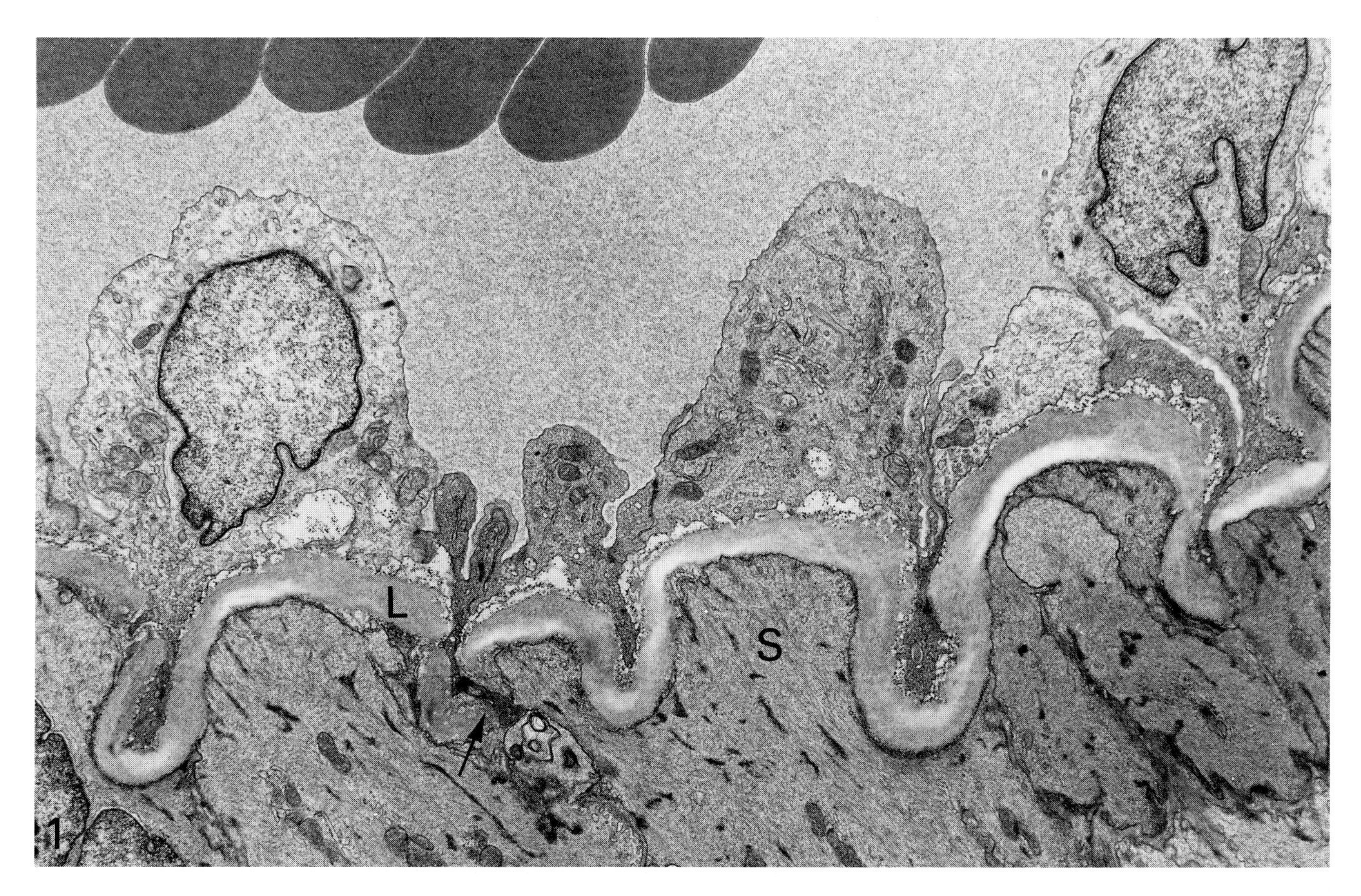

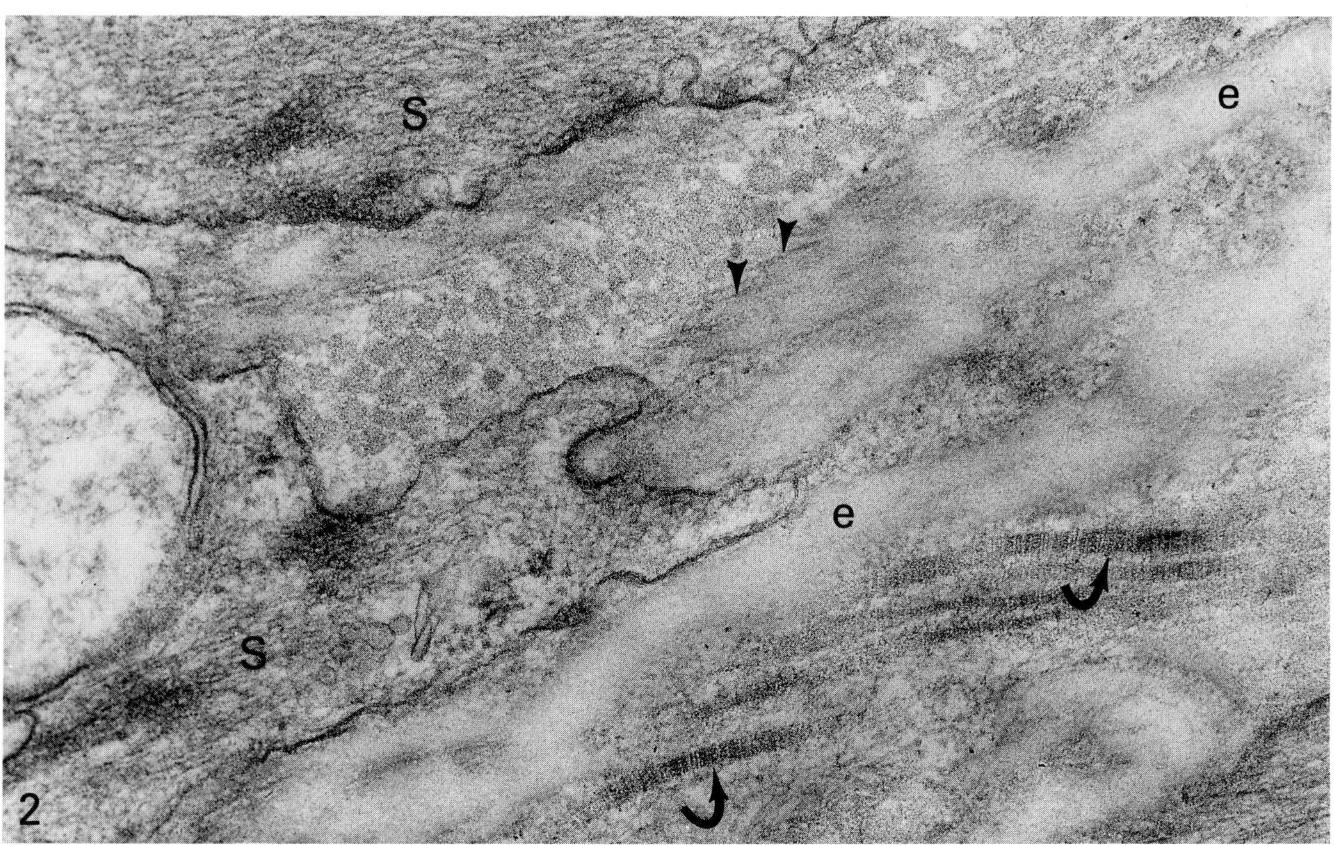

EM1: 7,500× EM2: 74,000×

BLOOD

HEMATOPOIESIS

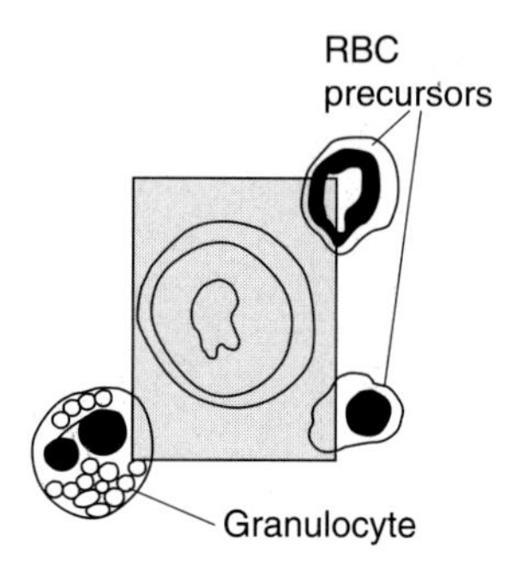

The development of blood cells **(hematopoiesis)** is a complex sequence of events in which a single totipotent stem cell gives rise to eight very different mature cell types with functions that range from carrying oxygen to producing antibodies. The process of hematopoiesis is continuous throughout a lifetime and operates to replace 3.7×10^{11} blood cells that are normally lost each day. In addition to its routine capacity, this system is very sensitive to increased demands related to environmental changes. For example, during severe infection, the number of granulocytes increases from 5,000 to 50,000/μl within a few days.

Many of the cells in bone marrow are similar to the one shown in the micrograph. These cells cannot be morphologically classified into any particular developmental lineage, but they contain features that are characteristic of a relatively undifferentiated **stem cell:** a large euchromatic nucleus (N) with a prominent nucleolus (n), and a cytoplasm packed with free polyribosomes. Activity is directed toward protein synthesis for internal use, such as for proteins involved in cell division, which is the most common event for stem cells. As differentiation proceeds, the nucleus becomes smaller and more heterochromatic and the cytoplasm acquires differentiation products such as hemoglobin or granules. Beginning with the **totipotent stem cell,** a hierachy of **pluripotent stem cells** exists. These progressively lose their ability for self-renewal as they become more restricted in lineage.

Early progenitor cells are not identified by their morphology but instead by the type of progeny they produce within clonal colonies. Cells referred to as erythrocyte-colony-forming units (E-**CFU**s) form only erythrocyte colonies. Progenitors of granulocyte (refers only to neutrophils in this case) and monocyte colonies are referred to as GM-CFUs. The growth and differentiation of CFUs in vitro is sensitive to specific growth factors, proteins including erythropoietin and many colony-stimulating factors (CSFs). Some of these factors have been produced using recombinant technology and are being used successfully in clinical trials. Results include (1) reduction of the need for transfusion in anemic patients when administered erythropoietin, and (2) the restoration of the white cell count in AIDS patients following GM-CSF treatment.

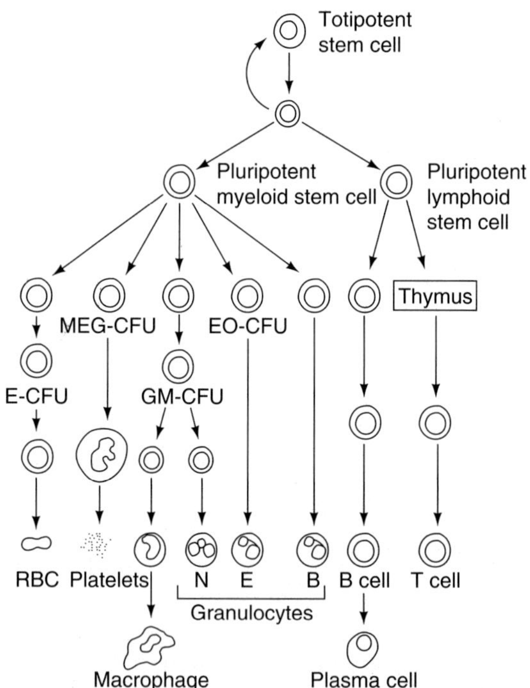

Modified from S. C. Clark and Robert Kamen, *Science 236*:1229(1987).

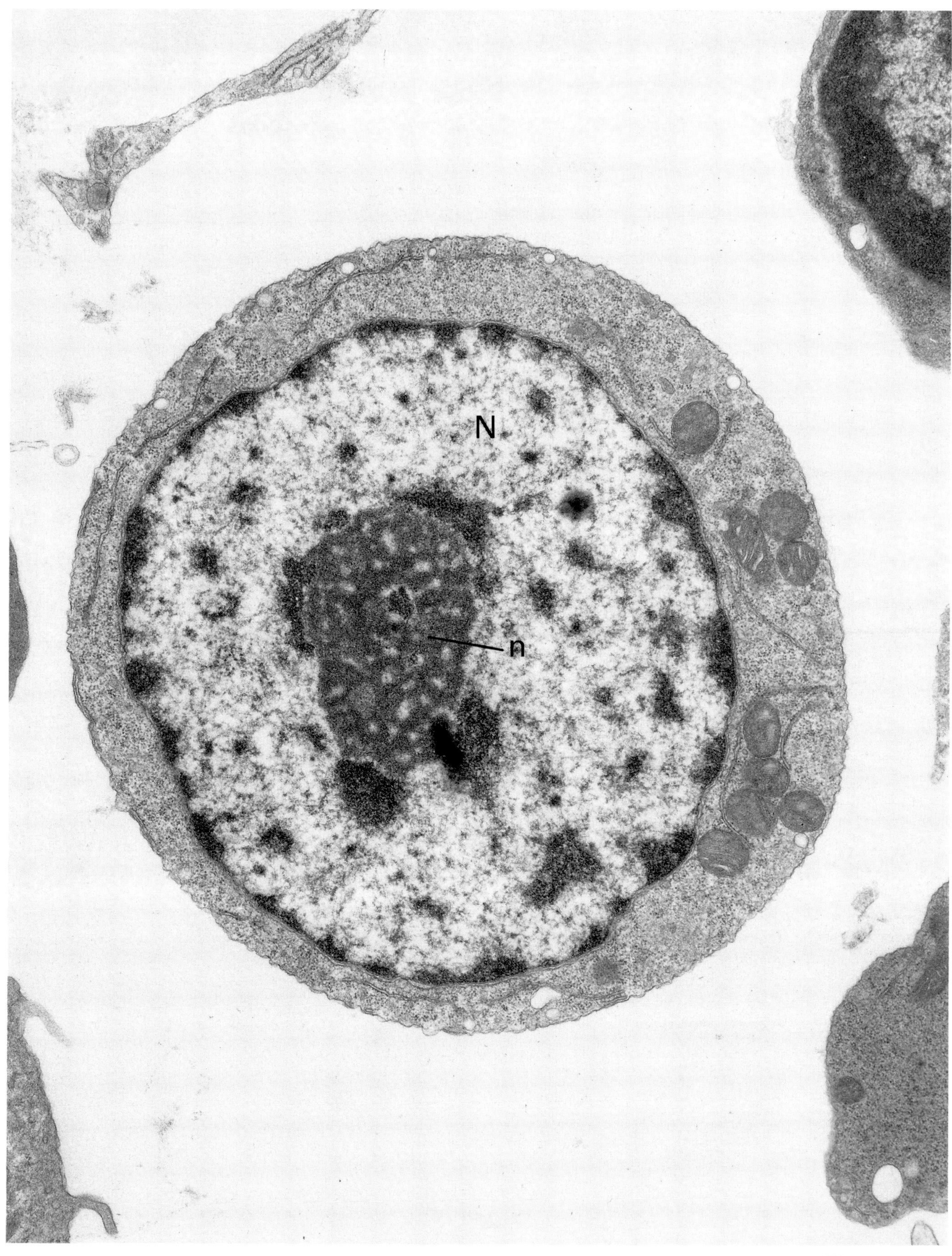

EM: 24,000×

ERYTHROPOIESIS: Early Development

Red blood cell development **(erythropoiesis)** occurs in 4–6 days and is characterized by the gradual synthesis and accumulation of hemoglobin to the exclusion of all organelles, including the nucleus. The large cell (P) on the micrograph is probably a **proerythroblast,** the first morphologically recognizable stage in the red cell line. The proerythroblast is formed from its precursor, apparently following stimulation by erythropoietin. Even though hemoglobin is being synthesized at this early stage, it is not concentrated enough to observe the electron density associated with iron. As differentiation proceeds to the **basophilic** (not shown in micrograph), **polychromatophilic** (PE, micrograph), and **orthochromatophilic** (OE, micrograph) erythroblast stages, hemoglobin concentration and electron density increase. Along with this, there is a reduction in free ribosomes and a progressive chromatin condensation from the euchromatic nucleus of the proerythroblast to the extreme condensation of the orthochromatophilic erythroblast.

Proerythroblast
Basophilic erythroblast
Polychromatophilic erythroblast
Orthochromatophilic erythroblast

The synthesis and assembly of **hemoglobin** is tightly controlled within different parts of the cell. Protoporphyrin is synthesized within the mitochondria (arrowheads, micrograph) in a complex series of reactions. Iron is transported into the mitochondria, where it combines with protoporphyrin to form heme. The globin chains (2 α and 2 β), synthesized on polysomes, are then assembled into the mature hemoglobin.

Iron–transferrin complexes are brought into the cell by receptor-mediated endocytosis. Iron is released and transferrin is recycled to the surface along with the receptor. In the early stages of red blood cell (RBC) development, such as the proerythroblast stage, the number of transferrin receptors within the cell membrane is large. With progressive development this number diminishes along with the other organelles important to hemoglobin synthesis.

Bone marrow preparations such as that used for this micrograph show developing cells isolated from one another and other supporting cells. Actually, all blood cells develop in microniches in intimate contact with some type of bone marrow **stromal cell.** Stromal cell contact is essential for long-term blood cell development in culture. In vitro studies and biopsy specimens show many developing RBCs in the same stage enclosed in processes of a single stromal cell, forming an **erythropoietic unit.**

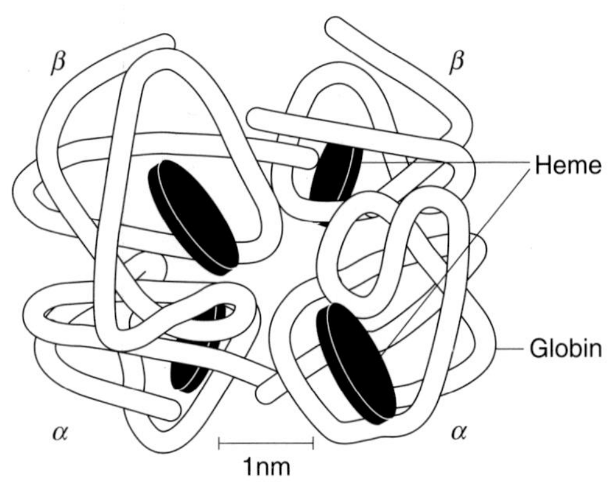

Modified from H. A. Harper et al., *Physiologische Chemie,* Springer-Verlag, New York, 1975.

Stromal cell

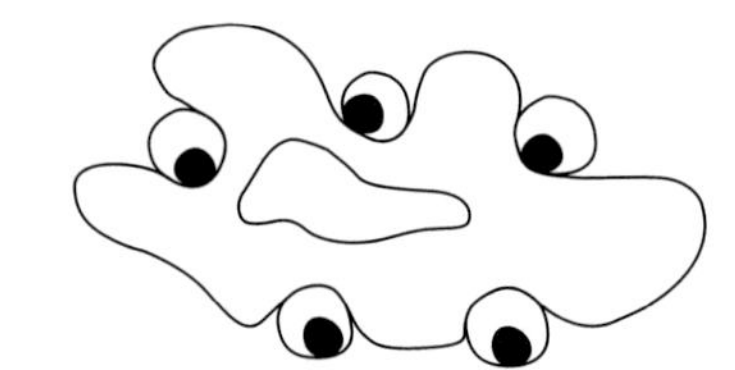

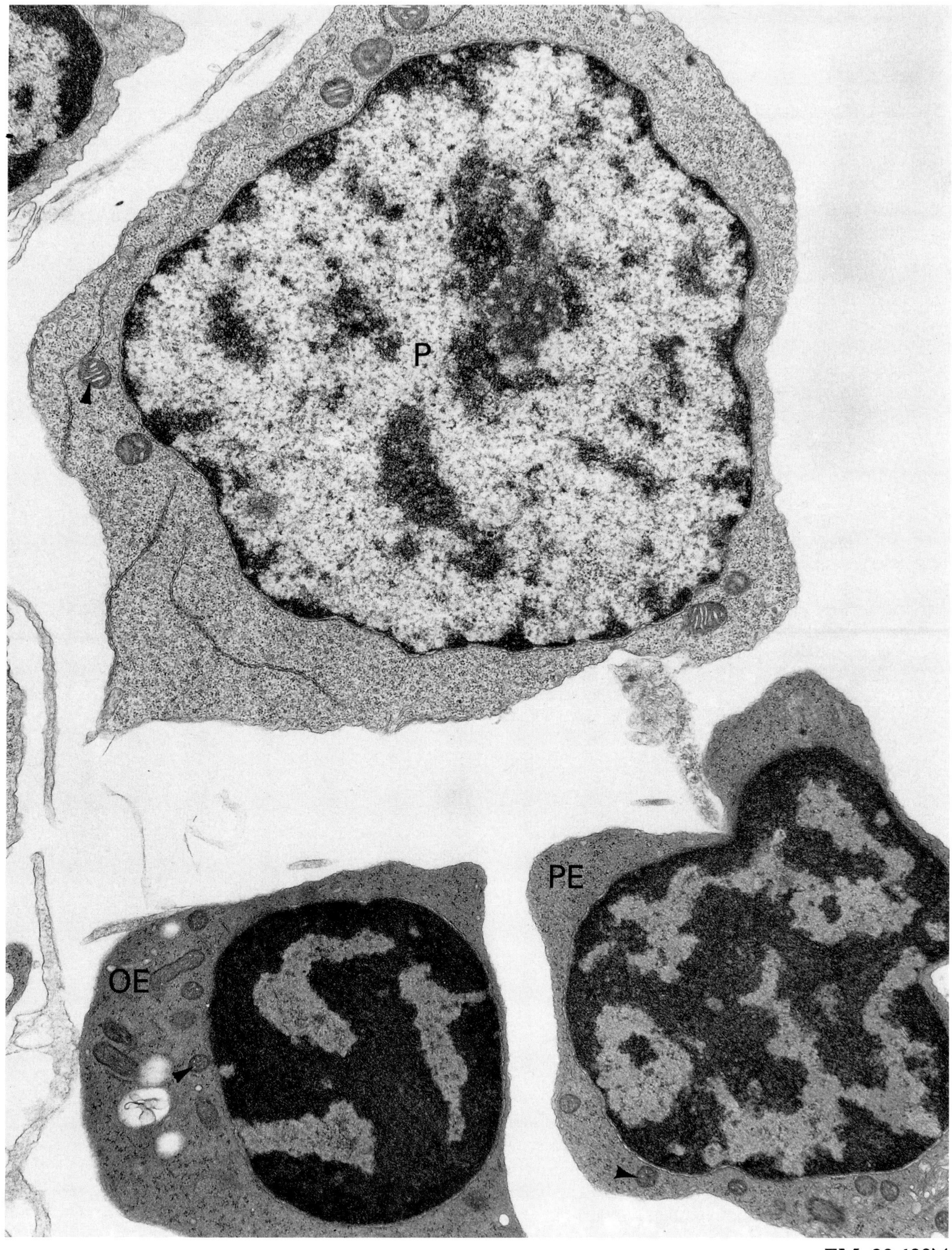

EM: 20,400×

Orthochromatophilic erythroblasts (normoblasts) extrude their nuclei, which are immediately phagocytosed by associated macrophages. The remaining cytoplasm repairs quickly to form an immature red blood cell, the **reticulocyte** (R, micrograph). Normally one third of reticulocytes leave bone marrow and complete their differentiation in peripheral blood. The reticulocyte in the micrograph is in the process of squeezing between endothelial cells (E) lining the bone marrow sinusoid. The polysomes (arrowheads) and mitochondria (m) still present continue to be involved in hemoglobin synthesis as the cell completes maturation. The gradual loss of these structures as reticulocytes mature does not occur within lysosomes, but occurs within the cytoplasm when the polypeptide ubiquitin binds to organelle proteins, initiating their enzymatic destruction. Reticulocytes develop in one to two days into mature RBCs, biconcave sacks tightly packed with hemoglobin.

Orthochromatophilic erythroblast

Reticulocyte

Mature RBC

The synthesis of the red blood cell **cytoskeleton** and associated unique membrane components occurs along with hemoglobin synthesis and is significant to RBC functioning. Spectrin, a protein dimer of two nonidentical rod-shaped polypeptides, binds to membrane proteins such as Band 3 and glycophorin via other proteins such as ankyrin and protein 4.1. This membrane–cytoskeleton network maintains the discoid shape that provides the large surface: volume ratio that is critical to the soft, pliable nature of the red cell. Mature cells are continually deformed as they travel through narrow capillaries. Even the reticulocyte demonstrates considerable flexibility as it enters the peripheral blood (micrograph).

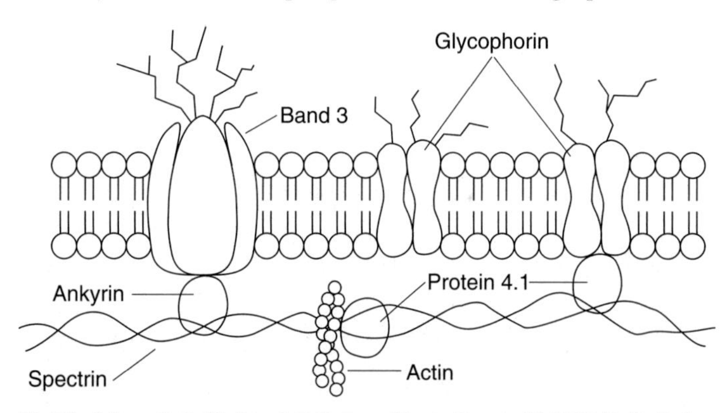

Modified from S. B. Shoket & S. E. Lux, *Hosp. Pract. 19* (1984). In L. Stryer, *Biochemistry,* Freeman, New York, 1988.

Erythroid cells depend upon an **attachment to fibronectin** in the stromal cell matrix for proper differentiation. Unattached cells grown in suspension are fragile and fail to assemble a stable cytoskeleton. As differentiation progresses, the number of fibronectin receptors on erythroid cells decreases, and cells normally detach at the reticulocyte stage. A constant number of reticulocytes leave the marrow and populate the peripheral blood as they detach from fibronectin.

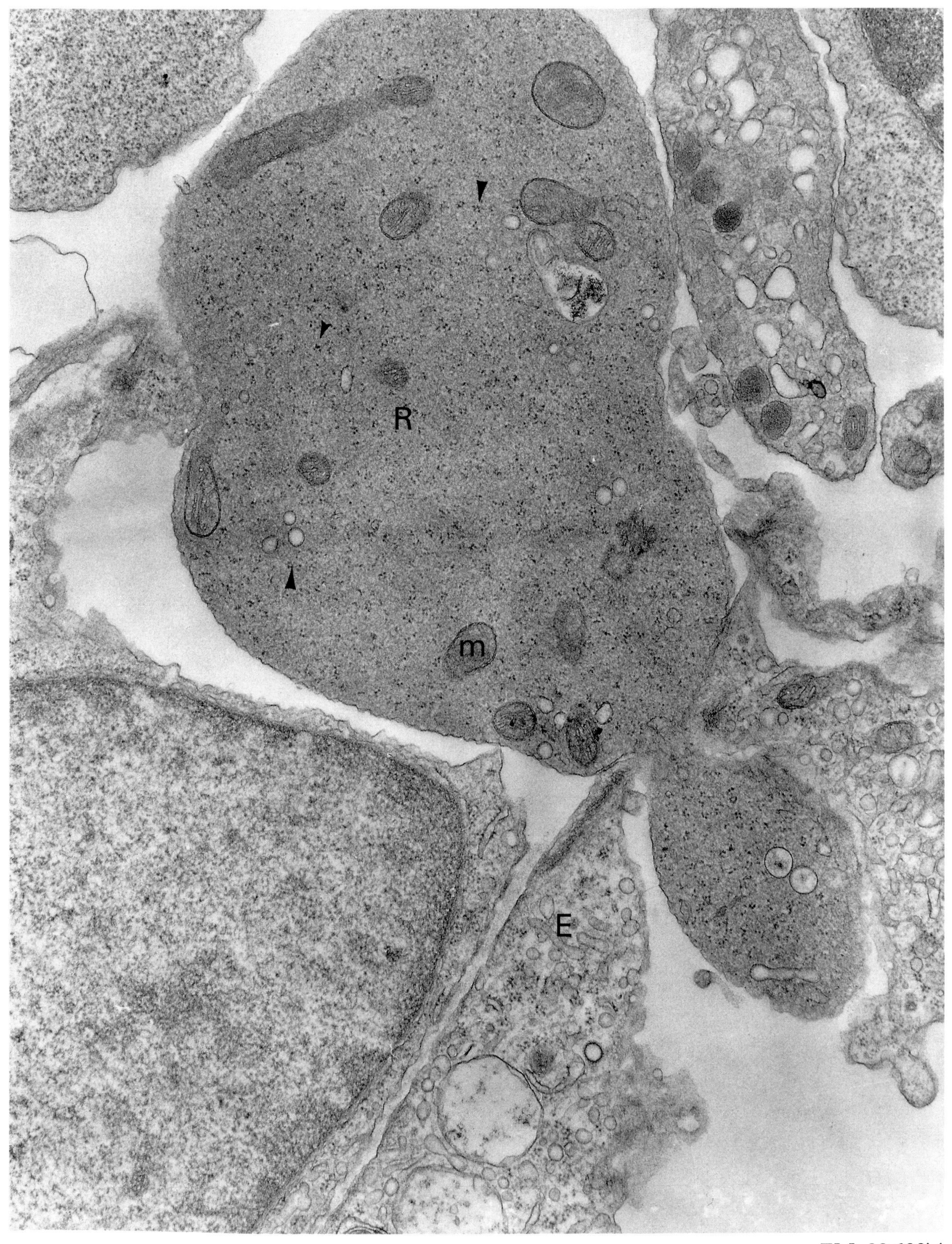

EM: 30,600×

GRANULOPOIESIS: Promyelocyte

Granulopoiesis is the process of differentiation of neutrophils, eosinophils, and basophils, white blood cells (leukocytes) characterized in morphology and function by unique granules. The immediate precursor to these granulocytes, the **promyelocyte** (P, micrograph), is a large cell with a euchromatic nucleus and well-developed nucleolus. The granules synthesized at this stage are called primary since they are the first to be synthesized during development, and are called azurophilic due to their affinity for azure dyes in light-microscope preparations. **Primary granules** produced in the promyelocyte stage are distributed to all granulocytes as development proceeds. **Secondary or specific granules** of mature granulocytes are synthesized in subsequent stages. Many of the large, dense primary granules (arrows, micrograph) are similar to lysosomes in their content of acid hydrolases (e.g., β-glucuronidase), but others contain components such as neutral proteases (e.g., elastase) and microbicidal enzymes (e.g., lysozyme, myeloperoxidase).

Promyelocytes, identified as a single cell-type in bone marrow smears, are actually different in their developmental potential. In culture some promyelocytes form colonies that give rise to (1) neutrophils and monocytes, (2) eosinophils, or (3) basophils. The development of these clones seems to depend upon separate colony-stimulating factors.

Myeloperoxidase (MPO), a heme-containing microbicidal enzyme synthesized during this stage and sequestered in the primary granules, has received considerable attention due to its role in the formation of hypochlorous acid and other toxic oxygen species that function during the respiratory burst of phagocyte killing (see Cell, page 20). This enzyme is synthesized as a large precursor on the rough ER, processed in the Golgi, and transported to primary granules. The gene that encodes MPO, localized to chromosome 17, is translocated (partially or completely) to chromosome 15 in acute promyelocytic leukemia. Such malignant promyelocytes exhibit increased numbers of abnormal granules and unusually high concentrations of MPO.

The processes of **stromal cells** (S) are seen in this micrograph adjacent to the promyelocyte. Stromal cell glycosaminoglycans adsorb colony-stimulating factors and concentrate them in specific microenvironments in which they act to stimulate clonal expansion of precursor cells.

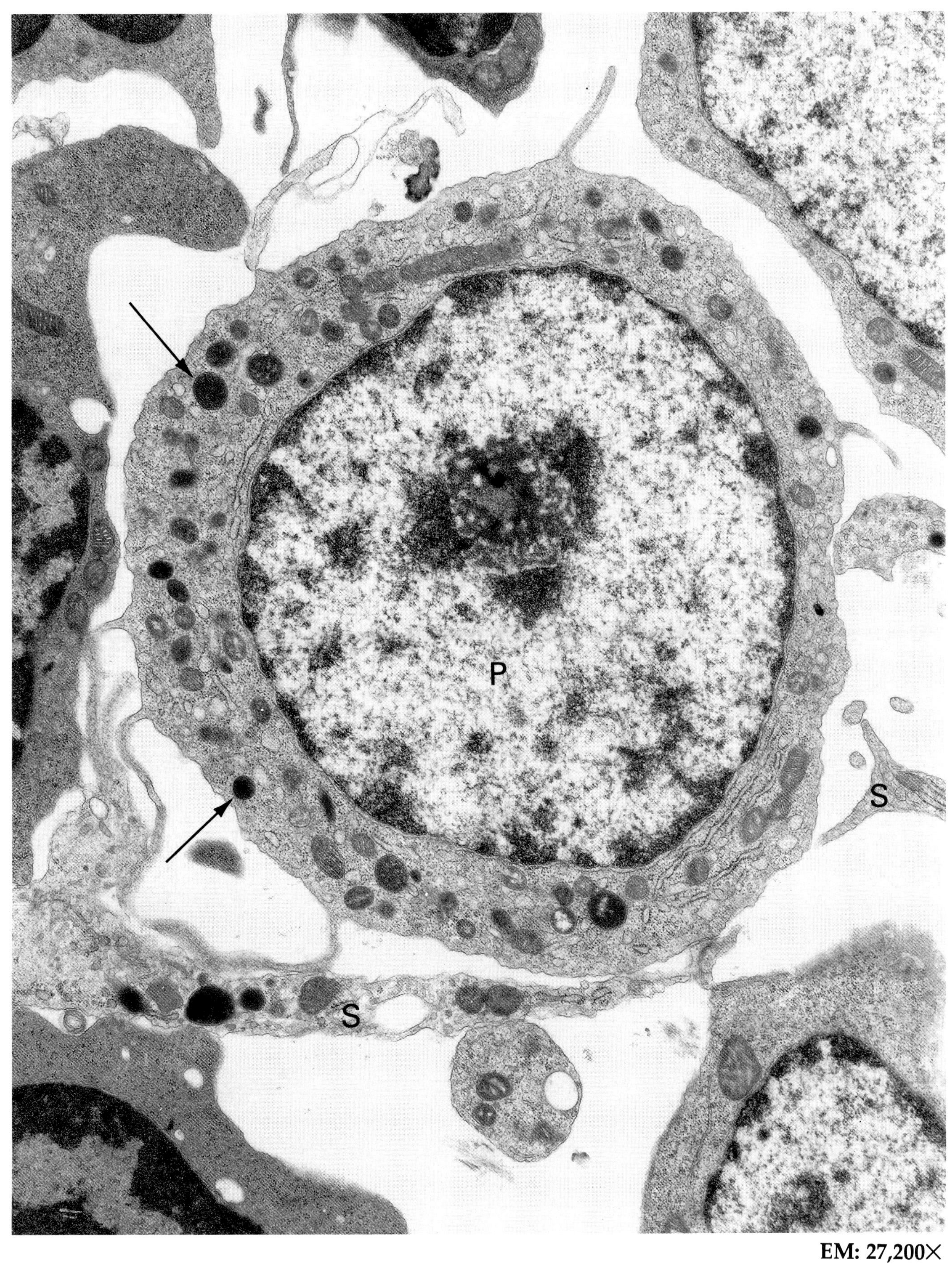

EM: 27,200×

Granulopoiesis: Myelocyte and Band Neutrophil

Promyelocytes divide and differentiate to form **myelocytes** (micrograph 1). Myelocytes and earlier stages are capable of division and form a mitotic pool of precursor cells within the bone marrow. Cells within this pool are only seen in peripheral blood in disease states. During the myelocyte stage, **secondary (specific) granules** are formed that contain at least one component unique to the granulocyte type (e.g., lactoferrin in neutrophils, histamine in basophils, major basic protein in eosinophils). In the neutrophil stages in the micrographs, the specific granules (arrows) are generally smaller and less dense than the larger primary granules (arrowheads) inherited from the promyelocyte. Specific granules, initially synthesized in the myelocyte stage, are also synthesized during subsequent stages. The prominent Golgi (G) in the **band neutrophil** in micrograph 2 is critical to granule formation.

Evidence suggests that several different kinds of specific granules are synthesized during neutrophil development, each containing different substances. In mature neutrophils they are released at different times and are under different control mechanisms. Contents of neutrophil specific granules include (1) lactoferrin, a glycoprotein that facilitates the formation of the hydroxyl radical in respiratory burst activity, (2) chemoattractants, opsonins, and activators of complement synthesis, and (3) collagenase, an enzyme important to the migration of neutrophils in loose connective tissue.

> The membranes of specific granules concentrate many receptors and enzymes. Following initial exposure to the bacterial chemoattractant N-formyl-methionyl-leucyl phenylalanine (FMLP), granule membranes fuse with the cell membrane, resulting in the immediate exposure of concentrated FMLP receptors ("up regulation"). An activated human neutrophil has up to 50,000 FMLP receptors.

During differentation into a mature neutrophil, the nucleus undergoes a series of changes, from large, round, and euchromatic to band-shaped and less euchromatic to a small, heterochromatic nucleus with 3–5 lobes in the mature cell. At the same time the cell becomes smaller. Both band and mature neutrophils are normally released into peripheral blood in a ratio of 1 : 3.

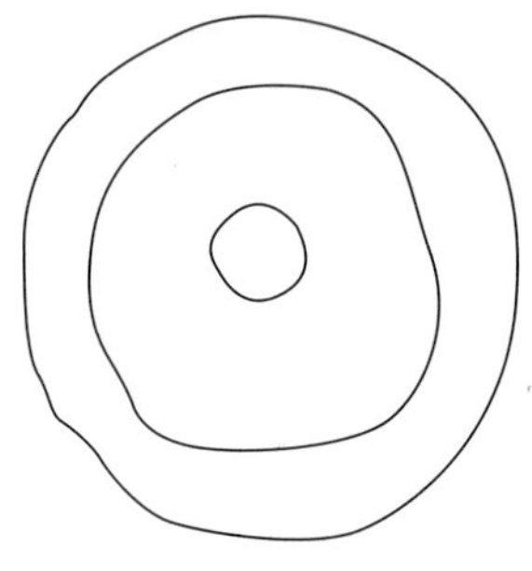

Promyelocyte

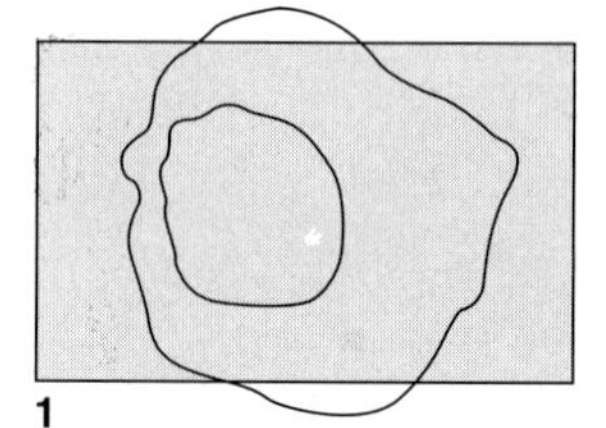

1

Myelocyte

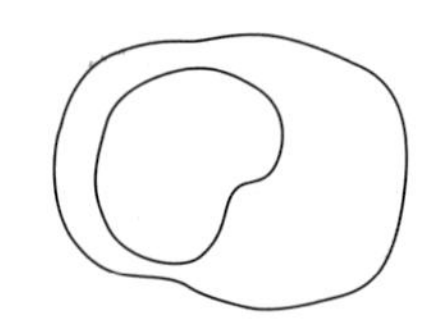

Metamyelocyte

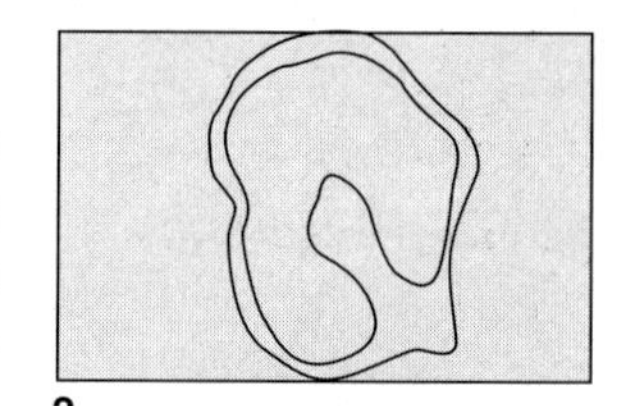

2

Band neutrophil

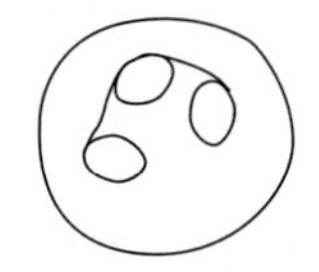

Mature neutrophil

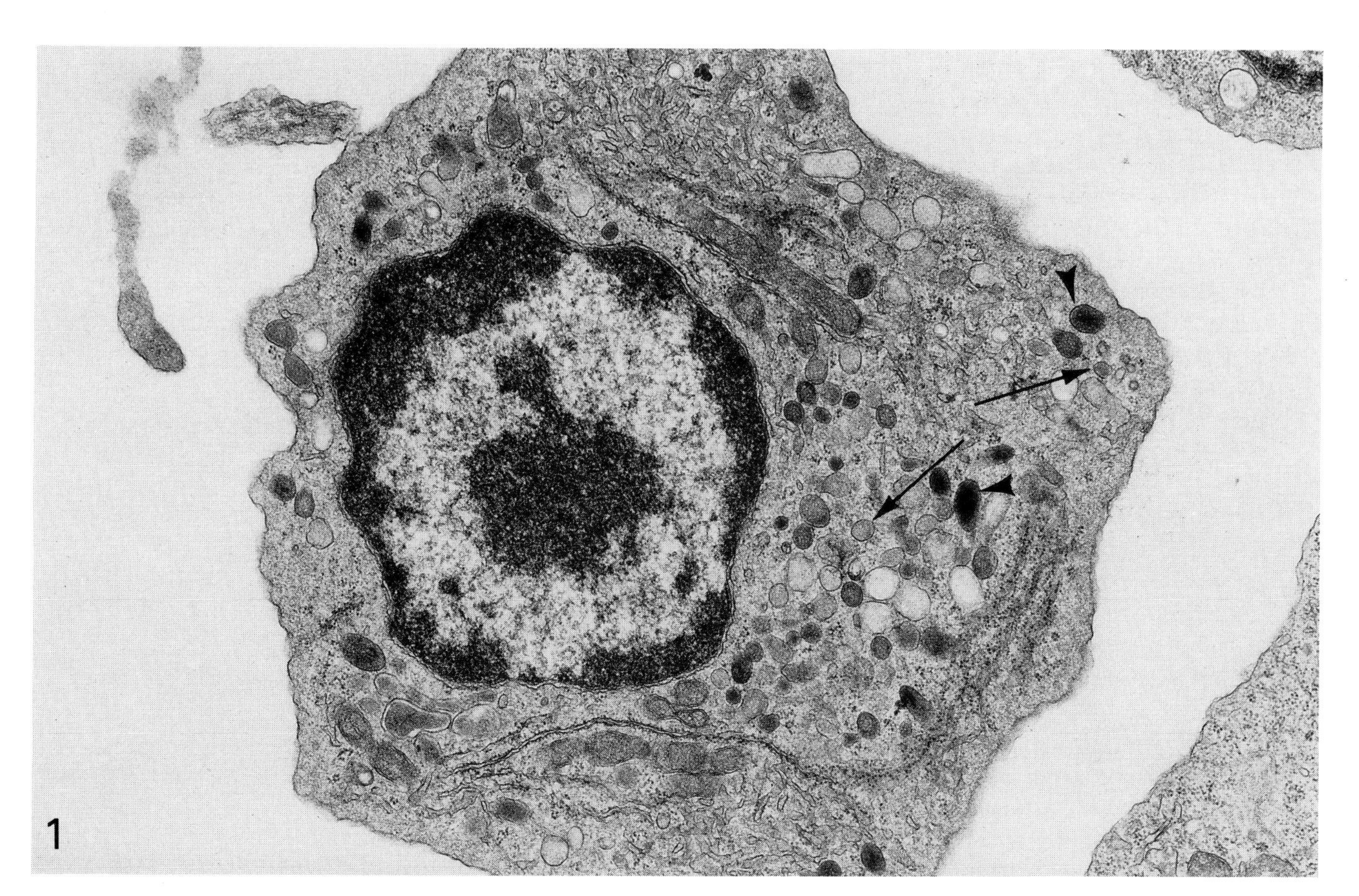

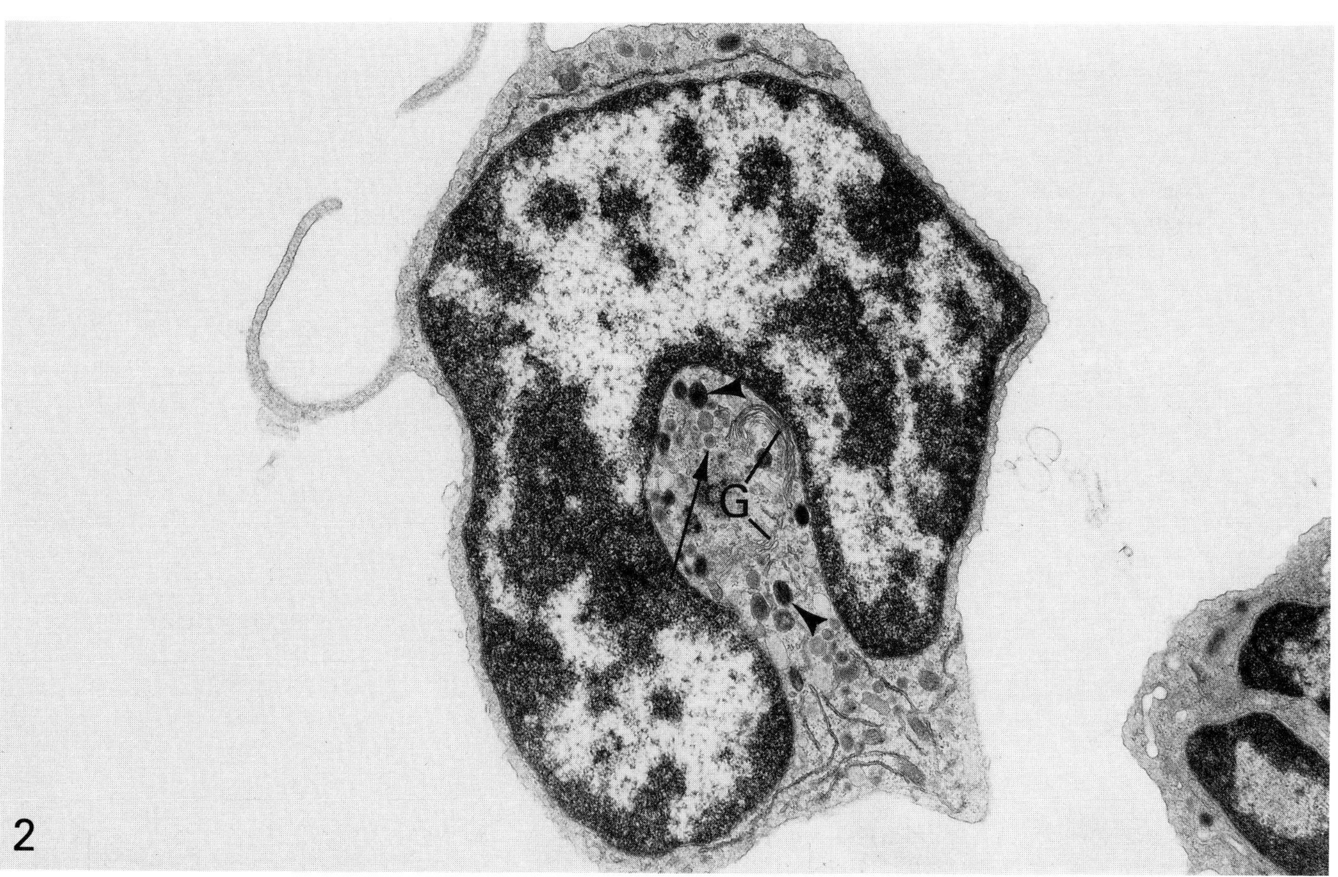

EM 1: 24,000× EM 2: 20,800×

GRANULOCYTE: Neutrophil

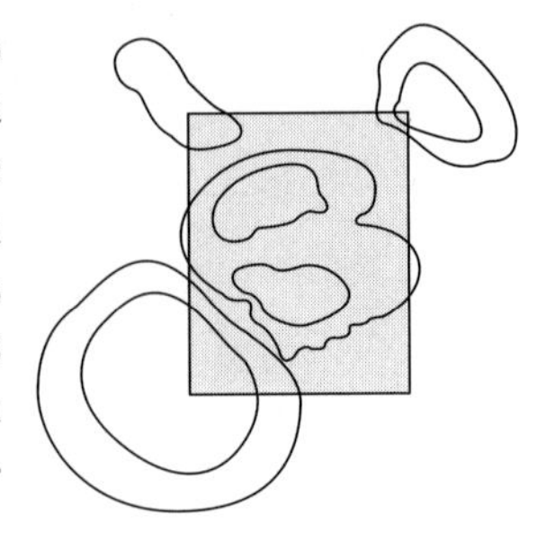

Neutrophils have a relatively heterochromatic nucleus (N) divided into three to five lobes (section shown in micrograph passes through only two) and a cytoplasm packed with **primary** and **secondary granules** (arrows). In the mature cell, the rough ER (arrowheads, micrograph) and Golgi (G, micrograph) remain active in the production and packaging of secondary granules. Following 10 days of postmitotic development in bone marrow, mature segmented neutrophils (segs, polymorphonuclear leukocytes, or PMNs) leave bone marrow and travel in peripheral blood for approximately 10 hours before they squeeze between endothelial cells (diapedesis) and migrate to infected regions.

Neutrophils are the initial phagocytes at the site of infection. In comparison to their companion phagocytes, the macrophages, neutrophils are quicker and more specialized. They reach a site of infection first and concentrate on the phagocytosis of opsonized bacteria. Macrophages arrive later and clear the battlefield remains. In both of these cell types, receptor contact with the coated antigenic material initiates increased enzyme activity in the cell membrane and granules.

A variety of types of granules are produced by neutrophils, but it is impossible to distinguish them by ultrastructure and even difficult using histo- or immunochemical techniques. Some of the granules are associated with the phagocytic role of neutrophils, while others release substances involved in migration and cell interaction. Many are involved in the respiratory burst that kills phagocytosed bacteria. When opsonized bacteria bind to neutrophil receptors, oxygen consumption increases 100-fold and toxic oxygen species are formed. One of these, hypochlorous acid, is capable of altering nucleotides and cytochromes as part of the killing action. How these events are coordinated in subcellular compartments is not precisely known; however, one of the initial steps involving NADPH oxidase occurs on the cell membrane.

In a rare X-linked disorder, chronic granulomatous disease, neutrophils are unable to undergo a respiratory burst, and affected individuals are subject to recurrent, sometimes fatal, infections. The defective gene in this disease has been found to code for a part of a cytochrome that may act as an electron carrier in association with NADPH oxidase activity.

Aside from the role of oxidants in defense against microorganisms, their release can also cause tissue injury and contribute to such diseases as adult respiratory distress syndrome and arthritis. In addition, products of phagocyte oxygen metabolism have recently been implicated in carcinogenesis.

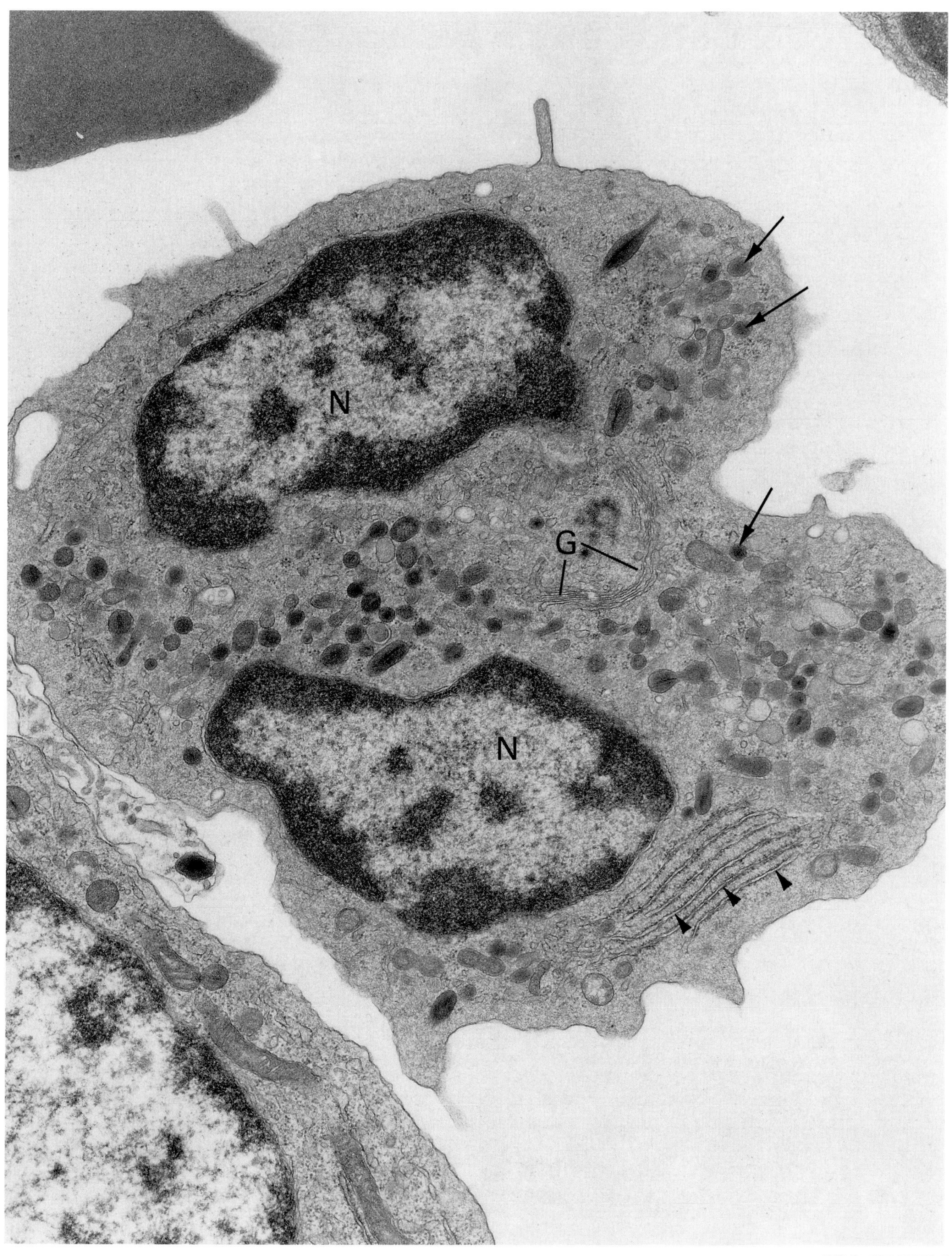

EM: 27,200×

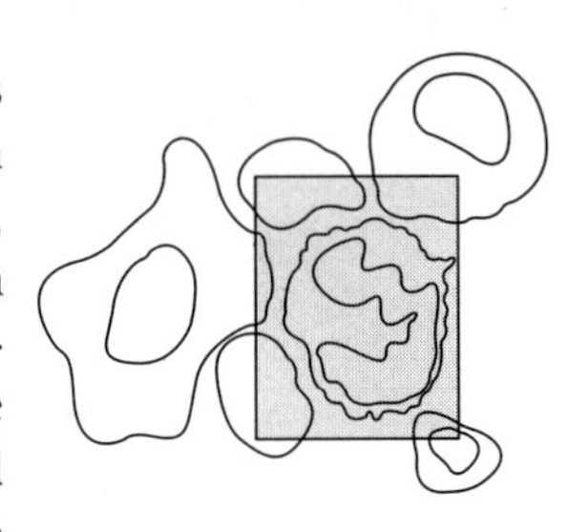

Eosinophils (micrograph) contain characteristic large specific granules (arrowheads). These specific granules contain several **cationic proteins,** of which the most abundant and well characterized are eosinophilic cationic protein, major basic protein, and eosinophil peroxidase. Major basic protein differs from eosinophilic cationic protein in many respects, including its tendency to aggregate and form the insoluble crystalline cores (arrows, micrograph) that give eosinophils their unique appearance on electron micrographs. The band-shaped nucleus (N, micrograph) shown here will separate into two to three lobes as the cell completes maturation.

Eosinophils are attracted to sites of infection and function to dampen the effects of mast cell degranulation (see Connective Tissue, page 78). In addition, they are the principal defense against **schistosomiasis,** a parasitic helminthic disease that kills more than 800,000 people per year throughout the world.

In schistosomiasis, the number of eosinophils in larval-infected tissues reaches $100{,}000/mm^3$. Eosinophil receptors attach to the Fc portion of antibodies that coat the larvae. Receptor binding activates the release of granule contents directly onto the larval surface. Both **eosinophilic cationic protein** and **major basic protein** are capable of causing damage; however, eosinophilic cationic protein seems to be the most potent and is effective in vitro at a concentration as low as 10^{-7} M. Eosinophilic cationic protein acts by altering membrane permeability by creating transmembrane pores. This type of membrane damage is also caused by complement and cytotoxic T cells, and may be a common killing mechanism in immune defense.

Like neutrophils and macrophages, eosinophils depend upon a respiratory burst for their functioning. Peroxidase activity within the eosinophilic granules is part of a sequence leading to the formation of toxic oxidants. In contrast to other cell types, however, the eosinophil peroxidase preferentially oxidizes bromide instead of chloride to form hypobromous acid (HOBr), a more toxic and faster acting agent. In addition to the role of peroxidase in killing, this enzyme, along with eosinophil histaminase, helps regulate the effects of mast cell activity at sites of antigen invasion.

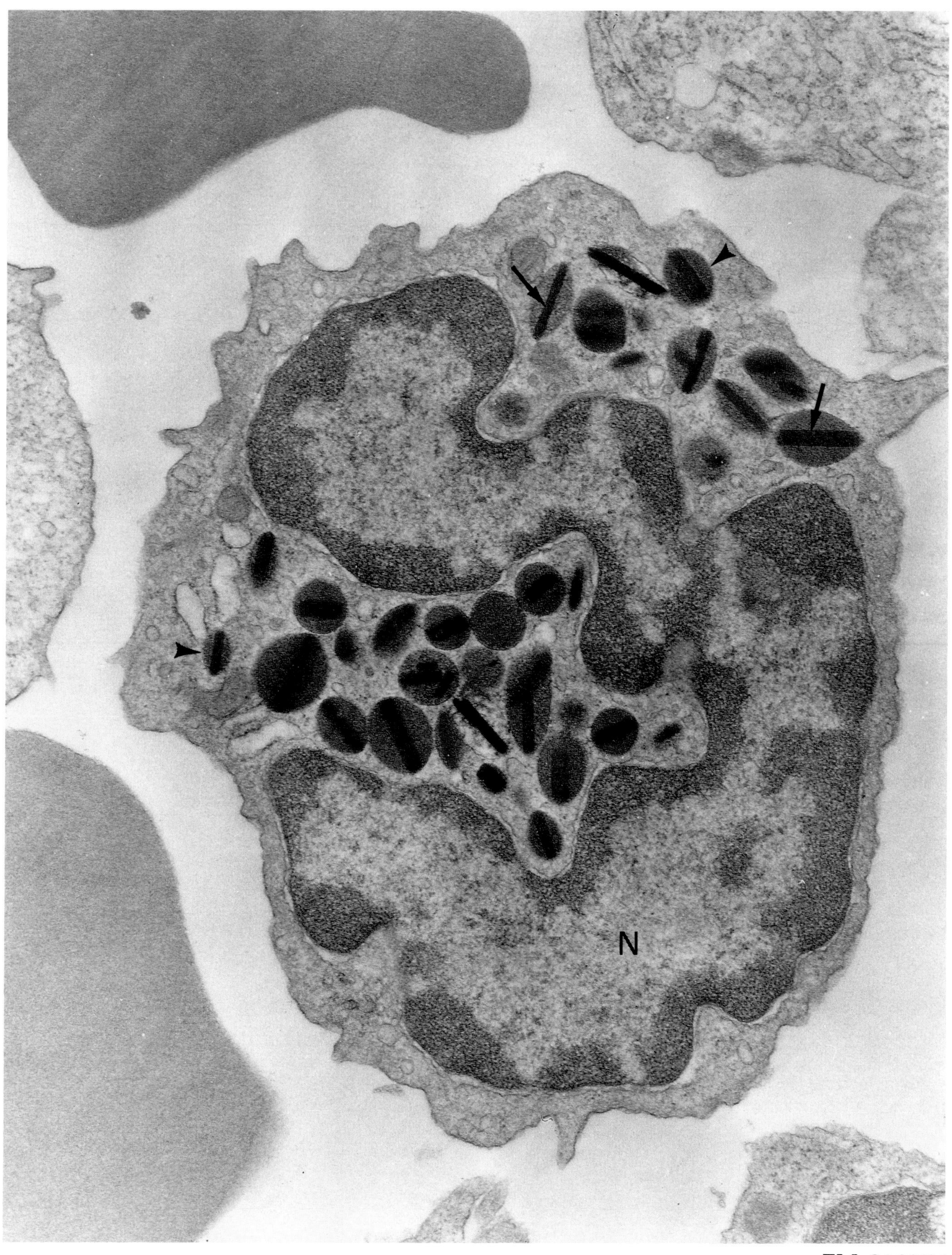

EM: 34,000×

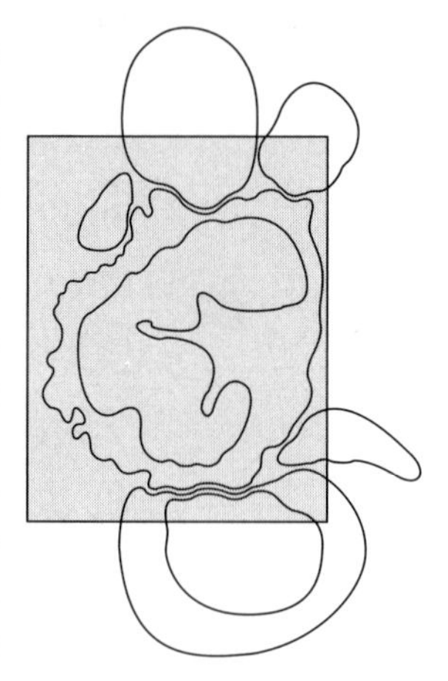

In contrast to granulocytes, a major part of the differentiation of the **agranular leukocytes** (monocytes and lymphocytes) occurs after they leave bone marrow. These cells leave the marrow as relatively immature cells capable of mitosis and travel to other tissues and organs where they divide and undergo critical differentiation changes.

Monocytes (M, micrograph), often recognized by their euchromatic indented nucleus (N, micrograph), are the largest cells normally found in peripheral blood. They circulate for approximately 14 hours and then migrate into tissues throughout the body where they differentiate into a variety of cell types. The most ubiquitous is the **macrophage,** which is found in many different regions where it functions in defense (secondary lymphoid regions, connective tissue, liver, lung) and in removing old and diseased red blood cells (spleen and liver) and platelets (spleen). Monocytes also fuse to form specialized cells such as osteoclasts in bone, giant foreign body cells in certain types of inflammation, and Reed–Sternberg cells in the malignant lymphoma of Hodgkin's disease.

Many of the organelles seen within the monocyte in the micrograph will become particularly important during the division and differentiation into these varied cell types. The centriole (c) and its associated microtubule-organizing centers (arrows) will play major roles as organizers of the mitotic spindle during division, and the primary lysosomes (arrowheads) will increase in number to perform their function either in phagocytosis (macrophages) or secretion (osteoclasts).

Monocytes are attracted to areas of injury by bacterial components, products associated with tissue injury (e.g., fragments of elastin and fibronectin, platelet-derived growth factor), and many cytokines released from activated cells involved in defense. One such cytokine chemoattractant is granulocyte–macrophage colony-stimulating factor (GM-CSF) produced by T lymphocytes at the site of injury. CSFs seem not only to promote growth during blood cell development in bone marrow but also to affect the differentation and the activity of certain blood cells during activation.

Monocytes (and macrophages) are important secretory cells. One significant secretion released in response to infection is interleukin 1, a protein with far-ranging effects. This protein acts locally on T cell activation and systemically on (1) the enhancement of liver cell synthesis of proteins associated with inflammation (the hepatic acute phase reaction) and (2) the release of glucocorticoids that regulate immune function during stress. Interleukin 1 released by monocytes stimulates the synthesis of corticotropin-releasing factor (CRF) in the hypothalamus, which, via adrenocorticotropic hormone (ACTH) from the pituitary, increases glucocorticoid release.

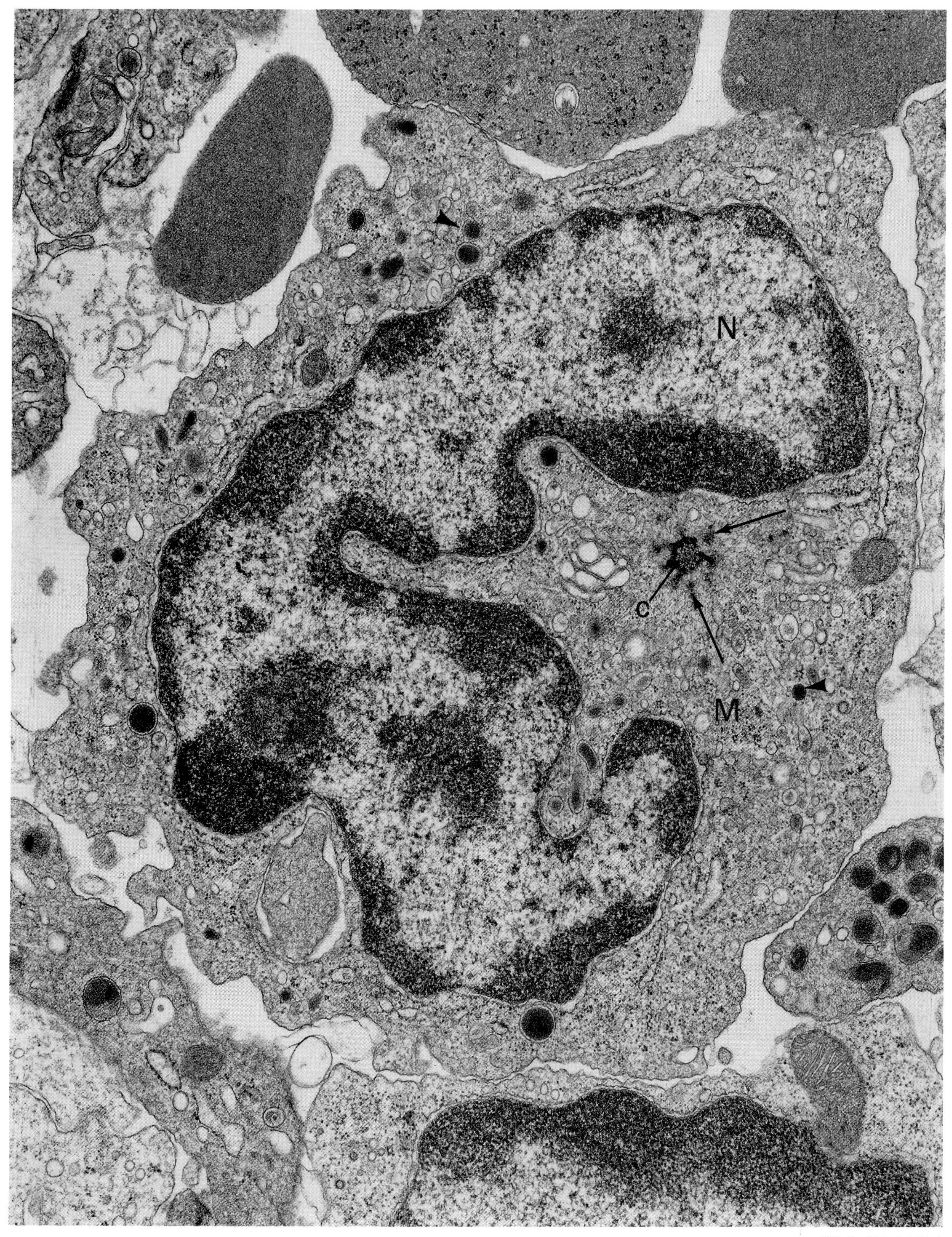

EM: 27,200×

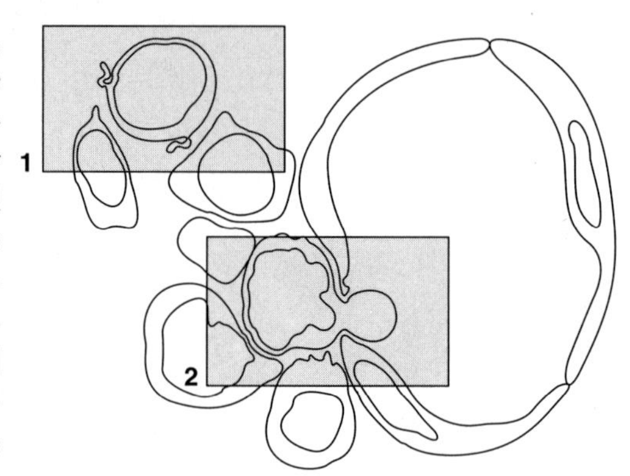

Most **lymphocytes** (L, micrograph 1) are small (10 μm), with a thin rim of cytoplasm and surface projections. These cells, like monocytes, leave bone marrow in a relatively undifferentiated stage and enter the peripheral blood, where they travel to lymphoid organs (spleen, lymph nodes, thymus) and tissues (Peyer's patches in the intestine and connective tissue underlying most luminal surfaces). The lymphocyte (L) in micrograph 2 is in the process of squeezing between two endothelial cells (E) to enter a bone marrow sinusoid (S). Peripheral blood lymphocytes represent less than 5% of the total lymphocyte pool. Since lymphocytes periodically enter and leave peripheral blood during their life span, this traveling 5% represents an extremely **heterogeneous population.** Some of those that have recently left bone marrow are not yet immunocompetent; others are memory cells that have undergone a previous activation by antigen in a peripheral organ and are migrating to another peripheral site.

Two major classes of lymphocytes develop from a pluripotent lymphoid stem cell in the bone marrow, the **B cells** involved in humoral immunity and the **T cells** involved in cell-mediated immunity. Lymphocytes destined to become T types leave the bone marrow and travel to the thymus, where they differentiate and develop immunocompetence. In contrast, B lymphocytes remain in bone marrow and seem to develop immunocompetence in this region before traveling to the peripheral lymphoid organs.

The first committed B precursors to be recognized contain small amounts of cytoplasmic immunoglobulin which is probably a type of IgM. There is no surface expression of this immunoglobulin at this early stage, but these cells are committed to respond to a single antigenic determinant. In the next stage of B cell development IgM is expressed on the surface. It is not clear whether this occurs in bone marrow or in peripheral lymphoid tissues and organs.

Acute lymphocytic leukemia, the most common leukemia in children, is a result of a deficiency of the enzyme adenosine deaminase in a bone marrow lymphocyte stem cell. Purine synthesis is impaired and a toxic metabolite, deoxyadenosine, accumulates. As in other leukemias, increasing numbers of abnormal cells crowd out other developing blood cells and spill into the peripheral blood.

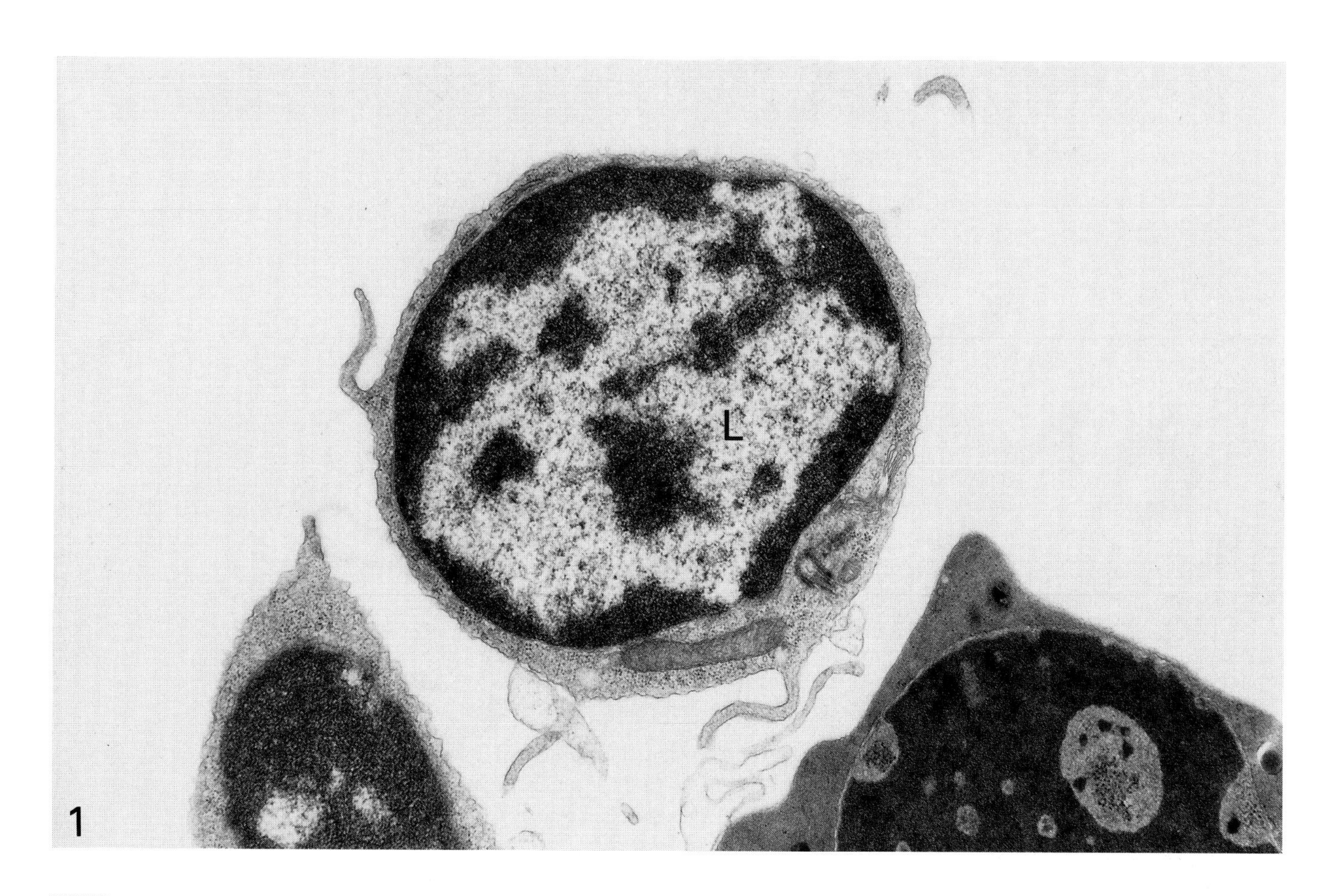

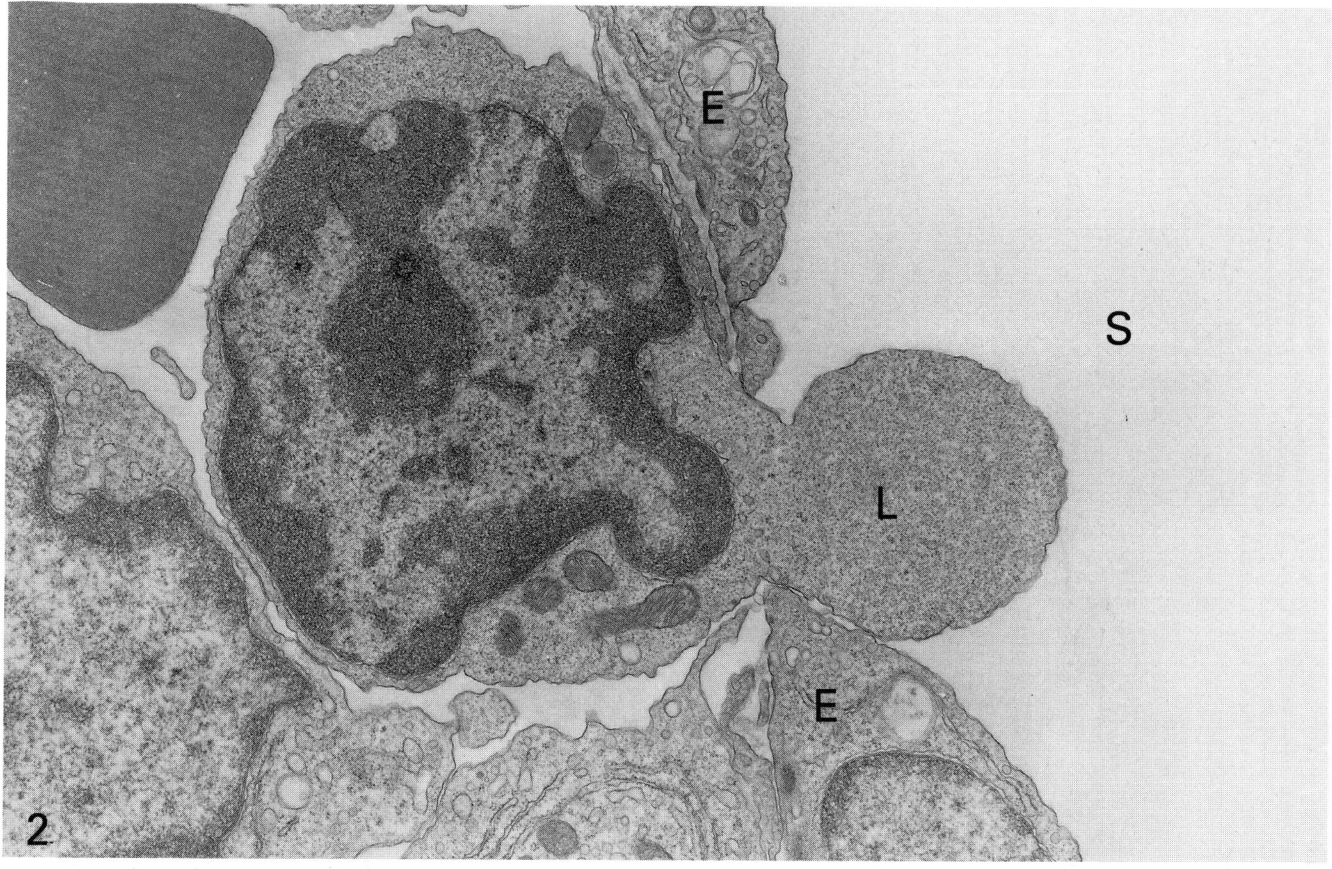

EM 1: 19,200× EM 2: 16,800×

PLATELETS: Megakaryocyte

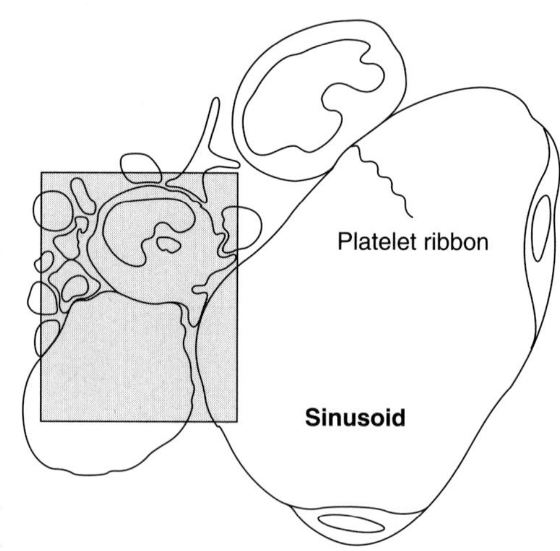

Megakaryocytes (M, micrograph) are large, polyploid cells that give rise to **platelets.** They develop from small pluripotent stem cells that undergo first cellular division, then a subsequent series of chromosome divisions without cytoplasmic division (endomitosis). During endomitosis several metaphase plates are formed and chromosomes separate and move apart. The section of the dividing megakaryocyte in the facing micrograph is probably a polar view through one of these metaphase plates (arrows designate chromosomes).

Following division the different chromosome groups come together enclosed by the same nuclear membrane. Endomitosis results in cells with large multilobed nuclei (N, micrograph) that are 8N, 16N, 32N, or 64N. The greater the number of chromosomes in a megakaryocyte, the greater the amount of cytoplasm that is synthesized and the greater the number of platelets formed. Megakaryocytes can form up to 8000 platelets.

During the cytoplasmic maturation of megakaryocytes, the cell membrane invaginates to form channels separating cytoplasmic islands about 3–4 μm in diameter. These **platelet demarcation channels** (arrowheads, micrograph) eventually coalesce to release platelets. During early cytoplasmic maturation the outer rim of megakaryocytes is frequently devoid of organelles and forms a kind of ectoplasm (e, micrograph).

Megakaryocytes typically rest next to bone marrow sinusoids (S, micrograph) and extend cytoplasmic projections between endothelial cells into the sinusoids. Many platelets appear to be released together as a "ribbon" from these extensions, with the actual formation of individual platelets usually occurring after the ribbon separates from the megakaryocyte.

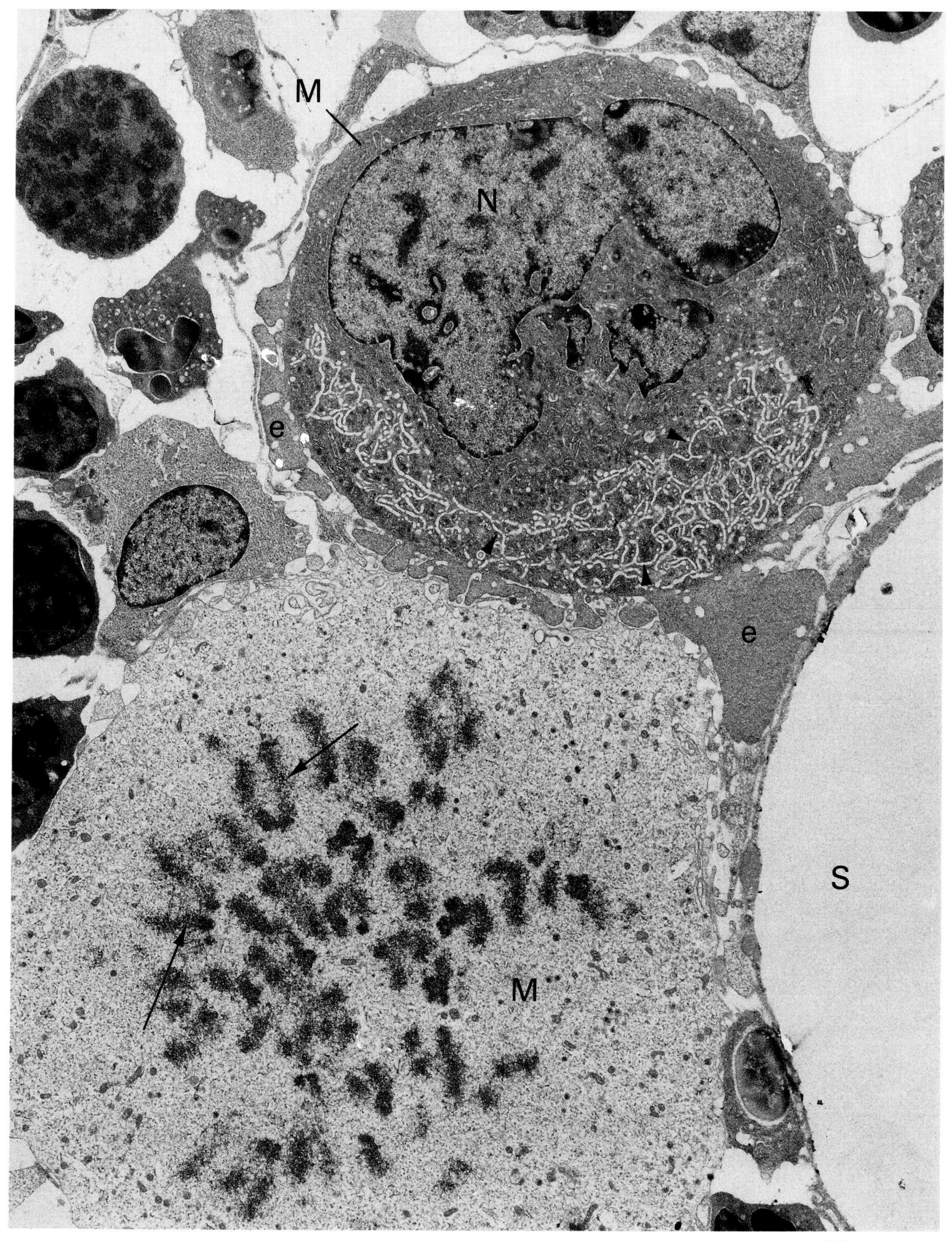

EM: 6,150×

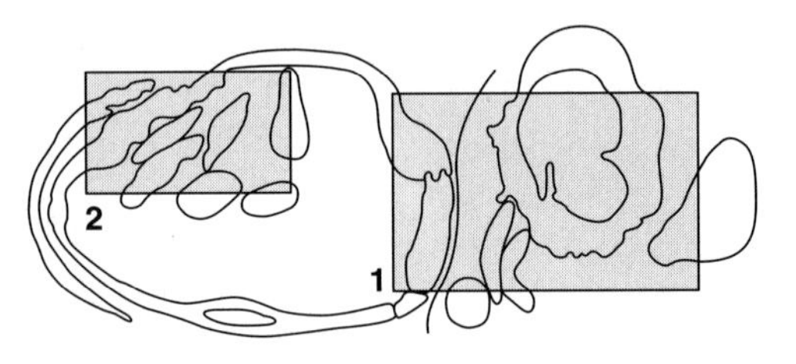

Platelets (P, micrographs 1 and 2) are small (2–4 μm) **biconvex cell fragments** that repair blood vessels. These cell fragments reach concentrations of approximately 250,000/mm^3. At a site of injury they respond to substances as diverse as thrombin, adenosine diphosphate (ADP), and basal lamina proteins. The glycocalyx covering platelets contains many receptors for such agonists. Once activated, platelets change shape, acquire surface stickiness, aggregate, release mediators (both newly synthesized and stored), and contract. This sequence, which usually repairs a damaged vessel surface and maintains blood homeostasis, can also lead to vessel wall thickening and result in disease states such as arteriosclerosis.

During the initial stage of activation platelets adhere to exposed basal lamina. Attachment activates the release of components from two types of granules, those with a dense core (arrows, micrograph 1) usually separated from the granule membrane, and those with variable form and less dense contents (arrowhead, micrograph 2). The **dense core granules** contain products, including adenine nucleotides and serotonin, that are picked up from other cells and temporarily stored. ADP is an important platelet aggregator, and serotonin causes the contraction of smooth muscle in damaged vessels, minimizing blood loss.

The less dense granules consist of at least two types, **lysosomes** with acid hydrolases, and **α-granules** that contain molecules active in coagulation including platelet factor-4 and fibrinogen. Alpha granules also contain platelet-derived growth factor, which, due to its effect on vascular smooth muscle proliferation and migration and its ability to attract monocytes, is often considered a key element in vascular disease.

When platelets are activated, both the α- and dense core granules release their contents either directly onto the surface or into the **open canalicular system,** which forms snakelike channels that open onto the platelet surface. One region where the open canalicular system is continuous with the platelet surface is shown at the curved arrow in micrograph 2. A second system of channels, the **dense tubular system** (open arrow), is a separate organelle derived from megakaryocyte smooth ER that concentrates calcium and plays an important role in platelet activation.

Even though actin is the most abundant cytoskeletal element in platelets, the most prominent in most electron micrographs is the ring of **microtubules** near the cell membrane, seen in cross section in micrograph 2 (circles). These microtubules maintain the integrity of platelets even when they undergo the pronounced shape change from discoid to the spiky spheres characteristic of activation.

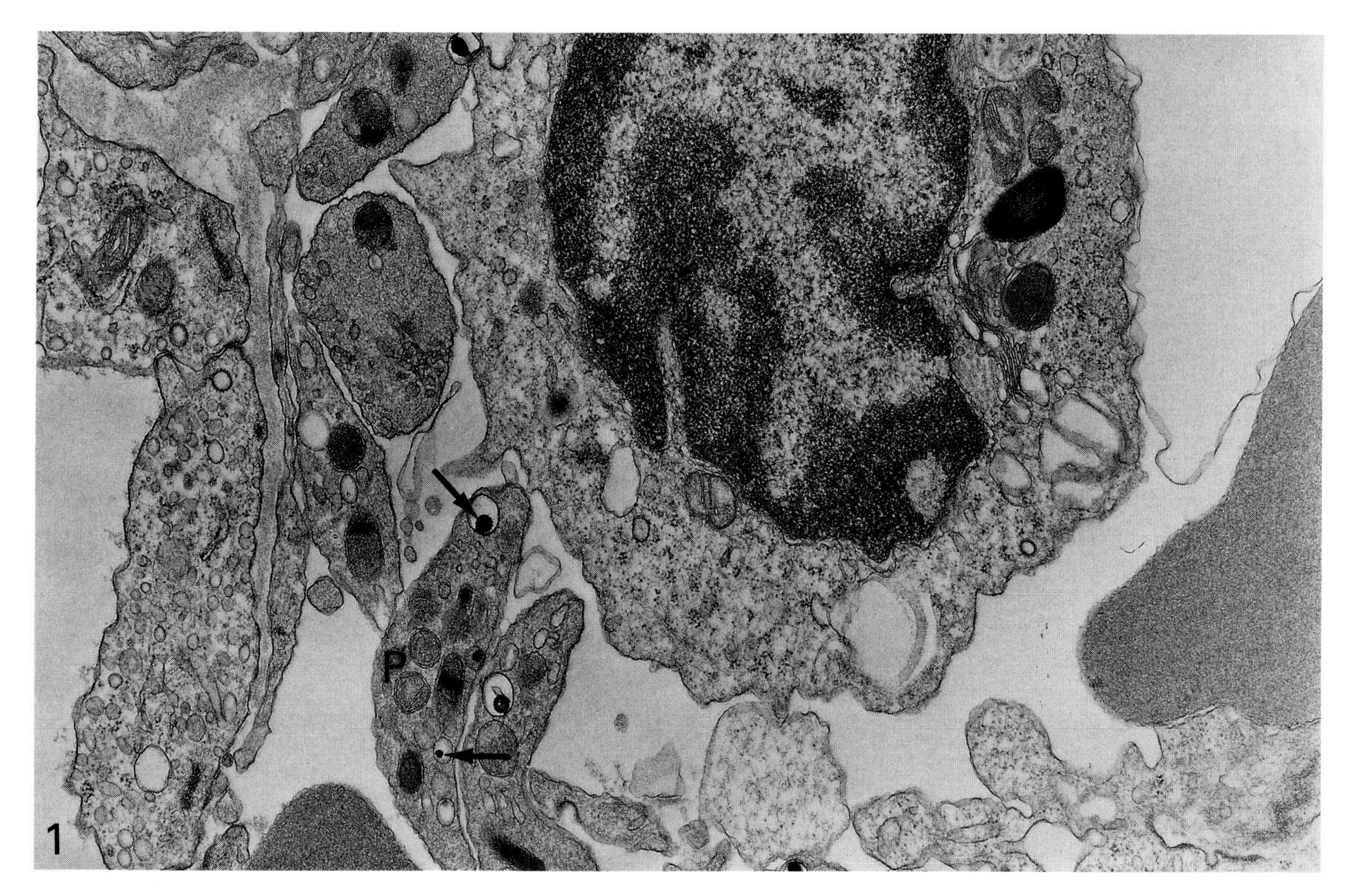

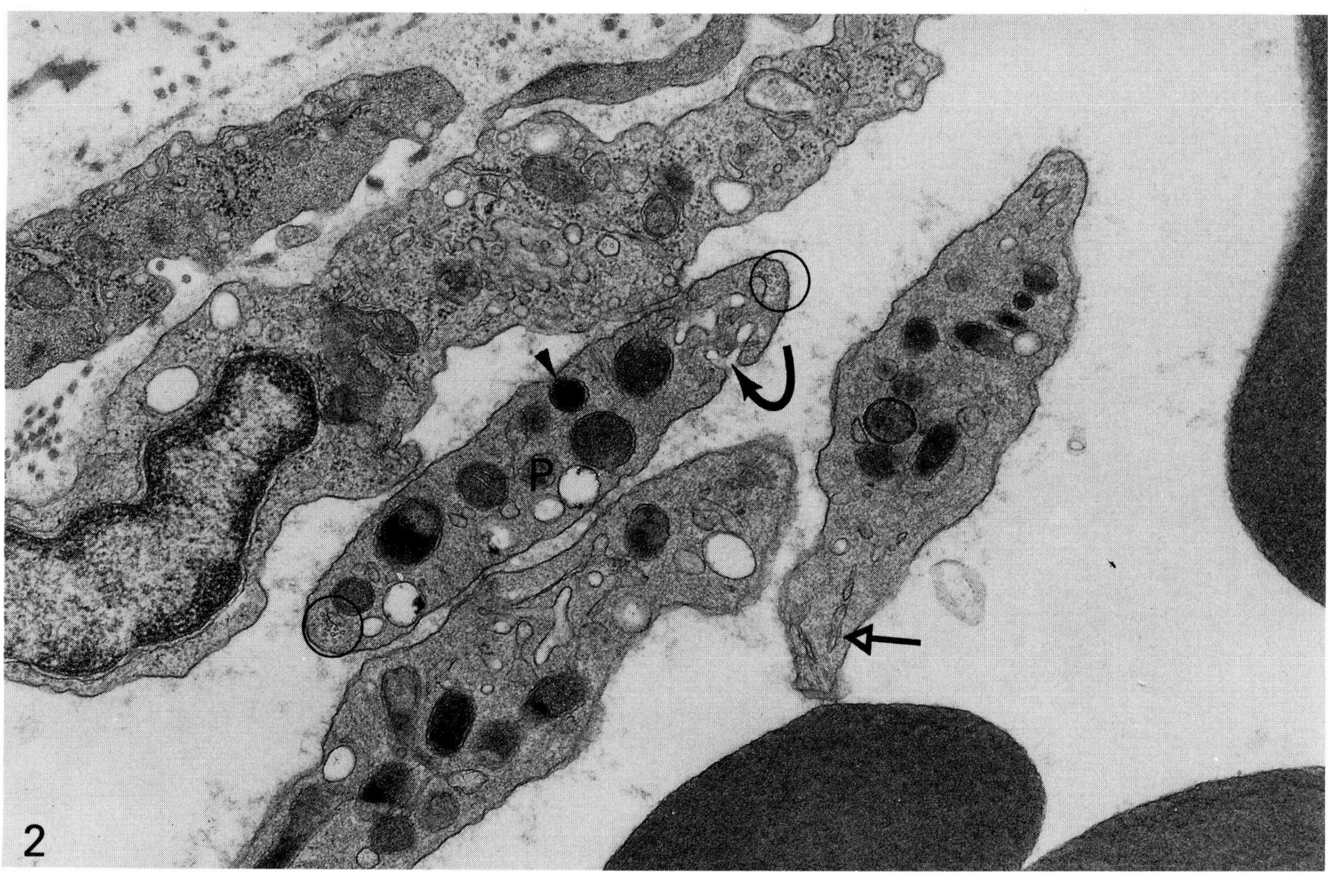

EM 1: 21,600× EM 2: 30,000×

Bone Marrow

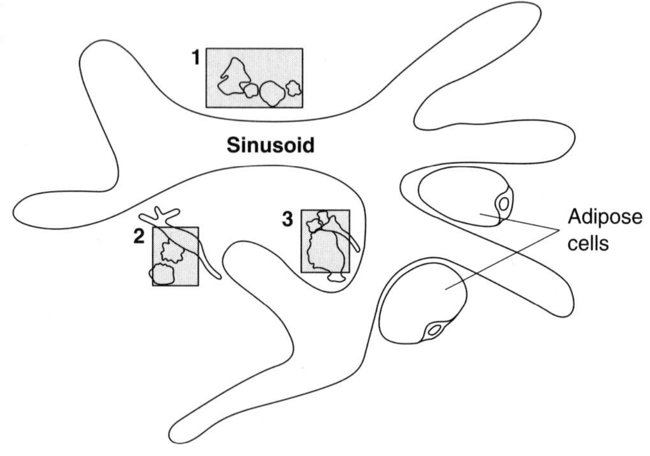

These low-magnification micrographs of bone marrow help to emphasize the diversity of cells in this organ. In micrograph 1, a striking comparison can be made between the small heterochromatic nucleus of the **orthochromatophilic erythroblast** (OE), the lobed nucleus of the mature **neutrophil** (N), and the small round nuclei of the **lymphocyte** (L) and **plasma cell** (P). The cytoplasm of these four cells is equally distinct. Each characteristic provides insight into the function of the cell, from the electron-dense hemoglobin in the RBC precursor (OE) to the bacteriocidal granules in the neutrophil to the thin rim of cytoplasm with surface extensions in the motile lymphocyte to the rough ER dilated with antibody in the plasma cell.

In micrograph 2, the **macrophage** (M), with its primary and secondary lysosomes and irregular euchromatic nucleus, stands out from the **polychromatophilic erythroblast** (PE) and the developing **granulocyte** (G). As shown in micrograph 3, macrophages (M), which in connective tissue are considered relatively large cells, are dwarfed in bone marrow by **megakaryocytes** (ME). With their small size and dense nuclei and cytoplasm, the polychromatophilic (PE) and orthochromatophilic (OE) erythroblasts in this micrograph provide even more contrast.

All of the cells identified in these micrographs originated from a single precursor cell during a rapid sequence of mitosis and differentiation. The severity of bone marrow disease relates to the far-reaching effects of a defect in a single stem cell. Acute myelocytic leukemia, one of the most severe bone marrow disorders, with a median survival time of two months, is probably a disease of the pluripotent myeloid stem cell (see Blood, page 164). This cell continues to divide, but differentiation is impaired and immature cells accumulate in bone marrow and peripheral blood. The reduction in the number of mature erythrocytes, granulocytes, and platelets results in anemia, infection, and bleeding disorders. In severe cases, some of the malignant precursor cells settle in blood vessels in the brain and form tumors, leading to cerebral hemorrhage.

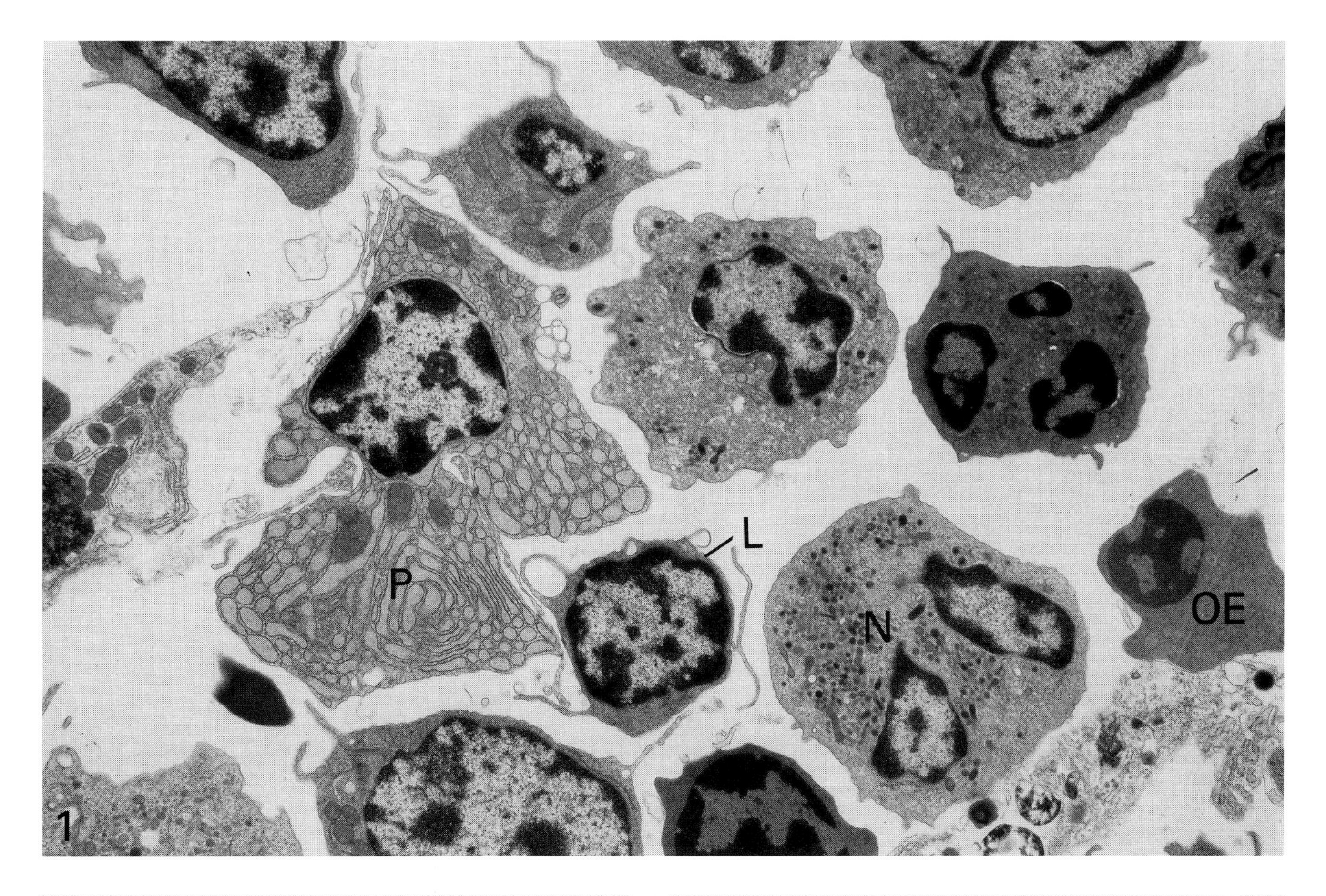

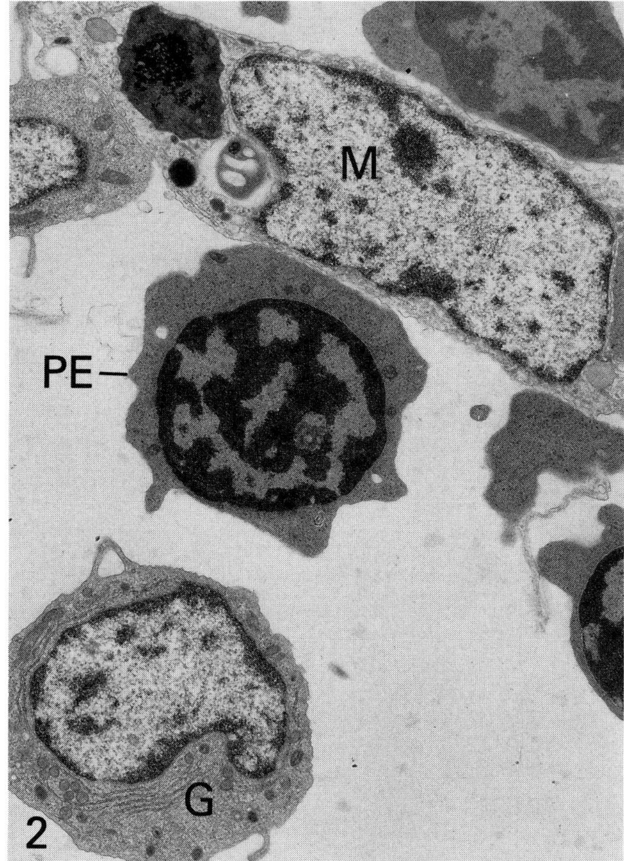

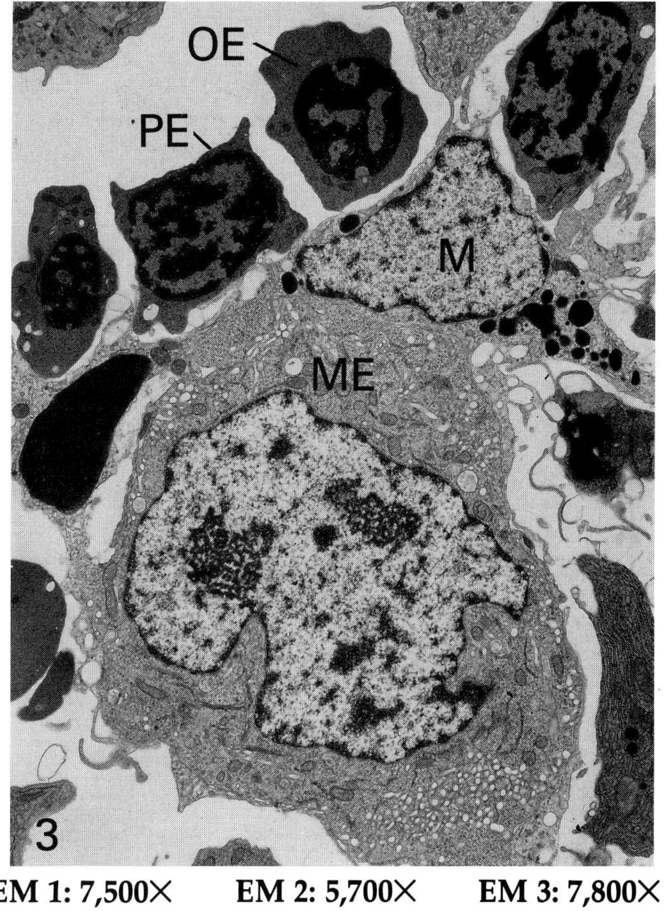

EM 1: 7,500× EM 2: 5,700× EM 3: 7,800×

IMMUNE SYSTEM

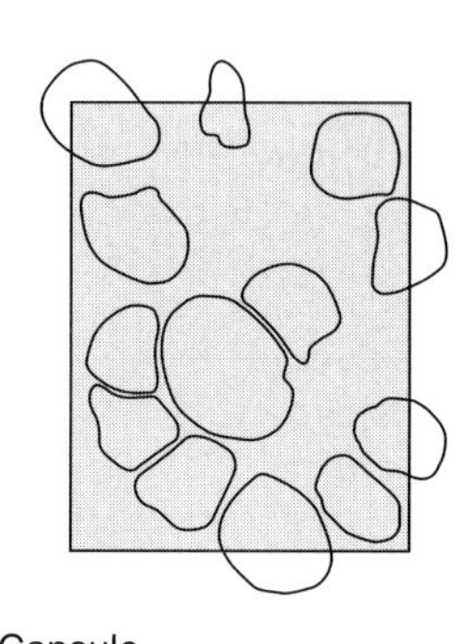

The thymic cortex is packed with T lymphocytes that are going through the **"thymic selection"** that renders them immunocompetent (able to respond to individual antigens). This selection begins in the outer part of the thymic cortex, where T lymphocytes undergo proliferation. In the micrograph, a dividing lymphocyte (c indicates chromosome) within the outer cortex is surrounded by many other developing T cells (T), known as thymocytes.

Thymocyte differentiation progresses as the cells move from the outer to inner cortex. With development, new surface proteins are expressed, including the **T cell receptor** (TCR) and **T cell coreceptors** (CD4, CD8).

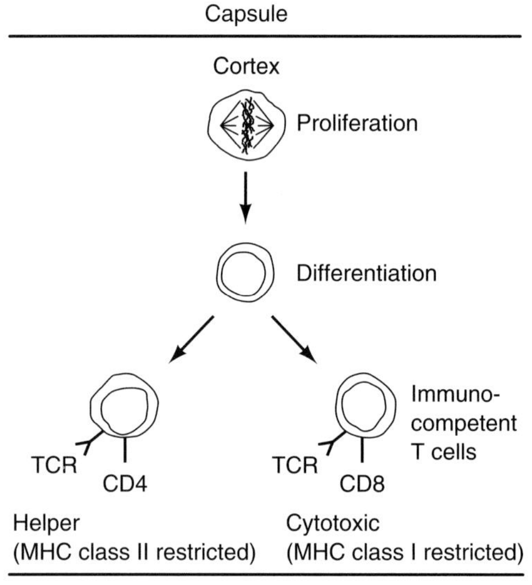

Unlike the B cell receptor, the T cell receptor does not recognize antigen alone; instead it recognizes only antigen bound by self **major histocompatibility complex** (MHC) molecules (i.e., T cells are **MHC restricted).** Most cells express MHC class I molecules; antigen-presenting cells also express MHC class II molecules. As T cells develop, they are programmed not only to respond to a single antigenic determinant, but also to respond to a foreign antigen only if it is in association with a particular class of MHC molecules, either I or II. Those T cells that function as **helpers** (T_H) are class II restricted (i.e., they respond only to antigen that is associated with class II MHC), and those that function as **cytotoxic** (T_C) cells are typically class I restricted (i.e., they respond only to antigen that is associated with class I MHC). The class restriction (I versus II) during thymocyte differentiation is a function of T cell coreceptors (generally CD4 for helpers, and CD8 for cytotoxic cells). If these proteins are blocked by monoclonal antibodies, T cell function is lost. Therefore, the response of T cells to antigen depends on both the T cell receptor and the coreceptors, CD4 and CD8, that define the T cell subsets.

During the process of thymic selection a population of T cells is formed that is capable, overall, of responding to almost any foreign antigen an individual may encounter. This process of T cell programming takes place in the absence of exposure to foreign antigen. These unexposed T cells leave the thymus and, along with B cells from the bone marrow, populate **secondary lymphoid regions** (e.g., lymph node, spleen, Peyer's patches). In contrast to the **primary lymphoid organs** (thymus and bone marrow), the secondary organs are specialized to facilitate lymphocyte exposure to foreign antigens.

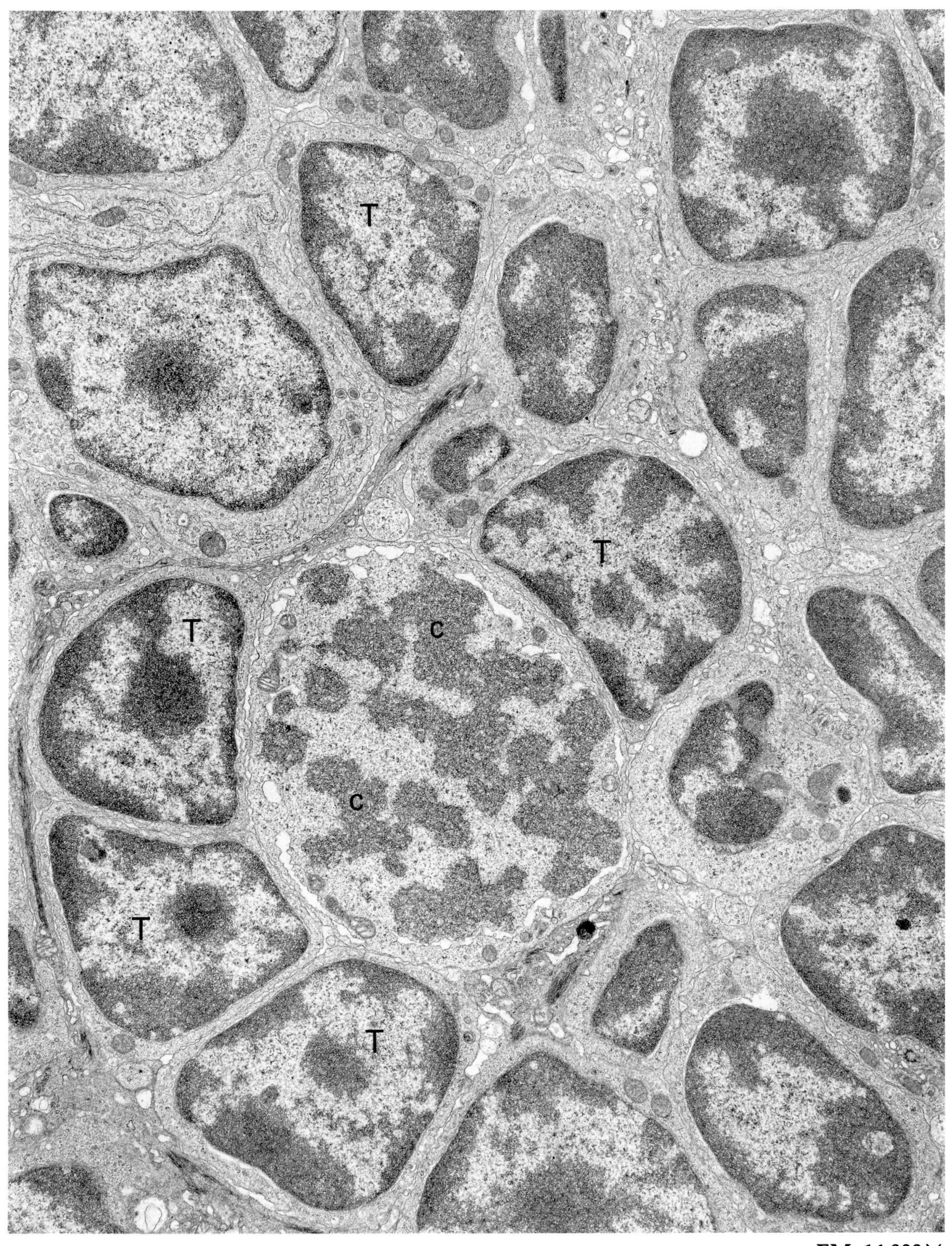

EM: 14,000×

THYMIC CORTEX: Support Cells

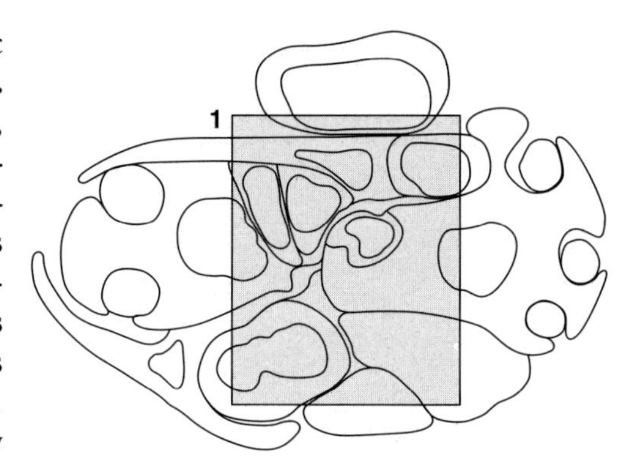

T lymphocytes (T, micrograph 1) develop in the thymic cortex in close association with **nonlymphoid stromal cells.** One of the most prominent, the **epithelial reticular cell** (E, micrograph 1), extends fine cellular processes (arrows, micrograph 1) that form a **cytoreticulum** that supports differentiating lymphocytes. The epithelial nature of stromal cells is evident by dense accumulations of tonofilaments (arrowheads, micrograph 1 and inset), that insert into **desmosomes** (circles, micrograph 1 and inset). Like the central nervous system, the thymic cortex does not contain connective tissue. Instead, the stromal cytoreticulum provides the necessary structural framework.

The development of immunocompetence by T cells in the thymus is extremely complex, since there is both a positive selection for the recognition of self MHC (T cells attack only foreign antigens in association with self MHC) and a negative selection against high-affinity reaction to self MHC (to prevent T cells from attacking self tissues as occurs in autoimmune diseases). Stromal cells have been implicated in the control and regulation of this T cell programming. Subsets of these cells express MHC proteins on their surface that seem to interact with T cell receptors. One model attributes thymic epithelial cells with exerting a positive selection and a special type of negative selection known as clonal anergy that functionally inactivates thymocytes. The influence of stromal cells occurs in a sequential manner as T cells differentiate. Evidence suggests that cortical epithelial cells affect positive selection on earlier T cell stages and medullary epithelial cells and dendritic cells affect negative selection on later T cell stages.

Nurse stromal cells (N, micrograph 1), situated close to the thymic capsule, appear to specifically bind to an early subset of differentiating thymocytes. Typically, thymocytes occupy deep recesses within the cytoplasm and remain associated with their nurse cells during experimental procedures used to isolate thymic cells.

Thymic stromal cells secrete cytokines, which not only act locally in the thymus to regulate thymocyte differentiation but also act at distal sites. Thymic secretions act in peripheral lymphoid regions to stimulate the growth and differentiation of B and T lymphocytes, and they also have far-reaching effects on entire hormonal systems (e.g., gonadal–adrenal–pituitary axis).

Immune disorders that occur with aging, including an increased incidence of malignancies and autoimmune reactions, have been associated with an observed reduction in circulating thymic cytokines.Studies suggest that synthetic thymic hormones may help to reverse certain age-related immunodeficiencies.

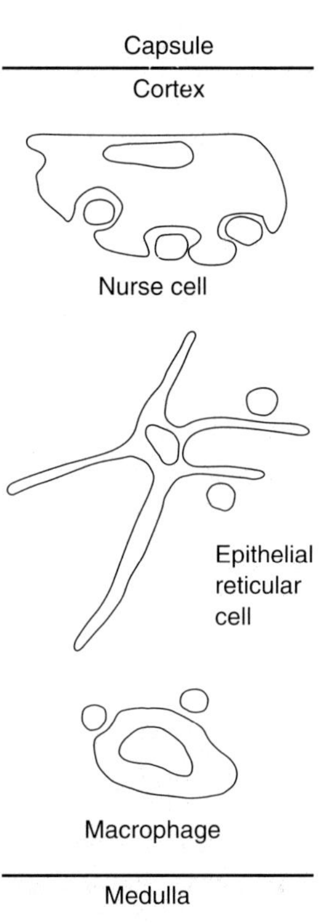

Modified from L. E. Hood et al. *Immunology,* 2nd ed. Benjamin Cummings, Redwood City, CA, 1984.

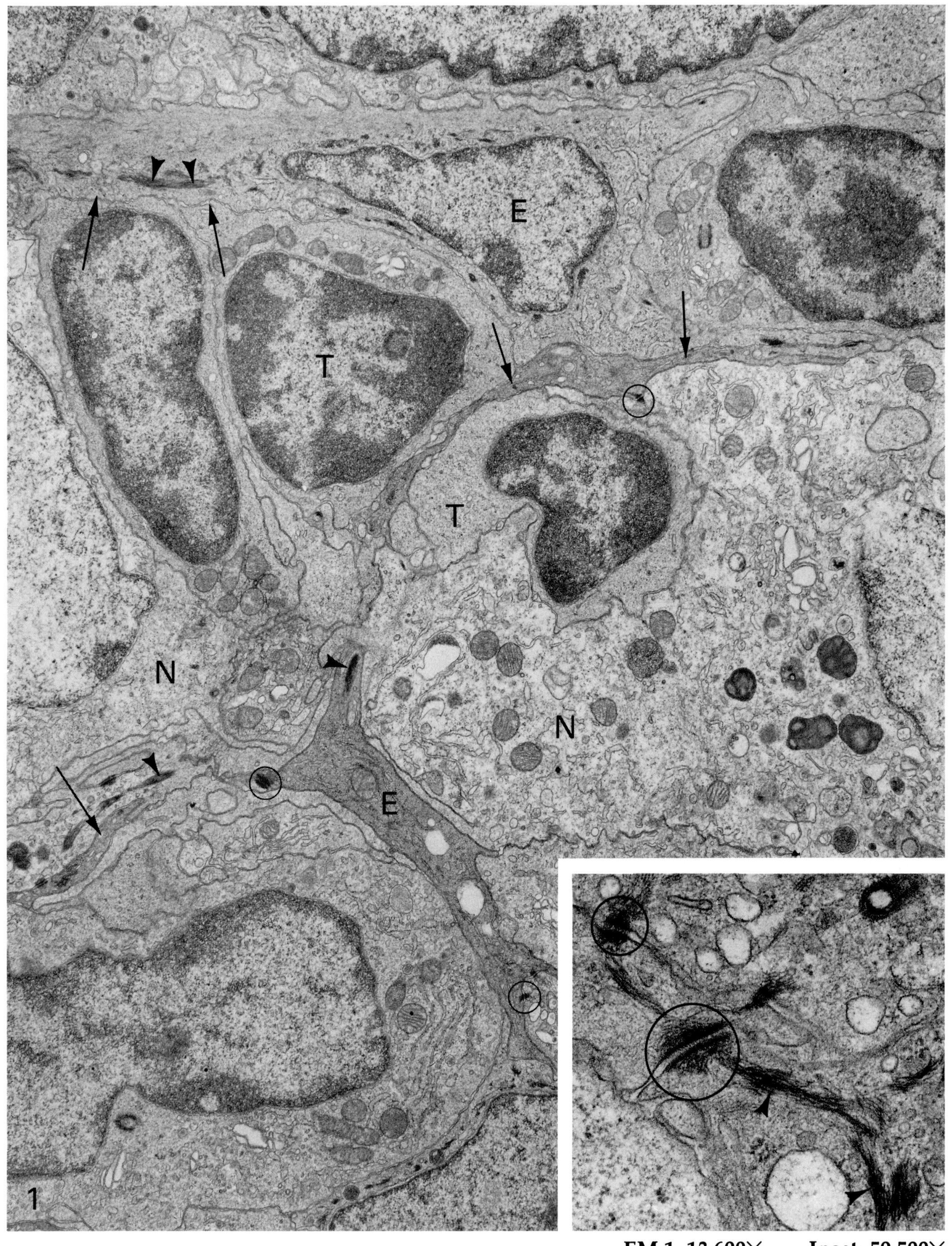

EM 1: 13,600× Inset: 59,500×

THYMIC CORTEX: Macrophage

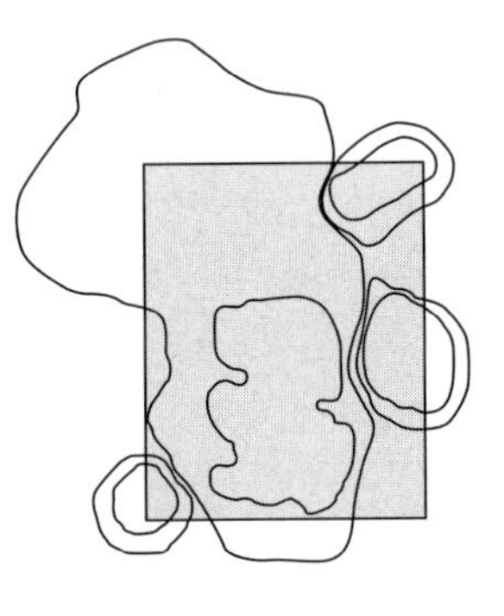

Macrophages (M, micrograph) are bone marrow-derived stromal cells that are relatively common in the thymus, particularly near the corticomedullary junction. These cells are frequently seen in the process of digesting lymphocytes (L, micrograph) in secondary lysosomes.

Very few of the lymphocytes that seed the thymus from the bone marrow actually complete development, selected as mature immunocompetent T cells. Ninety-nine percent die and are phagocytosed by macrophages. Evidence suggests that certain lymphocytes that die within the thymus would have reacted with a high affinity to self MHC and thus attacked and destroyed self tissues. **Tolerance to self,** therefore, appears to be largely a result of the destruction of T cells that would attack self (clonal deletion). However, certain populations of potentially self-destructive T cells survive and leave the thymus to populate the peripheral tissues. These T cells are somehow inactivated (clonal anergy) and therefore normally do not attack self.

The involution of the thymus at puberty is, in part, a result of an increased release of both gonadal and adrenal steroids. These hormones decrease thymocyte proliferation and cause thymocyte death. Certain adrenal steroids, glucocorticoids, specifically inhibit the synthesis of interleukin 2 by T cells. Interleukin 2 is needed for the growth of T cell populations, both in the thymus and peripheral lymphoid regions. The inhibitory effect of glucocorticoids is the basis for the well-known clinical use of these hormones as immunosuppressants (e.g., during organ transplantation).

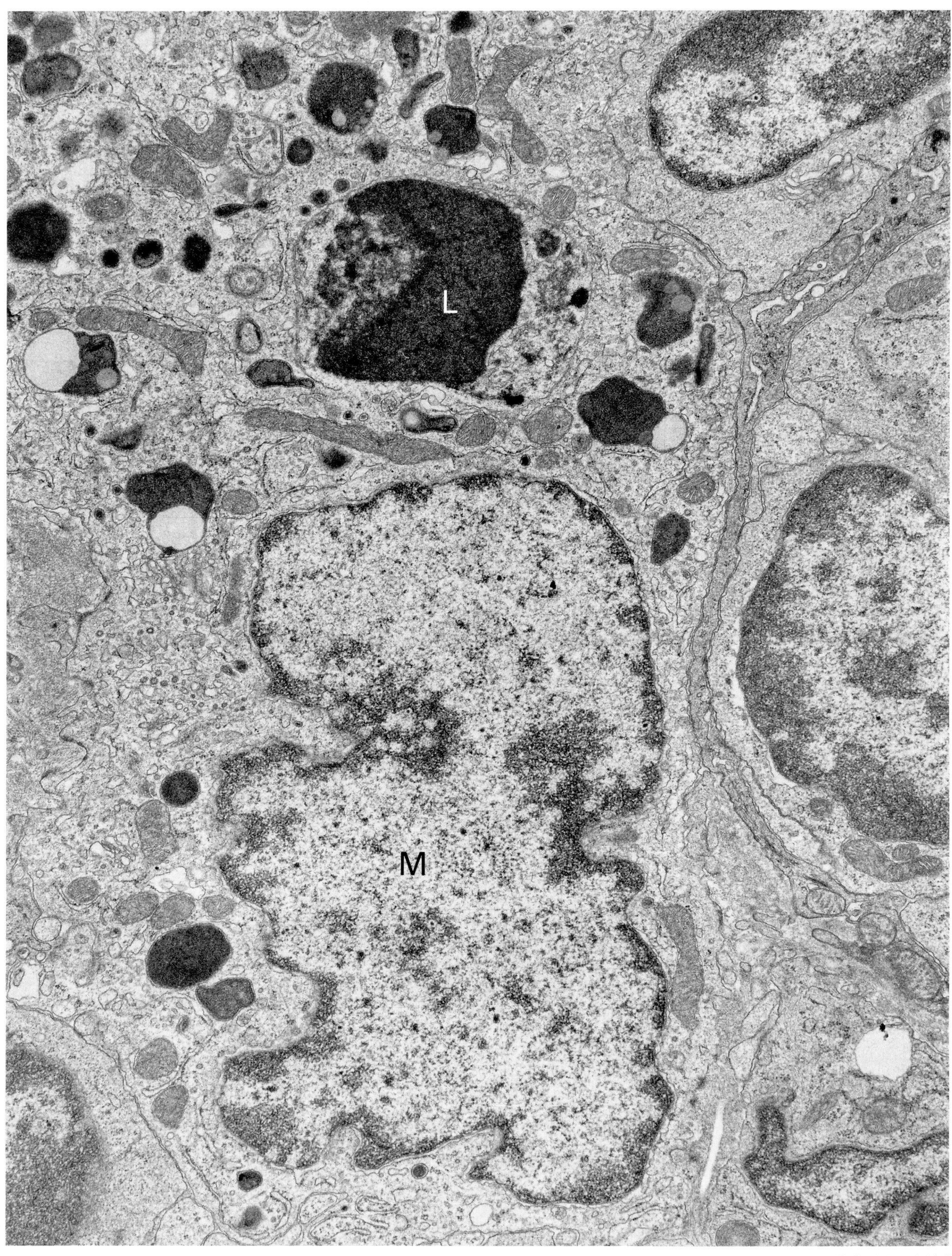

EM: 20,500×

LYMPHOCYTE ACTIVATION

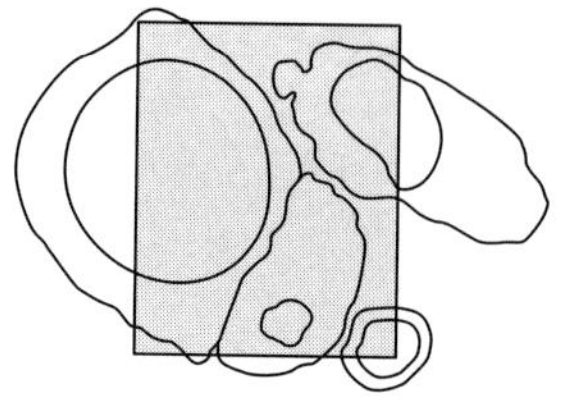

Lymphocytes are activated by antigen in **germinal centers** within secondary lymphoid regions. The basic immune response is similar, whether it is directed against foreign antigen entering via lymph (lymph node), blood (white pulp of spleen), or the lumen of the GI tract (tonsils and Peyer's patches). The micrograph, taken from a germinal center in a lymph node, illustrates the first major event in the activation of lymphocytes by antigen, the transformation of small lymphocytes (L, micrograph) into large **lymphoblasts** (LB, micrograph).

One pathway of **B lymphocyte activation** depends upon (1) antigen binding to B cell surface receptor (antibody) and (2) the activation of T lymphocyte helpers (T_H). T_H activation depends upon antigen presentation by **antigen-presenting cells,** or APCs (e.g., macrophages, endothelial cells, B lymphocytes). APCs bind antigen, and endocytose, digest, and process it. A resulting fragment of antigen, together with self surface protein (MHC class II), is presented to T_H cells. Helper T cells are activated and develop into lymphoblasts when their receptors bind the APC antigen–MHC II complex. T cell lymphoblasts release several proteins, lymphokines, that are essential to the growth and differentiation of B cells.

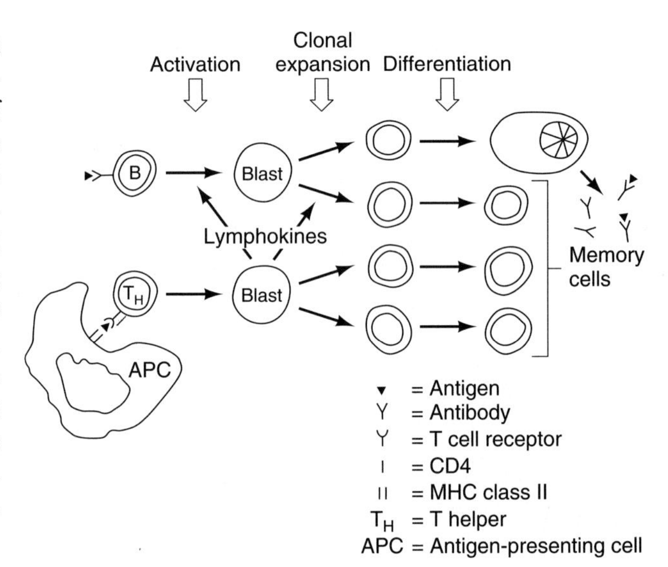

Lymphoblasts are characterized by large euchromatic nuclei and the predominance of free polysomes (arrowheads, micrograph) in the cytoplasm. Much of the gene activity during this stage is directed toward the synthesis of proteins used in mitosis during the clonal expansion of the activated lymphocyte. Following mitosis, B cells differentiate into (1) plasma cells (P, micrograph), the effector cell of B cell activation, and (2) memory cells, small lymphocytes that circulate, ready to respond to a second encounter with the same antigen. T_H memory cells are also produced following activation.

The B cell antigen receptor is an antibody with a hydrophobic tail inserted into the cell membrane. With the differentiation to plasma cells, mRNA coding for this antibody is spliced, removing the part coding for the hydrophobic region. As a result, antibody synthesized by plasma cells is no longer anchored in the membrane and instead is secreted.

As a part of the activation of B cells within germinal centers, a hypermutation may occur followed by a selection of B lymphocytes with the highest affinity for the activating antigen. Mesenchyme-derived follicular dendritic cells, the major stromal cell within germinal centers, concentrate antigen on their surface, interact intimately with activated B cells, and may be important in the selection that takes place following hypermutation.

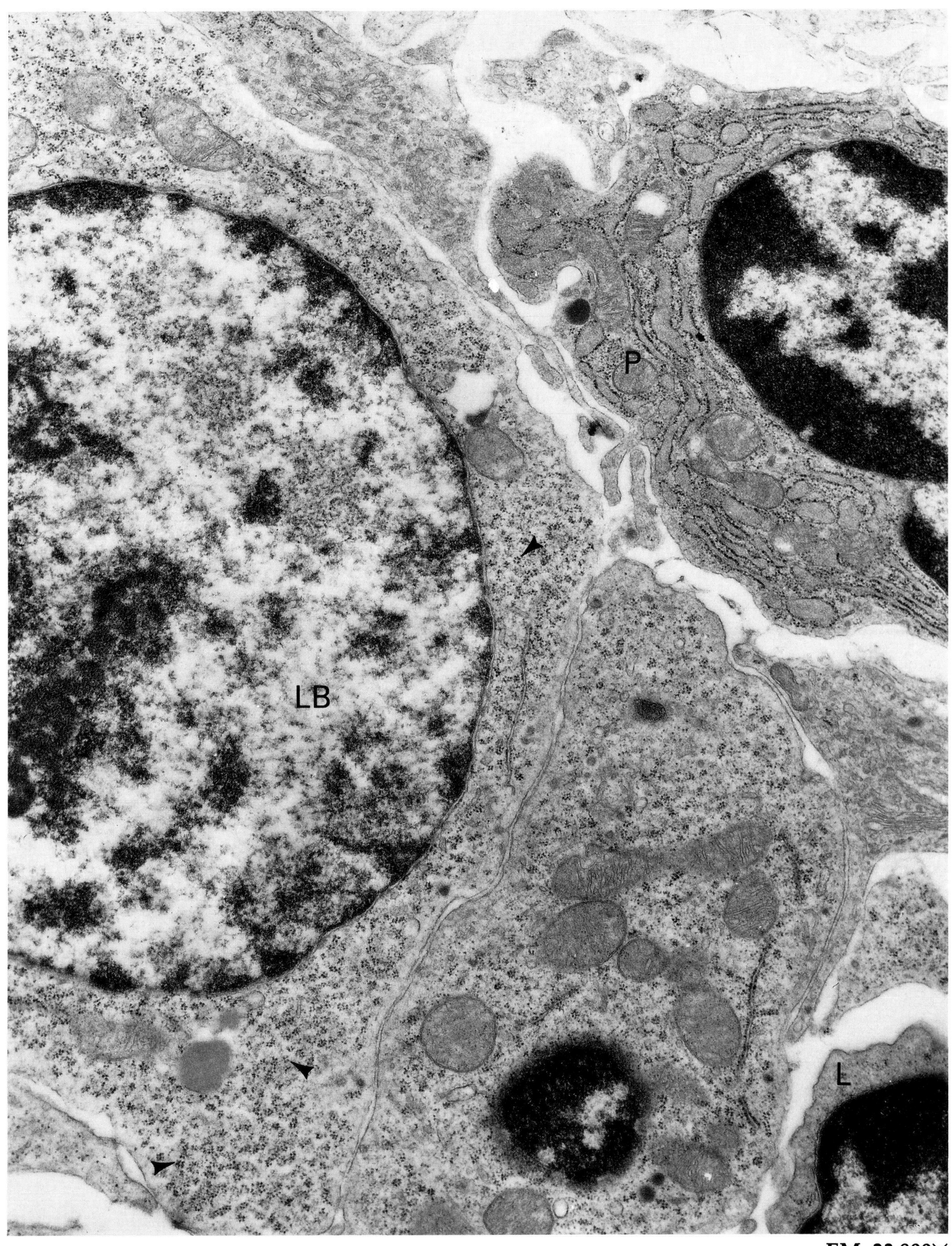

EM: 23,800×

PLASMA CELLS

Plasma cells differentiate from B lymphocytes following activation by antigen. Each plasma cell synthesizes and secretes **antibodies** that bind specifically to the antigen that initially activated the precursor B lymphocyte. Antigen/antibody binding is a major means of immune defense. Antibodies, synthesized within the rough ER (arrows, micrograph), are processed and packaged within the Golgi (G, micrograph) prior to secretion. In plasma cells, the heterochromatic nucleus does not reflect inactivity since the small part of the genome that is euchromatic is exceedingly active in maintaining the synthesis of many copies of a single antibody.

Antibodies are composed of two light (22 kd) and two heavy (55 kd) chains, each with a region that contains variable amino acid sequences and a region that contains constant amino acid sequences. The **variable regions** are the antigen-binding sites (Fab). The **constant regions** (Fc) define the functions of the individual classes and are identical within each antibody class.

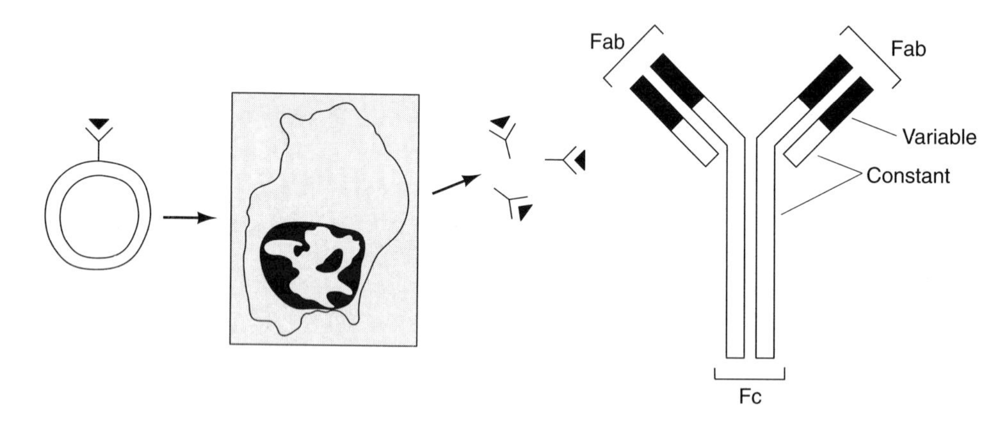

Five antibody classes have been identified.

IgG Major antibody of blood and lymph; complement activation; crosses placenta

IgM First antibody formed; complement activation; B cell receptor

IgA Major antibody of secretions

IgD Membrane antibody only

IgE Major antibody of allergic responses; Fc portion binds to mast cells and basophils

IgG is the major class of antibody. The Fab regions bind to and coat (opsonize) certain antigenic invaders (e.g. viruses, bacterial toxins) and prevent them from attaching to target cell receptors. The exposed Fc tails bind to (1) phagocytes, which stimulates phagocytosis, and (2) certain plasma proteins (complement), which initiates a cascade of complement interactions, eventually lysing foreign cells.

One of the actions of T cell lymphokines is to determine the class of antibody synthesized; interleukin 4 is associated with secretion of IgE and a certain type of IgG, and interleukin 5 with IgA.

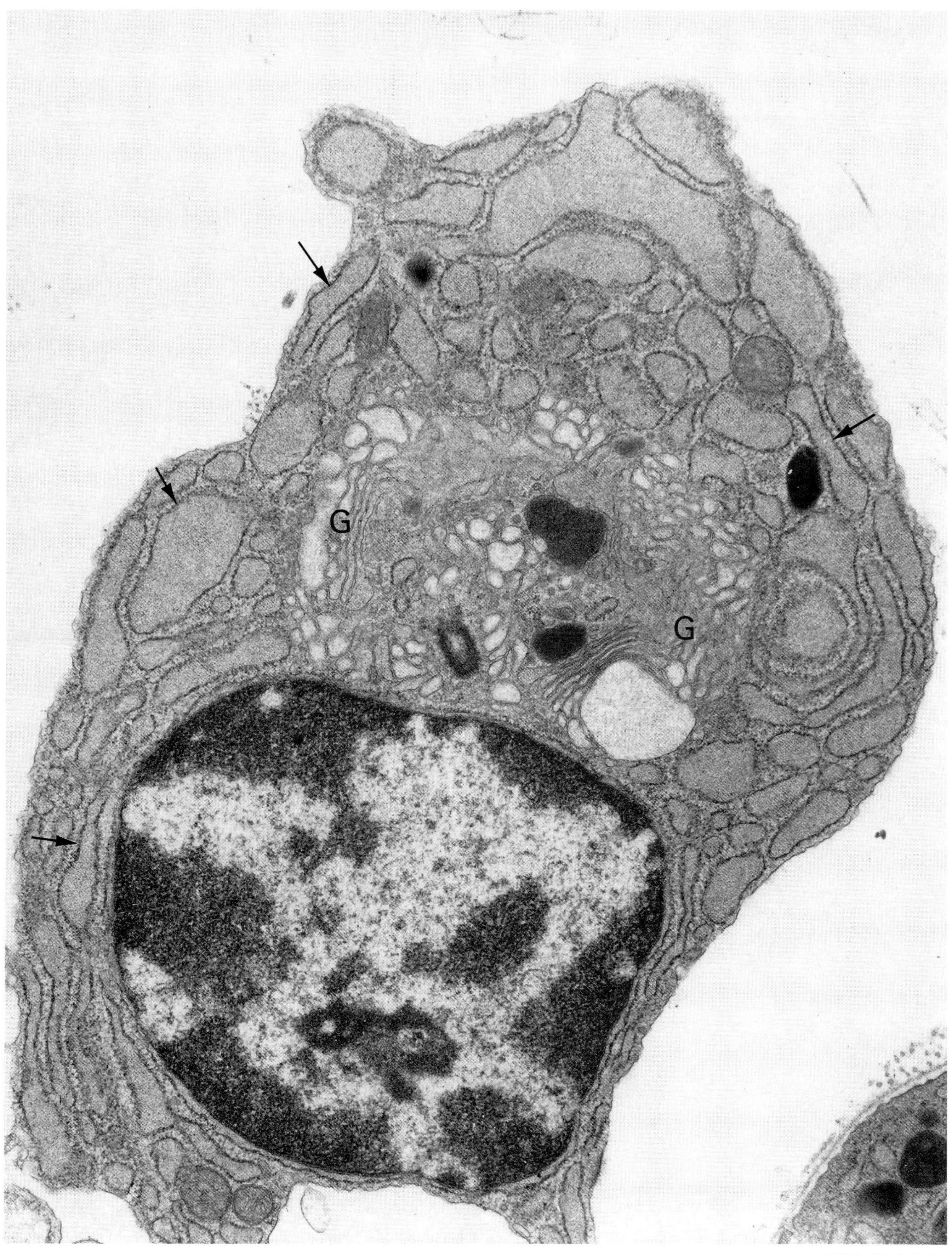

EM: 31,500×

HIGH ENDOTHELIAL VENULES

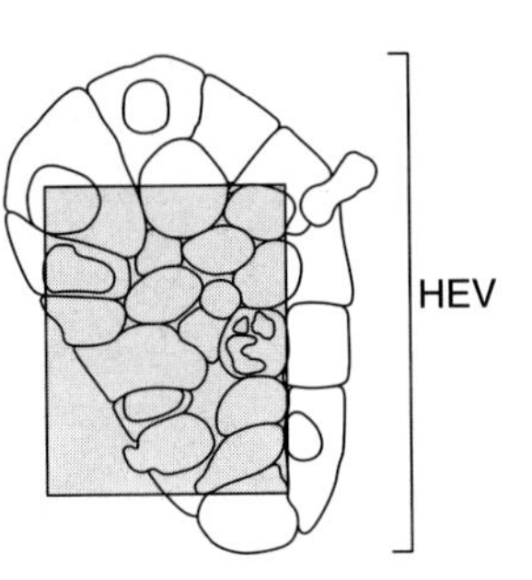

Lymphocytes move from the blood vascular system into lymphoid tissues across specialized venules lined with endothelial cells that express a unique protein on their surface. These specialized venules are frequently lined by cuboidal epithelium, as shown in the micrograph of a **high endothelial venule** (HEV) from a lymph node.

The lumen of the HEV in the micrograph is crowded with blood cells, including a red blood cell (R), lymphocytes (L), and a neutrophil (N). Two lymphocytes (L*) are seen migrating between the HEV endothelial cells (E) on their way to the tissue. One (arrowhead) has penetrated the HEV wall and is protruding into lymphoid tissue. If, in one to two days, lymphocytes do not encounter the antigen they are programmed to recognize, they enter the efferent lymphatic and circulate to another lymph node. This phenomenon of **lymphocyte recirculation** provides a type of surveillance essential to immune protection from foreign antigens entering the body via different routes. Each lymphocyte, programmed to respond to a specific antigen by expression of a unique receptor on its cell surface, travels to many different locations, thus increasing its chance of encountering its specific antigen.

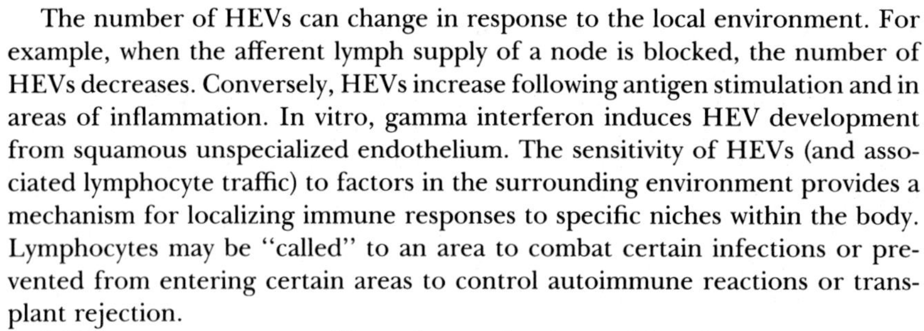
The number of HEVs can change in response to the local environment. For example, when the afferent lymph supply of a node is blocked, the number of HEVs decreases. Conversely, HEVs increase following antigen stimulation and in areas of inflammation. In vitro, gamma interferon induces HEV development from squamous unspecialized endothelium. The sensitivity of HEVs (and associated lymphocyte traffic) to factors in the surrounding environment provides a mechanism for localizing immune responses to specific niches within the body. Lymphocytes may be "called" to an area to combat certain infections or prevented from entering certain areas to control autoimmune reactions or transplant rejection.

HEVs express organ-specific surface molecules, and lymphocytes apparently use these to discriminate between different HEVs. This discrimination involves an interaction between unique surface molecules on (1) subsets of lymphocytes **(homing receptors)** and (2) organ-specific HEVs **(addressee ligands).** When lymphocytes and HEVs match, lymphocytes are transported across the venule wall into the tissue. This specific interaction determines the path of recirculation of subsets of lymphocytes. One such path taken primarily by B cells is via mucosal sites; another, taken primarily by T cells, is through peripheral lymph nodes. Some recirculation paths are extremely specific, such as the path taken by certain B lymphocytes that circulate only to the spleen. This subset of B lymphocytes is unique in not requiring T cell help for activation.

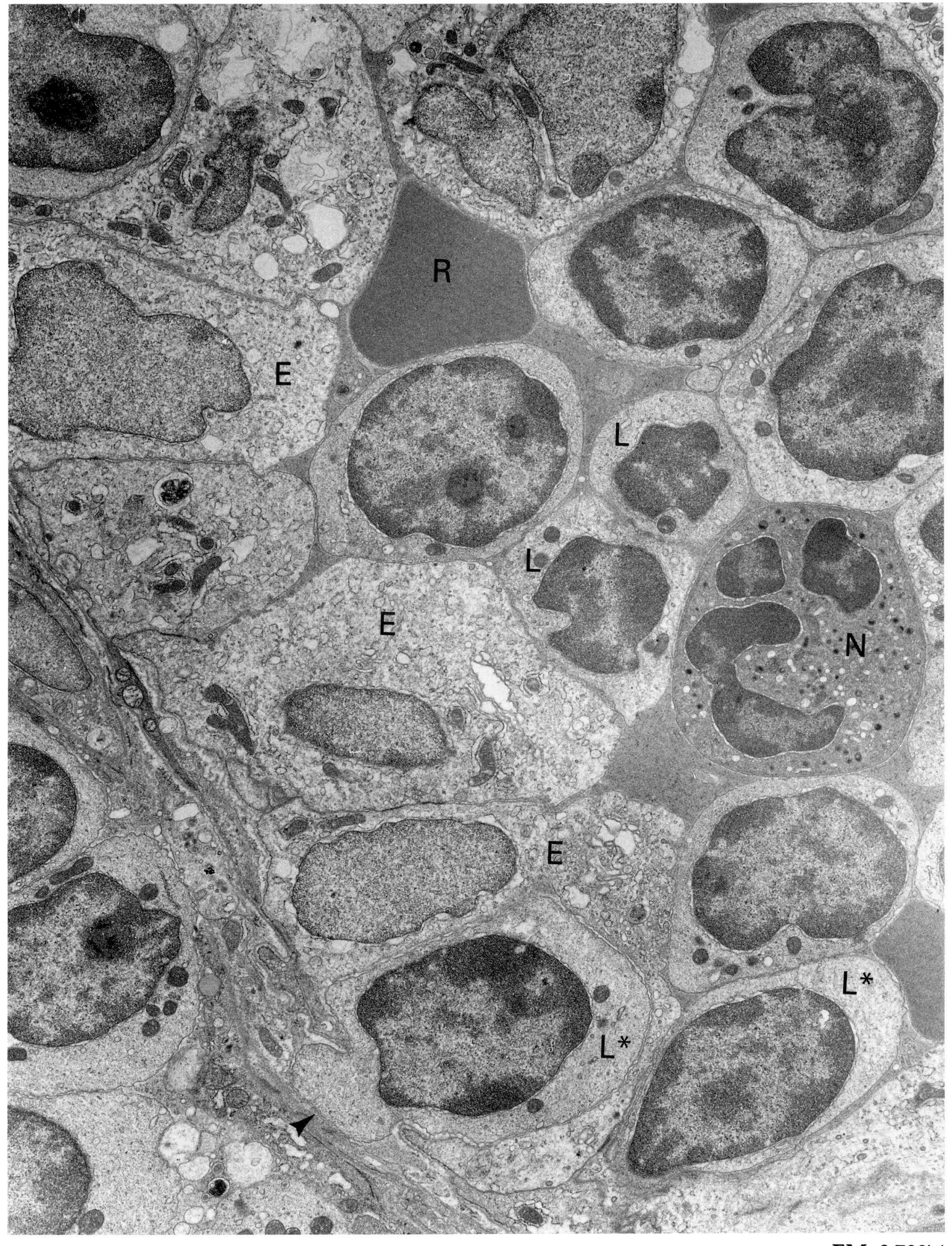

EM: 8,700×

Cytotoxic T Cells

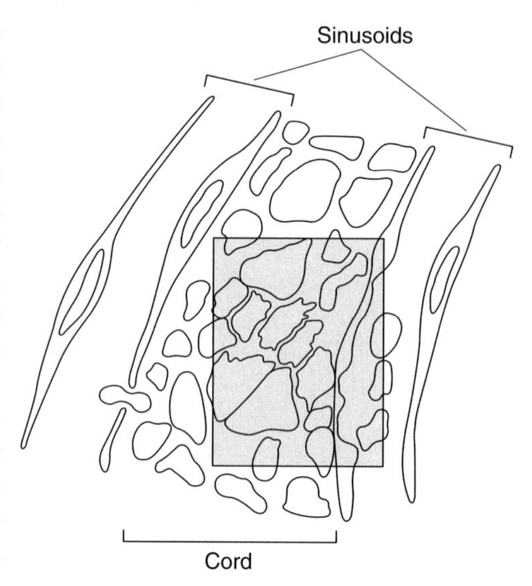

Secondary lymphoid organs (lymph nodes and spleen) are generally organized into regions in which (1) lymphocyte activation occurs and differentiation begins (germinal centers of follicles) and (2) the products of activation reside. In the spleen (micrograph) activation occurs in the white pulp, but the products of activation are predominantly found in the red pulp, which consists of cords (C) and sinusoids (S). The effector product of B cell activation, the plasma cell (P, micrograph), remains in the red pulp cords and releases antibodies into the blood sinusoids. In lymph nodes, plasma cells sit in the medullary cords and release antibodies into the lymph sinuses.

The principal effector cell resulting from T cell activation is the **cytotoxic T cell,** responsible for much of the direct cell-mediated immune defense. In routine electron micrographs cytotoxic T cells cannot be distinguished from other types of small lymphocytes. The population of small lymphocytes (L) sitting in the spleen red pulp cords in the micrograph probably includes cytotoxic cells as well as B and T memory and virgin cells. Many of these small lymphocytes will enter the sinusoid and recirculate to other lymphoid tissues.

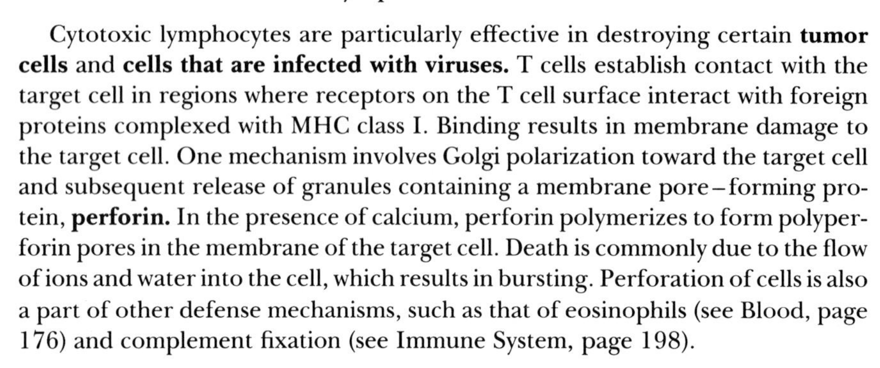

Cytotoxic lymphocytes are particularly effective in destroying certain **tumor cells** and **cells that are infected with viruses.** T cells establish contact with the target cell in regions where receptors on the T cell surface interact with foreign proteins complexed with MHC class I. Binding results in membrane damage to the target cell. One mechanism involves Golgi polarization toward the target cell and subsequent release of granules containing a membrane pore–forming protein, **perforin.** In the presence of calcium, perforin polymerizes to form polyperforin pores in the membrane of the target cell. Death is commonly due to the flow of ions and water into the cell, which results in bursting. Perforation of cells is also a part of other defense mechanisms, such as that of eosinophils (see Blood, page 176) and complement fixation (see Immune System, page 198).

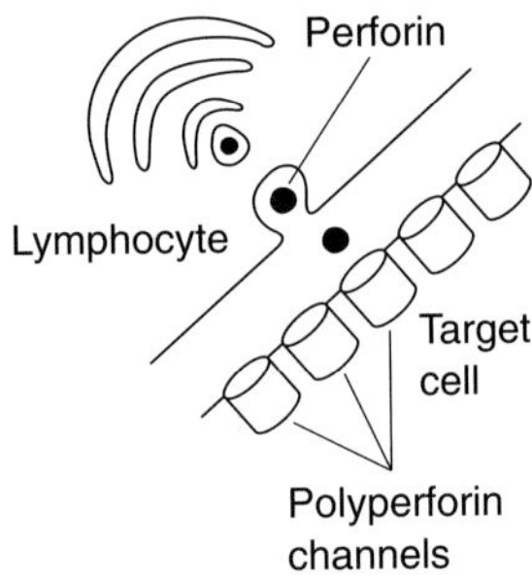

Macrophages (M, micrograph) are prevalent in the red pulp cords, where they phagocytose old RBCs (R, micrograph) as a part of the spleen's filtering function (see Immune System, page 206).

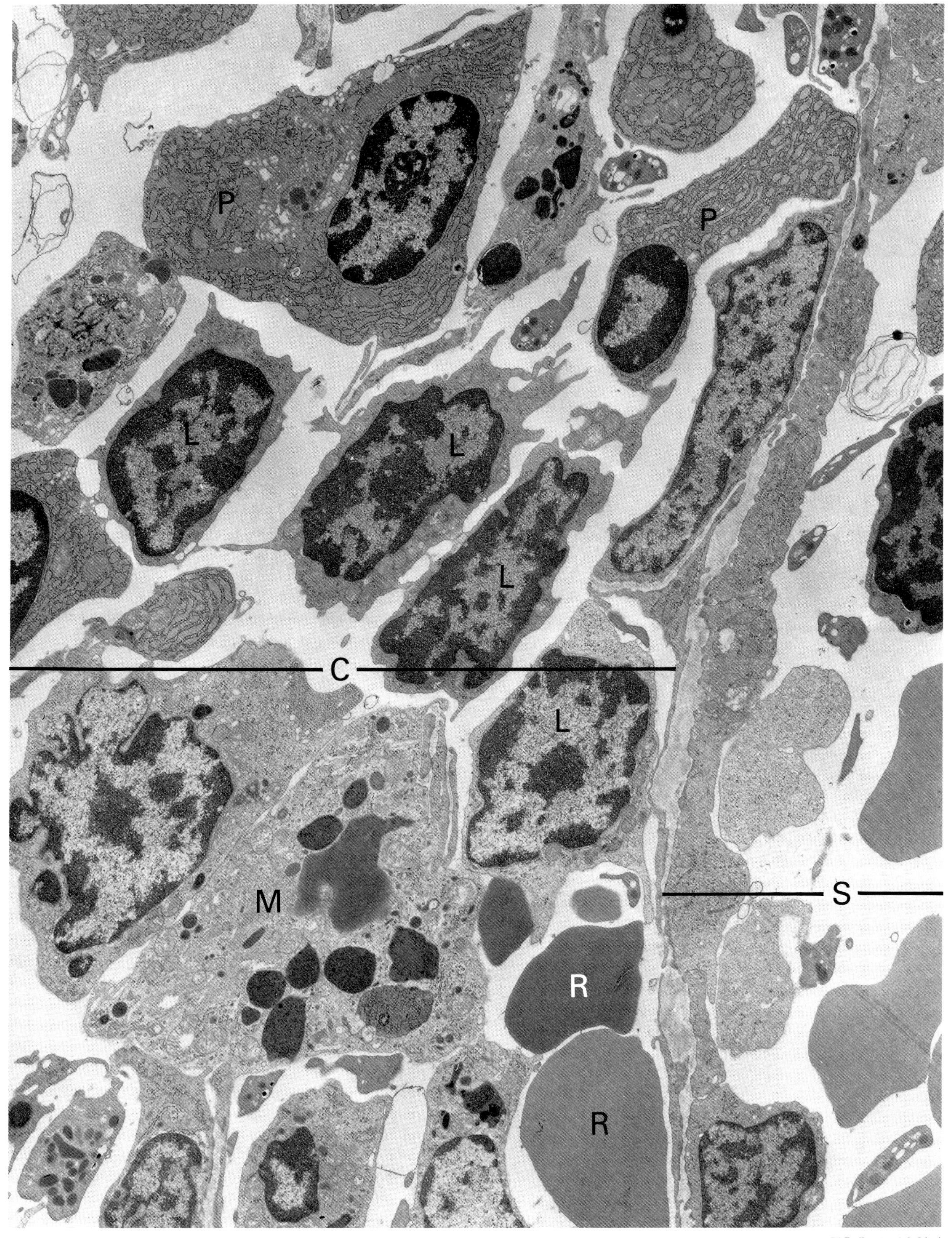

EM: 9,600×

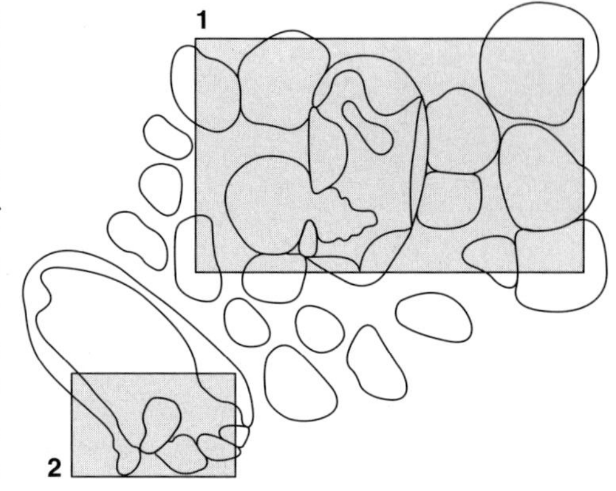

In addition to its immune function, the human spleen also functions in the adult as an important site of (1) blood cell storage, (2) red blood cell (RBC) and platelet destruction, and (3) reticulocyte maturation. These functions occur primarily in the red pulp cords (bracket, micrograph 1) and depend upon a unique and complicated circulatory system. Ninety percent of the blood entering the red pulp travels relatively directly into sinusoids (s, micrographs 1 and 2). Passage through this so-called **"closed" system** occurs in seconds, a rate equivalent to other organs. The remaining 10% passes through the cords before entering the sinusoids in the so-called **"open" system.**

Blood cells released into the cords migrate between a network of reticular fibers and the cellular processes of fibroblasts and macrophages. As they reenter the sinusoids, blood cells must squeeze between endothelial cells. This task is routine for white blood cells, such as the eosinophil (E) moving into the sinusoid in micrograph 1, and for healthy young RBCs such as the one entering the sinusoid (s) in micrograph 2. However, for aged and diseased RBCs that have lost their flexibility, this movement is not possible. Once trapped within the cords, these cells are phagocytosed by macrophages.

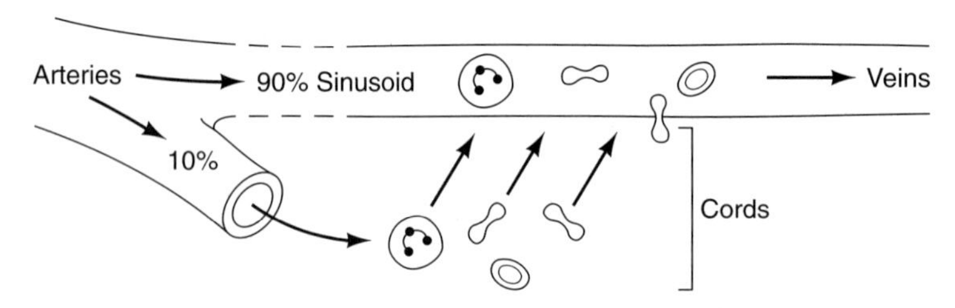

In micrograph 1 mature blood elements, including a neutrophil (N), eosinophil (E), and platelets (P), reside in the cords adjacent to many immature red blood cells such as polychromatophilic (PC) and orthochromatophilic (O) stages. These early blood cell stages are not normally seen in adult human spleens; however the spleen is a major hematopoietic organ throughout the life of many mammals (including mice, as in micrograph 1). In humans, the spleen functions in hematopoiesis only during fetal development and in the adult during disease states that interfere with normal production in bone marrow. In adults, blood cell formation outside of bone marrow is referred to as extramedullary hematopoiesis.

Stress fibers (encircled electron-dense region, micrograph 2) are commonly found within sinusoid endothelial cells near junctional regions. It has been suggested that these fibers not only stabilize the sinusoidal lining but also may shorten and play a role in the control of blood cell movement across the wall. RBCs apparently move across in bursts, possibly corresponding to a time when stress fiber changes open interendothelial slits.

Blood cells within red pulp cords represent a "pool" that is in dynamic equilibrium with cells in the general circulation. Only 3% of RBCs but as many as 30% of platelets are pooled in the spleen. Enlarged spleens can hold up to 72% of the total platelet mass, thus depleting the general circulation. This results in thrombocytopenia, with impaired vessel repair and consequent bleeding disorders.

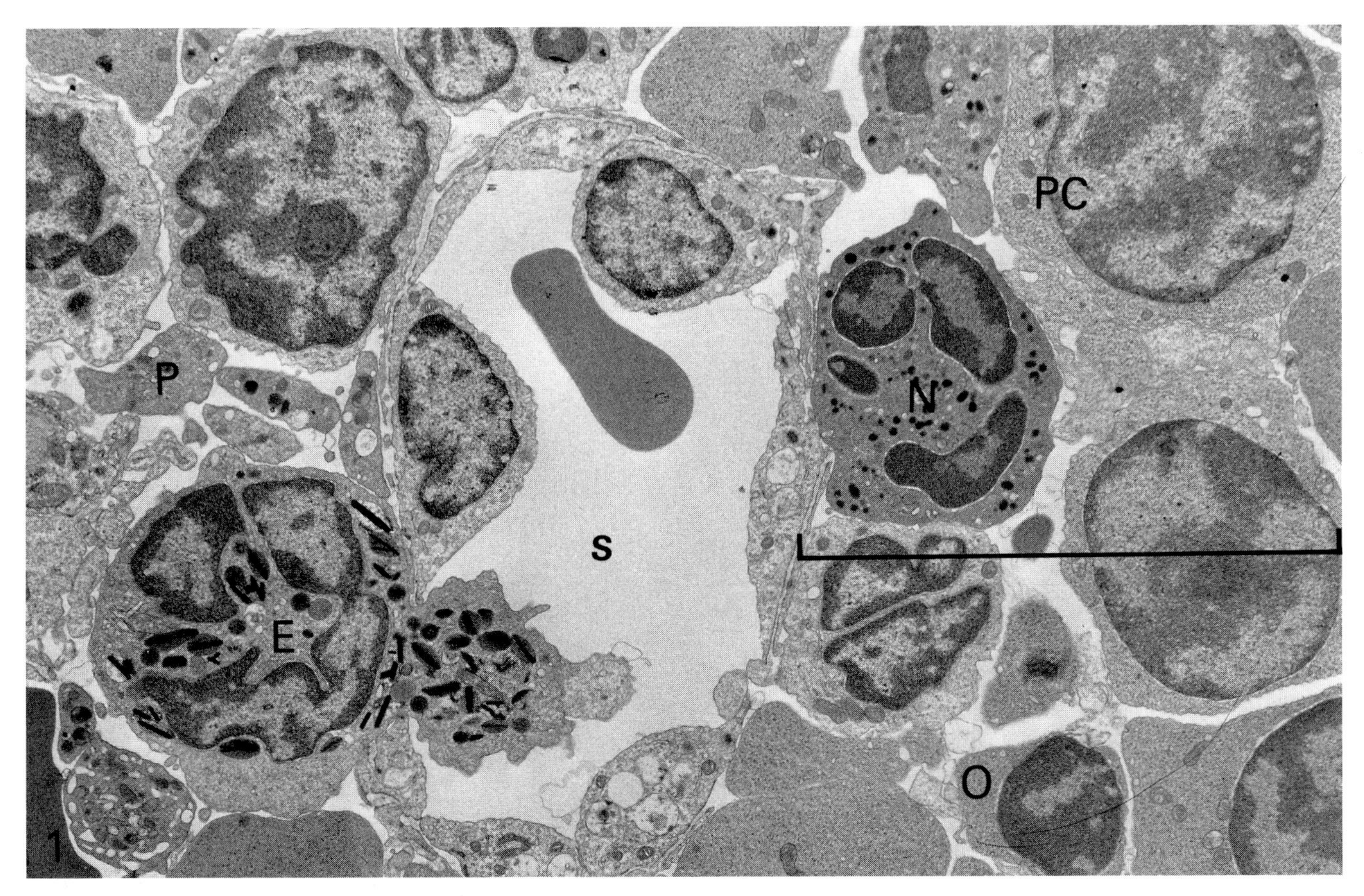

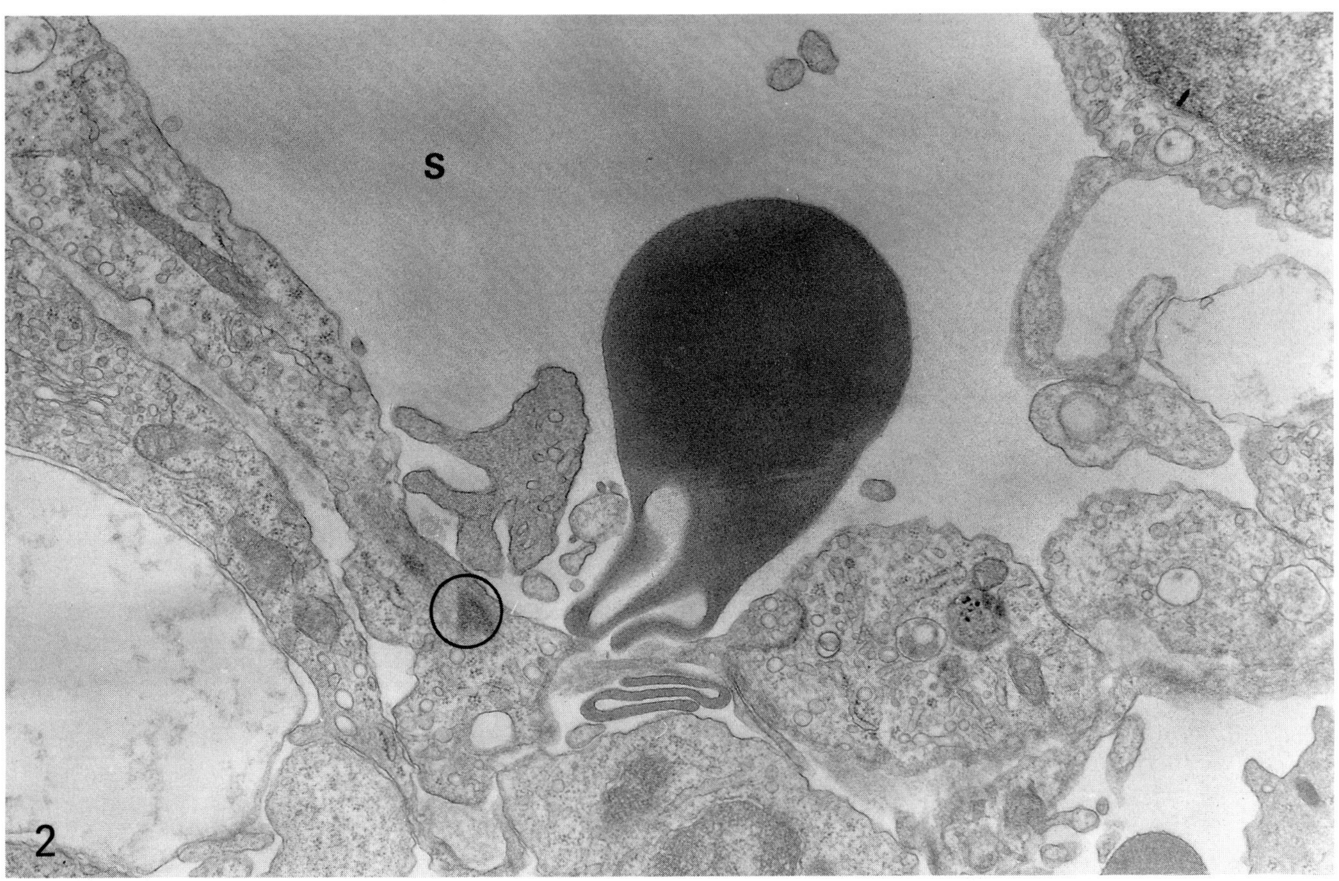

EM 1: 10,200× **EM 2: 22,500×**

SPLEEN RED PULP: Macrophage

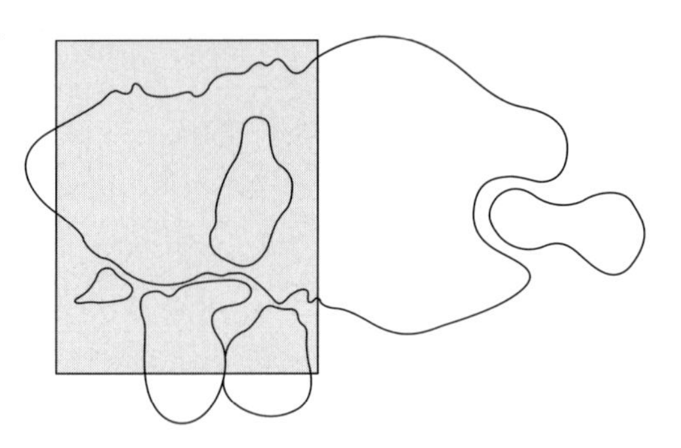

Macrophages (M, micrograph) are an important site of **red blood cell** (R, micrograph) and **platelet** (P, micrograph) **destruction** in the spleen. Several partially digested cell fragments are evident in two secondary lysosomes (ly, micrograph) and a residual body (b, micrograph) within this macrophage.

At the termination of both platelet (10 days) and RBC (120 days) life spans, surface changes occur that promote the removal of these blood elements from the circulation. One of these changes, the loss of sialic acid from specific cell membrane glycoproteins, results in (1) a reduction in surface negative charge and (2) the exposure of new antigenic sites that are immediately recognized as "foreign" and coated with antibodies. With the change in surface charge, the usual repulsion between cells is reduced and both platelets and RBCs are able to come into unusually close association with macrophages (arrows, micrograph). This intimacy facilitates both the binding of macrophage receptors with the Fc portion of the antibodies coating the blood elements and subsequent phagocytosis.

In RBCs, the surface change associated with aging follows a distinct sequence of events. In the absence of a nucleus and ribosomes, the lifetime of existing enzymes ends. As ATP is depleted, hemoglobin begins to denature. Hemoglobin is bound to Band 3, a predominant cell membrane glycoprotein that is a critical part of the spectrin–actin cytoskeletal network (see Blood, page 168). Evidence suggests that, as hemoglobin denatures, cross-links are formed between Band 3 glycoproteins, resulting in clustering of these membrane sites. In addition to being repositioned, Band 3 is altered by the removal of sialic acid. The newly exposed, clustered antigenic sites are coated with antibody. In addition to having reduced surface charge and being coated with antibodies, changes in the cytoskeleton reduce movement and flexibility. Macrophages have many opportunities to bind to the sluggish RBCs as they migrate through the red pulp.

The removal of aged and diseased RBCs by phagocytosis is just one role of the spleen in RBC homeostasis. The fate of RBCs as they pass through the open circulation varies, depending upon the unique properties of each cell. For example: (1) mature RBCs can be "treated" by a process known as **pitting,** in which inclusions such as malarial organisms and Heinz bodies (aggregates of denatured hemoglobin) are removed during transit between sinusoidal endothelial cells; (2) **reticulocytes,** immature RBCs released from the bone marrow, are retained in the red pulp for one to two days in a process of normal **maturation.**

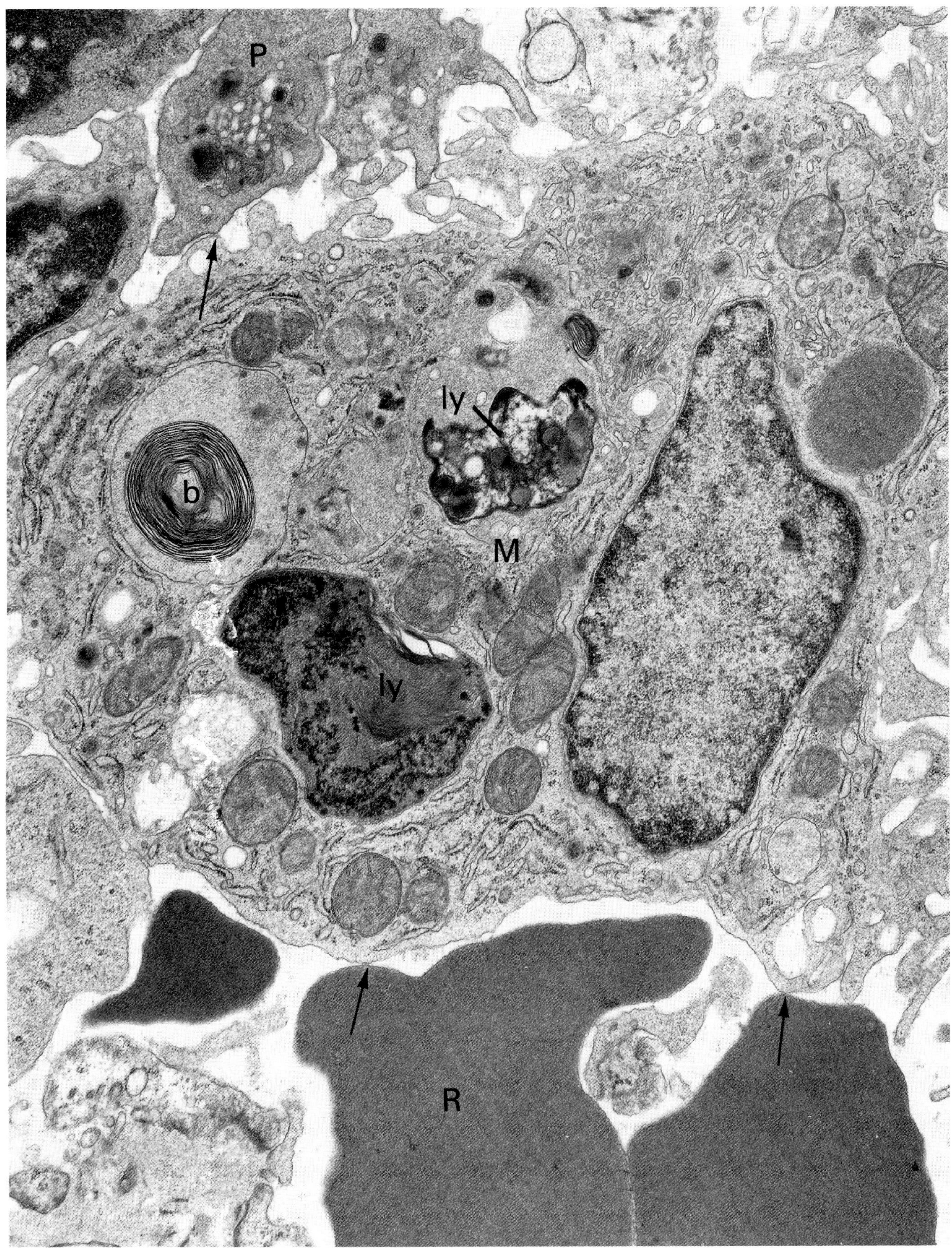

EM: 20,400×

Endocrine Glands

The **pituitary gland,** or **hypophysis,** includes an epithelial component, the adenohypophysis, derived from oral ectoderm, and a neural component, the neurohypophysis, which consists of a downgrowth of the hypothalamus. The **adenohypophysis** includes the pars distalis (anterior lobe), pars intermedia (intermediate lobe), and the pars tuberalis. The **neurohypophysis** includes the pars nervosa (posterior lobe), infundibular stalk, and median eminence (an extension of the hypothalamus). Secretions of the pars distalis and pars nervosa affect major physiological parameters by direct actions on target cells and/or indirect actions via second endocrine glands.

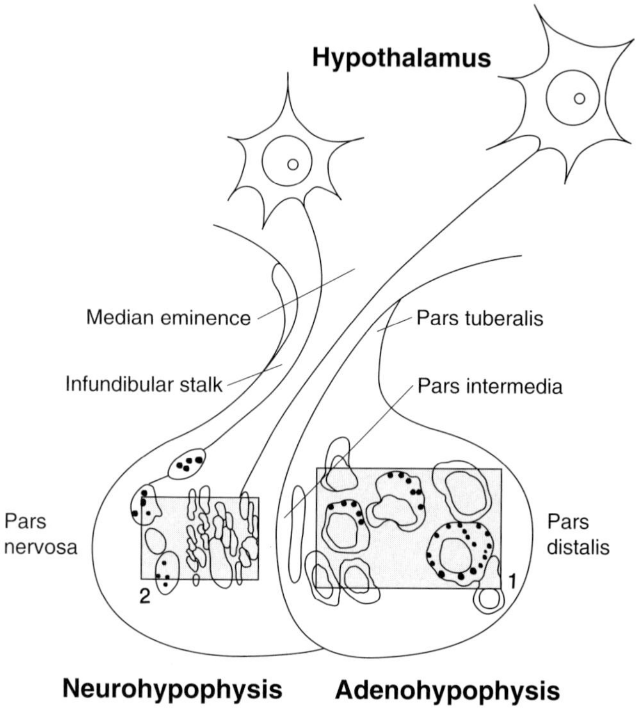

The **pars distalis** (micrograph 1) synthesizes hormones that fit into three categories: the **glycoprotein family** includes thyroid-stimulating hormone (TSH), follicle-stimulating hormone (FSH), and luteinizing hormone (LH), the **large single-stranded peptide family** includes growth hormone (GH) and prolactin (PRL), and the **pro-opiomelanocortin family** includes adrenocorticotropic hormone (ACTH), β-endorphin, and β-lipotropin. These hormones are synthesized by specific cell types that have been identified by immunocytochemistry: TSH is synthesized by thyrotrophs, FSH and LH by gonadotrophs, GH by somatotrophs, PRL by mammotrophs, and ACTH, β-endorphin, and β-lipotropin by corticotrophs. Correlations between immunocytochemistry and the morphology of granules on electron micrographs permit identification of cell type at the ultrastructural level. Three cell types, the somatotroph (S), mammotroph (M), and corticotroph (C), are shown in micrograph 1. In the pars distalis, the site of hormone synthesis and packaging (rough ER and Golgi) is directly adjacent to the site of storage within granules (g, micrograph 1), as is typical for most endocrine cells.

The **pars nervosa** (micrograph 2) secretes hormones of the **nonapeptide family** (oxytocin and the antidiuretic hormone, ADH). The secretory cells of the pars nervosa are modified neurons with cell bodies in the hypothalamus and axons (a, micrograph 2) that extend into the pars nervosa. In contrast to the pars distalis, the site of hormone synthesis is a considerable distance from the site of storage. Hormones stored within granules (g, micrograph 2) are actually synthesized in cell bodies within the hypothalamus and carried along the axons through the infundibular stalk to their final storage site.

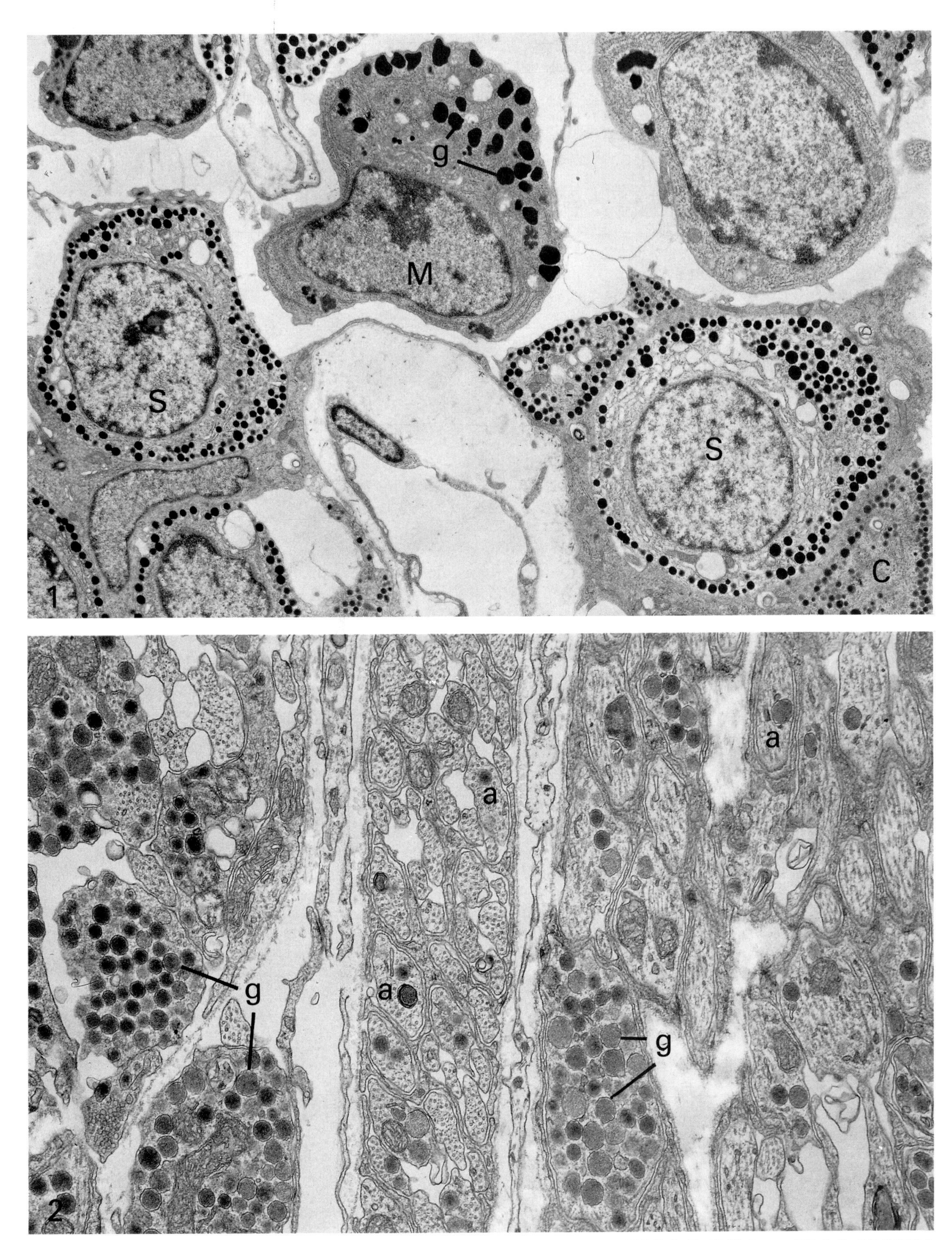

EM 1: 5,400× EM 2: 25,000×

PITUITARY GLAND: Pars Distalis, Portal System

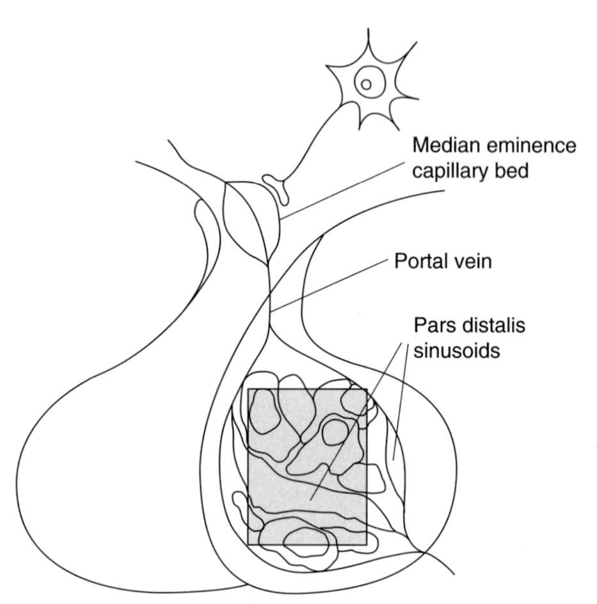

Cells of the **pars distalis** are surrounded by sinusoids (s, micrograph). Not only do the sinusoids collect hormones as they are secreted and carry them away from the gland, but they also bring factors that control hormone synthesis and secretion into the gland. These factors, present in the sinusoids, originate in both the central nervous system (hypothalamus) and periphery (target organs). The endothelium lining the sinusoids has fenestrations (positions indicated by arrows, micrograph), each covered by a thin diaphragm. The fenestrations are believed to facilitate the necessary bidirectional transport of protein hormones and regulating factors that is characteristic of many endocrine glands. The highly permeable endothelium is also important for cell-to-cell interaction within the pars distalis that has been shown to influence secretion.

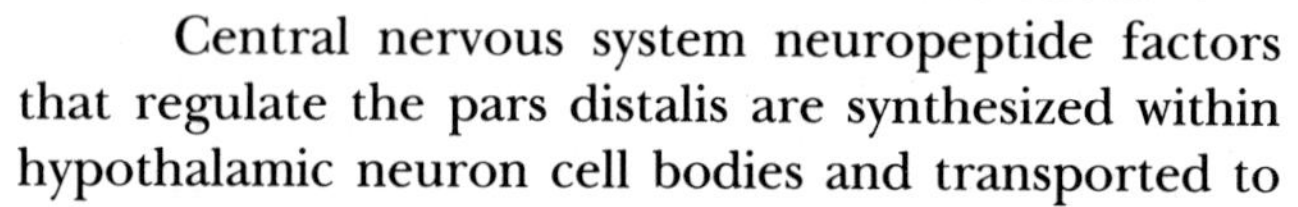

Central nervous system neuropeptide factors that regulate the pars distalis are synthesized within hypothalamic neuron cell bodies and transported to axon terminals in the median eminence. In the median eminence the neuropeptides are released from the axon terminals and carried to the pars distalis via the **hypophyseal – hypothalamic portal system (HHPS).** The first part of the HHPS is the capillary bed in the median eminence that initially collects the neuropeptides. A portal vein connects these capillaries to a second capillary bed, the sinusoids of the pars distalis. Each type of pars distalis cell, including the corticotrophs (C) and somatotrophs (S) in the micrograph, binds to and responds to a unique **neuropeptide-releasing factor.** Two anterior lobe cells, the mammatroph and somatotroph, also respond to specific **inhibitory factors.**

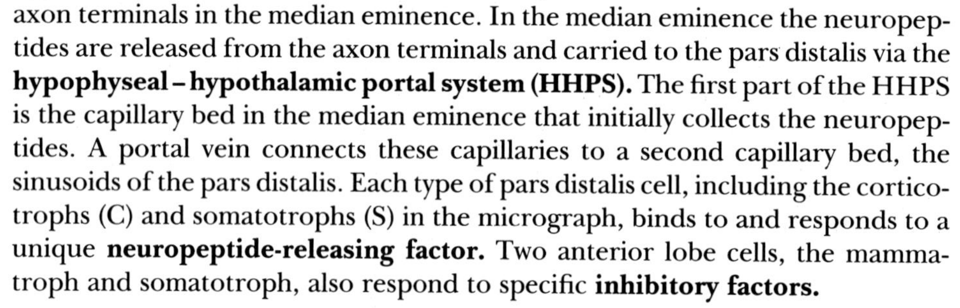

This portal system is an endocrine type of communication system, but due to the proximity of the site of release (median eminence) and site of action (pars distalis), it mimics certain characteristics of nerve–nerve communication: response occurs more quickly and requires the release of only small quantities of hormone. At the same time, however, the portal system does maintain the potential for exerting a broad range of effects. For example, certain hypothalamic neuropeptides act on more than one cell type. The stress-induced increase in secretion by corticotrophs and decreased secretion by gonadotrophs are both mediated by one releasing factor, corticotroph-releasing hormone.

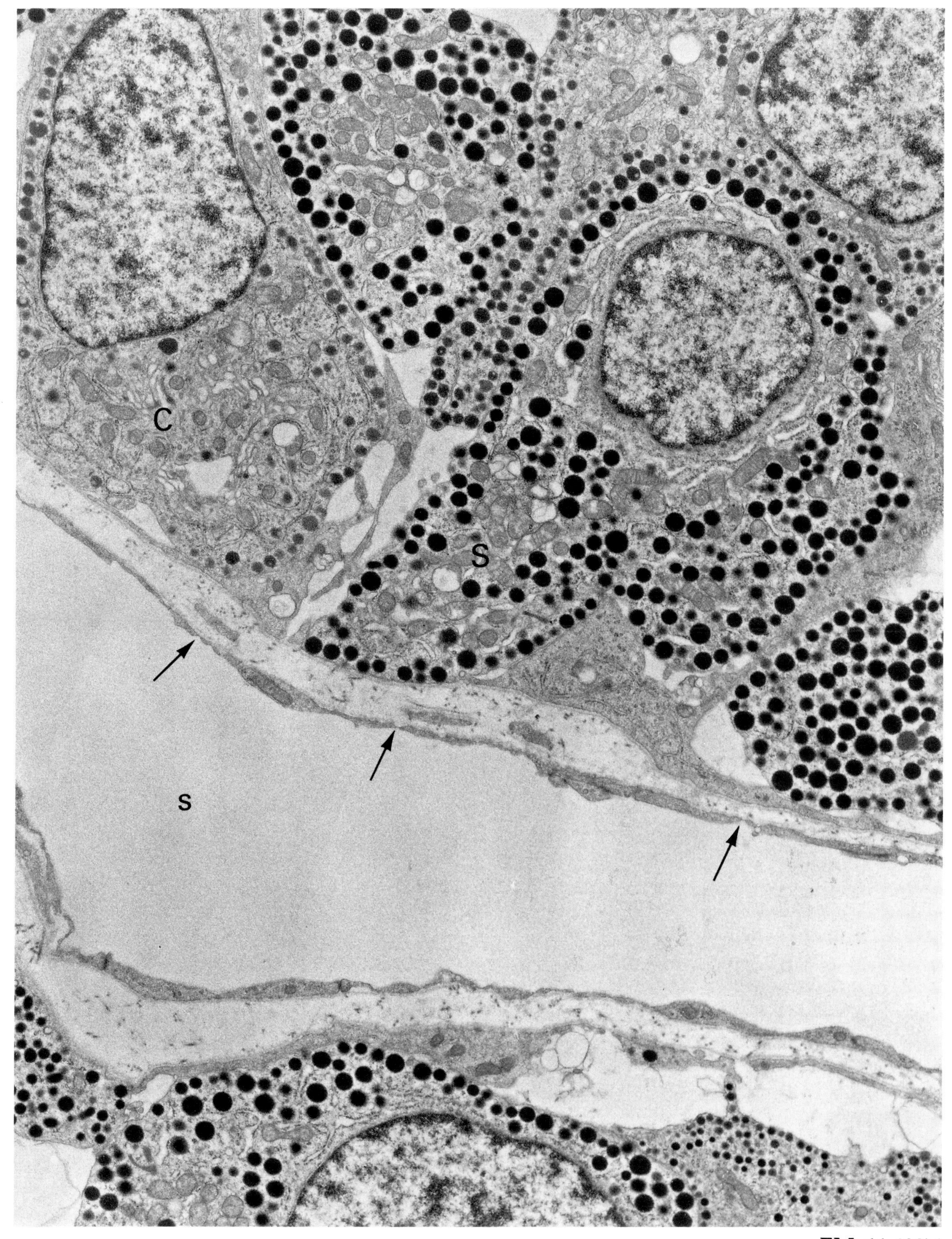

EM: 11,400×

PITUITARY GLAND: Pars Distalis, Regulated Secretion

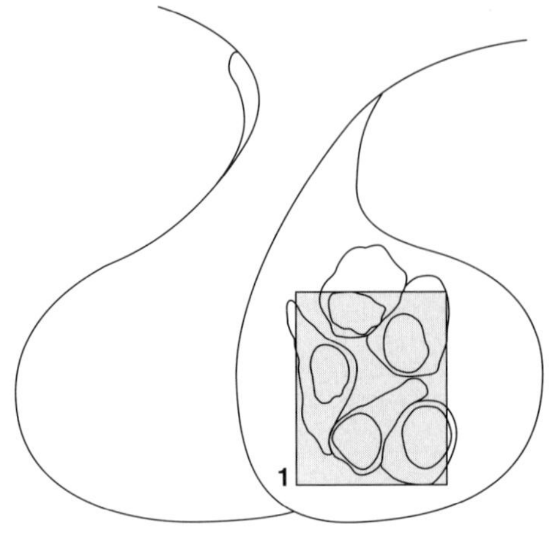

Cells of the pars distalis, in common with most endocrine and exocrine cells, carry out **two types of protein secretion,** regulated and constitutive.

Granules, as observed in the corticotrophs (C), somatotrophs (S), and gonadotroph (G) in micrograph 1, are characteristic of **regulated secretion,** in which the secretion is stored and only released upon stimulation.

Constitutive secretion does not involve storage within granules and is not as easily recognized in ultrastructure, even though small secretory vesicles do undergo exocytosis at the cell surface. Examples of constitutive secretion include the secretion of antibody by plasma cells, acetylcholinesterase by skeletal muscle cells, and collagen by fibroblasts. Constitutive secretion by anterior pituitary cells includes the production of laminin, a major component of the basal lamina (arrowheads, micrograph 1).

Regulated and constitutive proteins are sorted in the **trans region of the Golgi.** In one of the somatotrophs in micrograph 1, a large Golgi complex (outlined by arrows) is seen surrounding granules developing on the trans face. Regulated proteins destined for these secretory granules aggregate in clathrin-coated regions (arrow, inset) of the trans Golgi, where they are apparently recognized by a common receptor, just as lysosomal enzymes are recognized by mannose-6-phosphate receptors (see Cell, page 16). Different kinds of regulatory proteins are treated in the same way. When the gene for insulin is transfected into a corticotroph cell line, both ACTH and insulin are sorted into and stored within the same secretory granules. The first secretory granules leaving the trans region are coated (arrowhead, inset) and only lose their coat as the granule (g, inset) matures.

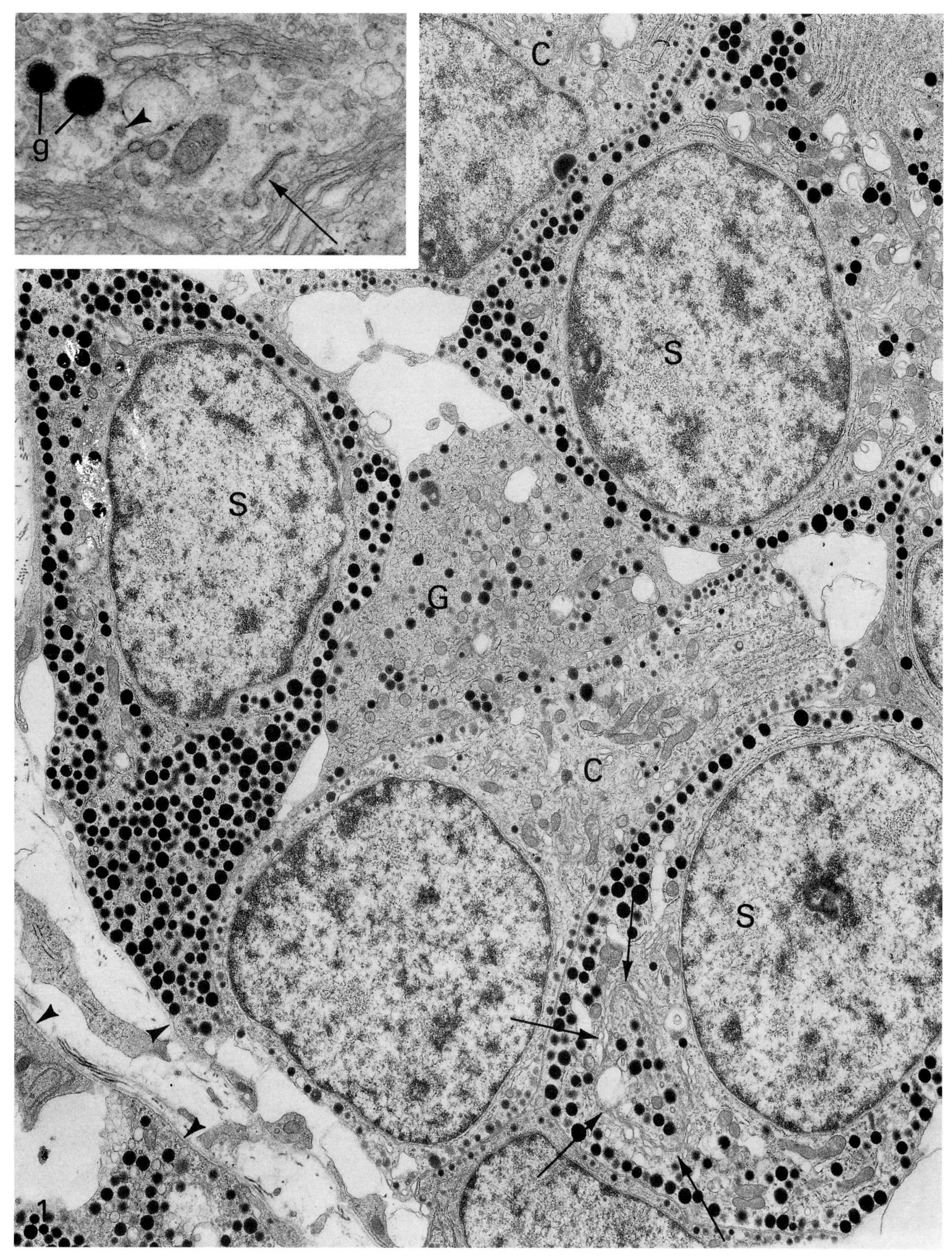

EM 1: 9,400× Inset: 37,000×

Pituitary Gland: Pars Distalis, Gonadotroph

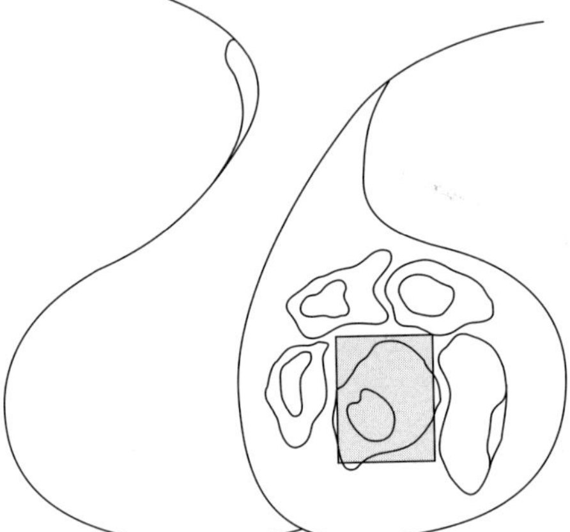

Gonadotrophs (micrograph) are recognized by relatively small (200 nm) granules of varying density scattered throughout the cytoplasm. Both the size of the cell and the number of granules vary with secretory activity. Most gonadotrophs synthesize, store, and release both **follicle-stimulating hormone** (FSH) and **luteinizing hormone** (LH). The action of gonadotropins in the female is on the ovary: FSH acts primarily on the granulosa cells to promote follicle growth and estrogen synthesis; LH acts primarily on the theca interna cells to promote ovulation and corpus luteum formation. Action in the male is on the testis: FSH acts on the Sertoli cells to promote sperm maturation; LH acts on the Leydig cells to promote testosterone synthesis.

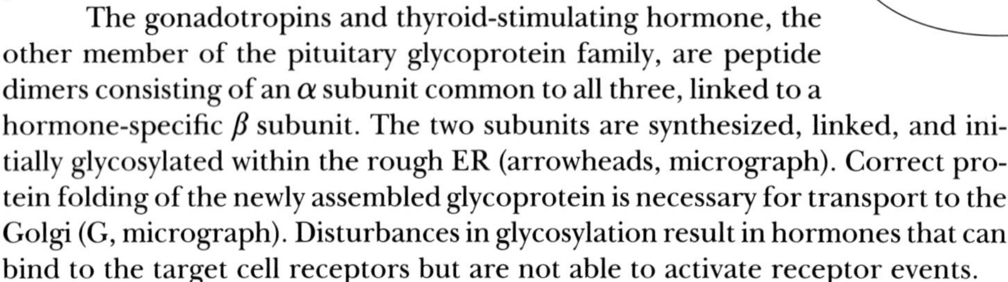

The gonadotropins and thyroid-stimulating hormone, the other member of the pituitary glycoprotein family, are peptide dimers consisting of an α subunit common to all three, linked to a hormone-specific β subunit. The two subunits are synthesized, linked, and initially glycosylated within the rough ER (arrowheads, micrograph). Correct protein folding of the newly assembled glycoprotein is necessary for transport to the Golgi (G, micrograph). Disturbances in glycosylation result in hormones that can bind to the target cell receptors but are not able to activate receptor events.

FSH and LH are typically found within the same granules. The ratio within the granules is determined by the dominant synthetic activity within a cell. This ratio is affected by many factors, including the frequency of the pulse with which gonadotropin-releasing factor is secreted and the action of specific gonadal peptides and steroids. One peptide, inhibin, produced by the ovary and testis in response to FSH, binds to gonadotroph cell surface receptors and, via second messengers, specifically inhibits the synthesis of FSH. The steroid hormones estrogen and progesterone cross the gonadotroph cell membrane, bind to receptors in the cytoplasm, and enter the nucleus, where they directly alter transcription.

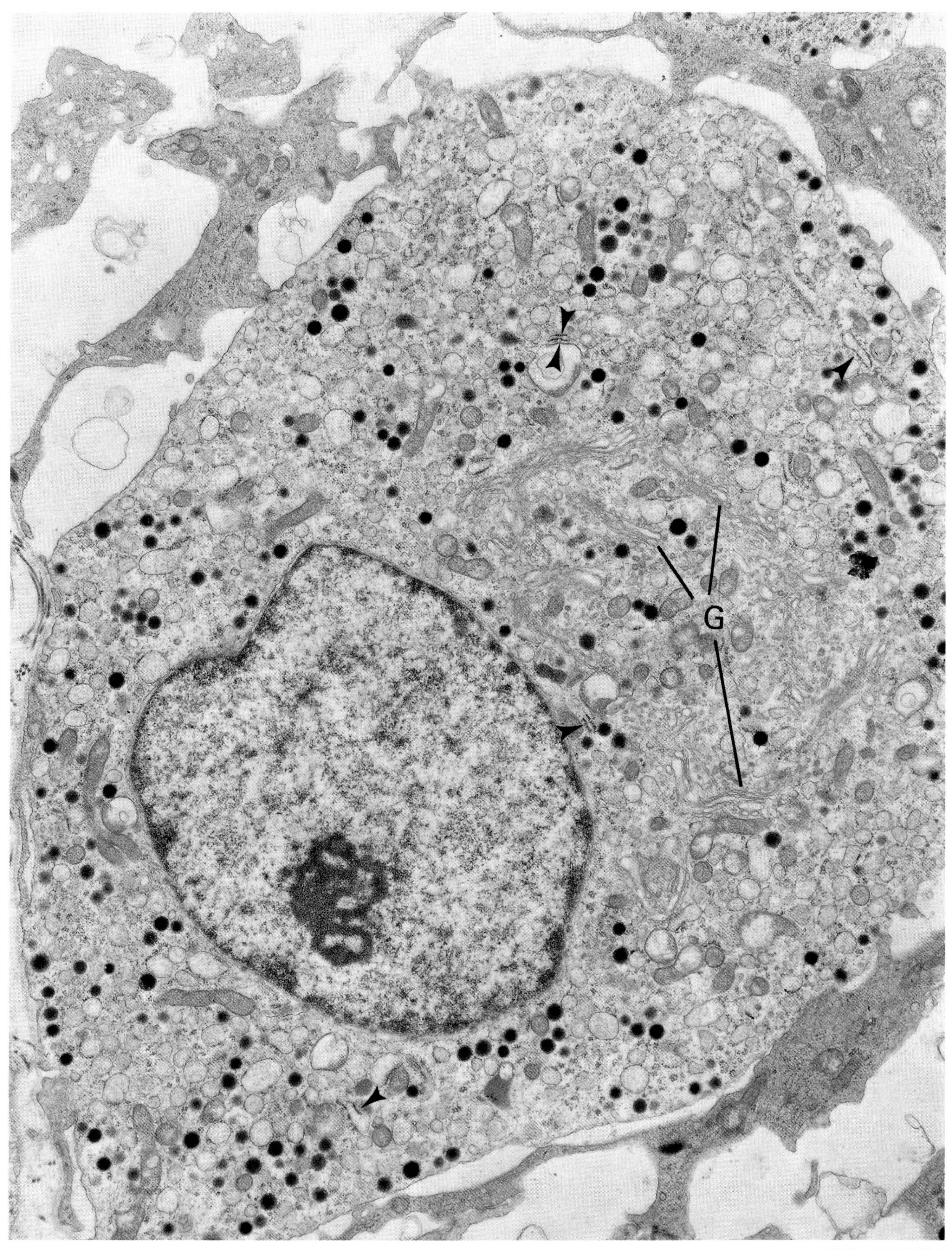

EM: 15,300×

PITUITARY GLAND: Pars Distalis, Corticotroph

Corticotrophs (C, micrograph 1) are recognized in micrographs of rat pituitary as elongaged cells with relatively small, peripherally situated granules (arrows, micrograph 1). **Adrenocorticotropic hormone** (ACTH), the major hormone secreted by corticotrophs, acts primarily to stimulate the inner adrenal cortex to produce glucocorticoids, which in turn have profound effects on carbohydrate and protein metabolism. The control of ACTH secretion is complex and includes not only a hypothalamic-releasing hormone and feedback from adrenal glucocorticoids, but also input from higher centers in the central nervous system associated with stress.

Like gonadotrophs, corticotrophs secrete more than one type of hormone. Unlike gonadotrophs, in which the final hormones result from combining different gene products (α and β subunits), the final secretory products in corticotrophs originate from cleavage of a single gene product, the prohormone **pro-opiomelanocortin** (POMC). This large (~285 amino acids) glycoprotein precursor is synthesized in the rough ER, glycosolated in the Golgi (G, micrograph 1 and inset), and concentrated into secretory granules at the trans face.

The **processing of POMC** occurs within the granules. Histochemical techniques have demonstrated that the smaller, less dense, clathrin-coated vesicles (arrowheads, inset) leaving the Golgi contain the entire prohormone POMC, whereas the mature, nonclathrin-coated granules at the cell periphery contain increasing amounts of the cleavage products of POMC. In the **pars distalis,** processing within the secretory granules typically progresses to form separate N-terminal glycopeptide, ACTH, and β-lipotropin.

PARS DISTALIS:	N-terminal glycopeptide	ACTH		β-Lipotropin	
PARS INTERMEDIA:	N-terminal glycopeptide	α-MSH	CLIP	γ-Lipotropin	β-Endorphin

Modified from P.A. Rosa et al., *J. Exp. Biol. 89*:217(1980), Company of Biologists, Ltd., Cambridge.

POMC is also synthesized in the **pars intermedia** of the pituitary, but in this region ACTH is usually not one of the final secretory products. Posttranslational processing within the intermediate lobe secretory granules is more extensive, and ACTH is split into melanocyte-stimulating hormone (MSH) and corticotropinlike intermediate lobe peptide (CLIP).

The function of the intermediate lobe in humans is unknown. In contrast to the anterior lobe, the intermediate lobe is poorly vascularized and richly innervated. Evidence suggests that secretion is controlled by direct neural input. POMC processing in the intermediate lobe, unlike that in the anterior lobe, is unaffected by corticotropin-releasing hormone or glucocorticoids, but is affected by certain neurotransmitter inhibitors.

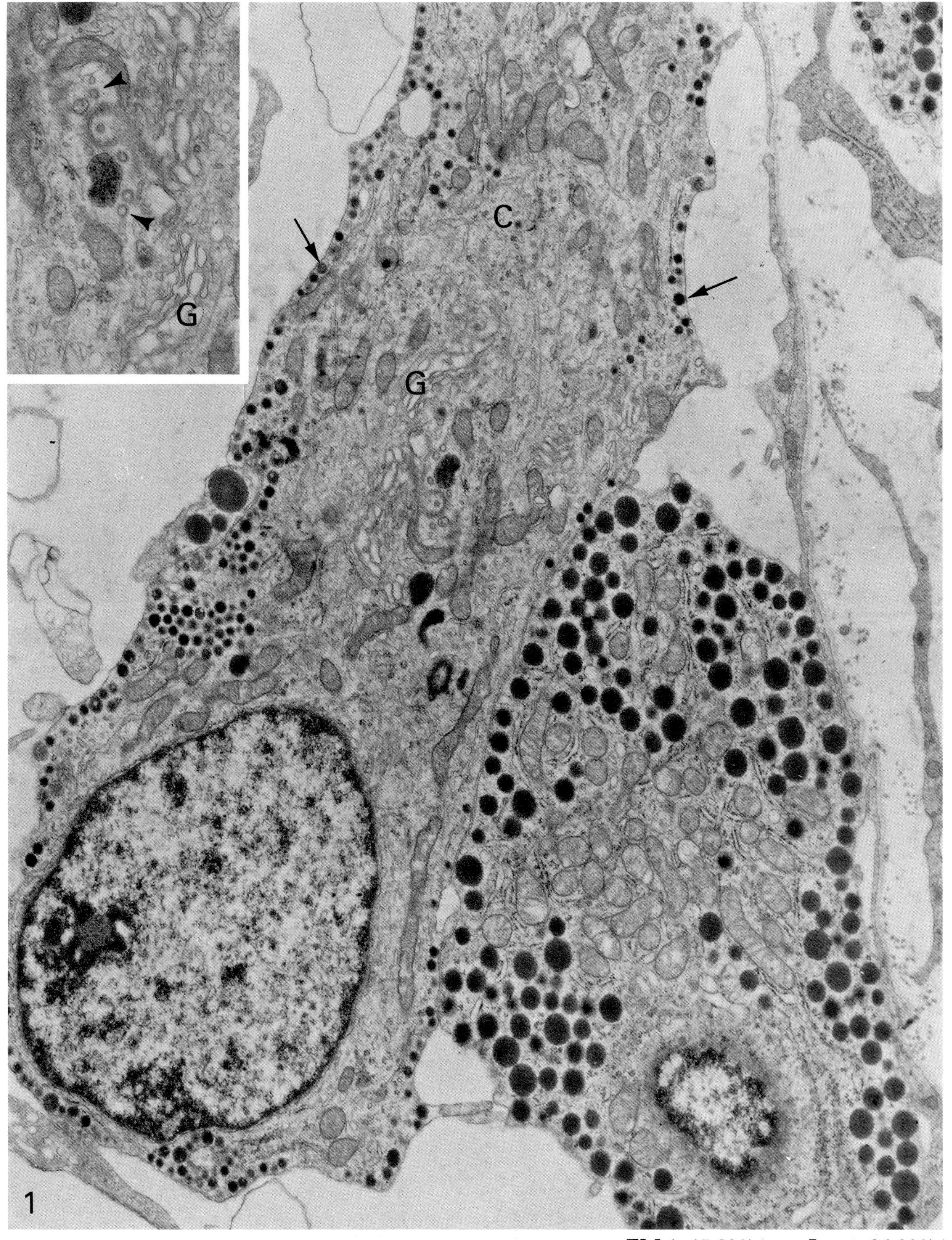

EM 1: 15,000× **Inset: 26,000×**

PITUITARY GLAND: Pars Nervosa, Overview

The **pars nervosa** secretes oxytocin and antidiuretic hormone (ADH or vasopressin). Oxytocin stimulates the contraction of (1) myoepithelial cells in the mammary gland, resulting in milk release, and (2) uterine smooth muscle during coitus and parturition (see Female Reproductive System, page 370). ADH acts primarily on the collecting tubules and ducts in the kidney to adjust water balance.

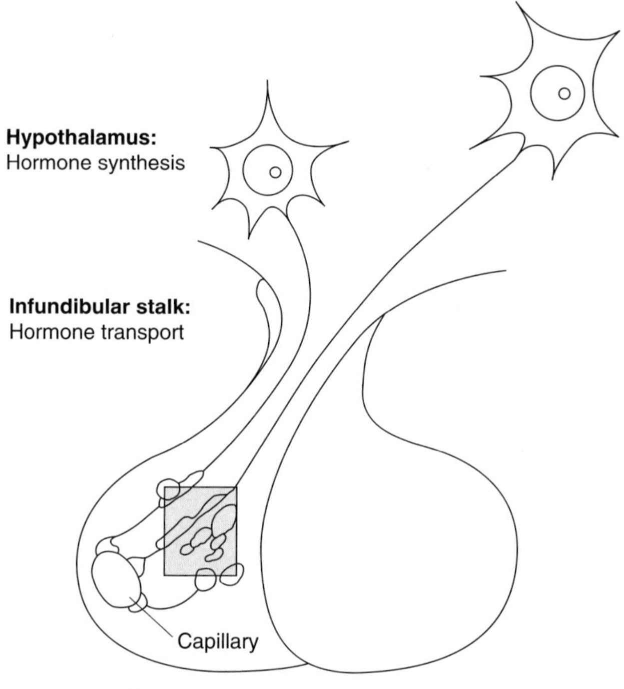

Oxytocin and **ADH** are peptides that differ by only two amino acids. They are synthesized as a part of large precursors in the cell bodies of separate neurons located both in the paraventricular and supraoptic nuclei of the hypothalamus. Precursors, synthesized on the rough ER, are shuttled through the Golgi and packaged into secretory vesicles that associate with axon microtubules. Microtubles (arrows, micrograph) move these granules (arrowheads, micrograph) down axons at the fast rate of 200 mm/day to the site of secretion, the pars nervosa. Over 100,000 axonal processes are packed together in the infundibular stalk, each transporting secretory granules that contain precursors of either oxytocin or vasopressin.

The secretory granules contain (1) the precursor molecule, which includes the hormone and its specific "carrier protein," or neurophysin, and (2) processing enzymes. During the transport of the granule to the posterior lobe, enzyme activity frees the hormone from its neurophysin. This processing continues even when the movement of granules is inhibited.

Secretion in the posterior lobe is controlled primarily by events originating in the hypothalamus. Sensory information from the periphery travels to the neuron cell bodies in the hypothalamus, where information is processed and an action potential is generated. The sensory information controlling oxytocin release is primarily from the mammary gland; information controlling ADH release is from neurons sensitive to decreased blood volume and increased osmolality. From the axon hillock, the action potential travels down the axons within the infundibular stalk to the posterior lobe, where it triggers granule exocytosis at the nerve terminal. Thus the major region of control of the final secretion (the hypothalamus) is located a considerable distance from the actual site of release (the pars nervosa). When the action potential carrying the stimulus is prevented from reaching the posterior lobe (i.e., when the infundibular stalk is severed), granule release does not occur.

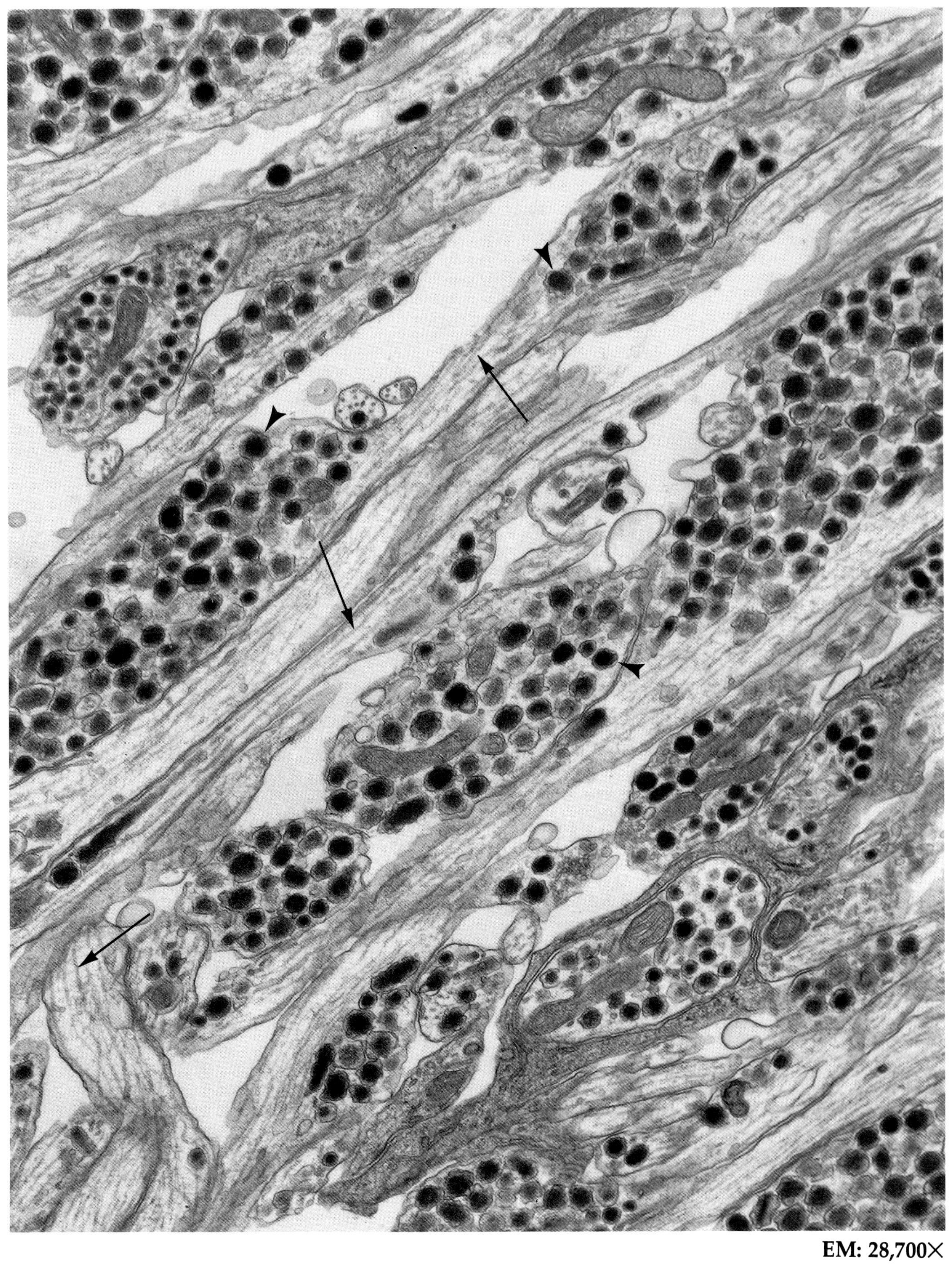

EM: 28,700×

Pituitary Gland: Pars Nervosa, Secretion

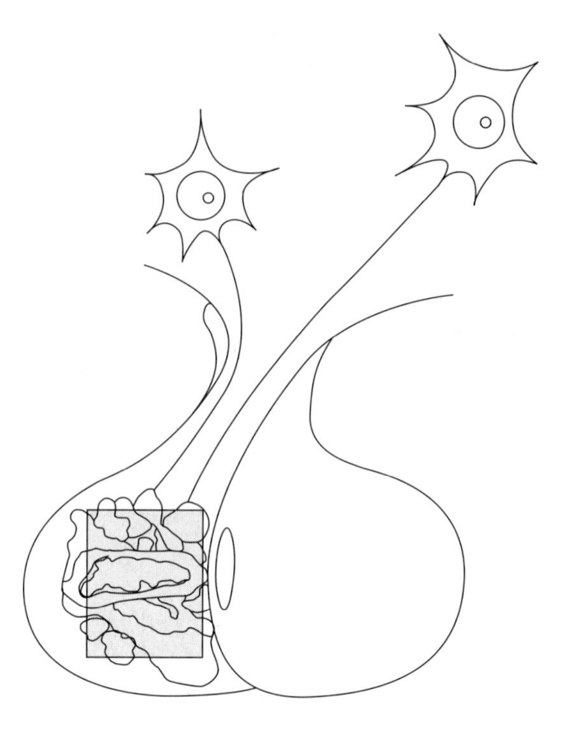

In the **pars nervosa,** axons of the secretory neurons terminate on a basal lamina (b, micrograph) that lines the perivascular spaces. The **axons** (a, micrograph) are attached to this basal lamina at synapticlike regions (circles). When an action potential reaches the axon terminal, an influx of calcium occurs and the **dense core granules** (arrows, micrograph) containing hormone, neurophysin, and other granule components undergo exocytosis and release their contents. Instead of traveling the short distance across a synapse, oxytocin and vasopressin cross the fenestrated endothelium (e, micrograph) and enter the vascular system to be distributed throughout the body. Dense core granules that do not discharge when they reach an axon terminal are shuttled to storage sites along the axon, where they accumulate as **Herring bodies** (H, micrograph). Granules within Herring bodies may be recycled to the terminal or fuse with lysosomes and undergo digestion. Lysosomes and secretory granules cannot be distinguished on routine electron micrographs.

Microvesicles (arrowheads, micrograph), packed within the axon terminals and similar in size and structure to synaptic vesicles of presynaptic terminals in other neurons, may be involved in the storage and release of nonpeptide neurotransmitters. Two different forms of secretory vesicles are also found in other endocrine cells. In pancreatic beta cells, large secretory granules containing the hormone insulin are accompanied by synapticlike microvesicles that appear to store the neurotransmitter GABA.

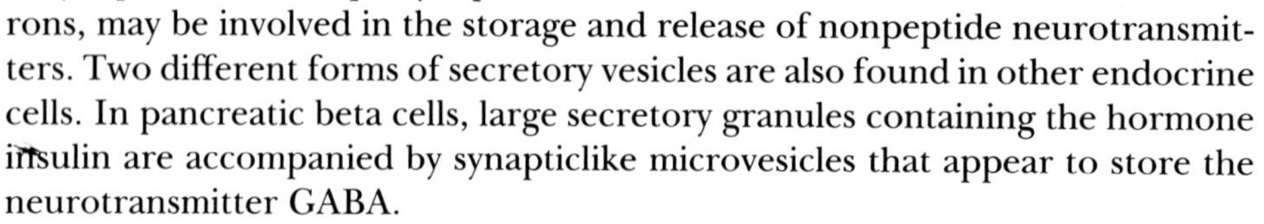

The posterior lobe, even though it is composed of axons originating from cell bodies located in the central nervous system, is unlike the central nervous system in two significant ways: (1) it is outside the blood-brain barrier, as shown by the presence of fenestrated capillaries, and (2) it supports neuron regeneration. When the infundibular stalk is severed and a posterior lobe is transplanted next to the severed stalk, axons from the hypothalamic neurons grow into the transplanted tissue and establish terminal contact with perivascular basal lamina. **Pituicytes** (P, micrograph), the glial cells of the posterior lobe, often recognized by their prominent lipid droplets (l, micrograph), play a significant role in this regeneration. Processes of the pituicytes attach to stretches of the perivascular basal lamina next to neuron terminals.

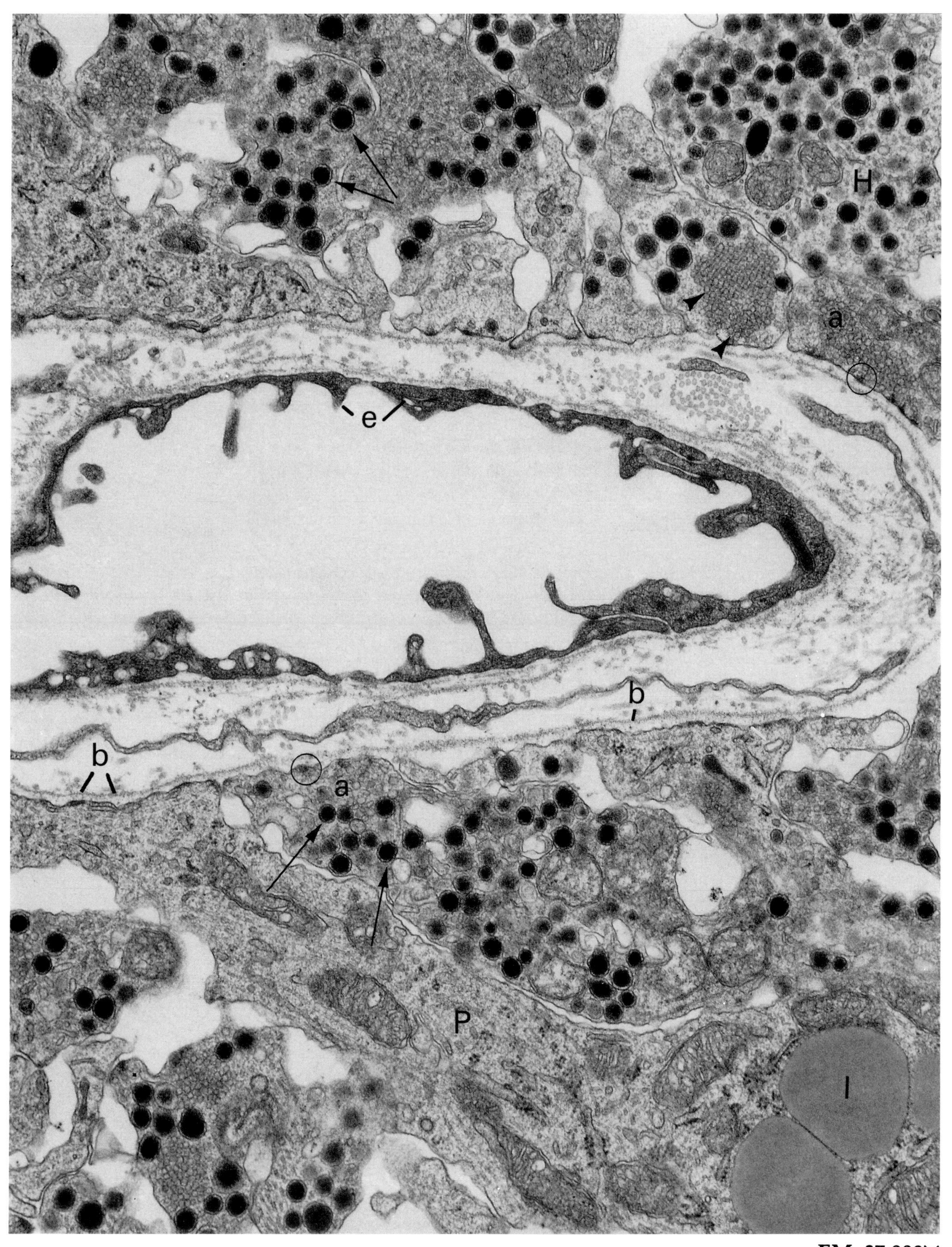

EM: 37,000×

PITUITARY GLAND: Pars Nervosa, Pituicyte

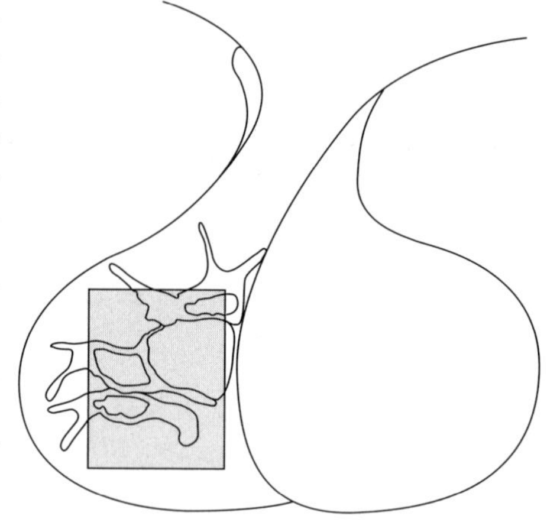

Unlike the hormone-secreting neurons, **pituicytes** (P, micrograph) are located entirely within the posterior lobe of the hypophysis. These glial cells contain **glial fibrillary acidic protein** (GFAP), a type of intermediate filament characteristic of astrocytes (see Nerve, page 142). Similar to other astrocytes, they extend processes (arrowheads) in many directions to contact blood vessels and envelop axons (a, micrograph). Processes of one pituicyte in the micrograph envelop a Herring body (H) that is larger than the pituicyte nucleus (N).

The degree of association between pituicytes and neurons varies with neuron secretory activity. With increased antidiuretic hormone secretion, such as occurs as a result of hemorrhaging or increased plasma osmotic pressure, **pituicyte processes retract** and appear to "free" the secretory neurons. As a result, more neuronal processes associate with the perivascular basal lamina surrounding capillaries. The pituicyte at the top of the micrograph contains **lipid droplets** (l). As the pituicyte processes retract during active secretion, the number of lipid droplets increases. It has been suggested that this is associated with membrane recycling during these dramatic shape changes.

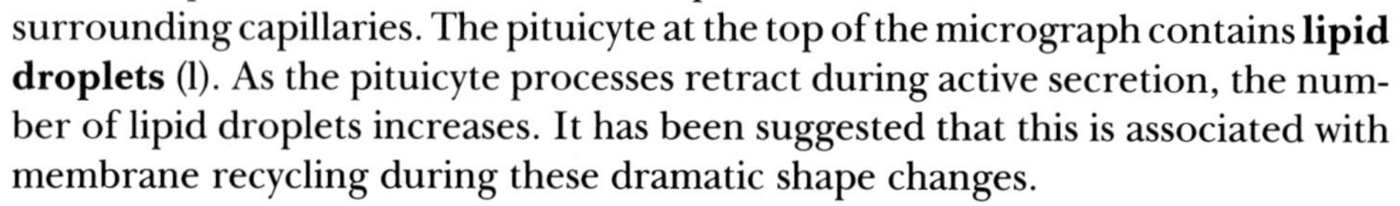

Even though the release of vasopressin and oxytocin is initially determined by the action potential generation at the axon hillock in the hypothalamus, it can be modulated in the posterior lobe by information acting on the nerve terminal. Some modulation effects are direct on the nerve terminal (e.g., effect of opioids) and others seem to be indirect via the pituicytes. Pituicytes have receptors for several neuroactive substances, including norepinephrine; norepinephrine has been shown to have a dramatic effect on pituicyte morphology in vitro.

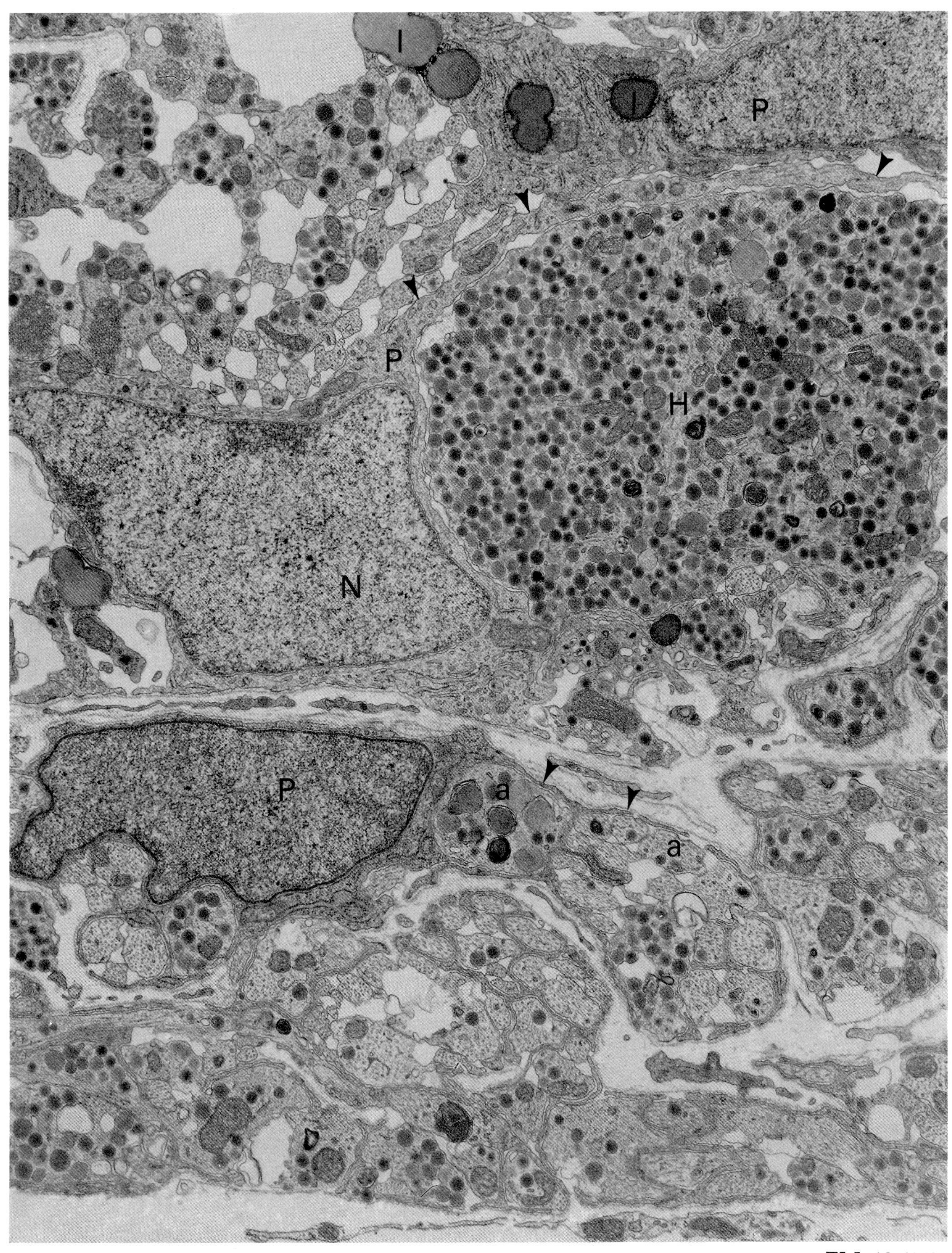

EM: 12,600×

THYROID GLAND: Hormone Production and Polarity

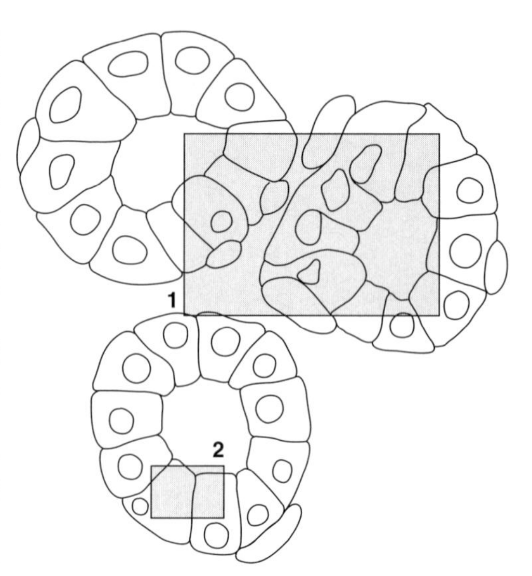

The **thyroid gland** secretes **triiodothyronine** (T3) and **tetraiodothyronine** (T4, thyroxine), tyrosine derivatives that control the metabolic activity of most cells. The synthesis and secretion of these hormones takes place in the epithelium (E, micrograph 1) lining spherical follicles. As in many other endocrine cells, a prohormone is synthesized and processed to form the final hormones. What is unique in the thyroid is that during processing the prohormone is exocytosed from the cell and **stored extracellularly** in the lumen (L, micrographs 1 and 2). This precursor, **thyroglobulin,** makes up greater than 75% of the weight of the thyroid, and represents enough stored precursor for 100 days of hormone supply.

The different stages in the production of T3 and T4 are highly ordered and depend upon the established **polarity of the follicular cells.** The apical surfaces (arrows, micrograph 1) lie adjacent to the lumen of the follicle, which contains thyroglobulin; the basal surfaces (arrowheads, micrograph 1) lie adjacent to an extensive network of capillaries (c, micrograph 1) that collect the final hormones as they are released from the cells. Red blood cells (R) are packed within the capillaries in micrograph 1.

T3 and T4 formation begins with the synthesis of the precursor thyroglobulin on the rough ER. Typically this large (>600,000 molecular weight) glycoprotein fills the expanded cisternae of rough ER (r, micrographs 1 and 2) so that at low magnification only thin strands of cytoplasm (C, micrograph 1) appear to remain. Glycosylation of the thyroglobulin begins in the rough ER, but is completed in the Golgi apparatus (G, micrograph 2). Secretory vesicles (curved arrows, micrograph 2), formed on the trans surface of the Golgi, coalesce and migrate to the apical surface where exocytosis occurs.

At the time of exocytosis or directly after, the tyrosine residues of thyroglobulin are **iodinated.** Iodide is actively pumped across the basolateral membrane into the cell. At the apical surface iodide is oxidized and bound to tyrosine residues of thyroglobulin. The **peroxidase** that catalyzes this reaction appears to be an integral membrane protein, specific to the apical microvilli (m, micrographs 1 and 2). Initially, one or two iodine molecules are bound per tyrosine to form monoiodothyronine (MIT) and diiodothyronine (DIT). Subsequent coupling occurs between these to yield T3 (MIT + DIT) and T4 (DIT + DIT) units, still a covalent part of the thyroglobulin molecule.

Polarity, such as the localization of the iodide pump to the basolateral membrane and the peroxidase to the apical membrane, is maintained by the tight junctions (circle, micrograph 2) between follicular cells. Other junctions seen in micrograph 2 link cells within a given follicle structurally (desmosome, d) and metabolically (gap junction, g). Secretory activity within a given thyroid follicle is typically uniform.

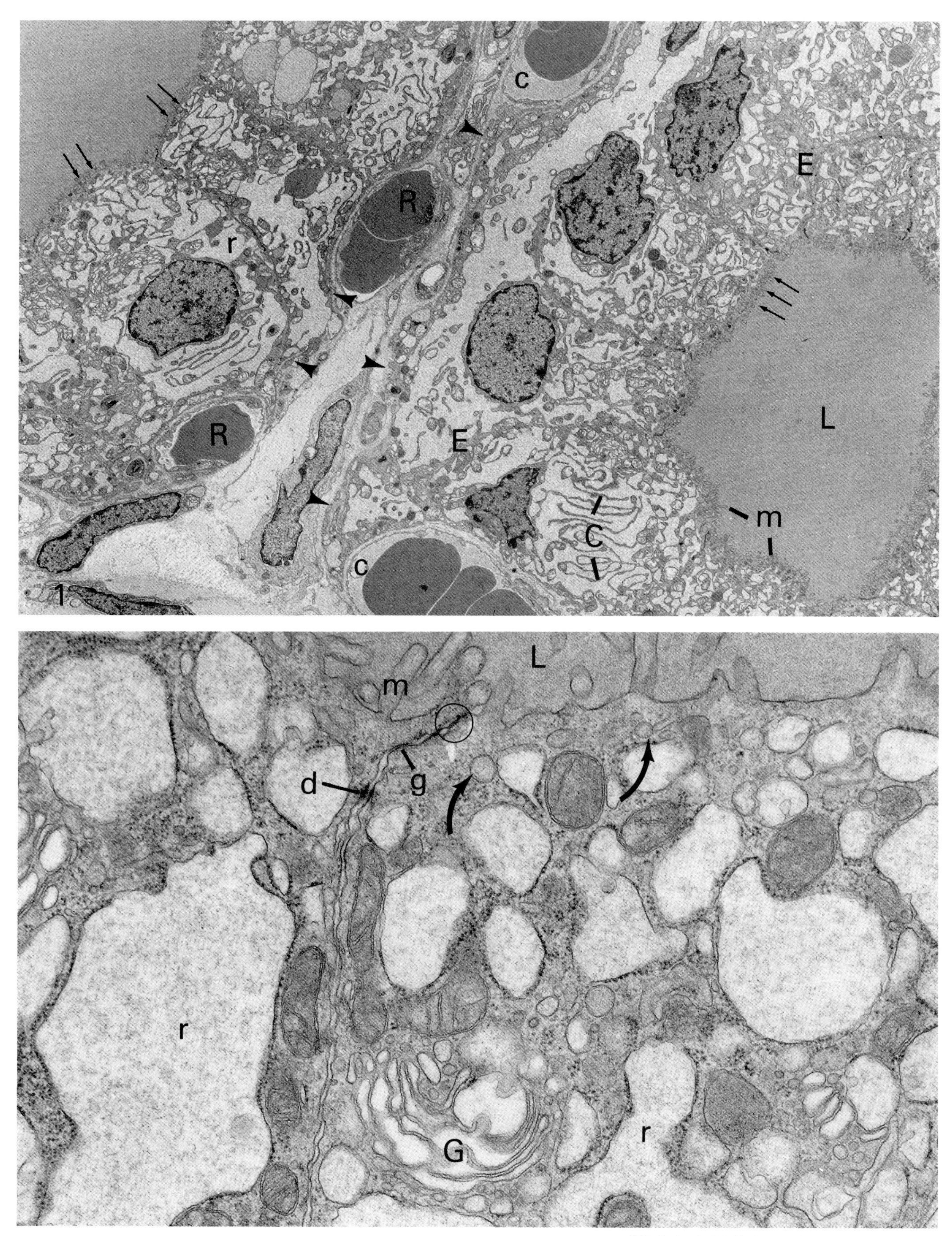

EM 1: 3,300×   EM 2: 35,200×

THYROID GLAND: Endocytosis and Hormone Processing

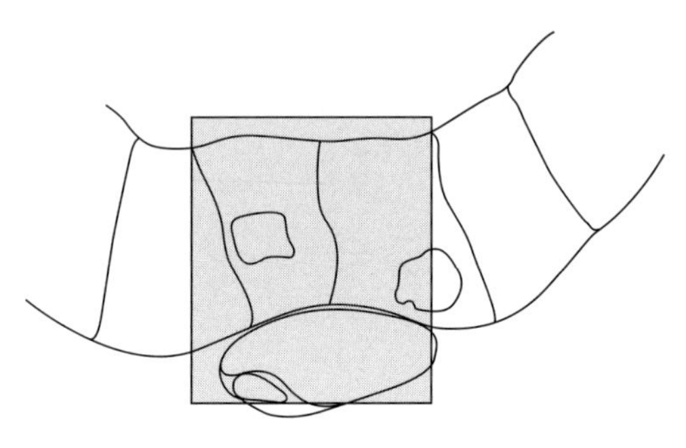

To form the final hormones, **thyroglobulin** is taken into the epithelial cells for processing. Uptake is via either **pinocytosis** or **phagocytosis.** During phagocytosis, pseudopods extending from the apical surface pinch off a section of the thyroglobulin, which is then incorporated into the cell. Phagocytotic activity and the resulting, sometimes large, phagosomes are characteristic of acute stimulation. Pinocytotic and phagocytotic vesicles that result from uptake fuse with lysosomes. The large dense vesicles (arrows, micrograph) near the apical surface are most likely secondary lysosomes that have formed following the fusion of smaller primary lysosomes with phagosomes.

Thyroglobulin is digested by proteases within the secondary lysosomes as they move from the apical to the basal surface. Monoiodothyronine and diiodothyronine are completely degraded, freeing iodide, which is recycled. The hormones triiodothyronine (T3) and tetraiodothyronine (T4) diffuse from the basally situated lysosomes (arrowheads, micrograph), leave the cell, and enter nearby capillaries (filled with red blood cells, R, micrograph). Even though more T4 is synthesized than T3, most of the biological activity of the thyroid gland resides in T3. Apparently T4 is converted to T3 in the peripheral tissues.

Thyroid-stimulating hormone (TSH) from the anterior pituitary binds to receptors on the basal cell surface and stimulates all phases of T3 and T4 synthesis and secretion. Initially, increases in blood TSH levels cause an increase in blood flow along with heightened thyroglobulin exocytosis, endocytosis, and digestion. With further TSH stimulation, the synthesis of both thyroglobulin and the microvillar peroxidase is increased. Chronically high levels of TSH result in cell division and cell hypertrophy. The enlarged thyroid gland is known as a goiter.

Graves' disease, one type of disorder associated with goiter, is characterized by the presence of autoantibodies that recognize the TSH receptor. When the antibodies bind to the receptor, they mimic the action of TSH and activate the follicular cells. The large amounts of T3 and T4 released do not inhibit the autoantibodies as they normally inhibit TSH in a negative feedback loop. This explains the paradoxical clinical findings of an overactive gland in the presence of low levels of TSH.

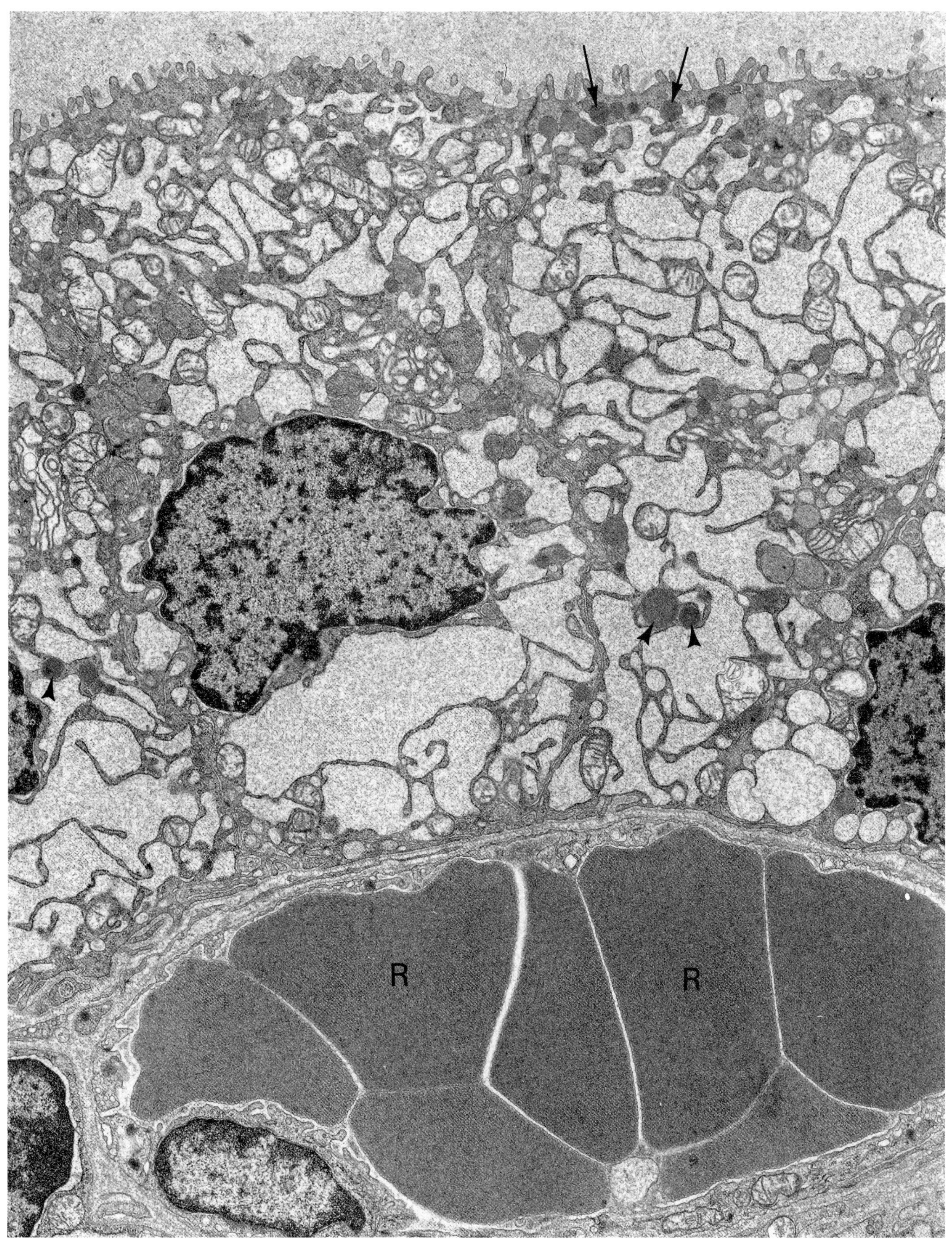

EM: 11,750×

Parathyroid Gland

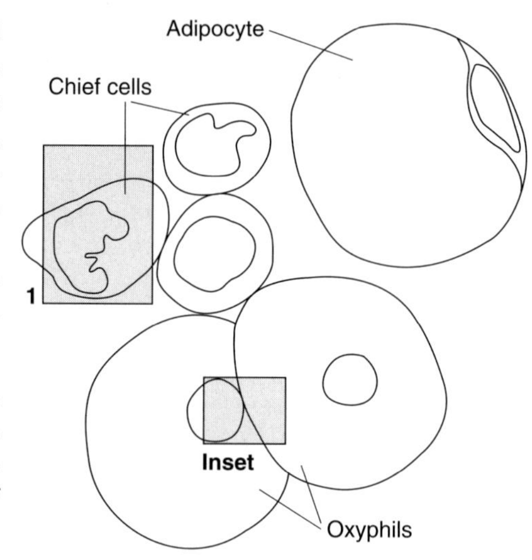

Chief cells (C, micrograph) of the parathyroid gland synthesize and secrete **parathyroid hormone** (PTH), a protein hormone that acts primarily on bone and kidney to increase circulating calcium levels. A PTH precursor is synthesized on the rough ER (r, micrograph 1) and processed to its final structure during its transport through the Golgi (G, micrograph 1) and packaging into secretory vesicles (arrows, micrograph 1). The secretion of PTH is controlled primarily by levels of **extracellular calcium;** high calcium inhibits and low calcium stimulates PTH secretion.

Evidence suggests that **lysosomal activity** is a pivotal control point in PTH release. At low extracellular Ca^{+2} concentrations, most storage granules are secreted. At high extracellular Ca^{+2} concentration, many storage granules, instead of being secreted, fuse with lysosomes, where PTH is digested. During this digestion process some of the lysosomes undergo exocytosis and PTH fragments are found in the circulation. Lysosomal digestion of secretory granules (crinophagy) occurs in other secretory cells, including the insulin-producing beta cells of the pancreas.

Beginning at puberty the number of chief cells begins to decline and adipocytes and another cell type, the **oxyphil,** increase in number. Oxyphils (inset) are larger than chief cells and have relatively small round central nuclei surrounded by cytoplasm packed with mitochondria. Oxyphils are not unique to the parathyroid gland and also occur in abnormal sites (e.g., thyroid carcinoma). Their function is unknown. Some evidence suggests that chief cells transform into oxyphils as they age, even though oxyphils bear no resemblance to chief cells and no intermediate stage has been observed.

The extreme proliferation of mitochondria in oxyphils may be a compensation for defective functioning; however the activity of certain mitochondrial enzymes (e.g., ATPase) appears to be normal.

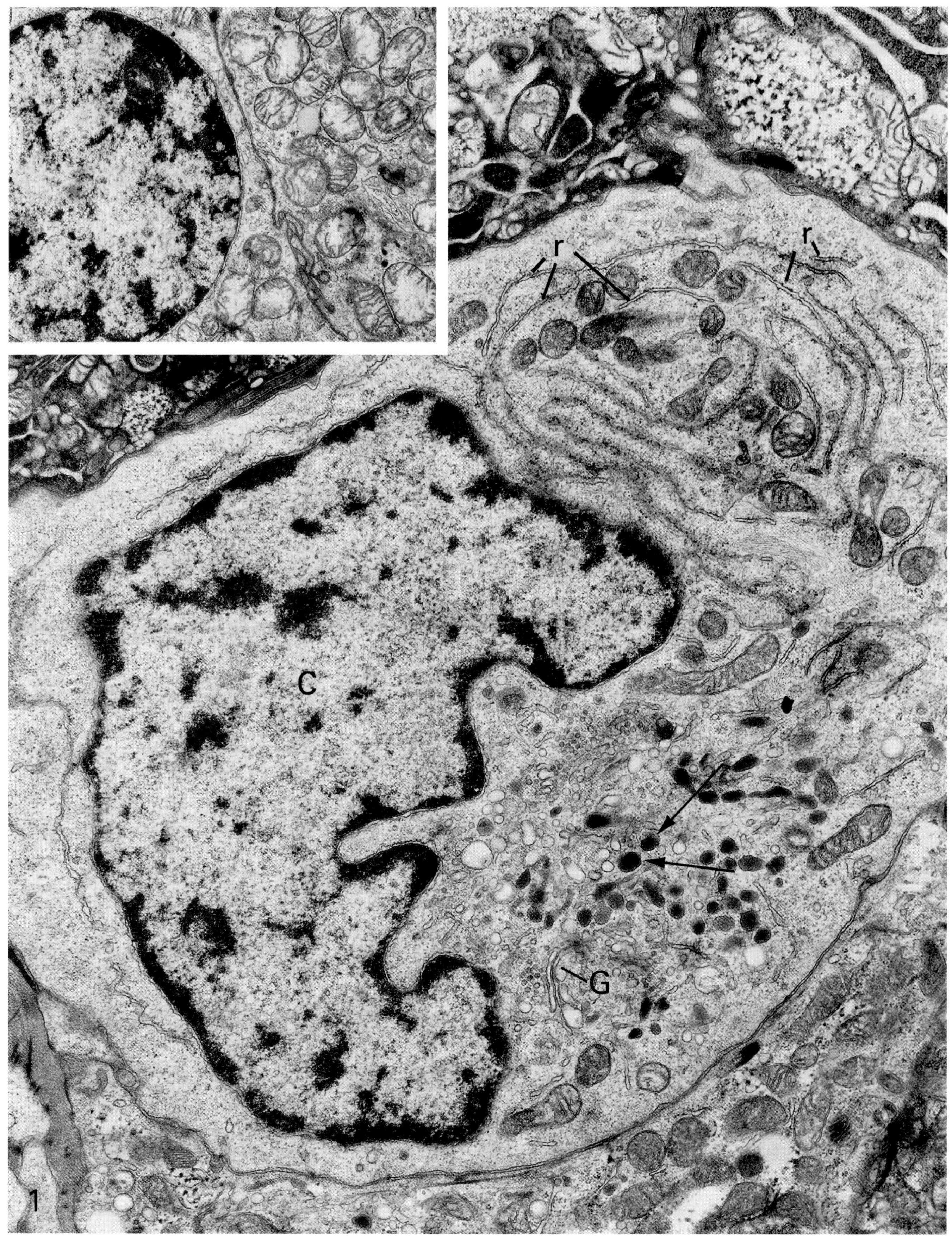

EM 1: 19,500× **Inset: 11,200×**

Islets of Langerhans

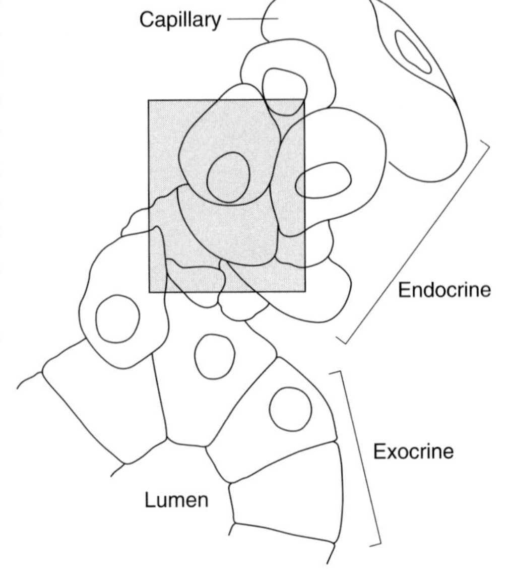

The endocrine portion of the pancreas is small (only 1% by weight) in comparison to the exocrine portion, and is concentrated in up to 1 million individual **islets of Langerhans** scattered between tightly packed exocrine acinar units. The function of the endocrine portion, like that of the exocrine portion, centers around securing and maintaining the nutrient supply of all cells. While the enzymes released from the exocrine pancreas free the nutrients from ingested material, the endocrine pancreas regulates the levels of nutrients, particularly glucose, available to individual cells.

Each islet consists of a network of heterogeneous endocrine cells, including at least four different types, each synthesizing a different hormone: (1) **β cells** (B, micrograph) synthesize insulin (promotes glucose storage), (2) **α cells** (A, micrograph) synthesize glucagon (mobilizes glucose), (3) **δ cells** (D, micrograph) synthesize somatostatin (inhibits alpha and beta cells), and (4) **F cells** (not shown) synthesize pancreatic polypeptide (regulates enzyme secretion by pancreatic acinar cells). Even though the cellular composition of islets varies, in most islets beta cells predominate, as they do in the micrograph. The appearance of mature granules in beta cells varies between species and the type of tissue preparation. In the rat (micrograph), insulin is concentrated in the center of the granule as a dense homogeneous circular mass surrounded by a relatively large light staining area. A portion of an exocrine acinar cell packed with rough ER (arrows, micrograph) is wedged between the endocrine islet cells.

Insulin is synthesized as a precursor, proinsulin, that contains two different amino acid chains joined by a connecting peptide. The drop in pH that occurs in coated vesicles leaving the trans Golgi (G, micrograph) activates an enzyme that clips the peptide from proinsulin to form insulin. As more insulin is formed, granules fuse and lose their coat. In humans, insulin precipitates and crystallizes as a rhombohedron of hexamers within the maturing granules (see Cell, page 16). Following exocytosis (triggered principally by a rise in blood glucose) the solubility of insulin increases in the relatively high pH and ionic strength of plasma.

A mutant form of proinsulin, differing by a single amino acid, has been found to follow the constitutive instead of the normal regulatory pathway of secretion. As a result, the mutant form does not mature and is continually secreted as the precursor, proinsulin. Proinsulin, as true for other prohormones, has only a fraction of the activity of the mature hormone. It is possible that certain disorders characterized by deficient insulin levels, such as diabetes Type I, are associated with altered guidance in the trans Golgi region.

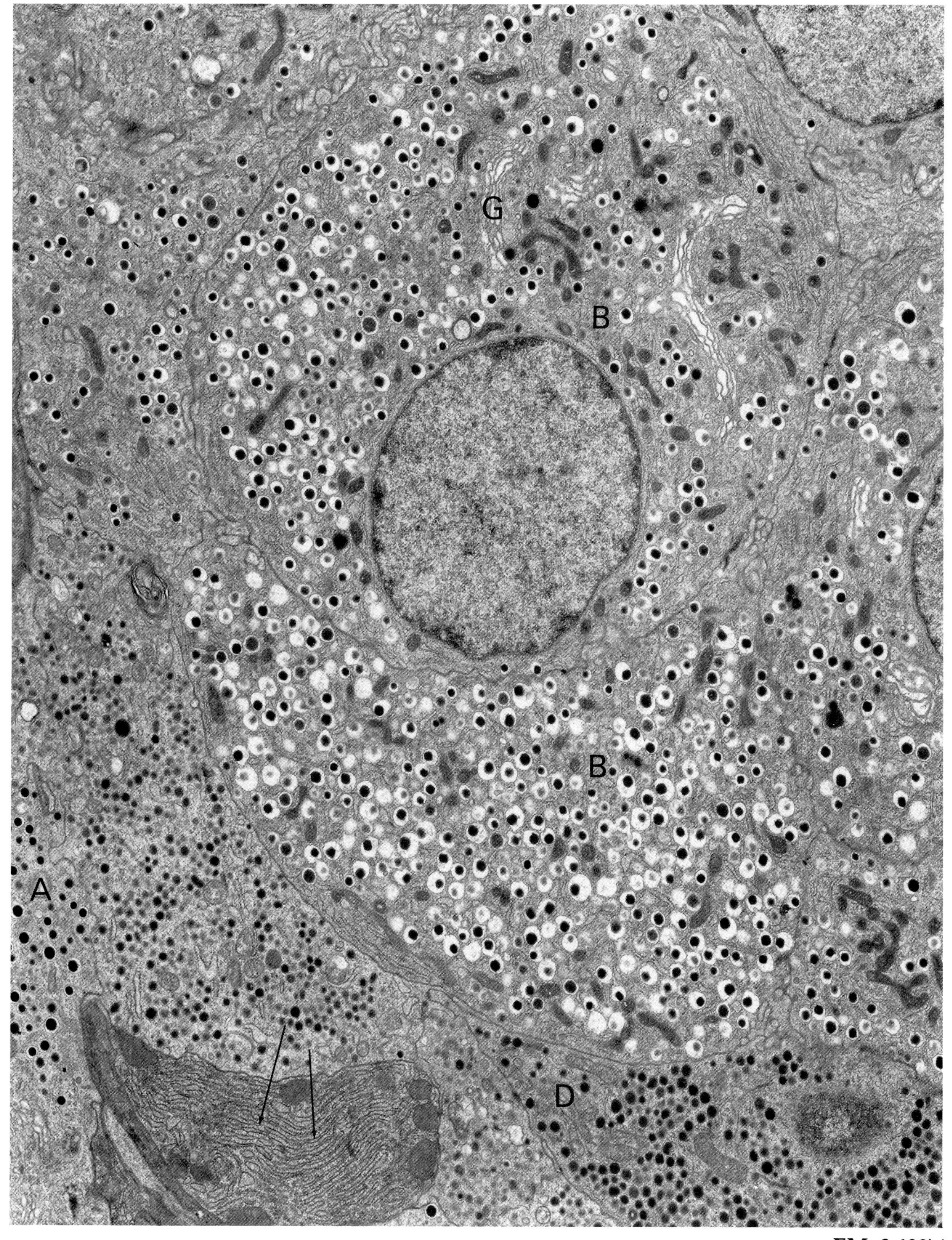

EM: 9,400×

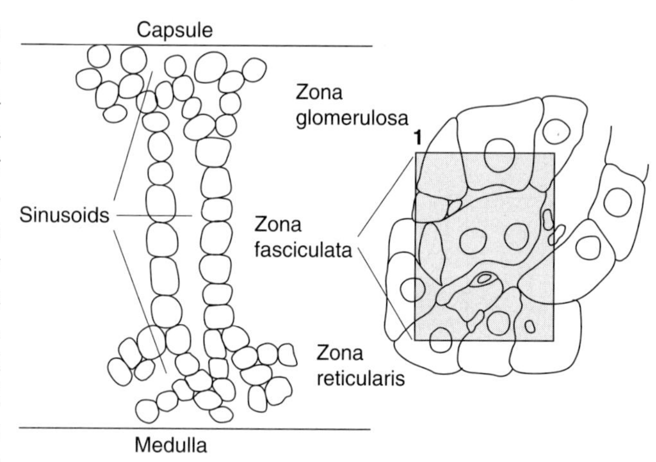

The **adrenal cortex** is divided into three zones, an outer **zona glomerulosa,** in which the cells are arranged in clumps, a **zona fasciculata** (micrograph 1), arranged in cords, and a **zona reticularis,** with cells arranged in an irregular network. Even though all three zones utilize cholesterol to synthesize steroid hormones and share some common biochemical pathways, typically there is a functional distinction between the zona glomerulosa and the fasciculata-reticularis zones. The glomerulosa secretes the mineral-corticoid aldosterone, which controls electrolyte and fluid balance, whereas the primary secretory product of the fasciculata and reticularis in humans is the glucocorticoid, cortisol, which regulates carbohydrate homeostasis. Both of these hormones are essential for life. In addition to having a very different secretory product, the zona glomerulosa is under separate control via the renin-angiotensin system, with angiotensin II acting directly on the glomerulosa cells. The remainder of the adrenal cortex, the fasciculata and reticularis, is under the control of ACTH and is an essential part of the response of the adrenal gland to stress.

In common with all endocrine cells, adrenal cortical cells are intimately associated with the blood vascular system. The association is so pronounced that the sinusoidal endothelium (arrows, micrograph 1) follows the contours of the adrenal cells as if forming a lining. The final steroid hormones, released from the cortical cells, cross the fenestrated endothelium and enter the sinusoids.

The ultrastructure of adrenal cortical cells clearly reflects their role in the synthesis of **steroid hormones.** The substrate, **cholesterol,** is frequently stored as esters with fatty acids in lipid droplets (l, micrograph 1 and inset; lipid is extracted in this preparation). Initially cholesterol is released from the lipid droplet by a hydrolase and transported to the inner mitochondrial membrane. This membrane contains the enzyme complex and electron transport chain that catalyze the side chain cleavage of cholesterol to form pregnenolone. This initial step is the critical rate-limiting control point in the synthesis of all adrenal cortical hormones. To accommodate this enzyme complex, the surface area of this inner mitochondrial membrane is greatly increased, forming **tubular cristae** (arrowheads, inset), a hallmark ultrastructural feature of steroid-synthesizing cells. Subsequent steps in hormone synthesis occur on enzymes in both the **smooth ER** (s, inset) and mitochondria. The increase in steroid hormone synthesis stimulated by ACTH occurs in parallel with an increase in surface area of mitochondrial cristae and smooth ER. A major point of action of ACTH is the formation of pregnenolone on the tubular cristae.

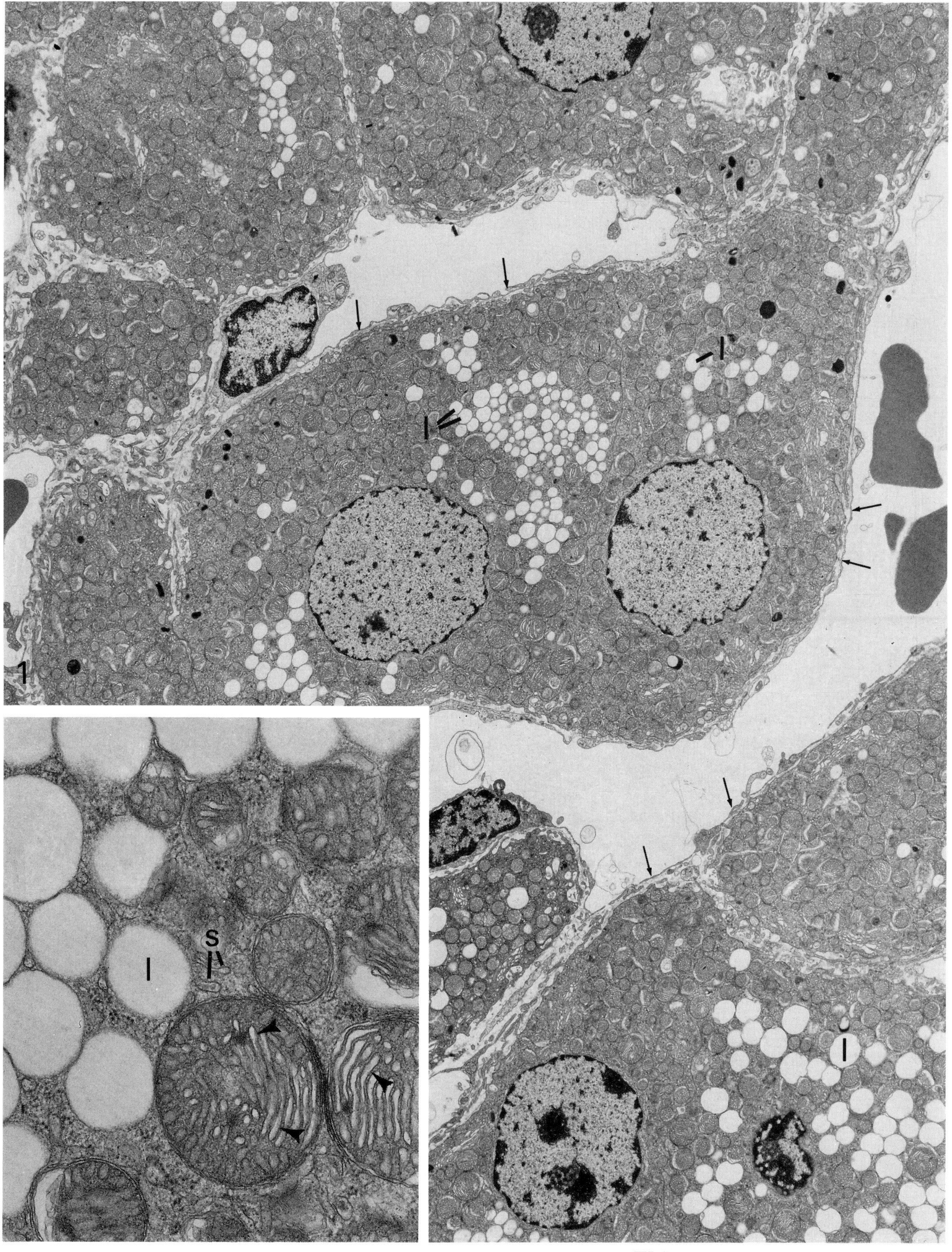

EM 1: 4,950× Inset: 33,000×

ADRENAL CORTEX: Cholesterol

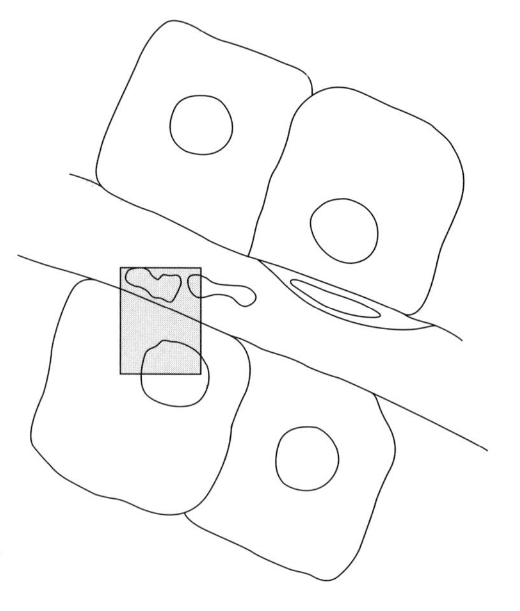

The close apposition of an adrenal cortical cell to an adjacent capillary (c) is well illustrated in the micrograph. In certain areas only a single basal lamina (b, micrograph) separates the fenestrated endothelium from the adrenal cell, further emphasizing the small distance that the steroid hormones have to travel following release from the cell. The **Golgi apparatus** (G, micrograph) is prominent in adrenal cortical cells and, like the other organelles involved in steroid synthesis in this gland, undergoes hypertrophy in response to ACTH. However, the classical role of the Golgi in the processing and packaging of secretory proteins is not applicable to steroid-secreting cells. The mechanism of steroid hormone release from the cells may not involve secretory vesicles and exocytosis.

The significance of the Golgi in steroid-secreting cells may relate more to the uptake of the hormone precursor **cholesterol** than to the secretion of the final hormone. Plasma low-density lipoprotein (LDL) particles are the major source of cholesterol for adrenal cortical hormone synthesis. LDL binds to surface receptors in coated pits and is taken up via **receptor-mediated endocytosis** into coated vesicles (circles, micrograph). Receptors are recycled, endosomes fuse with lysosomes (arrows, micrograph), and cholesterol is freed to enter the cell. The lysosomes, essential to the uptake of cholesterol, are formed within the Golgi that is frequently seen oriented toward the cell surface, as in the micrograph.

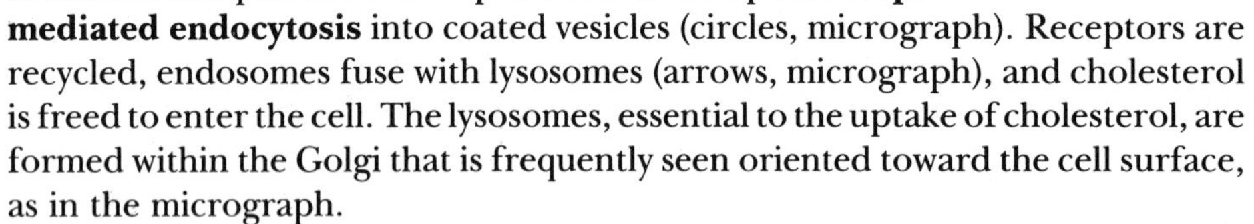

Following release from lysosomes, cholesterol is typically esterified to fatty acids that accumulate in the characteristic **lipid droplets** (l, micrograph). This storage form of cholesterol represents a readily available source of substrate for the rapid synthesis of adrenal hormones. Stored cholesterol is particularly prominent in the **zona fasciculata** and seems to reflect the dynamics of cells that synthesize cortisol.

When large doses of ACTH are administered, the glomerulosa switches from aldosterone to cortisol secretion. Following chronic stimulation, the glomerulosa cells, along with this functional change, accumulate lipid droplets and appear identical to fasciculata cells.

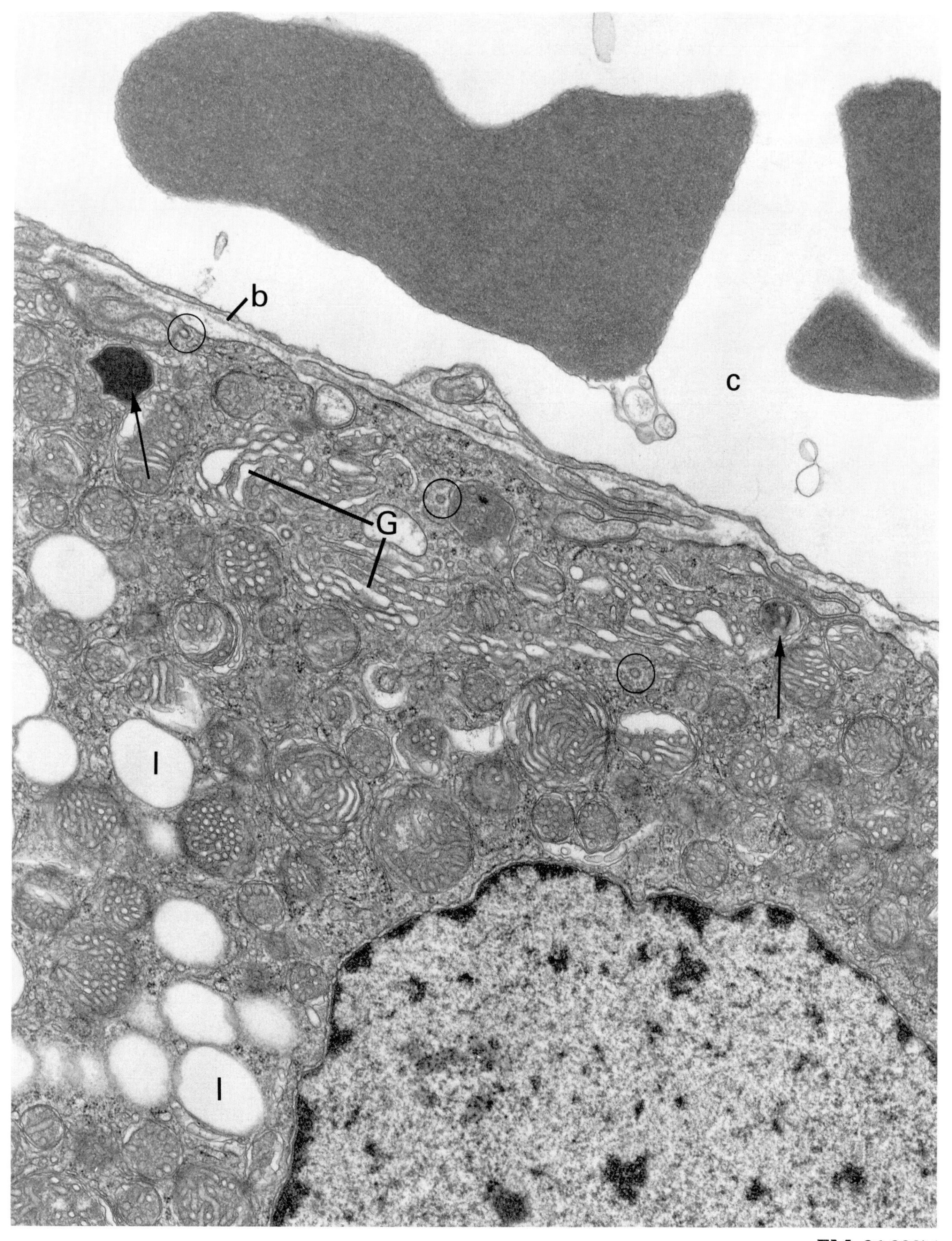

EM: 34,000×

ADRENAL CORTEX: Zona Reticularis

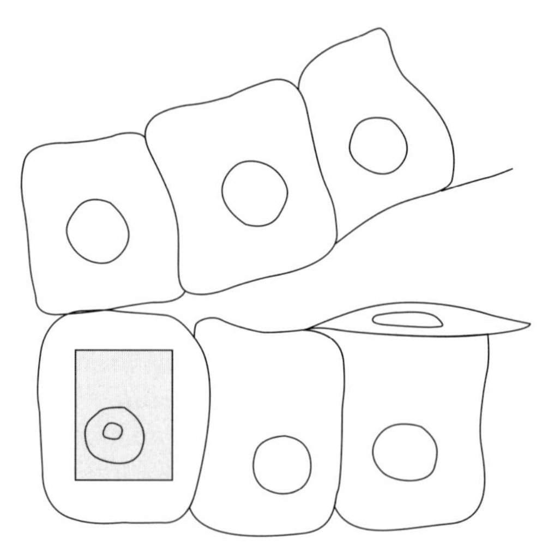

The cells of the **zona reticularis** (micrograph) are smaller and contain less lipid and fewer mitochondria than do cells in the other cortical zones. This morphology reflects a reduced responsiveness to ACTH and relatively low levels of hormone synthesis and secretion. The activities of many of the enzymes, including the enzyme needed in the cleavage of cholesterol to pregnenolone, are consistently lower than in the fasciculata. Even though the mitochondrial activity is relatively low, tubular cristae (arrowheads, micrograph), characteristic of the other cortical zones, are present.

The zona reticularis is in a unique position in the adrenal gland as the last to receive steroid-laden cortical blood and the closest to the medullary blood rich in catecholamines. Evidence suggests that it is also the "oldest" cortical zone since some differentiation and aging proceeds from the cortex periphery toward the center of the gland. The reticularis is distinguished

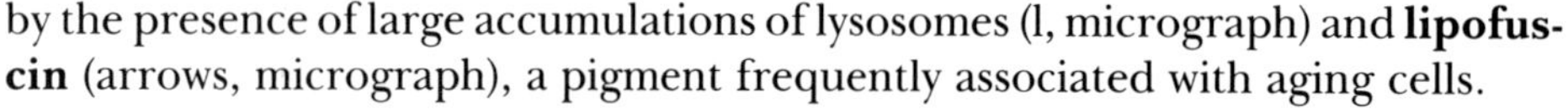

by the presence of large accumulations of lysosomes (l, micrograph) and **lipofuscin** (arrows, micrograph), a pigment frequently associated with aging cells.

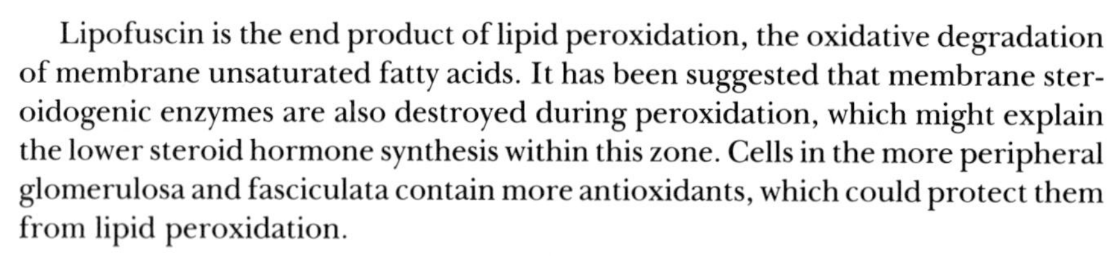

Lipofuscin is the end product of lipid peroxidation, the oxidative degradation of membrane unsaturated fatty acids. It has been suggested that membrane steroidogenic enzymes are also destroyed during peroxidation, which might explain the lower steroid hormone synthesis within this zone. Cells in the more peripheral glomerulosa and fasciculata contain more antioxidants, which could protect them from lipid peroxidation.

Androgens, in particular dehydroepiandrosterone and androstenedione, are produced by the zonas fasciculata and reticularis of the adrenal cortex. In addition, the reticularis may produce its own androgen, dehydroepiandrosterone sulfate. These androgens have only a fraction of the activity of the primary androgen, testosterone; however, they can be converted by peripheral tissues into more active androgens and also into estrogens.

When secreted in normal amounts, the levels of adrenal androgens and their conversion products usually have little effect. One exception to this occurs in obese individuals, in whom the conversion of androgens to estrogens by large numbers of adipocytes can have important clinical implications. Higher serum estrogen levels in obese women, created by conversion of adrenal androgens, can have both positive effects (counteract osteoporosis) and negative effects (increase incidence of endometrial cancer).

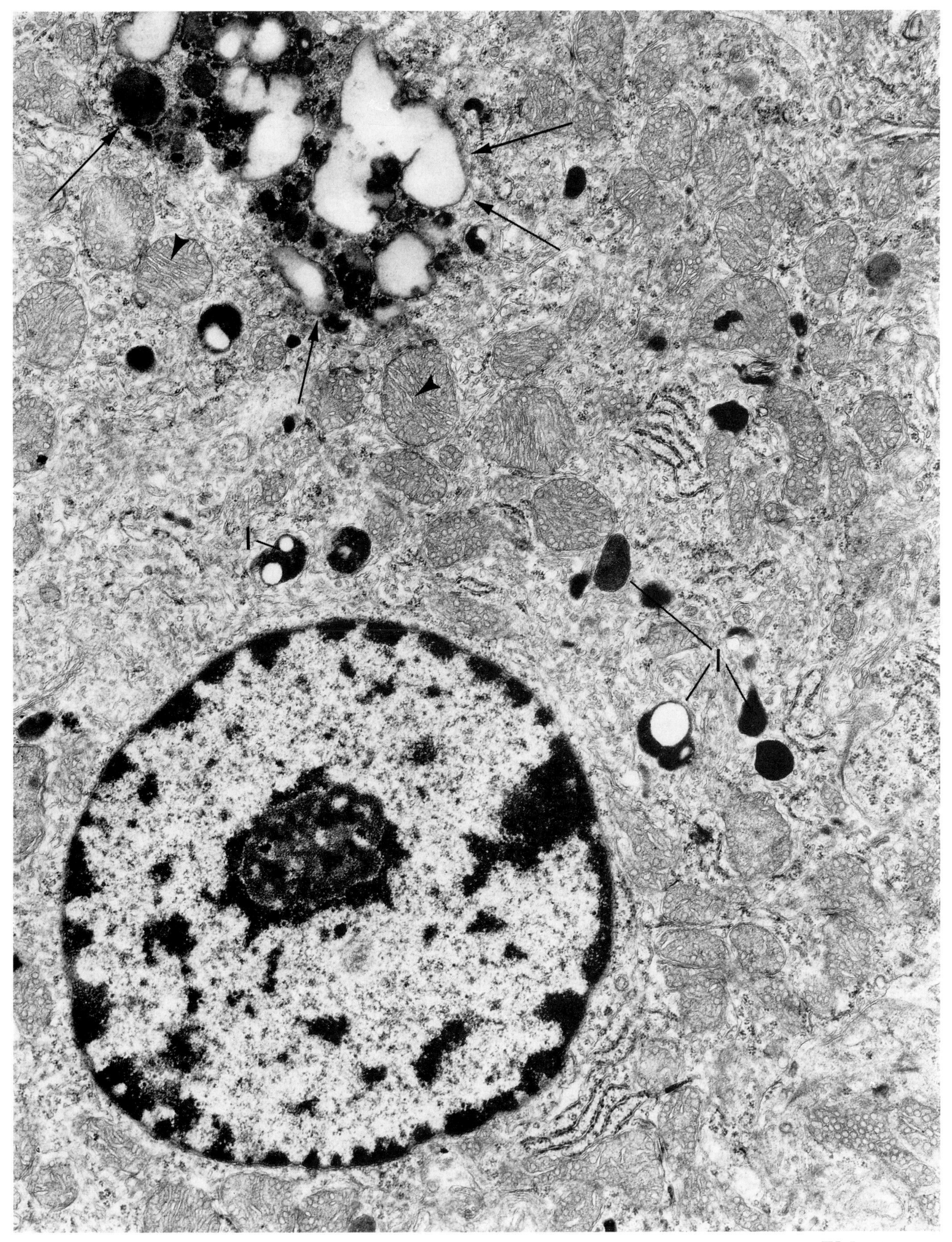

EM: 20,400×

ADRENAL MEDULLA

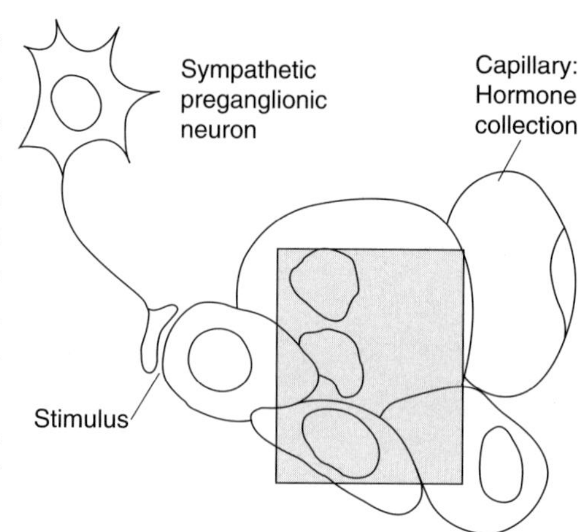

The catecholamines **epinephrine** and **norepinephrine** are the major hormones synthesized and secreted by the adrenal medulla. Norepinephrine is stored in large granules with a dense eccentric core (arrows, micrograph), whereas epinephrine is stored in smaller granules with a less dense central core (arrowheads, micrograph). Most of the medullary cells in humans store epinephrine. Both epinephrine and norepinephrine are tightly and efficiently packed in each granule by special molecules such as chromogranin A. Up to 3 million molecules of catecholamines are concentrated within a single granule. This large store of catecholamines "floods" the system within seconds after the onset of stress, exerting on most cells a variety of effects that prepare the body for "fight or flight."

Catecholamines are synthesized from **tyrosine** in a series of reactions that occur initially in the cytosol to form dopamine. Dopamine is transported into existing granules and converted within the granule to norepinephrine. Norepinephrine leaks out of the granule and is converted to epinephrine in the cytosol in a reaction catalyzed by phenylethanolamine-N-methyltransferase **(PNMT)**. The synthesis of this enzyme is induced by high concentrations of glucocorticoids that reach the medulla from the adrenal cortical capillaries. When epinephrine is synthesized, it is moved back into the granule for storage. The energy for movements of molecules into the granule is supplied by an **electrochemical gradient** much the same as operates across the inner mitochondrial membrane.

The granules in which catecholamine synthesis occurs originate from the classical rough ER-to-Golgi pathway. When the granules emerge from the trans Golgi, they do not contain catecholamines, but they do contain within their membrane certain enzymes necessary for catecholamine formation, and channels and pumps needed for the movement of substrates into and out of the granule as hormone synthesis progresses. The variation in granule morphology (particularly evident in the boxed region of the micrograph) reflects changes in content, as precursors and products vary during synthesis and packaging.

Cells in the adrenal medulla are innervated by preganglionic sympathetic neurons that control their secretion. Medullary cells are, in fact, **modified sympathetic postganglionic neurons** that develop, like other peripheral nervous system neurons, from the neural crest. Frequently, the sympathetic nervous system and the adrenal medulla are considered together as the **"sympathoadrenal medullary system."** The initial and fastest reaction to stress is sympathetic, via both direct innervation and indirect control involving adrenal medulla secretion.

In vitro, medullary cells send out processes typical of neurons. Attempts have been made to substitute adrenal medullary cells for damaged brain cells in Parkinson's disease. Dopamine synthesized by adrenal medullary cells can compensate for the deficiency of dopamine associated with this disease.

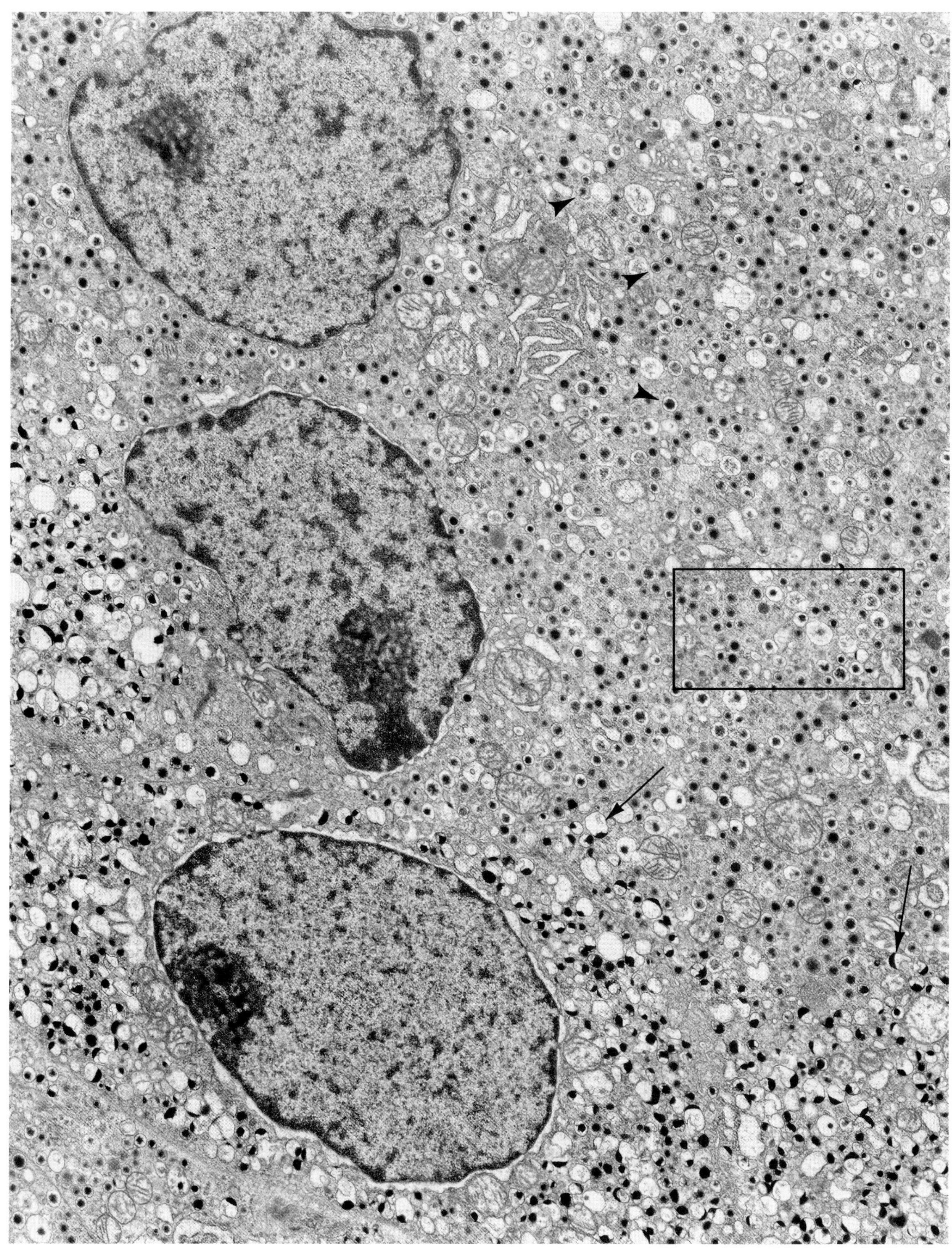

EM: 9,600×

CHAPTER TEN

SKIN

Overview

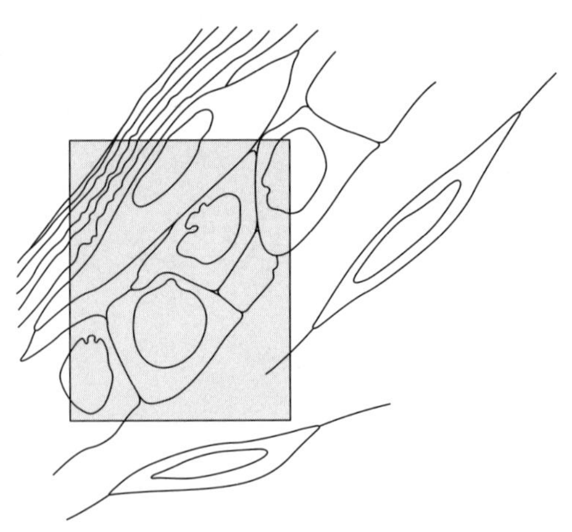

Skin, the largest organ in the body, is the major direct interface between the environment and the internal organs. Its components, an **epidermis, epidermal derivatives** (hair, glands, nails), and **dermis,** act as a protective barrier and at the same time serve as a critical means of communication between the external and internal environments.

The **epidermis** (E, bracket, micrograph) is a stratified squamous epithelium composed primarily of **keratinocytes** that undergo differentiation that culminates in a surface layer of dead cells packed with the protein keratin. Keratin provides the characteristic strength and inertness to the epidermis. As the keratinocytes mature and move closer to the skin surface, sequential expression (often in pairs) of genes within a family controlling keratin synthesis results in changes in the composition, size, organization, and quantity of keratin within the cell.

The **layers of the epidermis** are most easily defined by the changes that occur in keratin as differentiation proceeds. Stem cells in the stratum basalis (B, micrograph) and cells in the next stage of differentiation in the stratum spinosum (S, micrograph) synthesize keratin in the form of tonofilaments (arrows, micrograph). These filaments and their associated desmosomes (circles, micrograph) maintain the structural integrity of the epidermis. With age, the tonofilaments cross-link to form granules in the stratum granulosum (G, micrograph). In the final stages of differentiation, organelles are completely replaced with mature keratin that fills the highly ordered cells of the stratum corneum (C, micrograph). As cells are added from the bottom, other cells are lost or desquamated from the surface. Thus, optimal epidermal thickness is maintained. The progression from division to desquamation normally occurs in one month.

> The switch from basal cell mitosis to the beginning of programmed cell death occurs when keratinocytes detach from the basal lamina. Fibronectin, a prominent component of basal lamina, can regulate the terminal differentiation of keratinocytes.

Keratinocytes provide a framework that houses other cell types. Together, these cells carry out functions including protection of underlying tissue (from desiccation, foreign invasion, physical trauma, and UV irradiation), sensation, and the formation of a vitamin D precursor.

The **dermis** consists of a thin layer of loose connective tissue (papillary layer; p, micrograph) directly under the epidermis, and a deeper, thicker layer of dense irregular connective tissue (reticular layer; r, bracket, micrograph). The **papillary layer** (unusually thin in this micrograph) is the major region of cellular traffic associated with immune and inflammatory reactions. The **reticular layer,** recognized by dense masses of collagen (c, micrograph), provides strength and elasticity, is a conduit for nerves and the larger blood vessels, and, along with the hypodermis, houses the basal portions of hairs and the secretory regions of glands. Secretions of fibroblasts, the hallmark cell of loose and dense connective tissue, include not only the extracellular components of these tissues but also factors necessary for the normal division and differentiation of the epidermis. A codependency between dermis and epidermis operates at many levels.

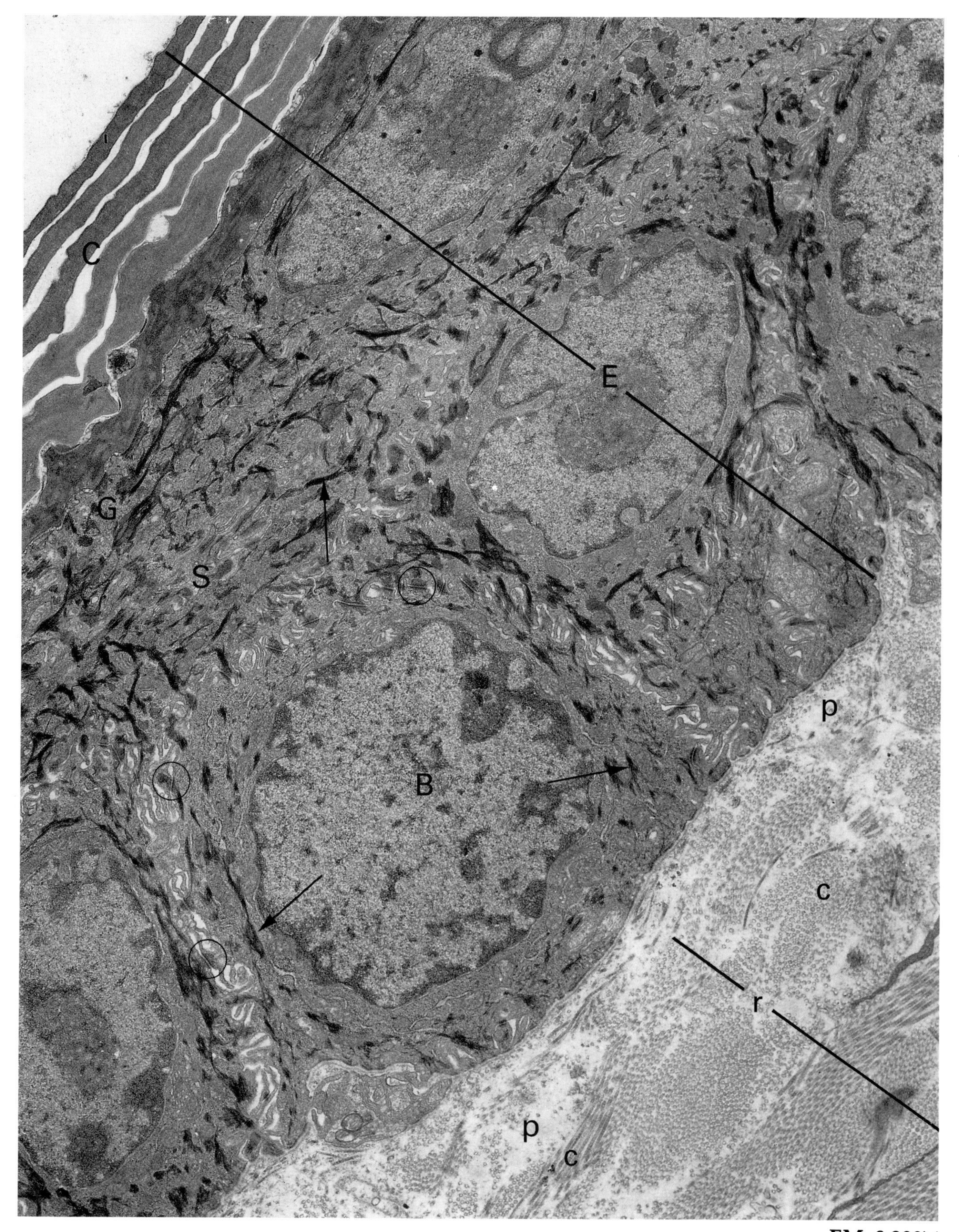

EM: 8,200×

EPIDERMIS: Stratum Granulosum

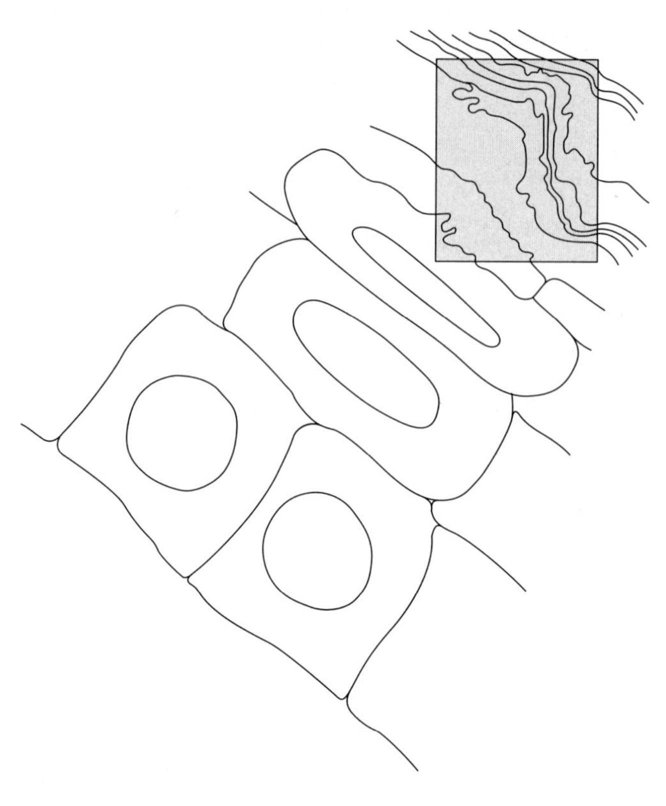

Critical events in the differentiation of the keratinocyte occur in the **stratum granulosum** (G, micrograph). The extent of these changes is clearly appreciated by comparing the structure of the relatively undifferentiated precursor cells in the stratum spinosum (S, micrograph) to the fully differentiated product, the stratum corneum cells (C, micrograph).

The straum granulosum is named for the presence of nonmembrane-bound **keratohyalin granules** (k, micrograph) that appear in conjunction with the tonofilaments (t, micrograph). These granules contain the protein **filaggrin,** which cross-links tonofilaments as keratin takes on a new form. Many of the ribosomes (arrowheads, micrograph) in the granulosum cells are concentrated around these granules and may be involved in the synthesis of filaggrin. The patchy occurrence of ribosomes reflects an overall alteration in the cytoplasm as these cells differentiate and lose many organelles, including nuclei. Even after the cells have lost their transcription ability changes in the keratin continue to occur (e.g., new disulfide cross-links, interaction with matrix proteins, molecular size changes) that are reflected by distinct ultrastructural changes during progression from the stratum granulosum outward (micrograph).

Some of the events related to cell death in the stratum granulosum begin to take place in the stratum spinosum. In the stratum spinosum the evenly dispersed ribosomes, in addition to supporting all of the metabolic events of this active cell, direct the synthesis of proteins that are essential to differentiation into the stratum granulosum, including the enzyme **transglutaminase** and the structural protein **involucrin.** As the spinosum cells differentiate into granulosum cells, calcium levels rise and activate transglutaminase. Once activated, transglutaminase cross-links involucrin to the inner leaf of the cell membrane, creating an impermeable barrier that isolates each cell from nutrients and accelerates cell death. The thickness of the involucrin envelope can be appreciated in certain areas (arrows, micrograph) in the stratum corneum.

Before being sealed and isolated from the environment, the cells of the stratum granulosum release a lipid secretion that occupies the spaces between corneum cells, providing cohesion and creating the permeability barrier of skin. Lipid can be seen in the cytoplasm as granules called **membrane-coating granules** (g, micrograph). After secretion lipid (l, micrograph) accumulates in the space between the stratum granulosum and the stratum corneum. The removal of sulfate from cholesterol sulfate, an important mortar component, is a necessary prerequisite for desquamation. Desquamation also depends upon the progressive loss of desmosomes, which are reduced to remnants (d, micrograph) in the stratum corneum.

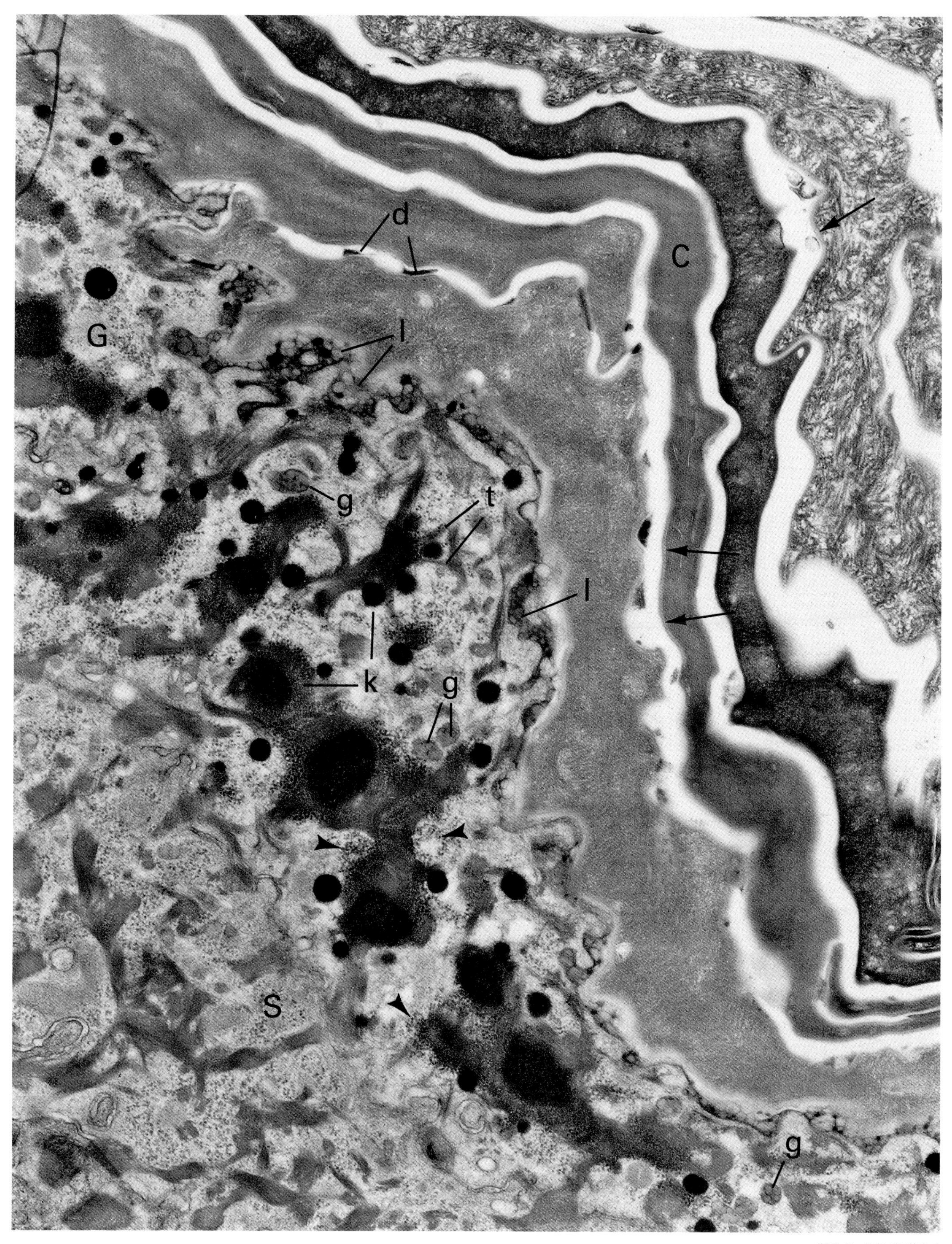

EM: 29,000×

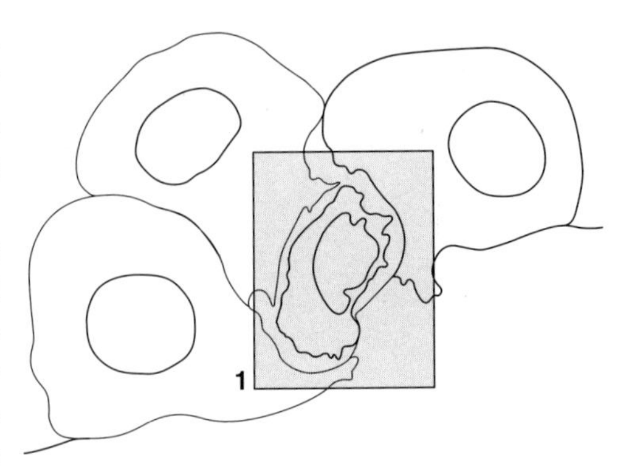

Dividing keratinocytes in the stratum basalis are vulnerable to mutation caused by exposure to short-wavelength UV radiation. A cap of melanin granules over the nucleus of these cells protects against chromosome damage by absorbing the potentially destructive radiation.

Melanin is produced by **melanocytes** (M, micrograph 1) that migrate from the neural crest to the epidermis during fetal development. Melanocytes enter the epidermis and remain attached to the basal lamina (b, micrograph 1), where they occupy regions between basal keratinocytes (K, micrograph 1). Melanocytes are easily recognized by (1) their tendency to separate from keratinocytes during tissue preparation, since these two cell types are not attached by desmosomes, and (2) the absence of tonofilaments (t, micrograph 1), which are obvious in keratinocytes. In contrast to keratinocytes, which move away from the basal lamina as they divide and differentiate, melanocytes maintain their position among the stem keratinocytes.

Melanin is synthesized within melanosomes (m, micrograph 1). Coated vesicles originating from the Golgi concentrate **tyrosinase,** the critical enzyme necessary for three steps in the single pathway to melanin formation. The coated vesicles become melanosomes as tyrosinase's substrates (tyrosine, 3,4-dihydroxyphenylalanine or dopa, and 5,6-dihydroxyindole) become available and the reactions take place to form melanin. Special melanosome matrix proteins provide a lamellar scaffolding (arrows, inset, courtesy of Dr. A. Breathnach) that aids in the spatial organization of the biochemical pathway associated with melanin formation. The protein framework that can be seen in the early melanosomes (e, inset) is obscured in mature granules (g, inset).

Melanosomes are donated to surrounding keratinocytes by the melanocytes that produce them. Melanosomes move to and concentrate within thin melanocyte processes (arrows, micrograph). The most common mechanism of intercellular transfer is phagocytosis of these processes by keratinocytes. The keratinocytes adjacent to the melanocyte in micrograph 1 have taken up several melanosomes (arrowheads). Most of the melanosomes within the keratinocyte cytoplasm will be moved to form a cap over the nucleus. Each melanocyte "feeds" a defined number of keratinocytes, and this number varies in different regions of the body. One melanocyte can feed as many as 36. Melanocytes do divide normally during routine turnover; however a set ratio is consistently maintained. A change in the ratio characteristic for a given region is an indication of a proliferative condition, either nonmalignant moles or malignant melanoma. In the absence of melanin, as in the albino condition, DNA is not protected from UV radiation and there is an increased incidence of DNA damage and epidermal cancer.

Differences in skin color are not related to differences in the number of melanocytes, but rather to the amount of melanin per cell, reflected by the size or number of melanosomes. The darkening of skin color associated with tanning initially involves a change in the configuration of existing melanin, but can subsequently involve increased synthesis. With prolonged sun exposure, both melanin structure and synthesis are altered, reducing the protection of DNA.

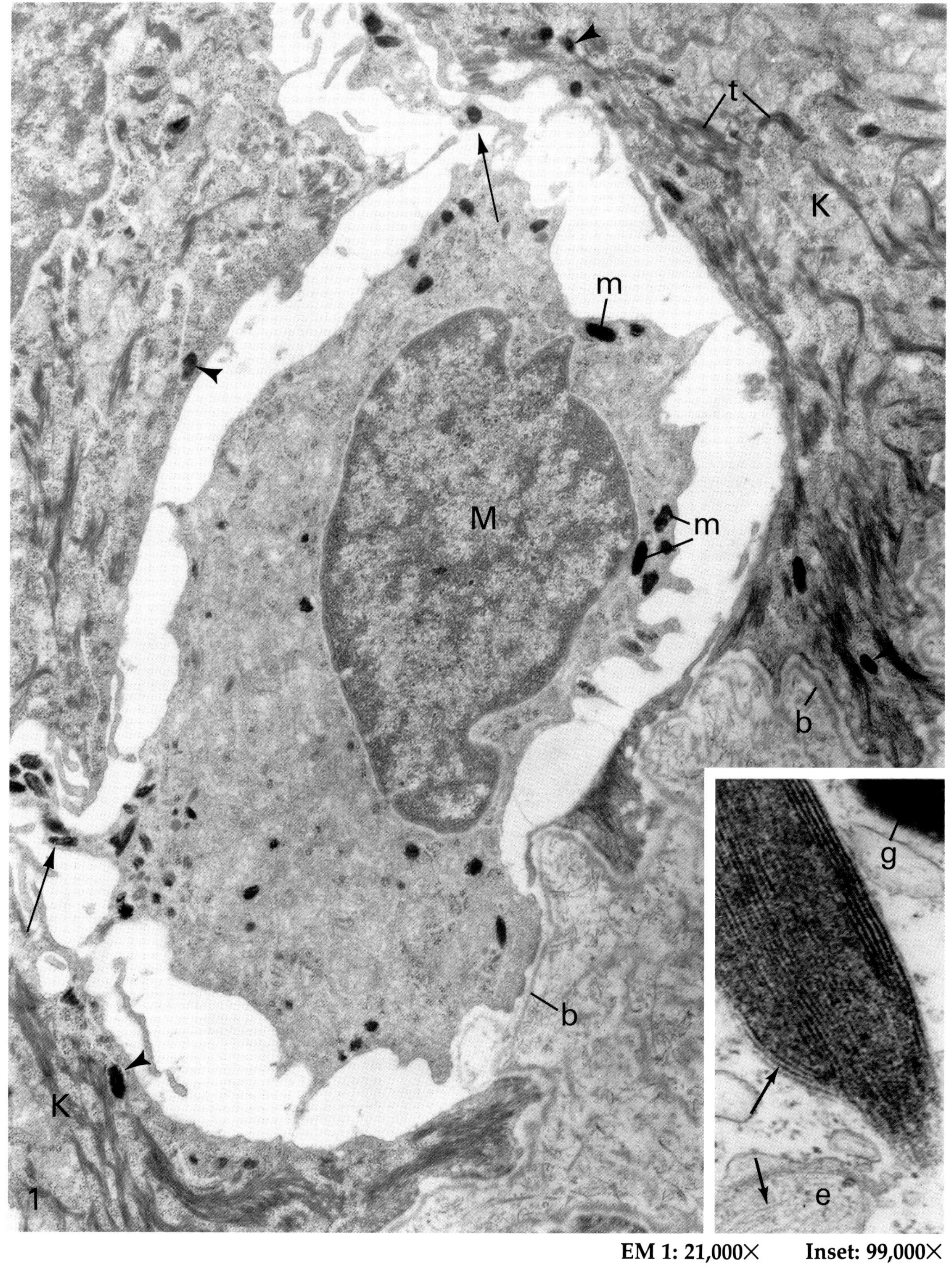

EM 1: 21,000× Inset: 99,000×

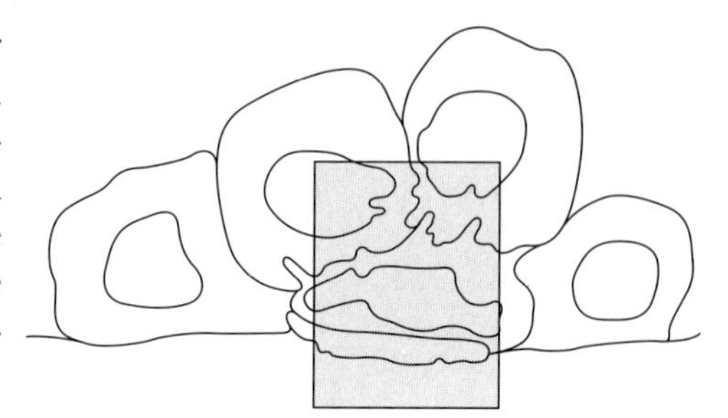

In addition to its obvious protective role, skin is a major sensory organ. Afferent neurons carry sensation to the central nervous system, where the information is processed. The sensory endings of these neurons terminate in both epidermis and dermis and also below the skin in the hypodermis. They are either naked (i.e., free) or associated with other cells that encase them (as in Pacinian, Krause, Ruffini's and Meissner's corpuscles) or, more simply, abut them (as with Merkel cells).

Sensory modalities, including pain, temperature, and touch, are sometimes matched with a particular structure. The **Merkel cell – neuron complex** in the micrograph is a mechanoreceptor localized within the epidermis, found predominantly in the most sensitive areas such as the lips and fingertips. Each Merkel cell (M, micrograph) rests in a cuplike depression formed by the terminal ending of its associated axon (A, micrograph).

Merkel cells have a structure characteristic of transducer sensory cells that act as intermediates between an initial stimulus and the neuron impulse. They contain: (1) **microvilli** (m, micrograph) seen projecting into depressions in adjacent keratinocytes (K, micrograph); (2) **granules** (arrows, micrograph) that contain peptides that are known neurotransmitters; and (3) **synapses** with their associated neurons (arrowheads in the micrograph indicate a possible synaptic region). Based on this overall ultrastructure, it is often proposed that the Merkel cell is a mechanoreceptor in which deformation of the microvilli causes granule release at the synapse and activation of the underlying neuron. However, electrophysiological studies have demonstrated that chemical synapse activity in Merkel cells is too slow to account for the generation of afferent impulses. In addition, the neuron ending itself is capable of responding directly as a mechanoreceptor.

There is evidence to suggest that even though the Merkel cell may not be the primary transducer, it can modify the neuron's response by affecting the threshold of response. In all of the other skin receptors, associated cells (and extracellular material) often modify the afferent nerve response but do not act as transducers of incoming messages as do classical receptor cells in other sensory modalities such as taste, vision, and hearing.

Merkel cells may also (1) provide essential metabolic support for the associated neuron (note the large numbers of mitochondria in the neuron, a characteristic of afferent as well as efferent nerve terminals); (2) be a "target" essential to the movement of neurons into the epidermis during development and nerve regeneration following injury; and (3) have effects associated with the release of neuropeptides, which, in addition to acting locally as synaptic transmitters, have far-reaching effects on autonomic nerves, blood vessels, and inflammatory cells.

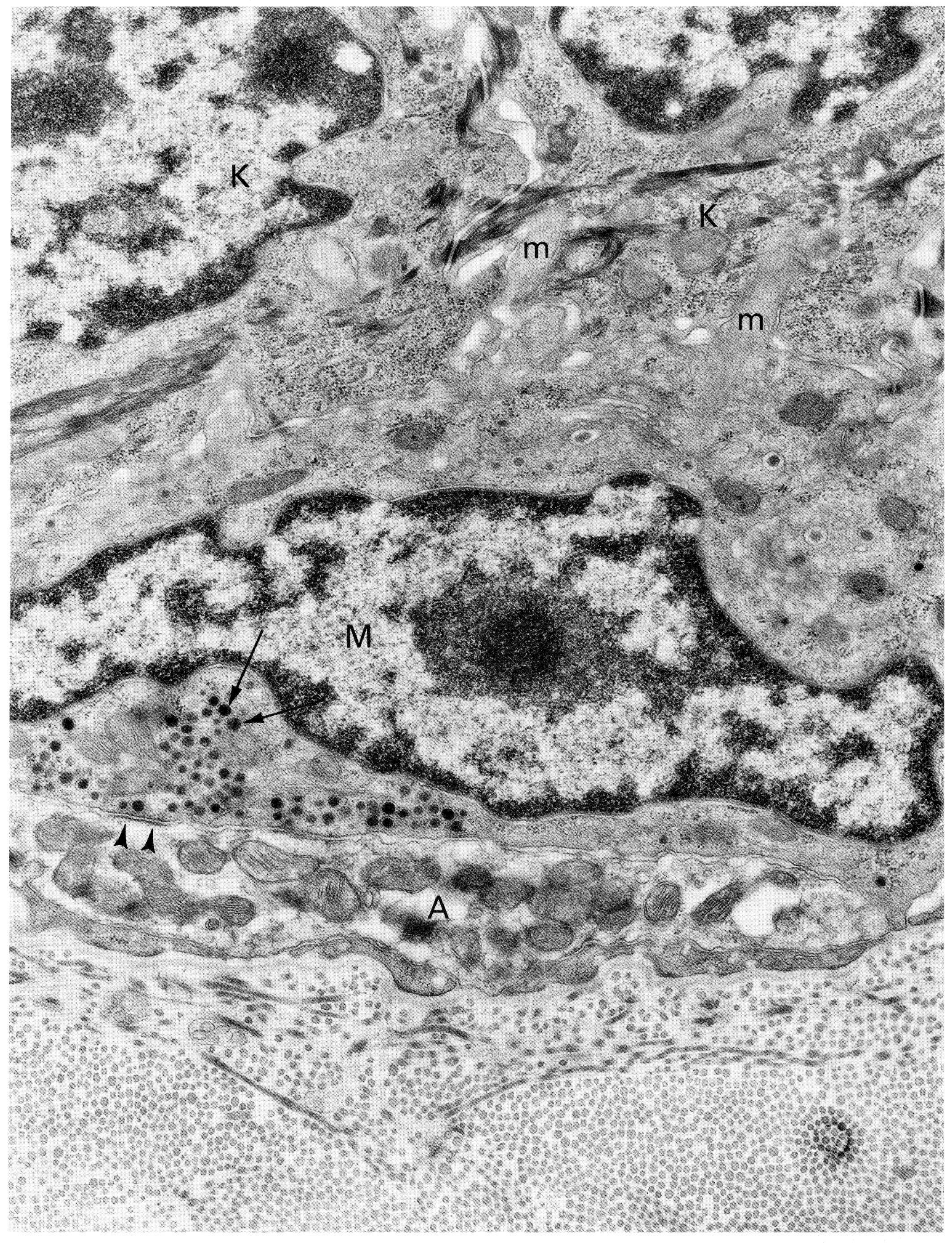

EM: 30,000×

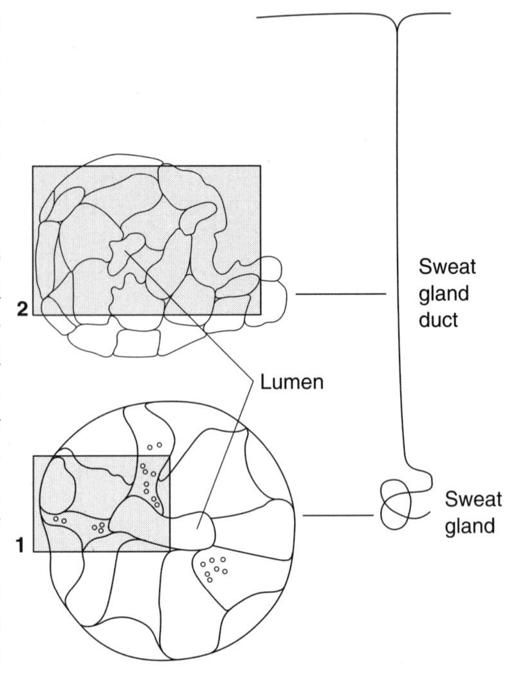

The surface of the epidermis is bathed with secretions released from eccrine (watery-serous secretion), sebaceous (lipid secretion), and apocrine (carbohydrate-rich secretion) glands. Of these, **eccrine sweat glands** (micrographs) are the most numerous and functionally significant. They help maintain homeostasis by their role in electrolyte balance, excretion, and thermoregulation.

The **secretory portion** (micrograph 1) of eccrine glands is found in the dermis (or hypodermis) at the end of a simple coiled tubular **duct** (micrograph 2). The major components of the secretion, ions (primarily sodium chloride) and water, are produced by **"clear" cells** (C, micrograph 1). Movement of NaCl across the cell into the lumen is driven by sodium/potassium exchange pumps located in the highly infolded basolateral membranes (arrowheads, micrograph 1). Water follows passively. Energy required for pumping activity is provided by mitochondria (m, micrograph 1).

The primary secretion released from the clear cells is isotonic with plasma. As it moves through the duct, sodium and chloride are reabsorbed by the lining cells in order to conserve these essential electrolytes. Sodium/potassium pumps in the highly infolded basolateral membranes (arrowheads, micrograph 2) of both cell layers of the duct move sodium out of the cell into the interstitium. Chloride follows passively through special channels in the membrane. Since the ductal lining cells are not as permeable to water as the secretory cells, the removal of ions from the lumen results in a final sweat that is hypotonic. Evaporation of this watery sweat acts to dissipate body heat, a significant mechanism of thermoregulation. Each individual has approximately 2 million eccrine sweat glands, situated in skin of nearly all regions.

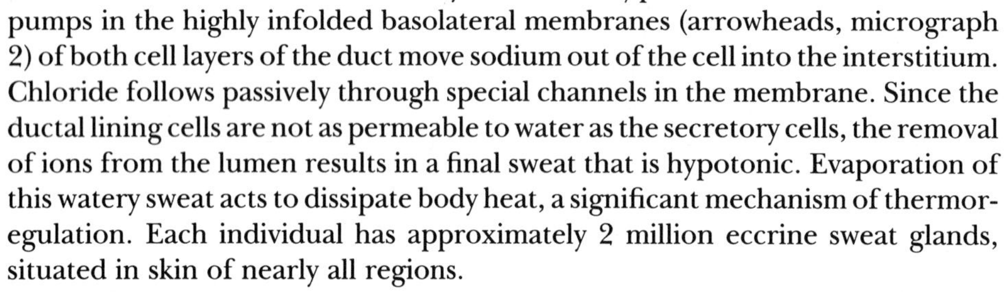

Cystic fibrosis is a fatal disease related to a single gene defect affecting the chloride channel. The reduced chloride permeability in the sweat gland ducts in individuals with this disease results in increased levels of NaCl in sweat, a characteristic frequently used as a diagnostic tool. A similar ion imbalance in respiratory epithelium leads to a thick, sticky mucus secretion, infection, and a subsequent life-threatening blockage of airways.

Dark cells (D, micrograph 1), named for their dense granules, are an integral part of the simple epithelium of the secretory portion, attached to the clear cells by junctional complexes (circle, micrograph 1). The granules contain glycoproteins that are released into the sweat gland lumen by exocytosis. Epidermal growth factor, a component of sweat, appears to be stored in the dark cell granules and may be involved in regulating the function of gland epithelial and **myoepithelial cells** (M, micrograph 1). Myoepithelial cells contract, as in other locations, to facilitate the expulsion of secretion.

The secretory cells make other significant contributions to the final secretion, including IgA (from plasma cells), which moves across the dark cells, and urea (from plasma), which moves across the clear cells. Eccrine glands share with the kidneys the role of urea excretion; elevated urea concentrations in sweat can suggest kidney failure.

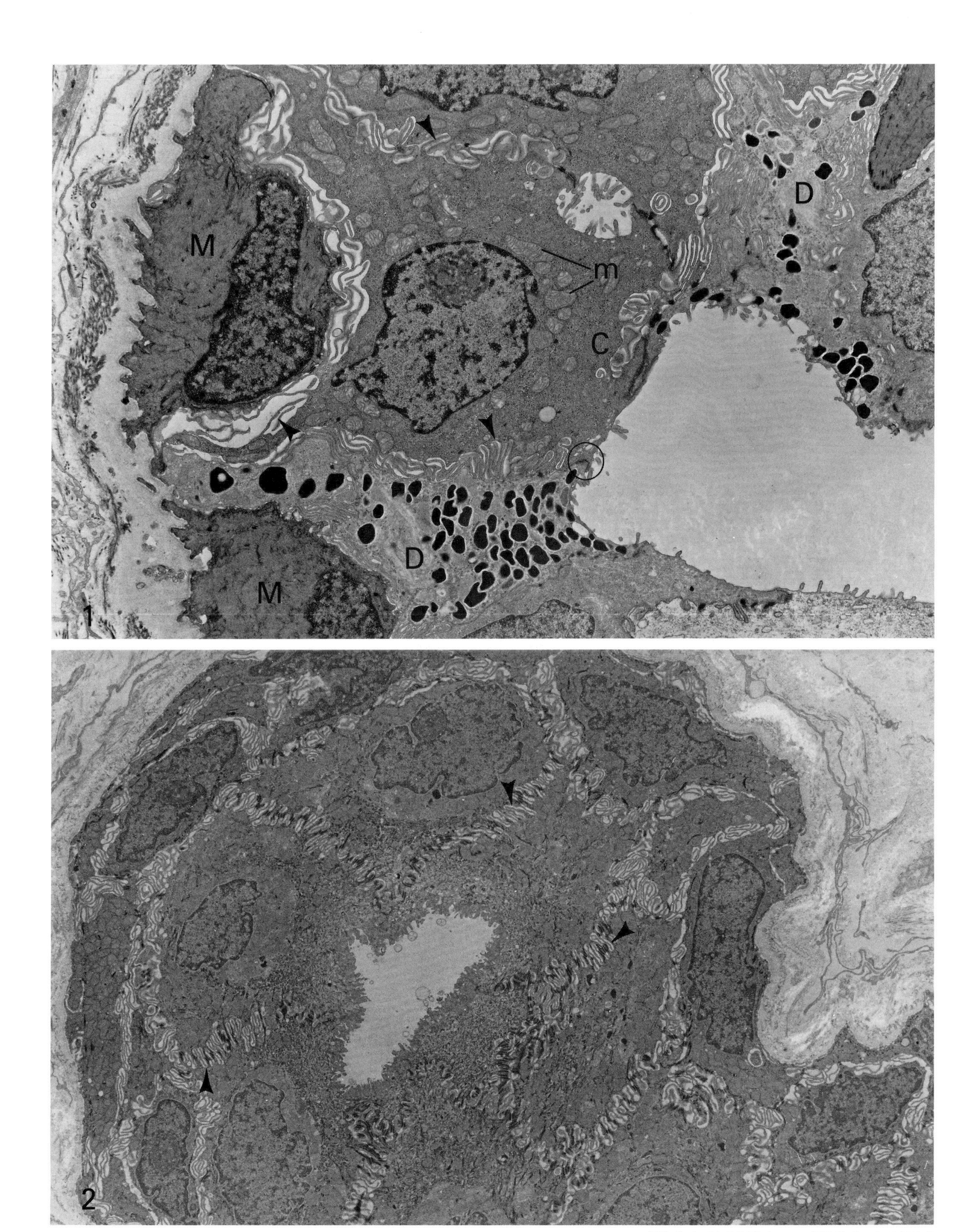

EM 1: 7,200× EM 2: 4,800×

EPIDERMAL DERIVATIVES: Hair and Sebaceous Glands

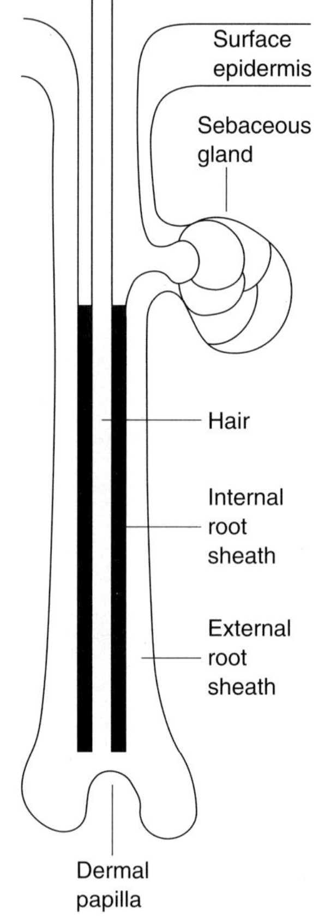

The micrographs illustrate the surface of **hair** (scanning electron micrograph 1), the **surrounding sheaths** (micrographs 2 and 3), and a **sebaceous gland** (micrograph 4). The adjacent diagram provides orientation to the relationship between these structures. An external root sheath grows down from and is continuous with the stratum basalis and spinosum of the surface epidermis. This sheath contains the stem cells that give rise to the hair, an internal root sheath, and sebaceous and apocrine glands. The hair and internal root sheath grow upward together associated by interlocking plates (surface plates of hair seen on micrograph 1) until the internal root sheath ends at the level of the sebaceous glands.

The **external root sheath** (e) is seen in micrographs 2, 3, and 4. The stem cells (S, micrograph 2) within the sheath, like stem cells in the gastrointestinal tract, have one of the most rapid mitotic cycles in the body, with the potential to replace damaged epidermis quickly and effectively. Two cell layers of the **internal root sheath** (i) are evident in micrographs 2 and 3. In both of these micrographs the outer layer (asterisk) is completely keratinized (except for two remaining nuclei, N, in micrograph 2), while the inner layer, in an earlier stage of differentiation, still contains granules (g). The granules, like keratohyalin granules of epidermis, are closely associated with keratin filaments (arrows, micrograph 3), are not membrane bound, and have been shown to contain the filament-linking protein filaggrin. At the point where the internal sheath degenerates, the space (S, micrograph 4) surrounding the hair becomes filled with the secretion of sebaceous glands.

The keratinocytes of the hair cortex and covering cuticle are packed with keratin filaments that are particularly insoluble, unreactive, and contain more disulfide bonds than does "soft" keratin. Formation of this "hard" keratin does not involve a keratohyalin granule stage, and the cells do not undergo regular desquamation. Hairs grow until the end of a growth cycle, after which the entire hair is replaced. On the scalp, hairs are replaced every two to six years. Adjacent hairs are in different phases of the cycle, so the dynamics of hair replacement is not obvious.

The development of sebaceous and apocrine glands is typically coupled to the development of hairs. **Sebaceous glands** secrete sebum, which contains fatty acids and is rich in squalene and wax esters, lipids not formed by the surface epidermis itself. As the gland cells differentiate, they increase in size and become packed with lipid droplets (L, micrograph 4). At their final stage of development they rupture and become the secretion. Sebaceous secretion contributes to the lipid coating of the epidermis and may act as a lubricant, but whether it has any added effect in preventing water loss over and above lipids produced by the surface epidermis is questionable. Sebaceous glands (like apocrine glands and certain types of hairs) undergo maximum development in response to rising androgen levels at puberty.

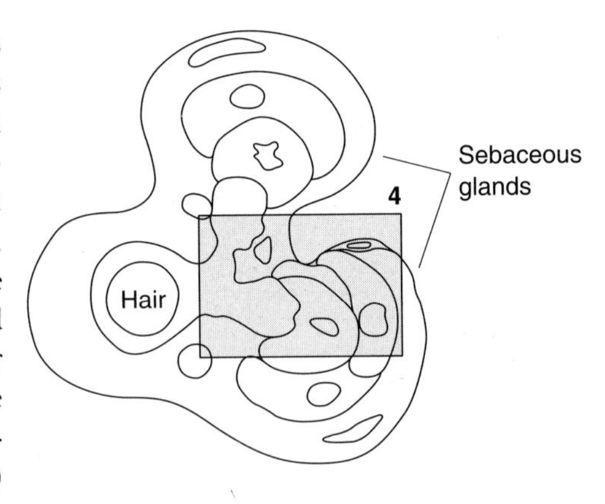

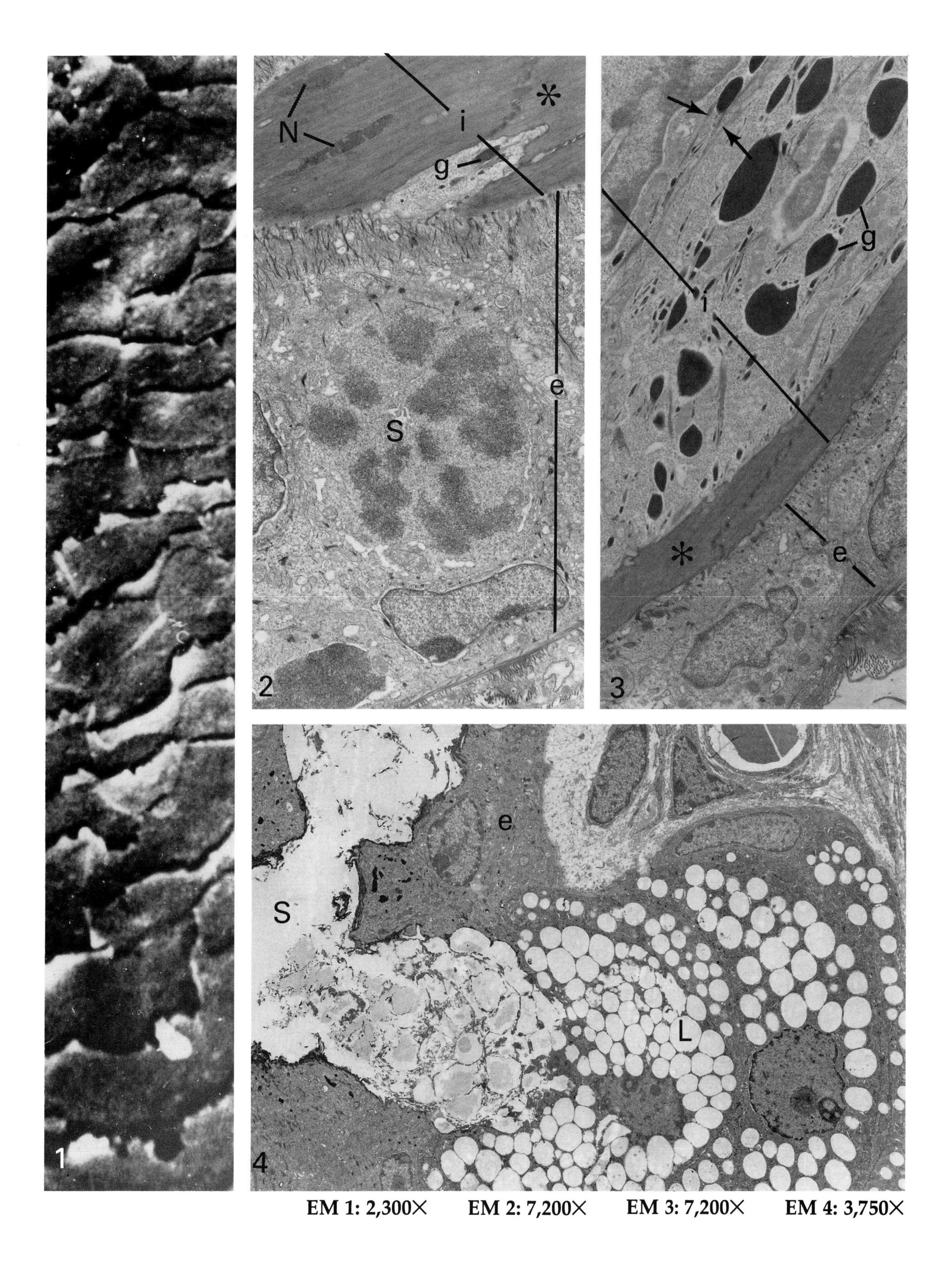

EM 1: 2,300× EM 2: 7,200× EM 3: 7,200× EM 4: 3,750×

DERMIS: Papillary Layer

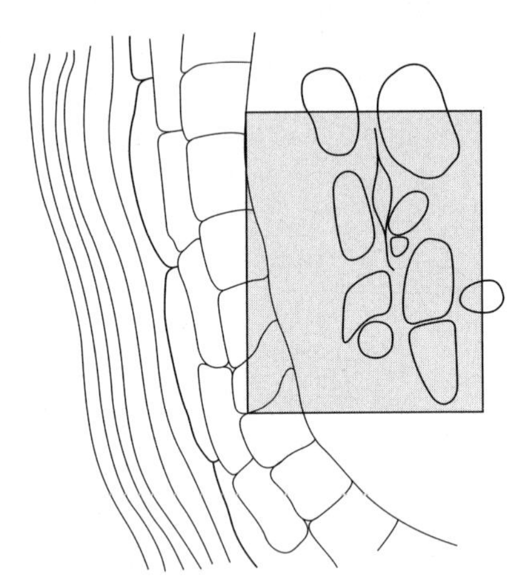

The **papillary layer** of the dermis is located directly under the epidermis (E, micrograph) and attached to it via a basement membrane (b, micrograph). A number of different types of cells are found within the loose connective tissue of the papillary layer, many of which migrate within the highly hydrated glycosaminoglycans synthesized by local **fibroblasts** (F, micrograph). Collagen (c, micrograph), another major secretory product of fibroblasts, is obvious, but considerably less concentrated here than in the underlying reticular dermis.

The papillary dermis houses an extensive network of **capillaries** (C, micrograph) that provide nutrition to the avascular epidermis. These small blood vessels are also the source of defense molecules and cells, including **lymphocytes** (L, micrograph). Some of the lymphocytes present in the connective tissue migrate into the epidermis, where, in conjunction with keratinocytes and epidermal macrophages known as Langerhans cells, they mount a specific epidermal immune response.

The amount of blood flowing through papillary capillaries is finely controlled and functions as one of the most important mechanisms of body temperature regulation. When the external temperature drops, body heat is conserved by reducing blood flow through these capillaries. Conversely, when the temperature rises, heat is lost through the epidermis via an increase in superficial blood flow. The arteriovenous shunts that adjust this flow are located in the papillary/reticular junction.

The thin, fenestrated wall of the capillary in the upper half of the micrograph differs in permeability from the thicker wall of the capillary below. The permeability of all types of papillary capillaries is influenced by both normal and abnormal events. Vasoactive substances, such as histamine released from **mast cells** (M, micrograph) as a part of a normal defense mechanism, increase vessel permeability. Dermal mast cells respond in the classic IgE-allergen manner, and also respond directly to neuropeptides released from neurons and Merkel cells. Permeability also increases when vessels are damaged, for example, as a result of immune diseases such as psoriasis or as a result of excess UV irradiation. The familiar blistering associated with sunburns is a consequence of plasma leaking through damaged capillaries and accumulating directly under the epidermis.

The thickness and composition of the dermis differs dramatically at different sites of the body. The presence of **skeletal muscle** (Mu, micrograph) is unique to regions where skin movement is under voluntary control, such as the face. The myelinated **nerve fibers** (N, micrograph) surrounded by Schwann cells (S, micrograph) may be those innervating the skeletal muscle, or they may be afferent fibers that originate from encapsulated touch receptors, Meissner's corpuscles. The bracketed region appears to be a glancing section through a Meissner's corpuscle.

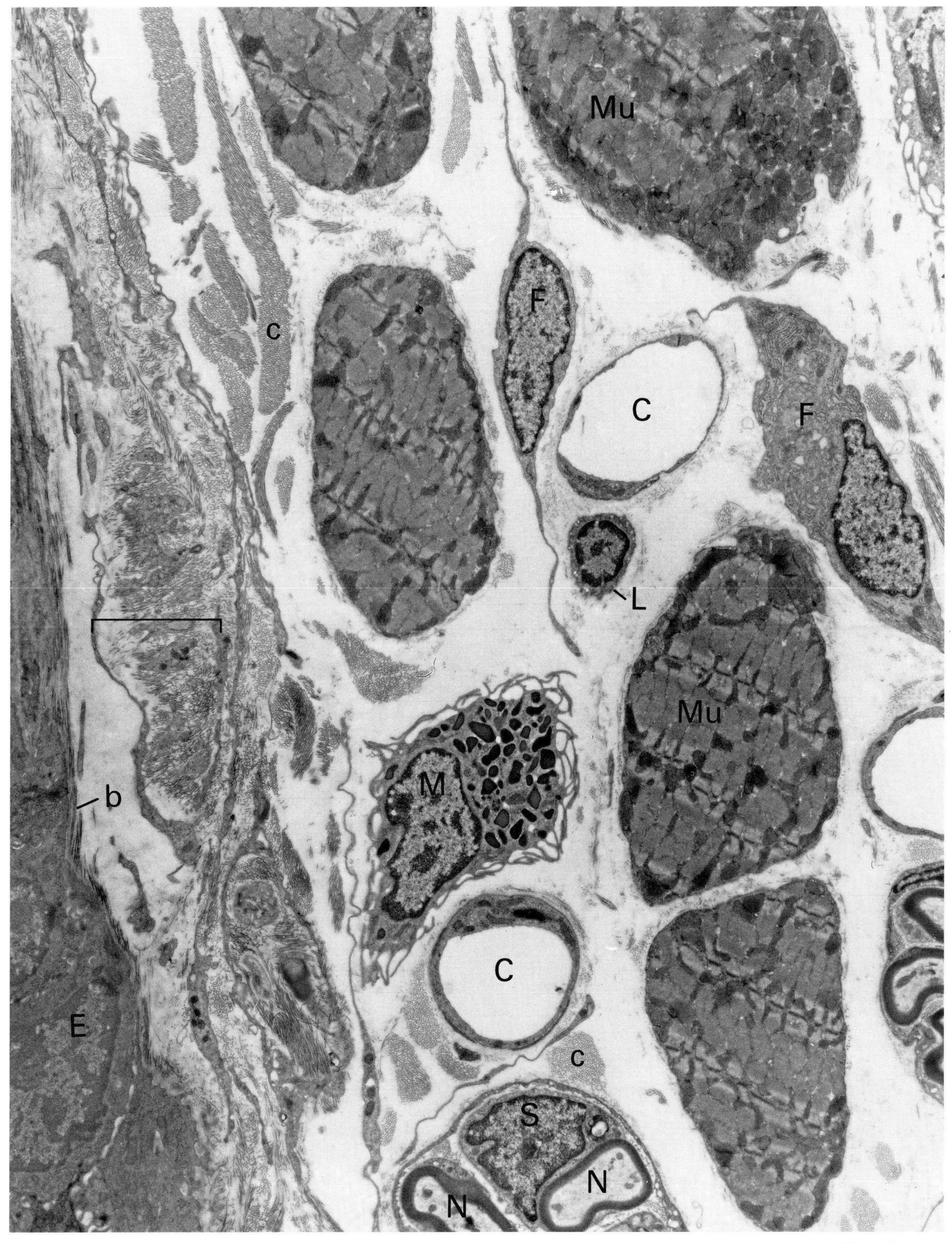

EM: 6,000×

MAJOR EXOCRINE GLANDS

SALIVARY GLAND: Serous and Mucous Cells

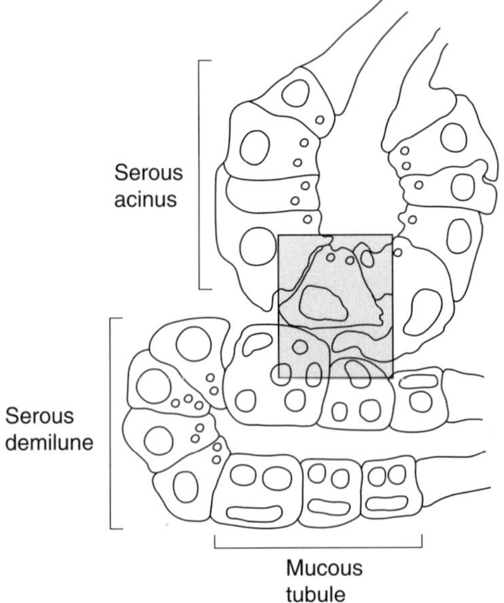

Three major salivary glands, the parotid, the submandibular (submaxillary), and the sublingual, secrete most of the saliva. In each, the primary secretion is produced by collections of glandular cells, **secretory end pieces.** Mucous cells are arranged into tubular end pieces, whereas serous cells occur as rounded **acini,** or as semicircular **demilunes** at the tips of mucous tubules. Basal laminas (arrowheads, micrograph) define the basal boundaries of the secretory end pieces and reflect their epithelial nature.

Serous cells (S, micrograph) secrete a watery component rich in enzymes important for digestion (e.g., amylase) and defense (e.g., lysozyme). These enzymes are packaged in electron-dense granules (g, micrograph) as inactive precursors, zymogens. The abundant rough endoplasmic reticulum (r, micrograph), well-developed Golgi (G, micrograph), and apically situated secretory granules are typical features of exocrine cells active in protein synthesis.

Mucous cells (M, micrograph) synthesize mucins, which are also stored in granules. Usually, however, these glycoproteins are partially extracted during tissue preparation, leaving a precipitate (*, micrograph) within the granule membrane. The synthesis of mucins begins in the flattened cisternae of rough endoplasmic reticulum in the base of the cell. Salivary mucus traps foreign organisms and provides the lubrication for swallowing and speech.

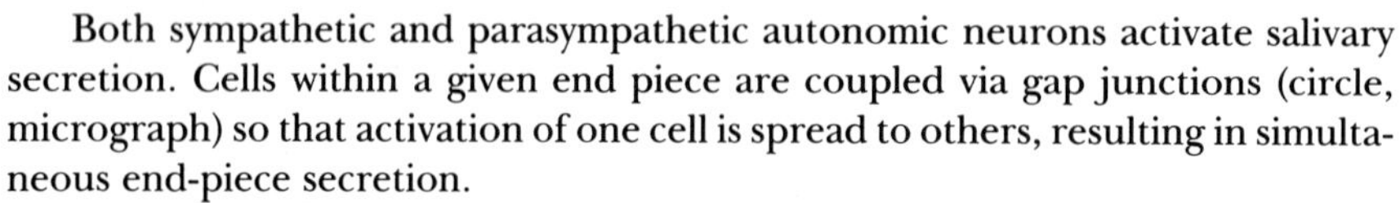

Both sympathetic and parasympathetic autonomic neurons activate salivary secretion. Cells within a given end piece are coupled via gap junctions (circle, micrograph) so that activation of one cell is spread to others, resulting in simultaneous end-piece secretion.

The primary salivary secretion is usually isotonic with plasma, which is partially a reflection of unusually leaky ($15\ \Omega \cdot cm^2$) tight junctions (arrows, micrograph). The electrolyte character of the secretion is altered as it passes through adjacent ducts.

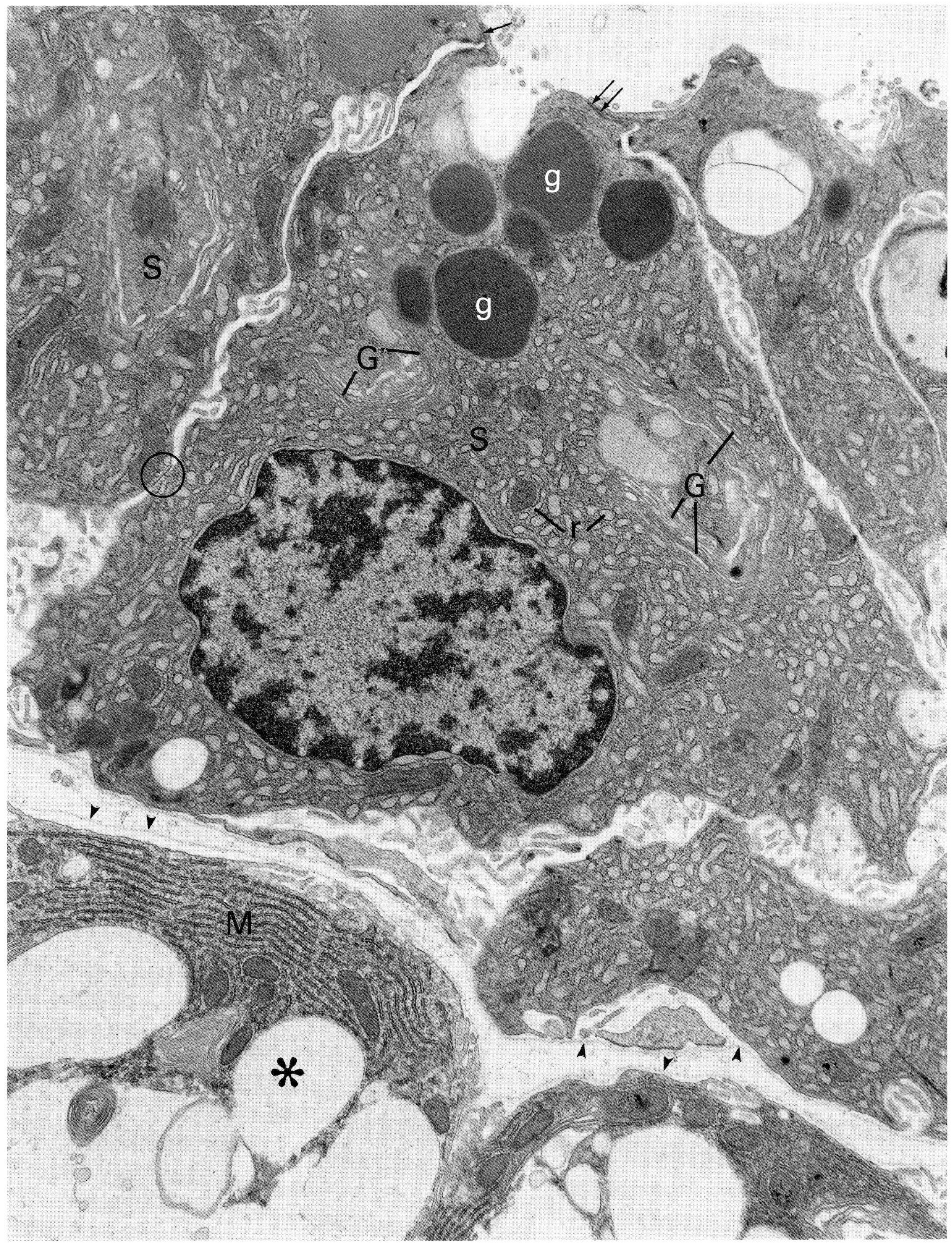

EM: 13,600×

SALIVARY GLAND: Striated Duct

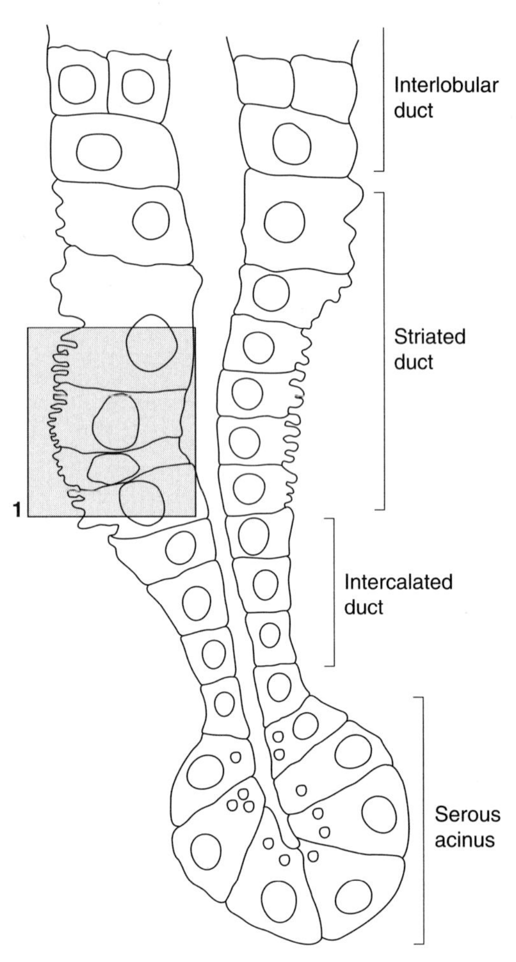

Salivary secretion passes through a **series of ducts** from the lumen of the secretory end pieces to the oral cavity. The electrolyte composition of the primary secretion is altered by active ion transport across the walls of these ducts. Ion exchange is most pronounced in the **striated ducts,** which connect small intercalated ducts to the larger interlobular ducts.

Striated ducts are composed of columnar cells with highly infolded basolateral surfaces (arrows, micrograph 1 and inset) and apically displaced nuclei (N, micrograph 1). The extensive membrane invaginations of each cell provide increased surface area for membrane pumps and channels important in ion transport. The most prominent proteins present within these membranes, $Na^{+}K^{+}$ATPases, split ATP and provide the energy for moving Na^{+} out of the cell. This active movement of Na^{+} across the basolateral surface pulls Na^{+} passively across the apical surface. In this way Na^{+} is removed from the primary secretion and returned to the circulation. The K^{+} that enters the secretion in exchange for Na^{+} reaches concentrations greater than that secreted by any other digestive gland, in some cases approaching intracellular levels. Mitochrondia (m, micrograph 1 and inset), packed between the membrane invaginations, provide the ATP substrate for the $Na^{+}K^{+}$ATPase activity.

The tonicity of the final secretion depends upon the amount of Na^{+} reabsorbed across the striated duct, since these ducts are relatively impermeable to water. The amount of Na^{+} reabsorbed is in turn dependent upon the rate of salivary secretion; faster rates, with less time for Na^{+} reabsorption, result in a saliva similar to the isotonic primary secretion, whereas slower rates, with more time for Na^{+} reabsorption, result in a more hypotonic saliva. The composition of the final salivary secretion is sensitive to the type of autonomic stimulation. Parasympathetic stimulation usually results in a copious, watery secretion, whereas sympathetic stimulation tends to result in small amounts of enzyme-rich secretion.

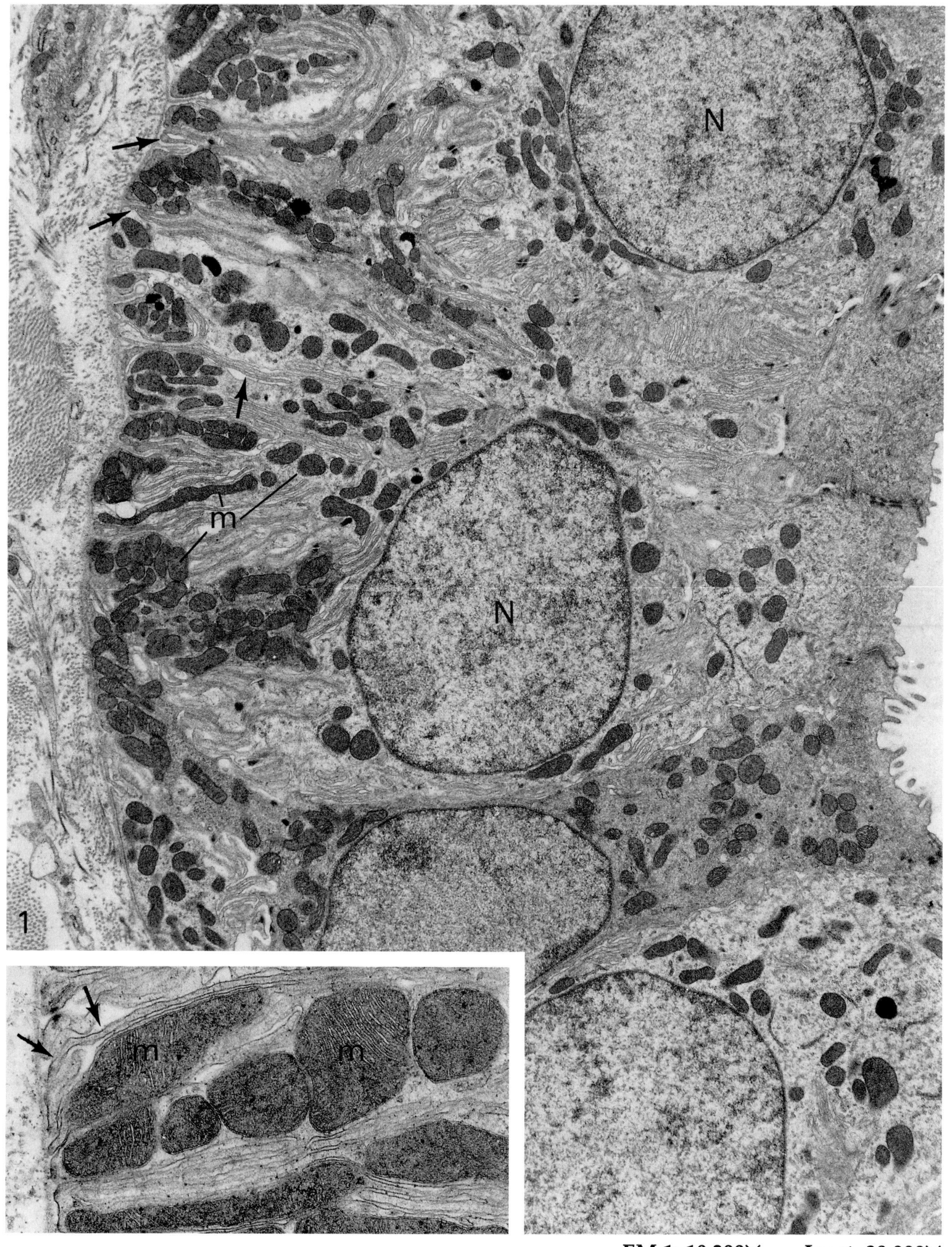

EM 1: 10,200×   Inset: 28,000×

Myoepithelial cells (M, micrograph) lie under the salivary secretory cells inside the basal lamina. Even though epithelial in origin, these cells are similar in ultrastructure and function to mesenchyme-derived smooth muscle; actin and myosin (note the fine filamentous appearance of the cytoplasm) function in contraction in association with dense bodies (arrowheads, micrograph). In contrast to the spindle shape of smooth muscle cells, myoepithelial cells have branching processes that form a basketlike network around the secretory end pieces and ducts. Both parasympathetic and sympathetic stimulation cause myoepithelial cell contraction, which squeezes the secretory units, forcing the secretion into the ducts.

Myoepithelial cells are frequent components of salivary gland tumors, and their presence is used by pathologists as a diagnostic tool. Recognition is facilitated by monoclonal antibodies directed against a unique subset of keratin intermediate filaments found in these cells.

Plasma cells (P, micrograph) are found under the epithelium of mucosal surfaces. Most of these plasma cells synthesize **IgA,** which crosses the basal lamina (arrows, micrograph) and is transported across the secretory cells (S, micrograph) and eventually secreted into the lumen. In the oral cavity, secretory IgA **(sIgA)** binds to and inactivates many pathogens.

The path of IgA secretion and transport to the lumen is illustrated in the diagram. Most IgA secreted by plasma cells is dimeric. Two IgA monomers are attached by a third type of chain, the **J-chain.** Following secretion, dimeric IgA binds to specific receptors on the basolateral surface of the secretory cells. These receptors are transmembrane glycoproteins known as **secretory component** (SC). IgA bound to SC is endocytosed and transported to the apical region of the secretory cell. During this transport, SC is cleaved and a part remains attached to and is exocytosed with IgA, now designated sIgA. The attached **secretory piece** is covalently linked to IgA and stabilizes this antibody after secretion. This mechanism of receptor-mediated transport in which the ligand remains permanently attached to the receptor is unique. IgA is secreted by the same mechanism across mucosal surfaces in other parts of the gastrointestinal, female reproductive, and respiratory tracts.

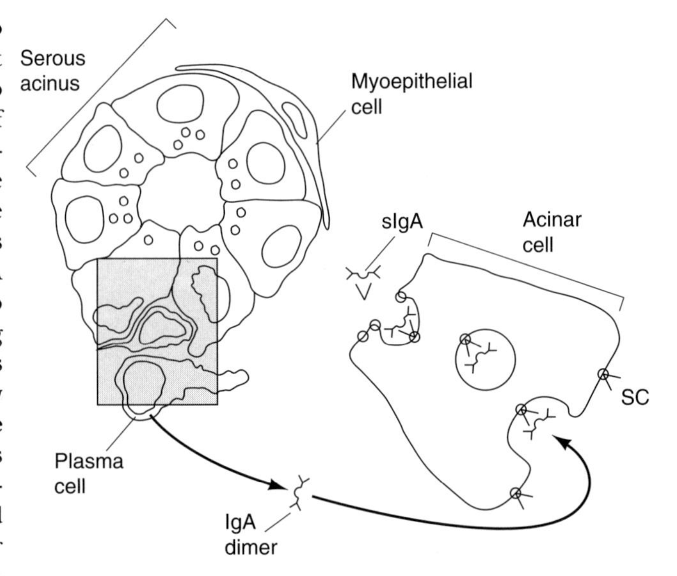

A J-chain is also present in pentameric IgM, and a small amount of IgM is transported across secretory cells by a similar mechanism. The presence of the J-chain in both IgA and IgM is necessary for receptor binding and transport; however, the J-chain may not be the actual binding site.

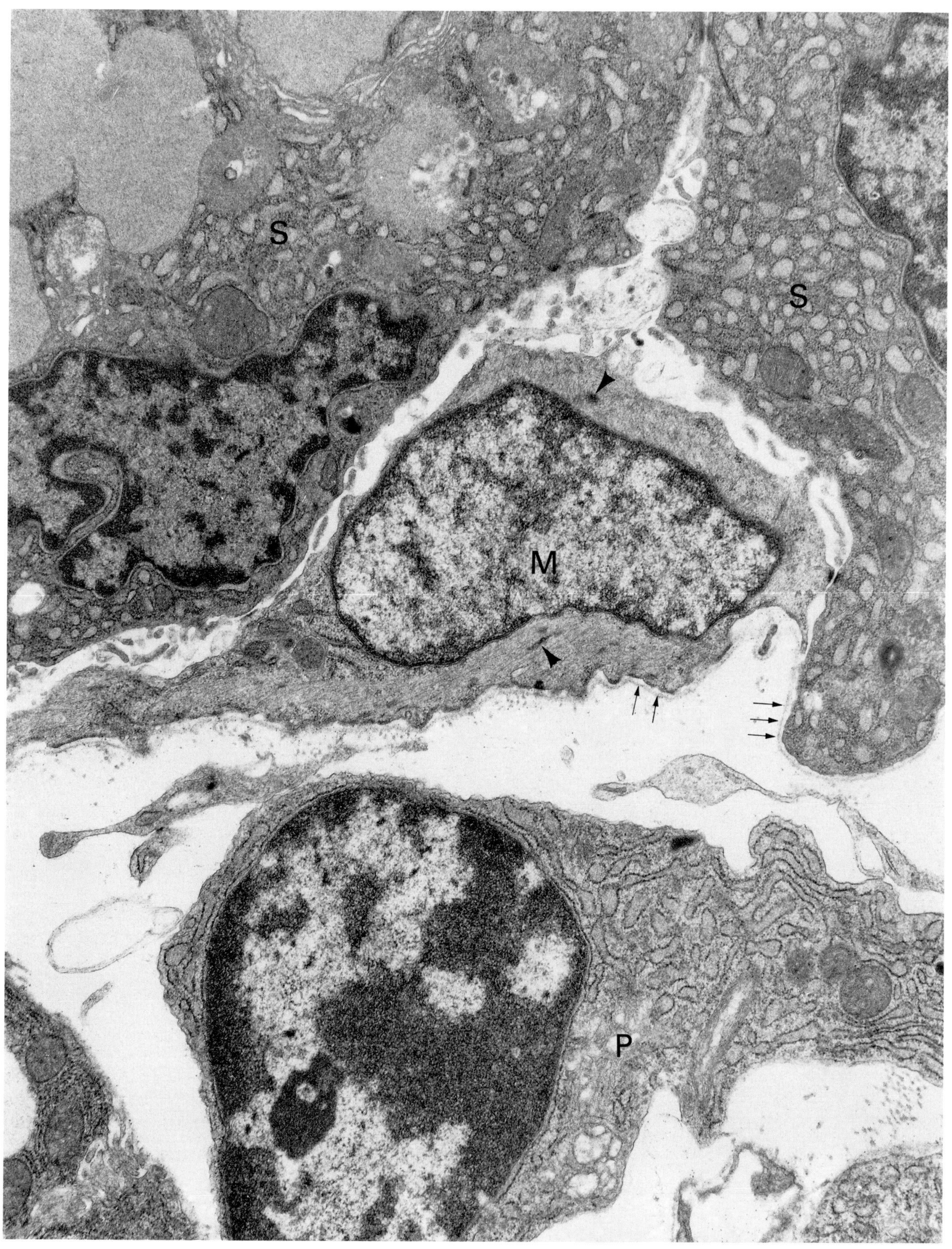

EM: 21,000×

SALIVARY GLAND: Activated Lymphocyte

Lymphocytes that are activated in follicles in gut-associated lymphoid tissue (GALT), such as tonsils and Peyer's patches, do not necessarily complete differentiation in this same region. Many **activated B cells** leave the region of activation and travel to new sites via the lymphatics and systemic circulation as they continue to differentiate.

The micrograph illustrates an activated B lymphocyte (L) leaving the circulation to enter a salivary gland. A small part of the cell has crossed the endothelium (E) of the venule and appears to rest on a pericyte process (P). The specific IgA that this cell is programmed to synthesize is already accumulating in the rough endoplasmic reticulum (arrows).

This distribution of activated lymphocytes "arms" a variety of exposed surfaces with antigen-specific effector cells.

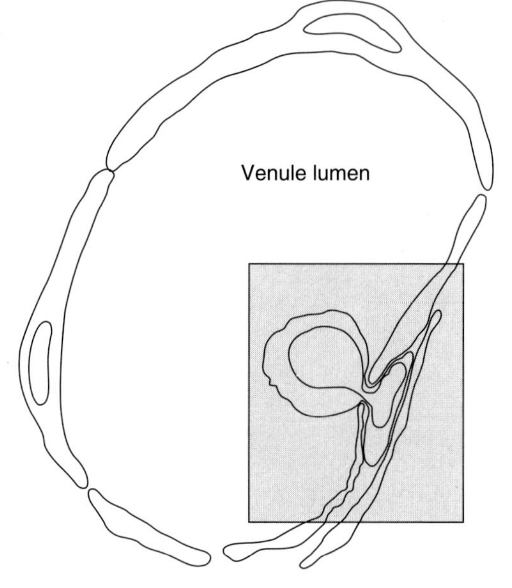

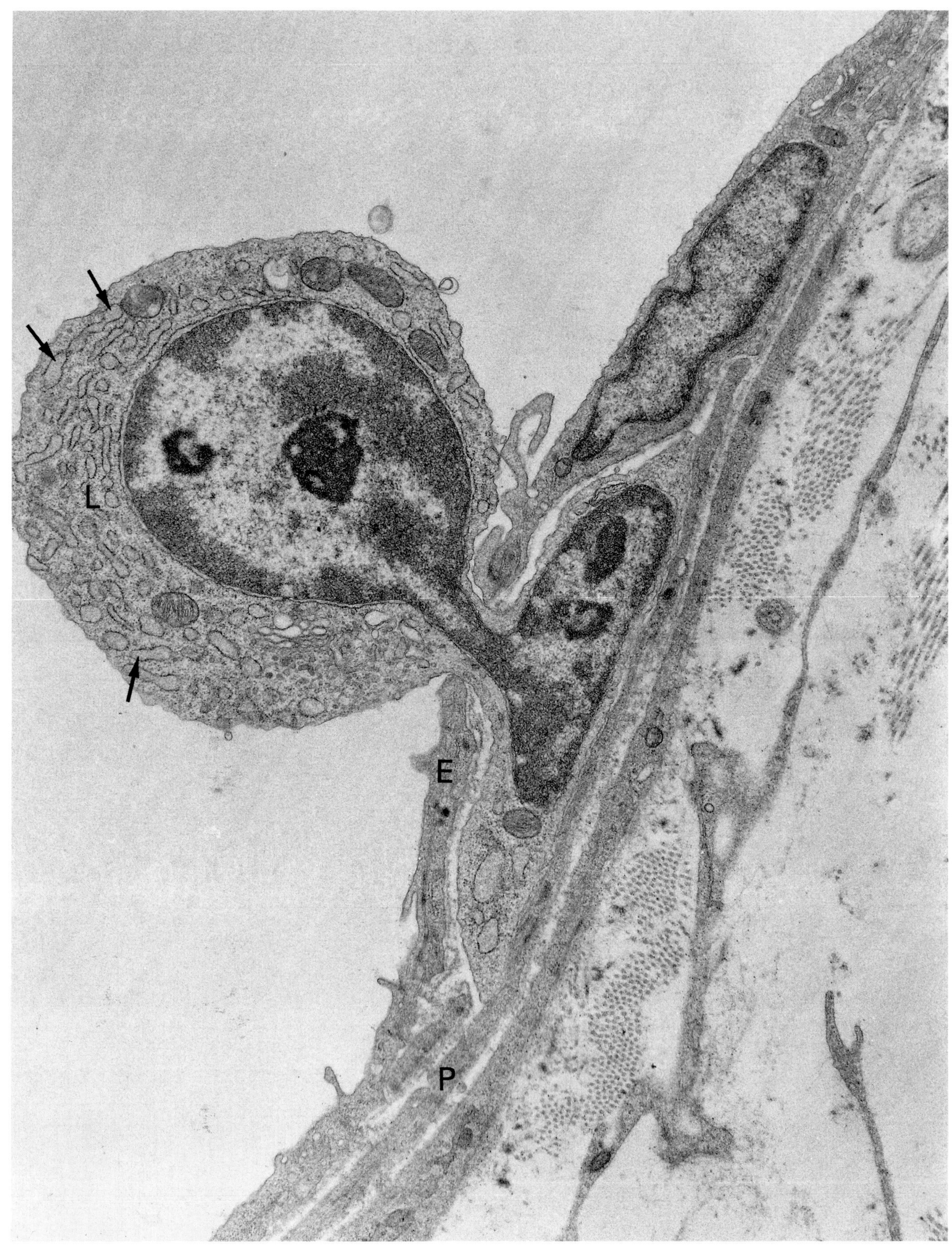

EM: 17,500×

PANCREAS: Acinar and Centroacinar Cells

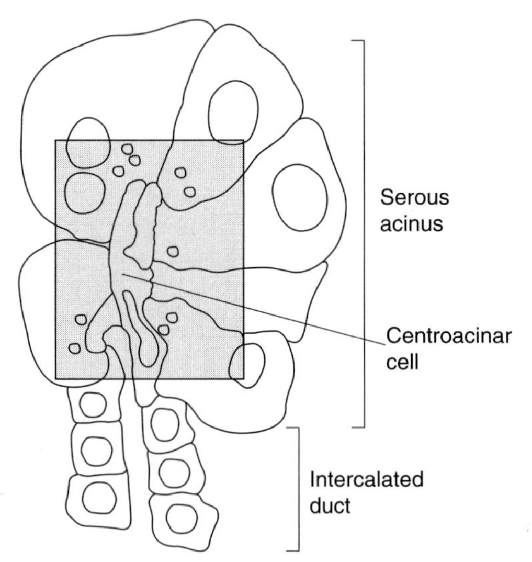

The exocrine pancreas is the major digestive gland in the body, producing more than 22 enzymes that digest all major classes of nutrients. The secretory units in the exocrine pancreas are arranged in acini, which secrete into intercalated ducts lined by cuboidal epithelium. Proximal ductal cells often appear to be in the center of the acinus, and are thus referred to as centroacinar.

In the micrograph, several **acinar** cells (A) surround paler-staining **centroacinar** cells (C). Proenzymes, stored within **zymogen granules** (g, micrograph), are the inactive precursors of digestive enzymes. These enzymes are synthesized on the extensive flattened cisternae of rough endoplasmic reticulum (arrows, micrograph) that pack the cytoplasm of the acinar cell. Pancreatic acinar cells were used in the original studies that "mapped" the ultrastructural path of protein synthesis and secretion.

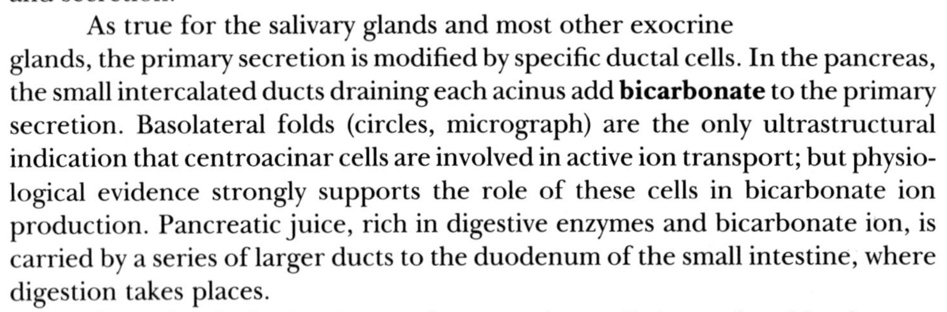

As true for the salivary glands and most other exocrine glands, the primary secretion is modified by specific ductal cells. In the pancreas, the small intercalated ducts draining each acinus add **bicarbonate** to the primary secretion. Basolateral folds (circles, micrograph) are the only ultrastructural indication that centroacinar cells are involved in active ion transport; but physiological evidence strongly supports the role of these cells in bicarbonate ion production. Pancreatic juice, rich in digestive enzymes and bicarbonate ion, is carried by a series of larger ducts to the duodenum of the small intestine, where digestion takes places.

Secretion by both acinar and centroacinar cells is regulated by the contents within the duodenal lumen. Following a meal, both bicarbonate and enzymes are released from pancreatic cells. The messengers between duodenal contents and pancreatic exocrine cells appear to be hormones released from endocrine cells in the wall of the duodenum. **Cholecystokinin,** a hormone released in response to products of digestion, acts primarily on acinar cells, whereas **secretin,** a hormone released in response to acid entering the duodenum from the stomach, acts primarily on centroacinar cells.

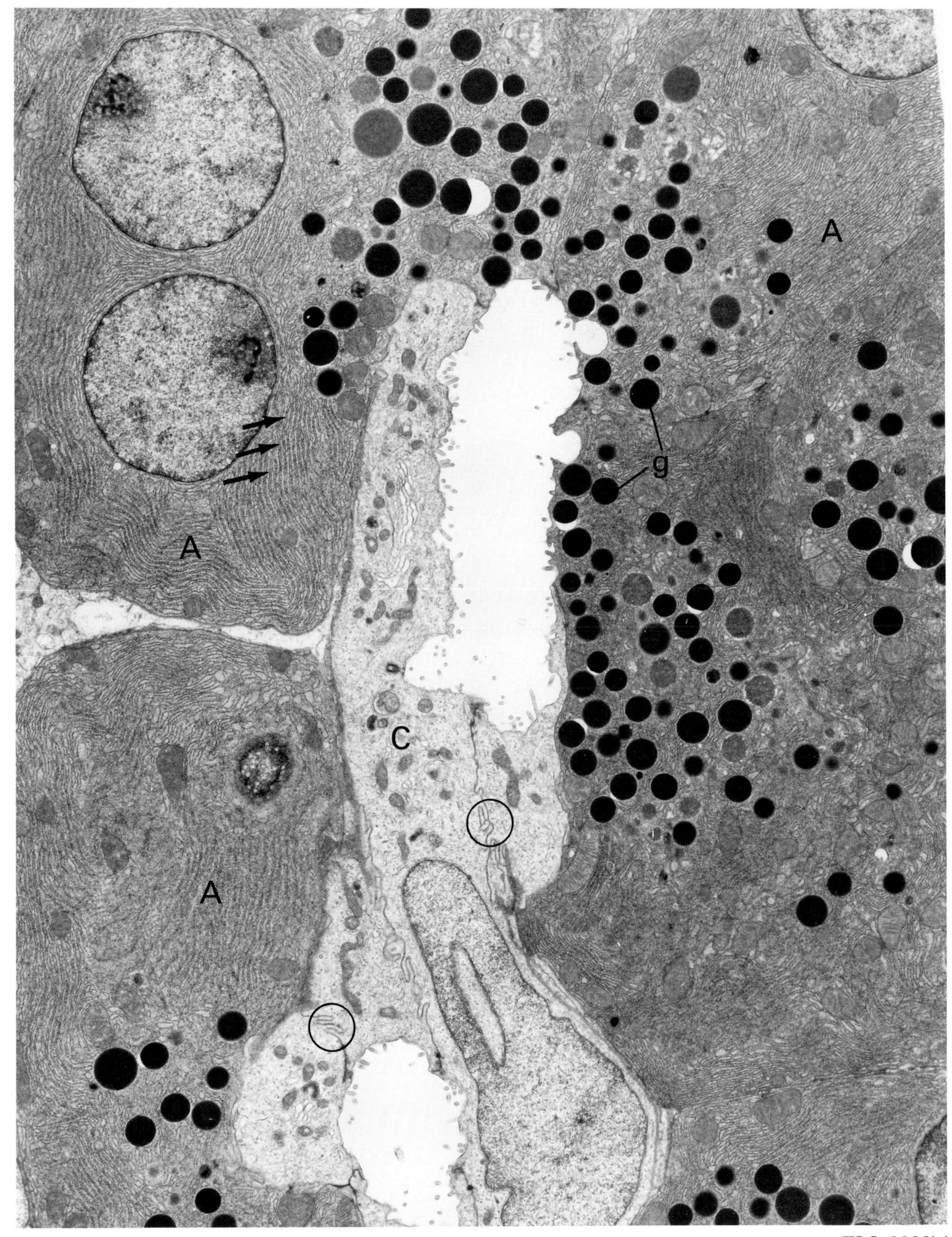

EM: 8000×

PANCREAS: Acinar Cells

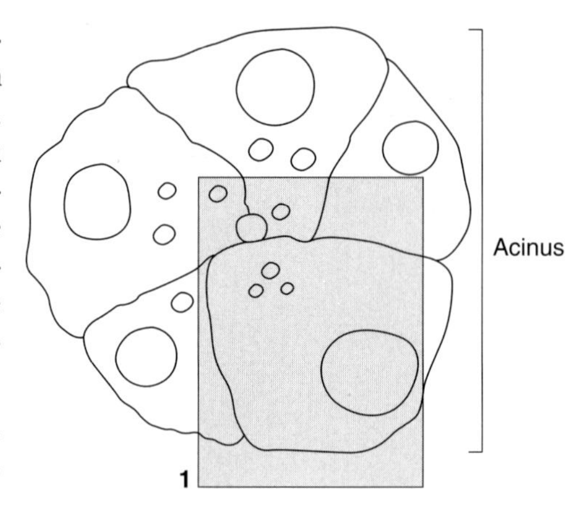

Within each **pancreatic acinar cell,** inactive **proenzymes** are synthesized on the rough endoplasmic reticulum (arrows, micrograph 1) and passed to the Golgi (arrowheads, micrograph 1), where they are processed and concentrated in condensing vacuoles (v, micrograph 1) to form proenzyme-containing **zymogen granules** (g, micrograph 1). The inactive proenzymes become active digestive enzymes within the duodenum. Premature activation of these enzymes occurs in the disorder pancreatitis, and results in autodigestion of pancreatic tissue, along with hemorrhaging and infection.

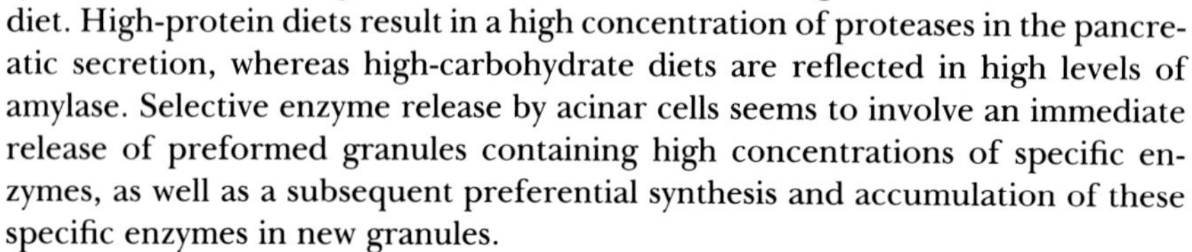

Each zymogen granule appears to contain all of the pancreatic enzymes; however, the concentration of individual enzymes varies between granules and is sensitive to changes in diet. High-protein diets result in a high concentration of proteases in the pancreatic secretion, whereas high-carbohydrate diets are reflected in high levels of amylase. Selective enzyme release by acinar cells seems to involve an immediate release of preformed granules containing high concentrations of specific enzymes, as well as a subsequent preferential synthesis and accumulation of these specific enzymes in new granules.

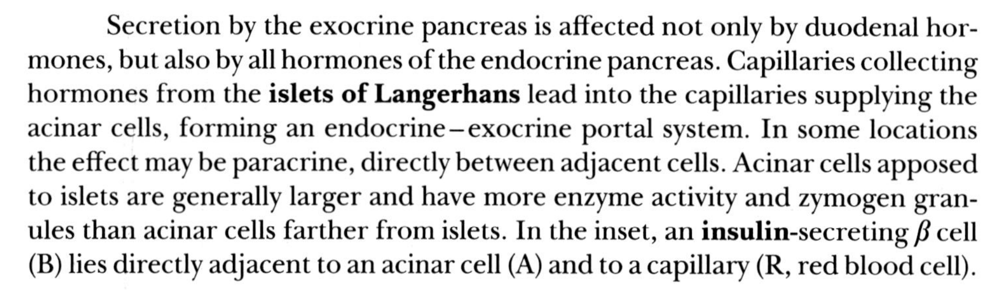

Secretion by the exocrine pancreas is affected not only by duodenal hormones, but also by all hormones of the endocrine pancreas. Capillaries collecting hormones from the **islets of Langerhans** lead into the capillaries supplying the acinar cells, forming an endocrine–exocrine portal system. In some locations the effect may be paracrine, directly between adjacent cells. Acinar cells apposed to islets are generally larger and have more enzyme activity and zymogen granules than acinar cells farther from islets. In the inset, an **insulin**-secreting β cell (B) lies directly adjacent to an acinar cell (A) and to a capillary (R, red blood cell).

Insulin receptors have been identified on acinar cells, and insulin has been shown to increase both amylase mRNA synthesis and the release of amylase from these cells. Carbohydrate digestion and absorption following amylase release increases glucose availability. Thus insulin has another role in glucose homeostasis, in addition to its well-known effect on the uptake of glucose by cells.

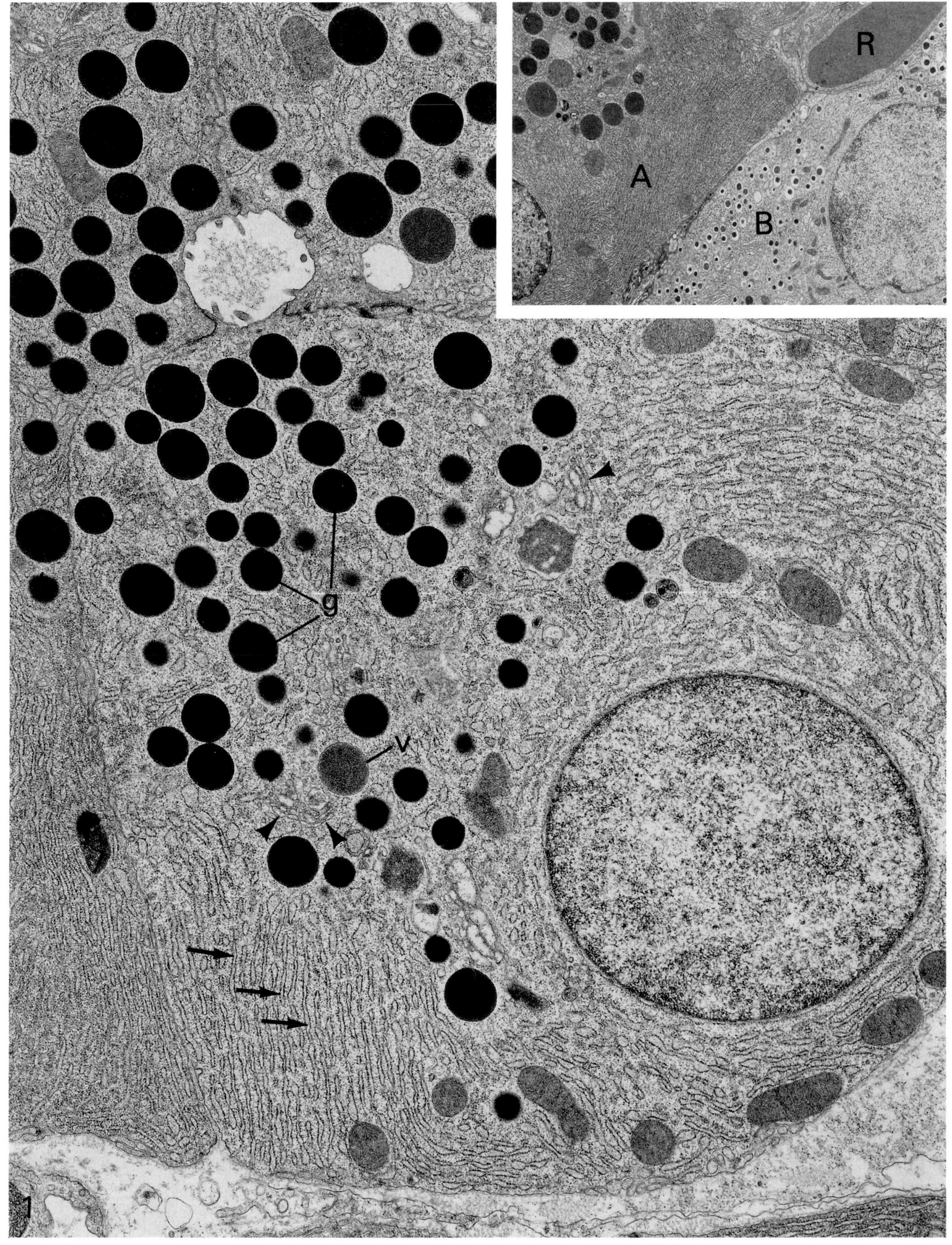

EM 1: 15,600× Inset: 4500×

The **classic liver lobule** is a hexagon with portal triads at each corner. **Triads** contain branches of the hepatic portal vein (V, diagram), hepatic artery (A), and bile duct (B). Oxygenated blood from the hepatic artery and nutrient-rich blood from the portal vein empty into sinusoids between rows of hepatocytes. The sinusoids, carrying secretions (e.g., glucose, very low-density lipoproteins, plasma proteins) from the hepatocytes, drain into a central vein. Exocrine secretion (bile) travels in the opposite direction through small canaliculi between individual hepatocytes and exits via the bile duct in the portal triad.

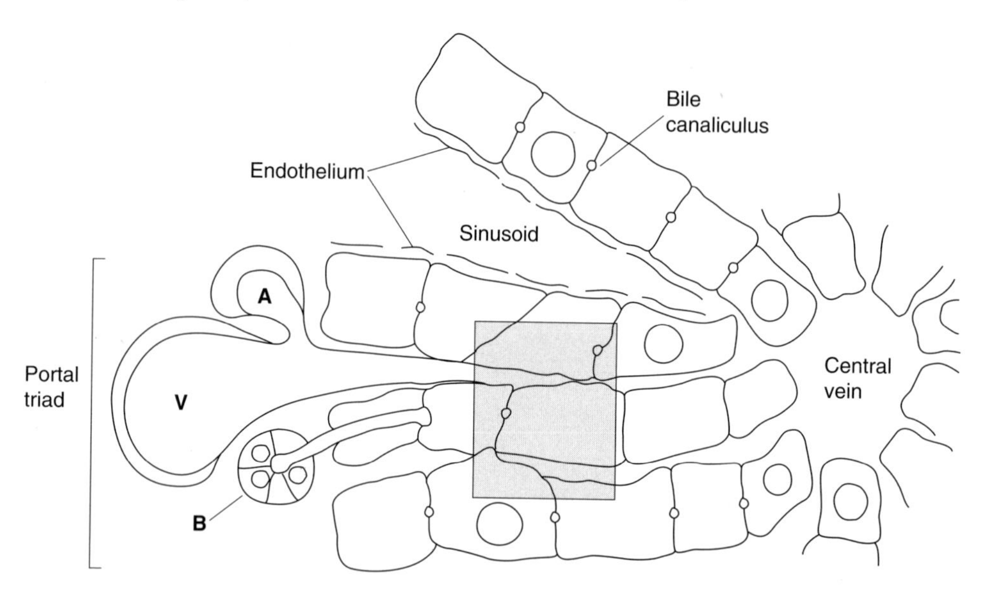

The micrograph depicts three rows of **hepatocytes** (H) separated by blood **sinusoids** (s). Each hepatocyte has three distinct surfaces: apical (arrowheads), basal (curved arrows), and lateral (short, thick arrows).

The **apical surface** lines a small lumen between adjacent cells, defined by junctional complexes (circles, micrograph). Bile, secreted into the lumen or **bile canaliculus** (c, micrograph), is transported to bile ducts and eventually to the duodenum of the small intestine. Bile functions in fat absorption and the excretion of certain substances (e.g., cholesterol, bilirubin).

Lateral surfaces contain the junctions that define cell polarity (tight junctions), bind hepatocytes together (desmosomes), and provide communication between cells (gap junctions). The **basal surface** is directly adjacent to the sinusoidal endothelium (e, micrograph). Plasma within sinusoids freely diffuses through open endothelial fenestrations to directly bathe both the basal surface in the **space of Disse** (D, micrograph) and the lateral surface between adjacent cells. Numerous microvilli (long, thin arrows, micrograph) project from the basal surfaces of each cell into the space of Disse, thus amplifying the surface area exposed to plasma.

Eighty percent of the blood entering the liver originates from the hepatic portal vein that drains the intestine. Many nutrients and toxins crossing the intestinal lining are shuttled to the liver in this portal system before entering the general circulation. The intimate association of each hepatocyte with the blood vascular system reflects a major role of the liver as a filter that monitors and adjusts plasma constituents.

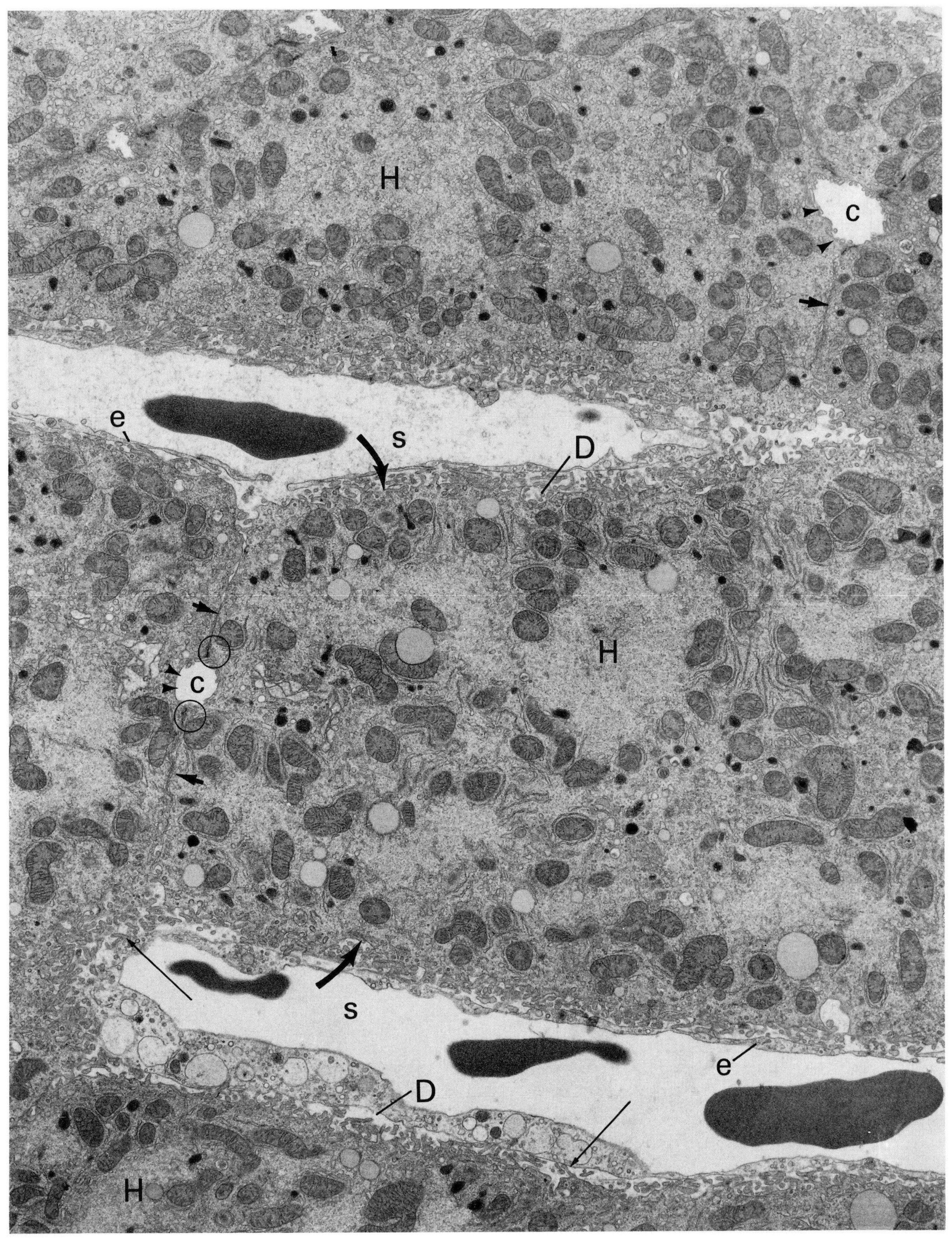

EM: 6800×

Blood entering the liver lobules through the hepatic portal vein is often rich in glucose, which is stored in **hepatocytes** in the form of **glycogen.** Glycogen within hepatocytes is a major source of glucose for other cells of the body. The liver contains 70–80 grams of glycogen, arranged as **rosettes** (circles, micrograph) of many individual particles. Large spaces (l, micrograph) represent sites where lipid was stored prior to its extraction during tissue preparation.

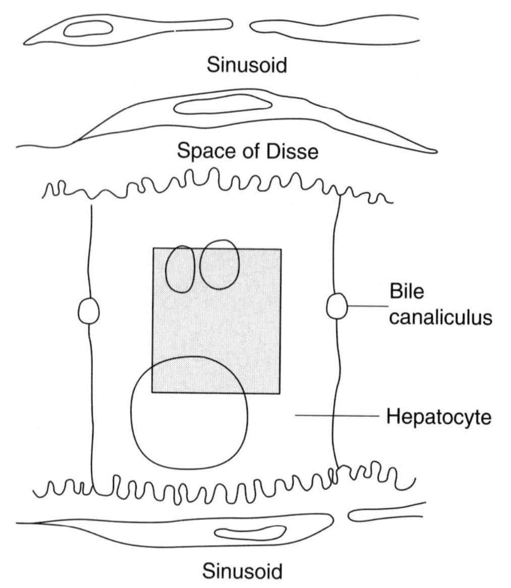

During meals with high glucose intake, insulin increases the ability of hepatocytes to synthesize glycogen. As blood glucose levels drop, glucagon, and to a lesser extent epinephrine, increase the ability of the hepatocyte to degrade glycogen. The product of degradation, glucose-6-phosphate, is dephosphorylated in the smooth ER (curved arrows, micrograph), and glucose diffuses from the cell into the general circulation.

Enzyme deficiencies associated with glycogen breakdown can result in **storage diseases;** in Type II glycogen storage disease the lysosomal enzyme α-1,4-glucosidase is missing and glycogen accumulates in lysosomes. The swollen lysosomes interfere with normal cellular metabolism, sometimes resulting in liver failure and death.

Ribosomes, both free (arrowheads, micrograph) and attached to ER (long arrows, micrograph), synthesize the **enzymes** needed for lipid, carbohydrate, and protein metabolism. In addition, the rough ER synthesizes protein components of **lipoprotein particles** and most of the **plasma proteins.** Plasma proteins include albumin, which is an important carrier molecule for substances such as hormones and enzymes and also maintains the osmotic pressure within the vascular system, fibrinogen and prothrombin, necessary for blood coagulation, and several complement factors that function in the immune response.

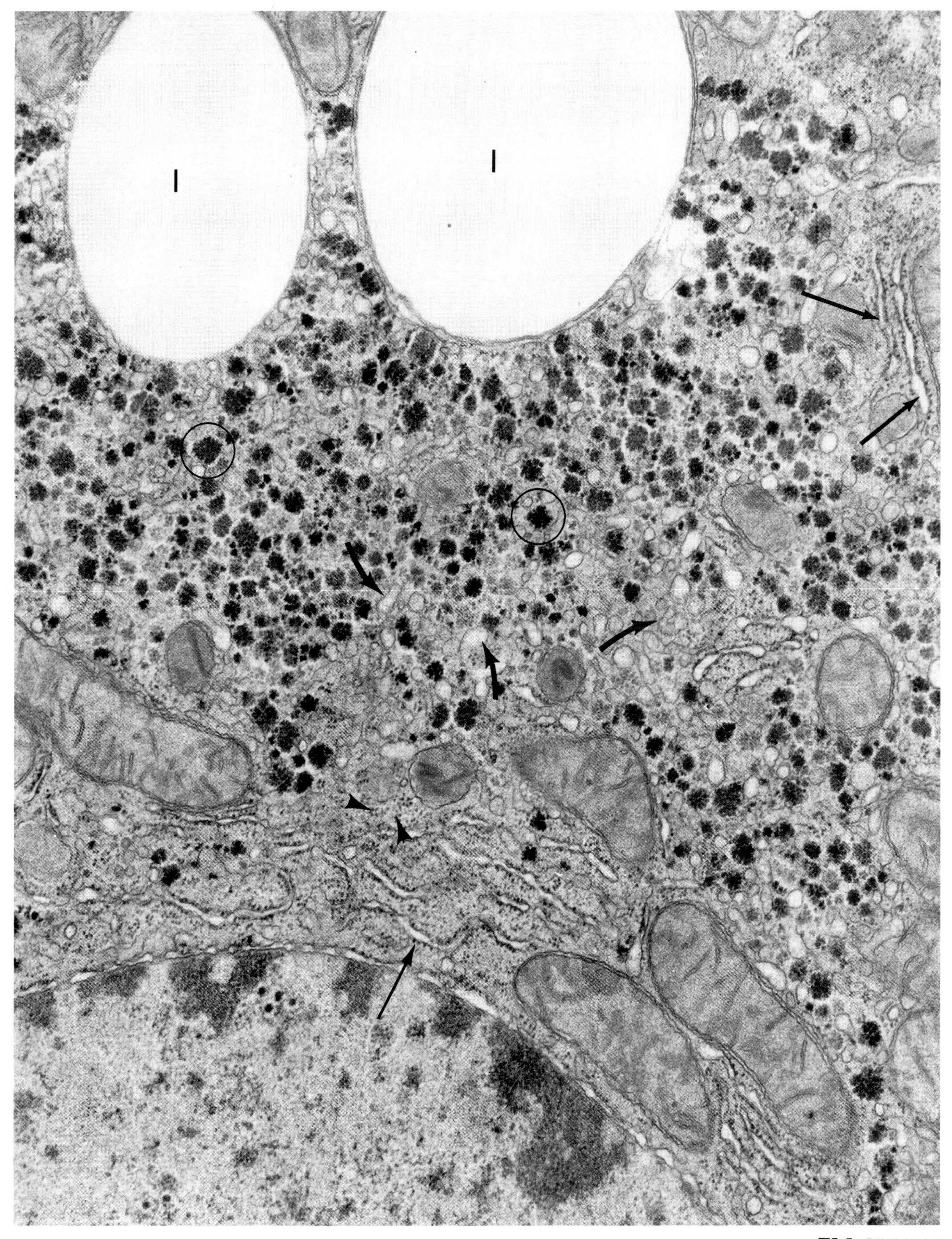

EM: 35,000×

The liver is the most important site of regulation of both lipid synthesis and degradation. **Hepatocytes** synthesize **triglycerides** (TG), **cholesterol** (CH), and **phospholipids** (PL) on the smooth ER (arrows, inset). Some of these lipids are stored as large lipid droplets (l, micrograph 1) in the cytoplasm; however, much of the lipid synthesized in the hepatocyte is packaged with protein synthesized on the rough ER (arrowheads, inset) and released into the circulation as **very low-density lipoprotein** (VLDL), one of several classes of lipoproteins found in plasma. VLDL particles are assembled in junctions between the rough and smooth ER and transported to the Golgi, where they are modified and directed to the space of Disse (D, micrograph 1) for secretion. VLDLs, like chylomicrons synthesized in intestinal enterocytes, contain a core rich in triglycerides, and are the main source of fatty acids for all cells.

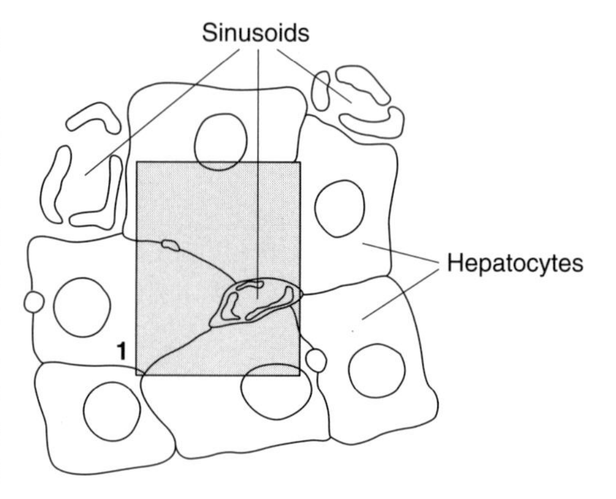

In addition to the liver's role in VLDL synthesis, it is also important in some aspects of the regulation of all other types of **lipoproteins** (table). These lipoproteins are continually undergoing synthesis, interconversion, and degradation. Hepatocytes are involved in these processes in many ways. They (1) synthesize the enzyme essential for the formation of cholesterol esters in high density lipoprotein (HDL), (2) remove chylomicron remnants from the circulation, and (3) are the indirect source of cholesterol-rich low density lipoproteins (LDLs), which are formed in plasma from VLDLs depleted of fatty acids.

Type	Origin	Size (Å)	Approx. density (g/ml)	Lipid Content	Function
Chylomicron	Intestine	800–5000	<0.95	90TG/8PL/5CH	Dietary TG source
VLDL	Liver	280–800	1.00	60TG/20PL/17CH	Endogenous TG sources
LDL	Plasma	200	1.05	60CH/30PL/10TG	CH source for most cells
HDL	Intestine/ Liver	50–150	1.10	50PL/32CH/10TG	Collects excess CH

Modified from R.M. Glickman and S.M. Sabesin, Lipoprotein Metabolism, in *The Liver, Biology and Pathobiology,* 2nd Ed.Press, New York, 1988.

Cholesterol, an essential component of cell membranes and certain hormones, is produced mainly in the liver from ingested carbohydrates, proteins, and fats. Typically, even though not essential, CH is also consumed in the diet. The liver responses in many ways to increases in cholesterol in the diet, including stopping the synthesis of cholesterol, adding more cholesterol to VLDL particles, and secreting more cholesterol into the bile. Bile secretion via the bile canaliculus (c, micrograph 1) is an important route for **cholesterol excretion.** Micelles of bile salts (cholesterol derivatives themselves) and phospholipids sequester cholesterol and solubilize it during its excretion.

Balancing the lipoprotein levels and cholesterol content in the circulation has proven to be a critical factor in vascular disease. In general, low levels of LDL, which brings CH into the system, and high levels of HDL, which clears CH out of the system, reduce coronary artery disease.

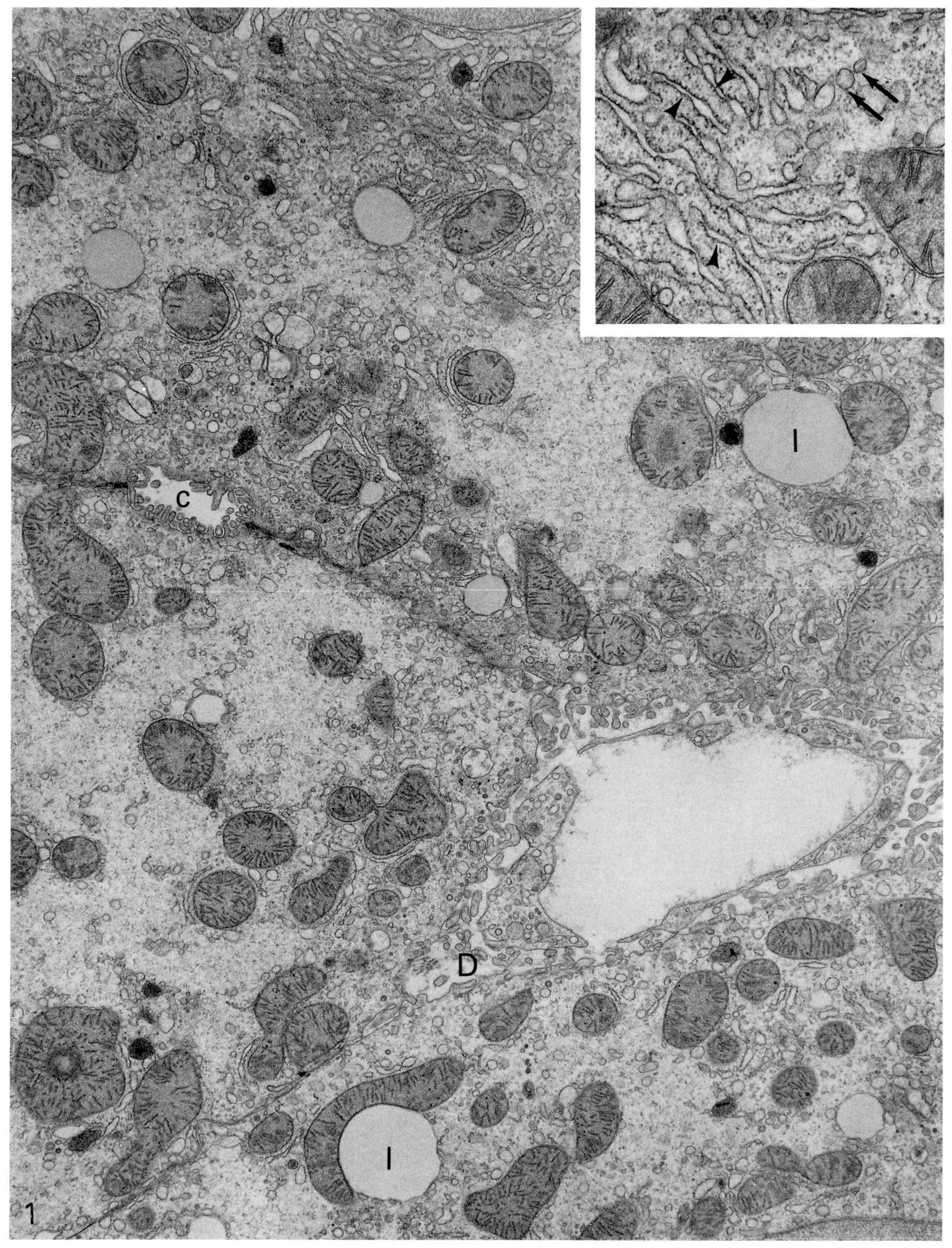

EM 1: 10,200× Inset: 30,000×

LIVER: Kupffer Cell

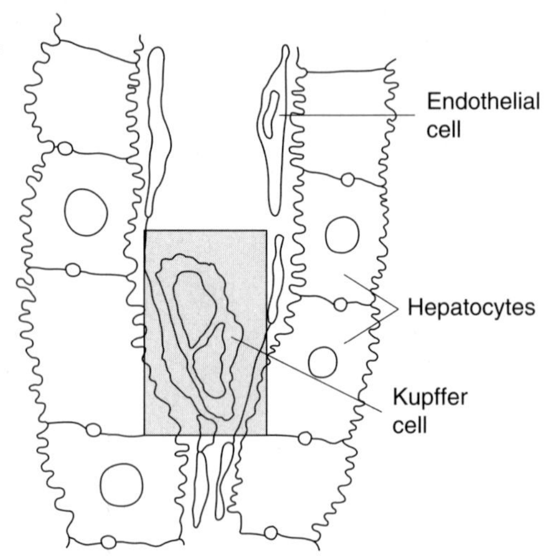

Kupffer cells (K, micrograph) are **macrophages** that inhabit liver sinusoids. Their major role is the **phagocytosis of particulate matter,** such as viral particles, clot complexes, damaged erythrocytes, immune complexes and bacteria. The Kupffer cell in the micrograph has recently phagocytosed an RBC (R) and is in the process of degrading it within a secondary lysosome.

Kupffer cells, endothelial cells (E, micrograph), and hepatocytes (H, micrograph) establish junctional contacts with one another, and this attachment has been shown in vitro to prolong hepatocyte survival. The Kupffer cell in the micrograph is associated with another cell at a junctional region (encircled). One other function of these junctions may be to anchor Kupffer cells in liver sinusoids.

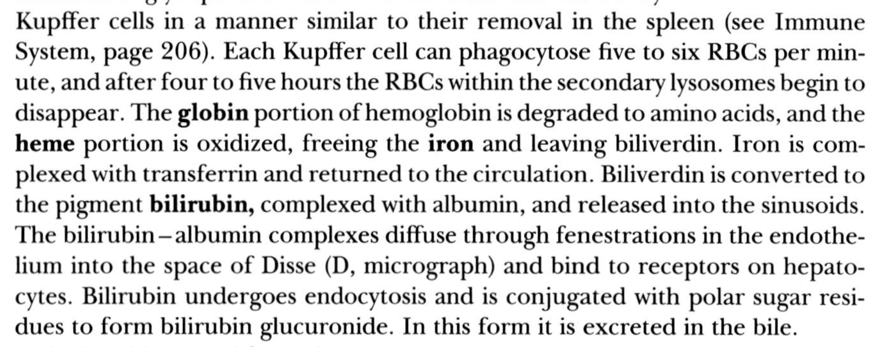

Blood cells that have lost **sialic acid residues** from certain membrane glycoproteins are cleared from the blood by Kupffer cells in a manner similar to their removal in the spleen (see Immune System, page 206). Each Kupffer cell can phagocytose five to six RBCs per minute, and after four to five hours the RBCs within the secondary lysosomes begin to disappear. The **globin** portion of hemoglobin is degraded to amino acids, and the **heme** portion is oxidized, freeing the **iron** and leaving biliverdin. Iron is complexed with transferrin and returned to the circulation. Biliverdin is converted to the pigment **bilirubin,** complexed with albumin, and released into the sinusoids. The bilirubin–albumin complexes diffuse through fenestrations in the endothelium into the space of Disse (D, micrograph) and bind to receptors on hepatocytes. Bilirubin undergoes endocytosis and is conjugated with polar sugar residues to form bilirubin glucuronide. In this form it is excreted in the bile.

Sialic acid removal from glycoproteins is not only a message for the phagocytosis of certain blood cells by the Kupffer cell, but it also signals the removal of plasma glycoproteins from the circulation by hepatocytes. Exposed galactose residues of asialoglycoproteins bind to specific receptors on the basal surface of the hepatocyte. Endocytosis follows, thus clearing asialoglycoproteins from the circulation.

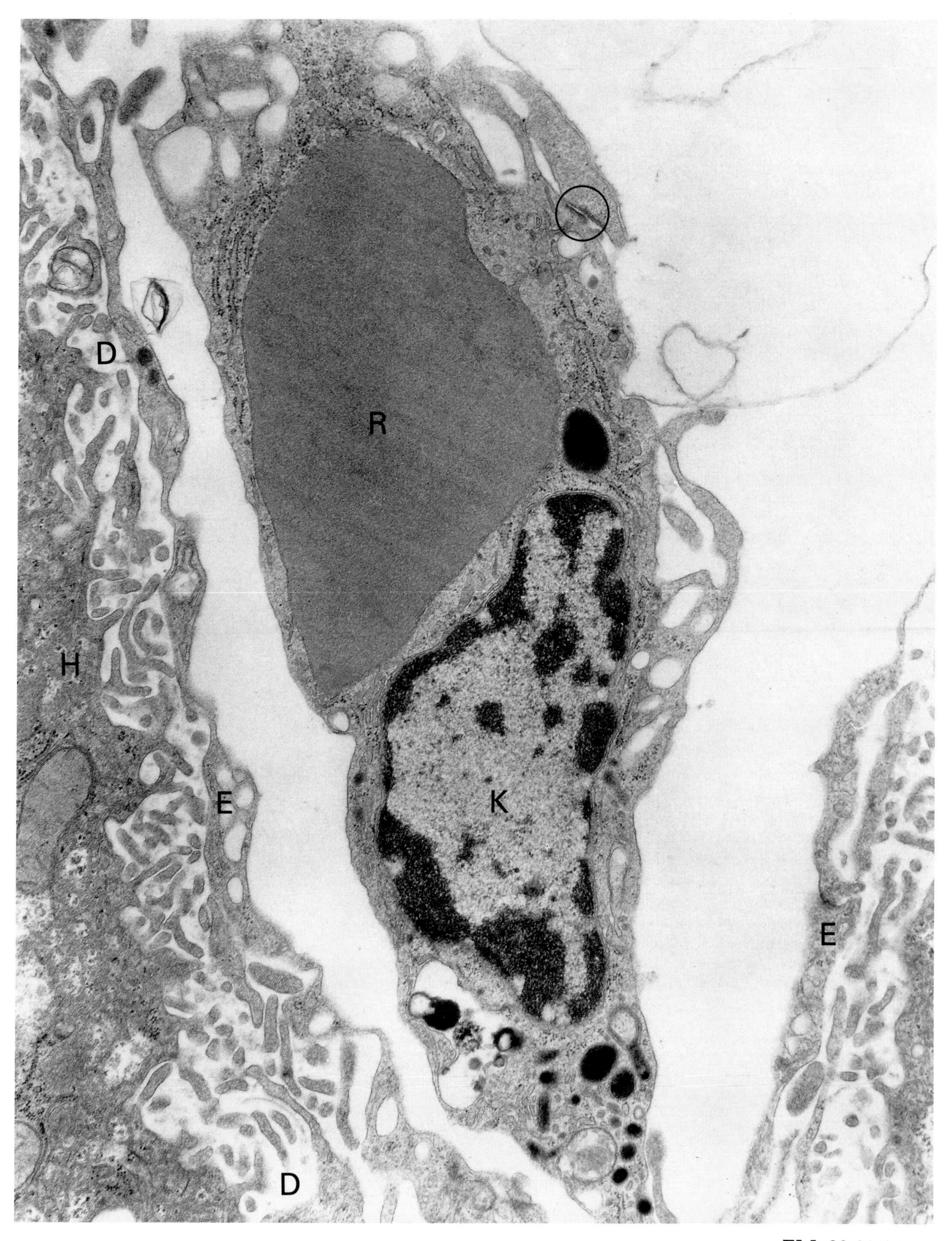

EM: 22,000×

GALLBLADDER EPITHELIUM

The gallbladder **concentrates and stores bile** between meals. Water from the lumen (L, micrograph 1) is transported across the simple columnar epithelium and enters blood vessels (V, micrograph 1) in the underlying connective tissue. The epithelial cells are attached to one another by junctional complexes that separate an apical surface with microvilli (arrows, micrograph 1) from a basolateral surface with membrane folds (arrowheads, inset). These specializations provide extra surface area for ion transport. The concentration of bile involves the **active movement of Na^+** from bile into the lateral intercellular space. Water follows, causing the intercellular space to swell. The close apposition of the adjacent lateral membranes in micrograph 1 and inset indicate that this epithelium is not in the process of actively concentrating bile.

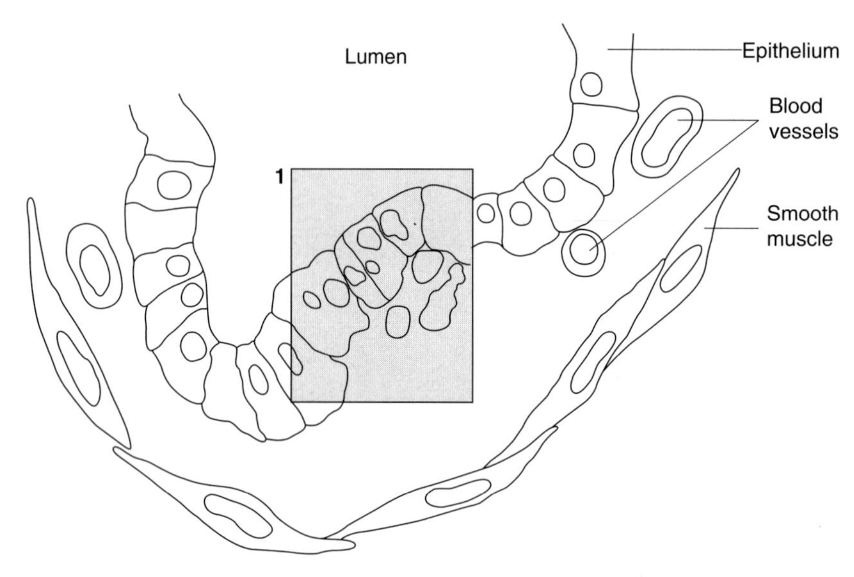

The daily secretion of bile in humans is 1 liter; the gallbladder holds ~50 milliliters. During meals, partially digested nutrients activate the release of cholecystokinin, a hormone synthesized by endocrine cells in the duodenum. One of the actions of this hormone is to cause smooth muscle in the gallbladder wall to contract, forcing bile into the duodenum. In the duodenum, bile salts emulsify and facilitate the digestion of fats. Bile salts then pass to the ileum, where most are reabsorbed, enter the portal vascular system, return to the liver, and are secreted again. Bile salts can undergo two cycles of this enterohepatic circulation during the digestion of one meal.

Gallstones are precipitates of bile components that occur in the gallbladder or bile ducts. The most common type are cholesterol stones caused by the secretion of excess cholesterol by hepatocytes. Bile salts and phospholipid micelles typically contain some cholesterol, but precipitate when overloaded.

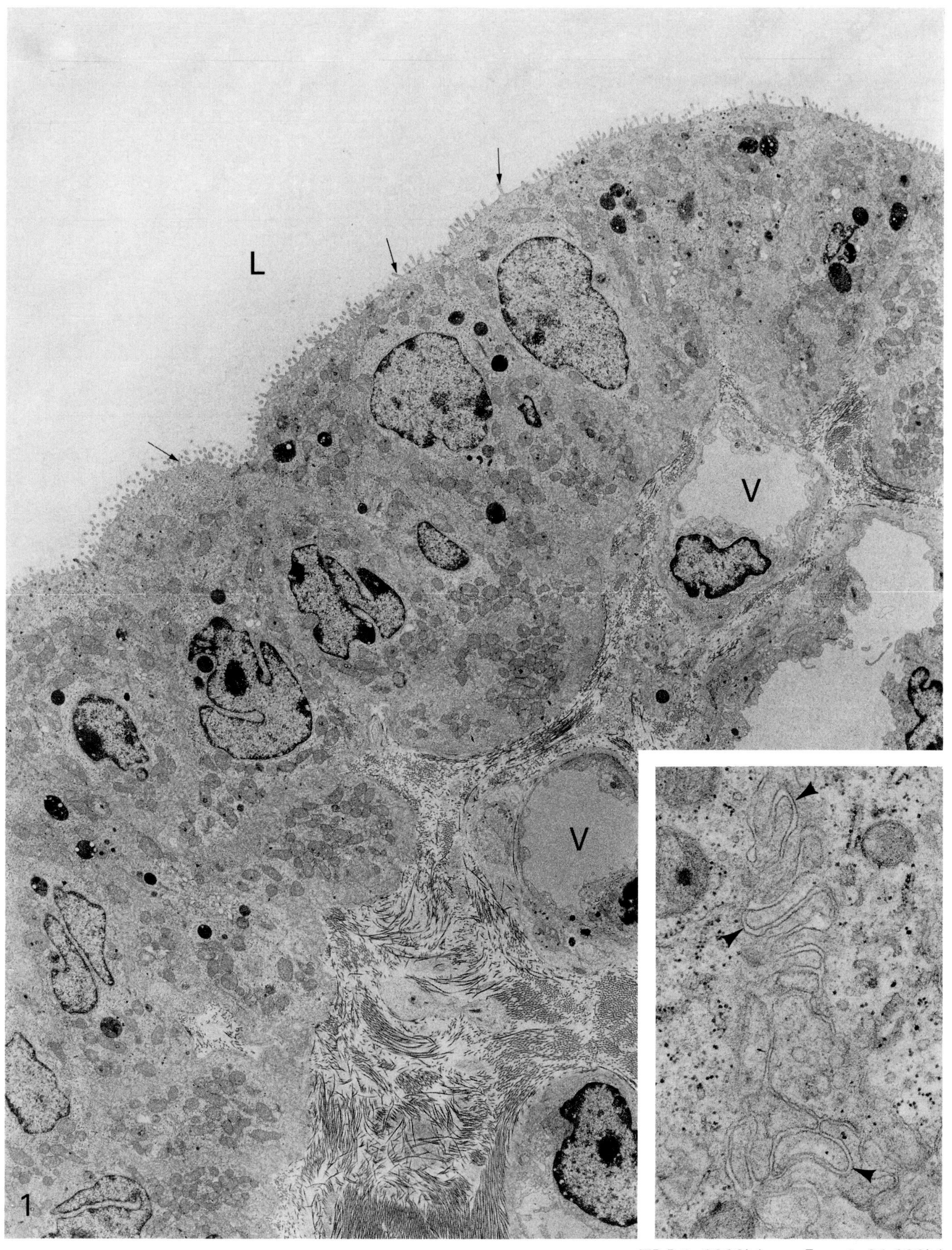

EM 1: 3800× **Inset: 31,000×**

CHAPTER TWELVE

GASTROINTESTINAL TRACT

Pit
Stem
Glands
Parietal

The stomach is a glandular organ in which the surface regularly dips down into pits that open into glands. The epithelial lining of the surface, pits, and glands is routinely replaced every three to four days by the division, differentiation, and migration of stem cells located at the junction of the pits and glands.

The major exocrine cell within these glands is the **parietal cell** (P, micrograph 1). Parietal, or oxyntic, cells secrete HCl in quantities sufficient to maintain an H^+ concentration in the stomach equivalent to 0.17N HCl, with a pH as low as 1. In addition to killing invading bacteria, this acid environment is necessary for the protein digestion that occurs in the stomach.

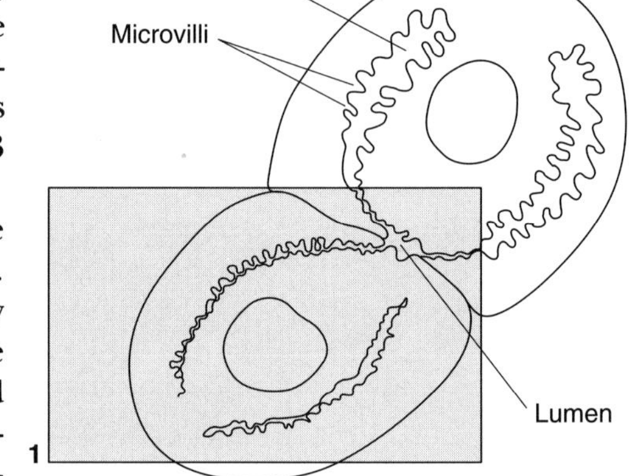

The apical surface of the parietal cell is deeply indented, forming a **canaliculus** (c, micrographs 1 and 2) around the nucleus. Densely packed **microvilli** project from the cell surface into the canaliculus. Hydrogen ions pumped across this extensive surface reach a concentration in gastric juice 2–3 million times that in plasma.

The membrane pumps that accomplish this feat are **H^+K^+ATPases,** which are packed in the microvillar membrane. These highly concentrated pumps create an electron density (arrows, micrograph 2) not seen in other membranes within the cell. The numerous mitochondria (m, micrographs 1 and 2) and extensive cell surface elaborations in the parietal cell are consistent features of cells involved in energy-dependent ion secretion. Fifteen hundred calories of energy are required for every liter of gastric juice secreted against the concentration gradient. The membranes of an extensive **tubulovesicular system** (arrowheads, micrographs 1 and 2) also contain H^+ K^+ ATPases that could provide extra membrane pumps during increased acid demand. It is unclear whether these tubules are continuous with the apical surface at all times or only during intense acid formation.

The surface area of the basal membrane is also increased by folds (circles, micrograph 1). In this region the membrane accommodates carrier systems such as the Cl^- and HCO_3^- exchange channel that brings Cl^- into the cell. The HCl released into the canaliculus creates an osmotic gradient that pulls H_2O across the cell to form the final secretion.

Several mediators act directly on parietal cells to stimulate acid secretion: histamine released from neighboring cells, acetycholine released from vagal nerve endings, and gastrin, a hormone secreted by enteroendocrine cells in the pyloric stomach. Each mediator binds to a separate receptor on the basal surface of the parietal cell and appears to trigger separate intracellular events that lead to acid secretion.

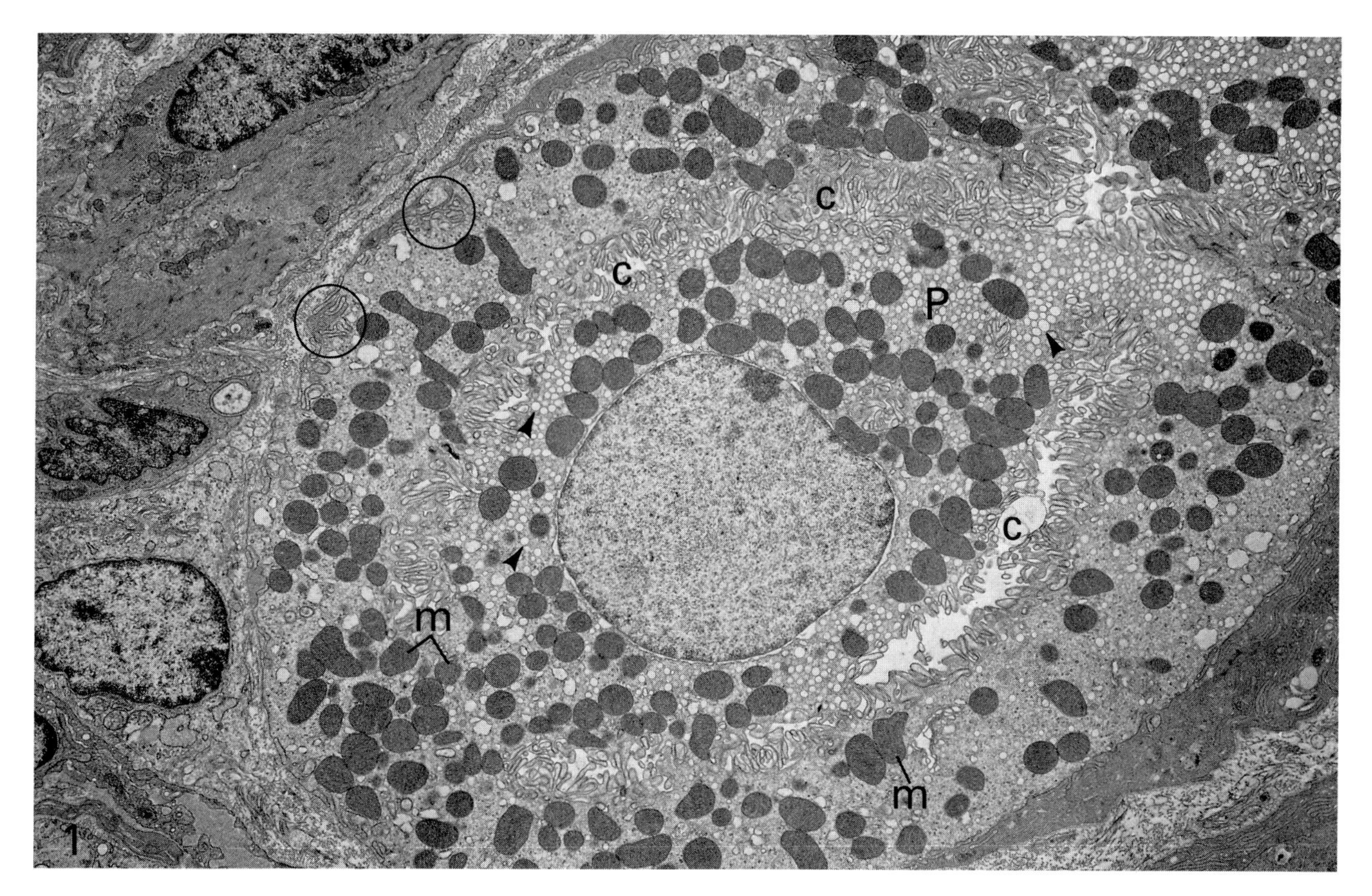

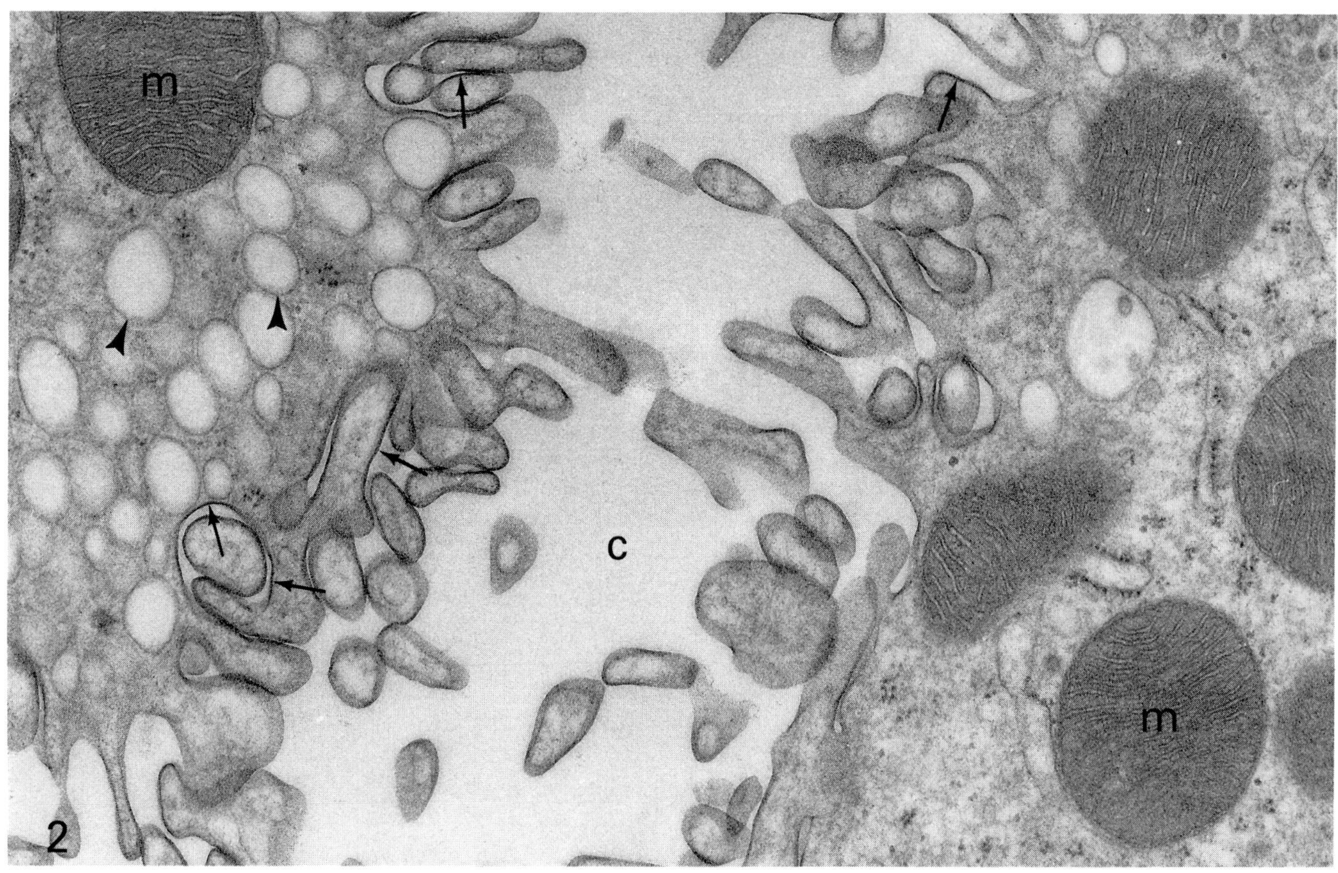

EM1: 4,900× EM2: 37,000×

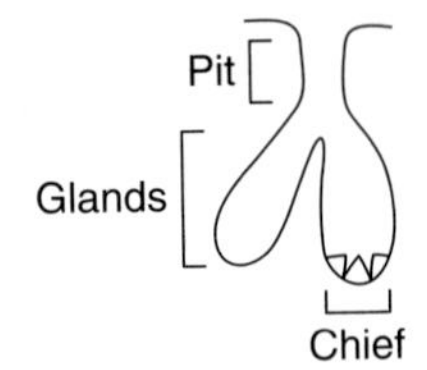

Parietal cells (P), concentrated in the necks of the gastric glands, are occasionally found near the base of the glands next to the chief cells (C), as shown in micrograph 1. Parietal and chief cells, bound by junctional complexes (curved arrows, micrograph 1), surround a central gastric gland lumen (L, micrograph 1).

In addition to the secretion of acid, **parietal cells** synthesize **intrinsic factor** (IF), a glycoprotein that binds to Vitamin B_{12} and is essential to its absorption. Newly synthesized intrinsic factor has been localized within parietal cells in the cisternae of the nuclear envelope and in the relatively sparse rough ER (r, inset) scattered between the tubulovesicular system (t, inset). Intrinsic factor secreted into the lumen of the stomach does not bind to B_{12} until a protective coat is removed by a pancreatic protease in the upper small intestine. The IF – B_{12} complex then travels to the lower small intestine, where receptors on surface epithelial cells bind to intrinsic factor and the complex is endocytosed.

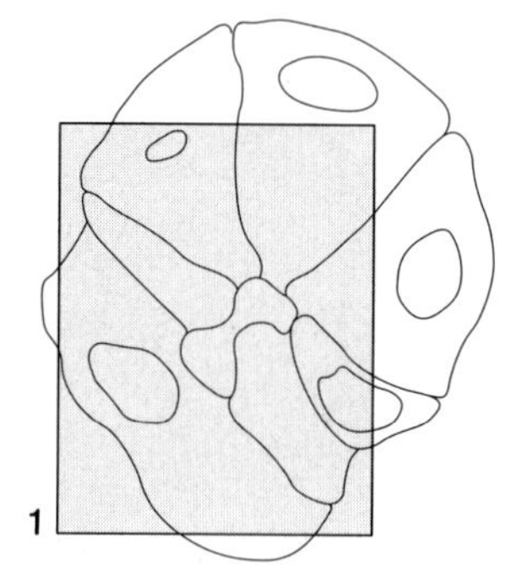

In autoimmune gastritis, antibodies are produced against the acid-producing H^+K^+ATPases in the parietal apical cell membrane. In addition to reduced acid in the stomach lumen and subsequent increased bacterial invasion, parietal cells fail to produce intrinsic factor. The resulting B_{12} deficiency further complicates the disorder by disrupting the synchrony of DNA synthesis and cell division in early red blood cell precursors. A subsequent reduction in normal RBCs leads to pernicious anemia.

Chief cells secrete enzymes that digest protein in the stomach lumen. The extensive rough ER (arrowheads, micrograph 1) in the base of these cells and the large secretory granules in the apical region reflect the role of these cells in protein synthesis and secretion. Proteases are synthesized, stored, and secreted as proenzymes, **pepsinogens.** On stimulation by acetylcholine, histamine, or gastrin, secretory granules fuse with the apical cell membrane and pepsinogens are released. Exocytosis is particularly rapid as a result of granule-to-granule fusion (straight arrows, micrograph) known as **compound exocytosis.**

Inactive pepsinogens are converted into active **pepsins** in the acid environment of the stomach with the removal of an N-terminal sequence. Pepsins are capable of partially digesting most types of proteins. Peptide bonds adjacent to aromatic amino acids are cleaved, and the resulting peptides of varying sizes pass to the small intestine, where further digestion takes place.

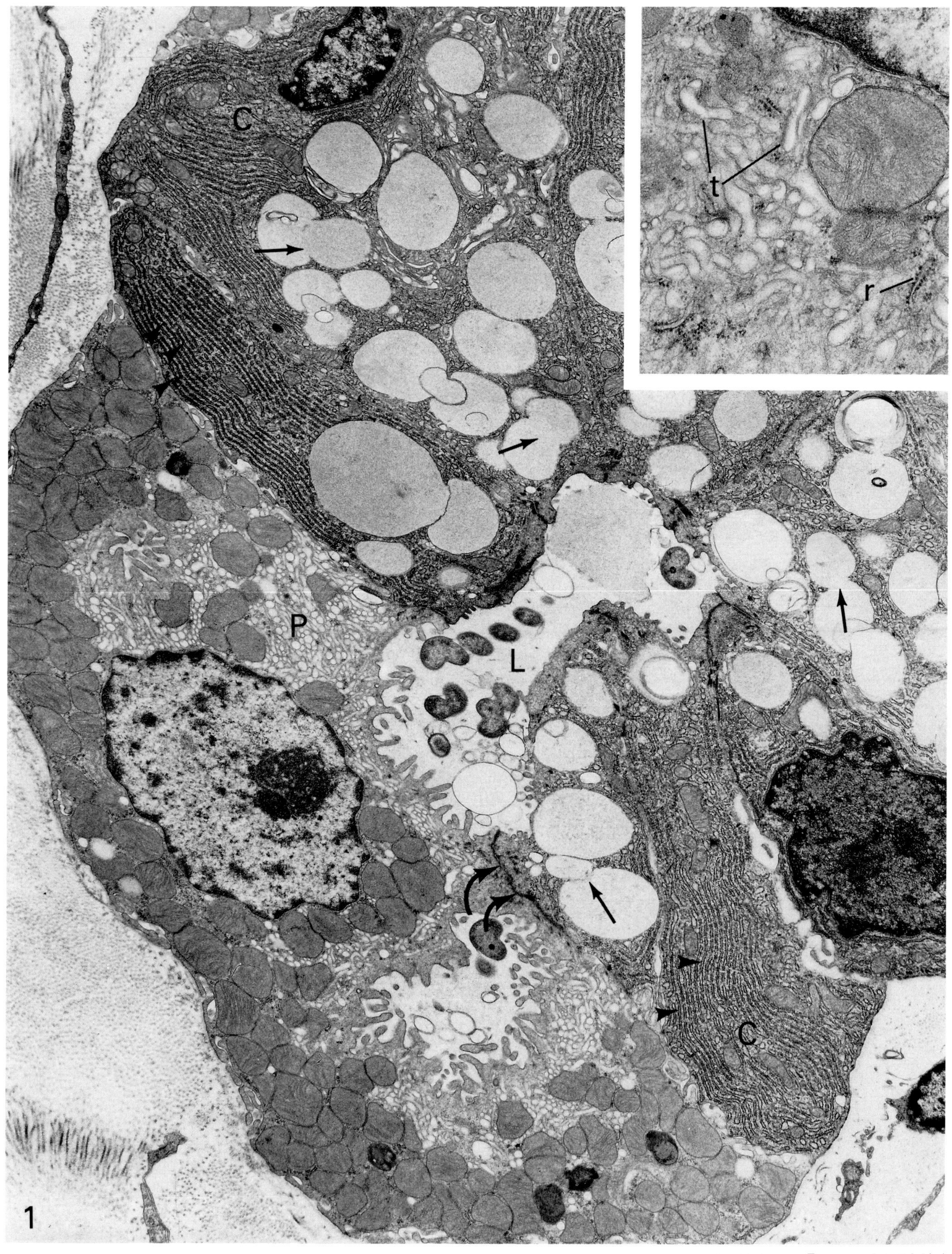

EM1: 10,200× **Inset: 31,500×**

STOMACH: Surface Mucous Cell

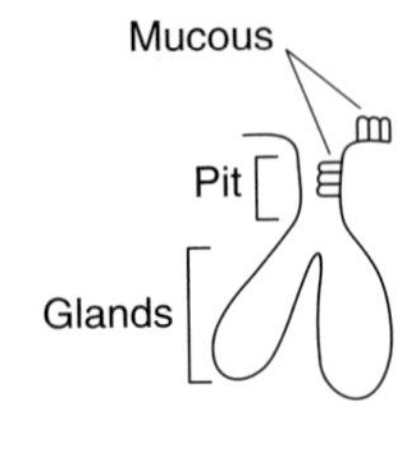

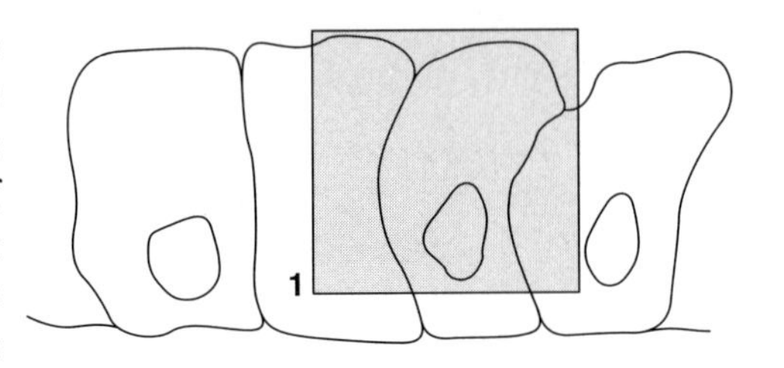

Mucins and associated water are packed in **mucous granules** (g, micrographs) in the apical region of the cells lining the stomach surface and pits. The release of mucus occurs in response to (1) the amount and character of ingested material and (2) the concentration of stomach secretions in the lumen. For example, distention of the stomach wall, concentrated acid, and amino acids all stimulate mucus secretion.

Mucins, high-molecular-weight glycoproteins, are synthesized by epithelial cells throughout the gastrointestinal tract. A mucous layer that is 5% **mucins** and 95% **water** coats the epithelial lining, forming a continuous **insoluble gel** that adheres to and protects the surface from chemical and physical injury. In the stomach, this viscoelastic coat is one of the factors responsible for the mucosal "barrier" that protects the surface lining cells from concentrated acid in the lumen.

Mucins in the thick (up to 100 μm) mucous layer formed on the surface of stomach cells retard the flow of H^+ but do not by themselves protect the stomach lining from digestion by acid. Hydrogen ions that flow through the mucus are neutralized by a gradient of $\mathbf{HCO_3^-}$ within this layer. As a result, the high concentration of H^+ (pH 1) above the mucous layer in the stomach lumen is reduced to normal (pH 7) at the cell surface. The localization of carbonic anhydrase in the surface mucous cells and the large number of apical mitochondria (m, micrograph 1) reflect bicarbonate ion formation and its energy requirements.

Substances that cause stomach injury and ulceration, such as aspirin, do so in many cases by altering the mucous–bicarbonate layer. The thinning of the mucous layer and reduced HCO_3^- formation following administration of aspirin is reversed in vitro when prostaglandins are added to the medium. This restoration of the thickness and bicarbonate character of the mucus, with an increase in pH adjacent to the cells, results in a **cytoprotective effect.** Aspirin is known to inhibit prostaglandin synthesis, and this may account for its ulcerative effects.

Even during normal mucus and bicarbonate production, the neutralization of acid is not always complete and some H^+ penetrates the tissue. Most is quickly removed by the underlying blood vessels, but remaining H^+ does unavoidable damage to surface cells. These cells are replaced by an immediate **epithelial restitution** of the surface involving the migration and spreading of adjacent cells into the damaged area.

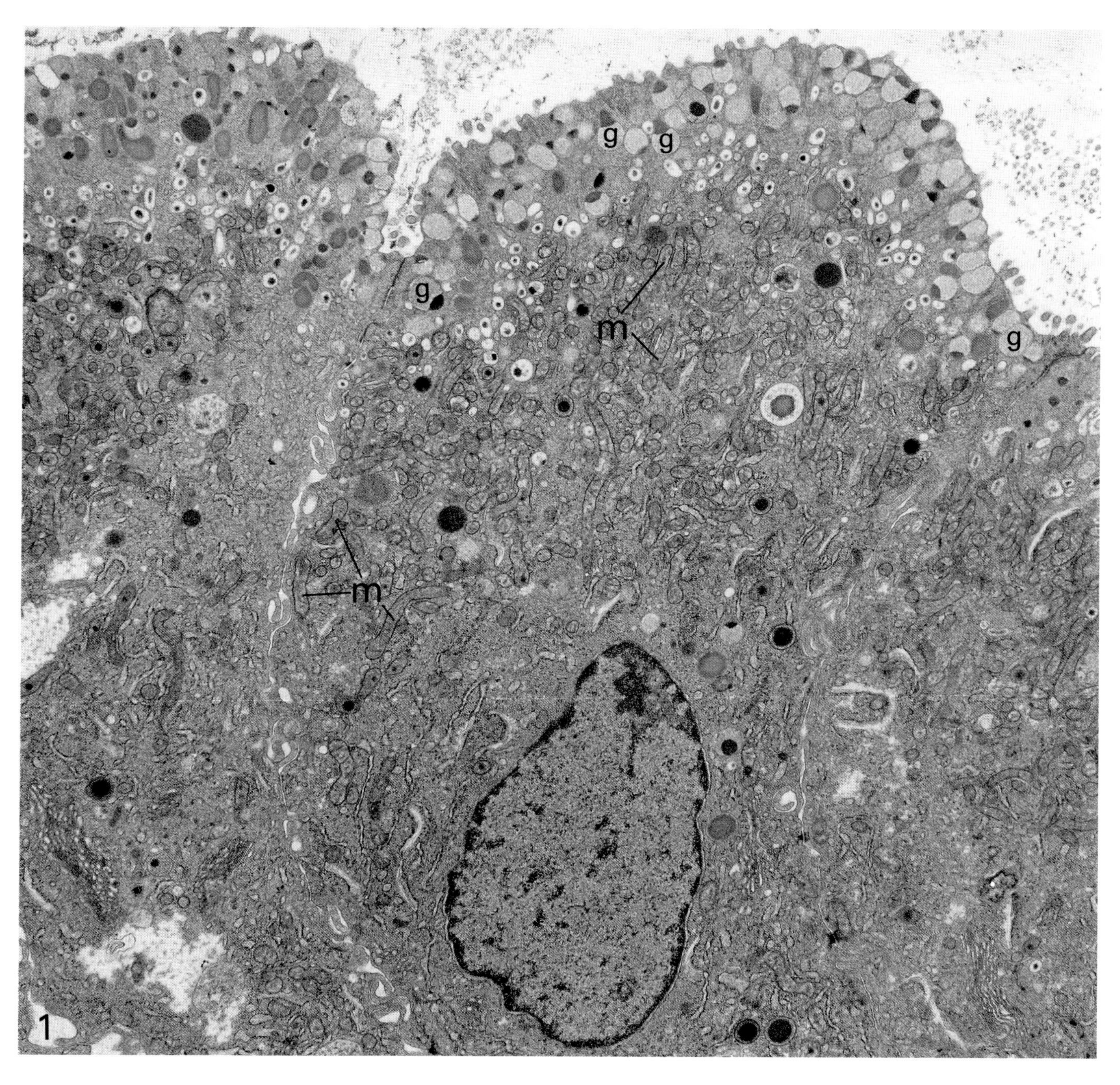

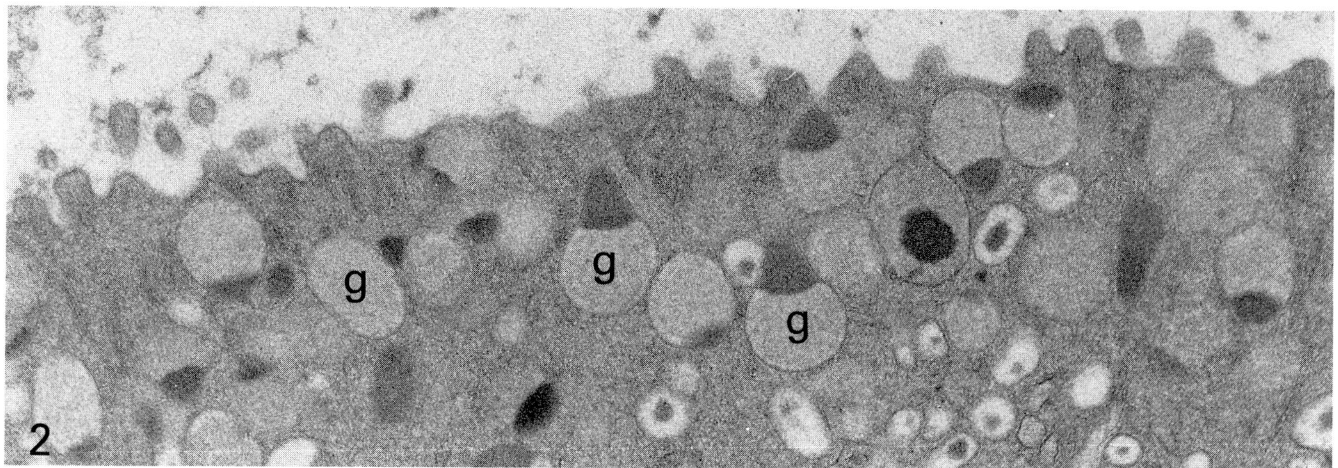

EM1: 10,800× **EM2: 28,500×**

ENTEROENDOCRINE CELLS

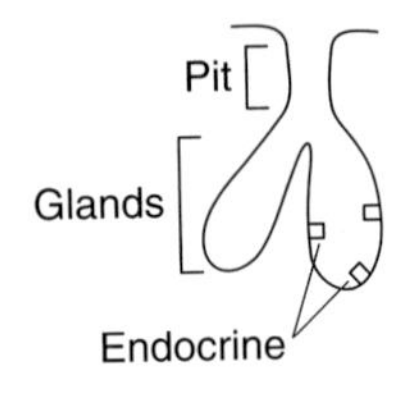

Enteroendocrine cells are scattered individually among the cells of the exocrine glands from the stomach to the colon. These endocrine cells, along with the nervous system, control and coordinate the muscular and secretory activity of the gastrointestinal tract. More than 30 different hormones are synthesized by the gastrointestinal (GI) tract. "Classical" endocrine cells release hormones that travel in the blood and act at distant sites, while the secretions of paracrine cells act locally.

Micrographs 1 and 2 illustrate an endocrine cell (E) between exocrine cells in the stomach glands. Endocrine cells in the GI tract have a distinct orientation, with basally localized granules that displace the nucleus toward the apical region. In the pyloric region of the stomach, the **gastrin-producing cell** (G cell) is the most common endocrine cell. Big (34 amino acids) and little (17 amino acids) gastrins are released into capillaries (c, micrograph 2) and travel in the blood to the gastric stomach, where they have equal potency in stimulating acid secretion by parietal cells.

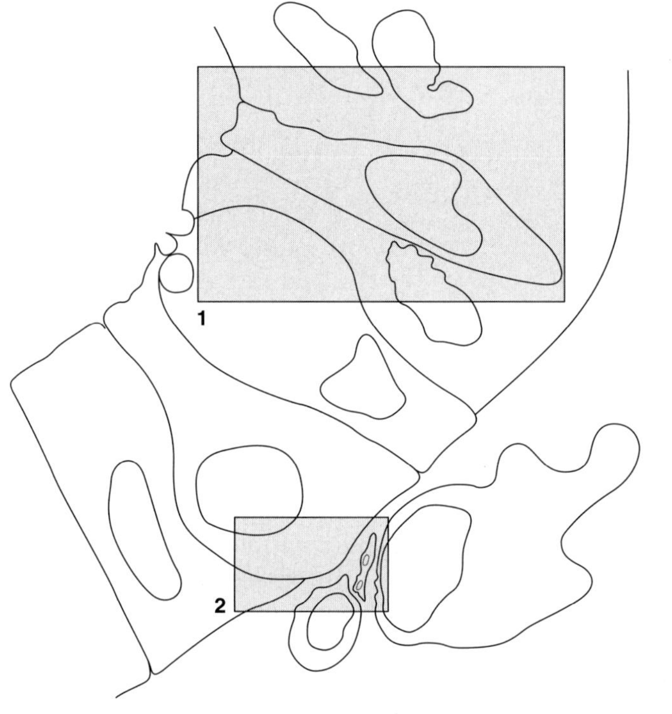

Endocrine cells typically extend a process to the digestive tract lumen (L, micrograph 1). This process is an important communication site to monitor the luminal contents. Dietary amino acids are the most important stimulants, and evidence suggests that their effect is direct. Lipid-soluble amino acids enter G cells via apical projections that extend to the lumen. These amino acids are decarboxylated to form amines that stimulate gastrin release. Many endocrine cells, in addition to G cells, are members of this amine precursor uptake and decarboxylation (APUD) system.

Gastrin release is also affected by secretions from nearby neurons and epithelial cells. Axons (a, micrograph 2) are frequently seen near the basal lamina adjacent to endocrine cells. Neuropeptides (arrows, micrograph 2) are found in many mucosal neurons, and one such peptide, gastrin-releasing peptide, acts specifically on G cells to stimulate gastrin secretion. Somatostatin, released by adjacent epithelial cells that extend thin processes toward the G cells, inhibits gastrin secretion. This mechanism is a part of the negative feedback associated with rising H^+ concentration. A thin process from a nearby epithelial cell (arrowheads, micrograph 2) is seen directly under the endocrine cell in micrograph 2.

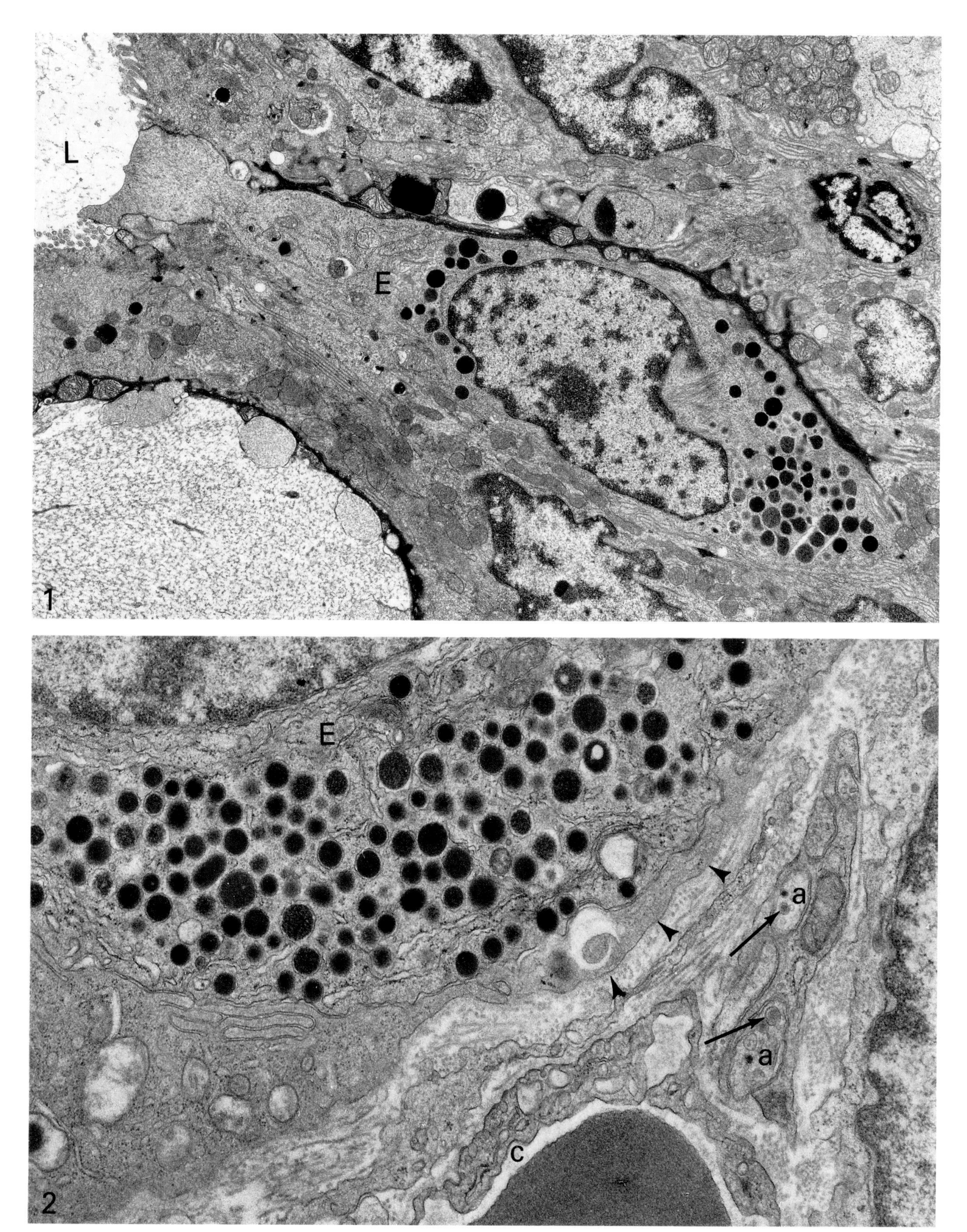

EM1: 8,100× EM2: 18,700×

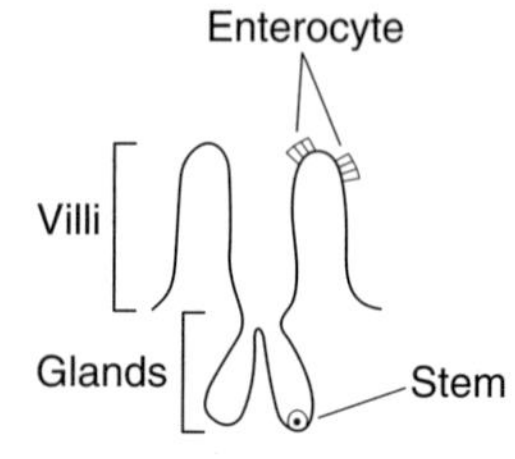

In contrast to the smooth surface of the stomach, the surface of the small intestine has unique projections, villi, that optimize interaction between the contents of the lumen and the cellular lining. Intestinal glands, the crypts of Lieberkühn, open at the base of the villi. In the intestine, stem cells are located at the base of the glands.

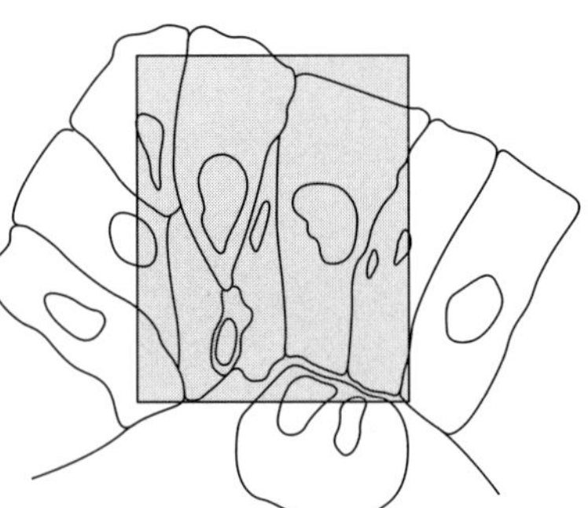

Enterocytes (E, micrograph), the most numerous cell covering the villi, are the site where nutrients cross from the external environment (lumen) to the underlying vessels (e.g., c: capillary, micrograph) and travel to all cells of the body. Their highly infolded apical surface of microvilli create a **"brush border"** (BB, micrograph) where the final stages of digestion and absorption occur (see Epithelium, page 60). The energy requirements for many of these functions are met by mitochondria (m, micrograph) throughout the enterocyte cytoplasm.

Ingested carbohydrates, proteins, and fats are processed in some manner by the enterocyte as they are taken up and transported to the underlying vessels. **Carbohydrates,** initially digested within the lumen by pancreatic enzymes, diffuse to the brush border, where digestion of oligo- and disaccharides is completed by enzymes (e.g., lactase, maltase, sucrase–isomaltase) that are integral membrane proteins. The resulting monosaccharides cross into the cell via membrane carrier systems in the brush border.

The significance of epithelial cell polarity is particularly obvious in enterocytes. Junctional complexes (circle, micrograph) isolate apical from basolateral domains, and not only segregate digestive enzymes to the apical luminal side but also maintain the position of channels important in the vectoral transport of some nutrients, such as glucose, across the cell.

An ATP-dependent Na^+K^+ pump localized in the highly infolded basolateral membrane (arrows, micrograph) of the enterocyte moves Na^+ out of the cell, thus maintaining an Na^+ gradient that favors the entry of Na^+ across the apical brush border. Glucose is cotransported across the apical region of the cell with Na^+. The transport of glucose out of the cell follows its concentration gradient through channels in the basolateral membrane. From the intercellular space glucose enters blood capillaries in the lamina propria and is carried in the portal system to the liver.

In this micrograph a lymphocyte (L) is situated between the basolateral surfaces of two enterocytes. Lymphocytes are one of the few cell types that penetrate the basal lamina to associate intimately with epithelial cells. They represent a large population (20 per 100 epithelial cells) of specialized immune cells that are involved in regulating the immune response to intestinal antigens.

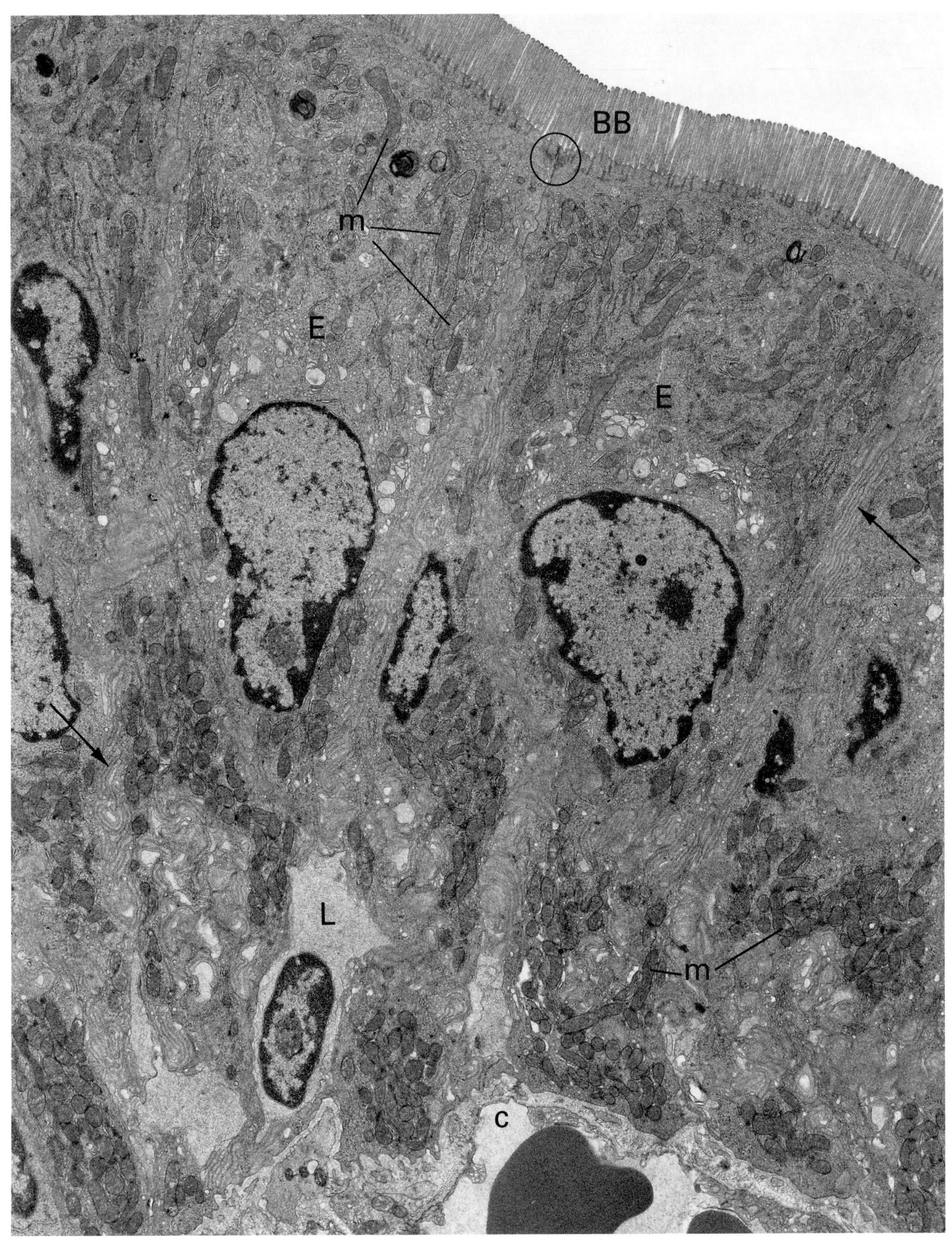

EM: 8,500×

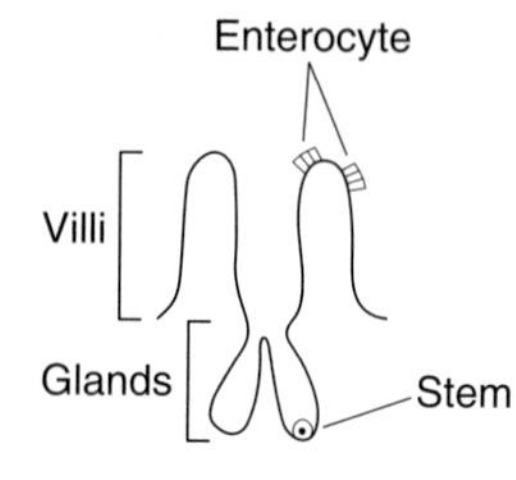

Like carbohydrate digestion, **protein digestion** in the small intestine occurs in two stages, luminal (under the action of pancreatic enzymes) and membranous (on brush border enzymes). However, unlike carbohydrate digestion, both stages depend upon brush border enzyme activity. Pancreatic proteases do not function until they are activated by the brush border enzyme enterokinase. This enzyme converts pancreatic trypsinogen to trypsin, which in turn activates all other pancreatic proteases. Small peptides resulting from digestion by pancreatic proteases in the lumen are further broken down by brush border enzymes.

Digestive enzymes are continually synthesized within the rough ER (straight arrow, micrograph) of **enterocytes** and inserted into the brush border membrane. The active sites of these enzymes are a part of the **glycocalyx** (curved arrows, micrograph) that appears as a fuzzy coat covering the microvilli. During the three to four days between the formation of enterocytes from stem cells at the base of the glands and their migration to the tip of villi, the activity of brush border enzymes increases. This activity remains even after cells are sloughed from the villus tip with normal turnover.

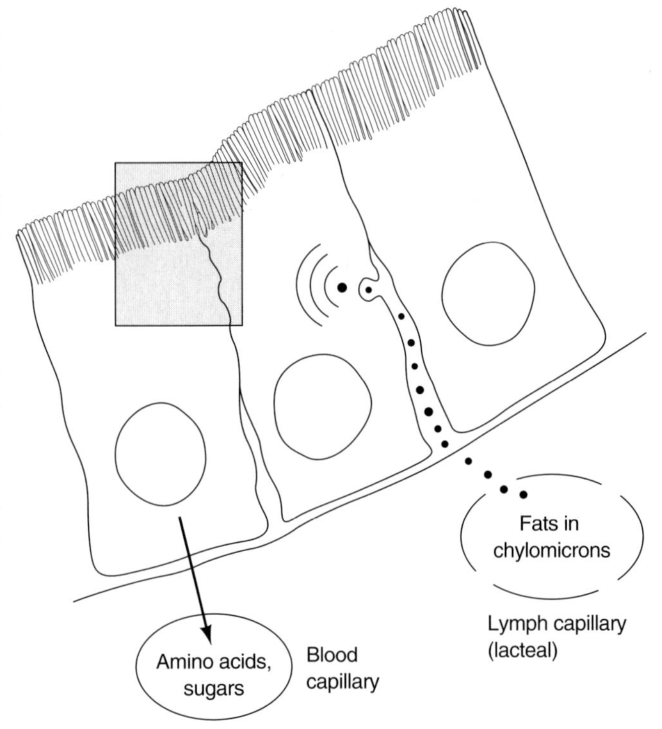

The digestion of **fats** occurs completely in the lumen in emulsified micelles, under the action of pancreatic lipase. Triglycerides, the most abundant form of dietary fat, are broken down to fatty acids and glycerol, which diffuse across the microvilli. In the apical region of the enterocyte, triglycerides are resynthesized within the smooth endoplasmic reticulum (arrowheads, micrograph). The triglycerides are then packaged with other lipids and covered with a surface of 80% phospholipids and 20% protein. The Golgi seems to be the site where these protein–lipid packages, **chylomicrons,** are assembled and directed to the basolateral surface of the cell. Following their exocytosis into the interstitial space, chylomicrons pass into the blind-ending lymphatic capillaries called **lacteals** within each villus. In contrast to blood capillaries, lymphatic capillaries apparently have loose cell-to-cell associations that allow the large, 120-nm chylomicrons to cross the endothelium. Following a fatty meal, the amount of smooth ER in enterocytes increases and the lacteals dilate with chylomicrons.

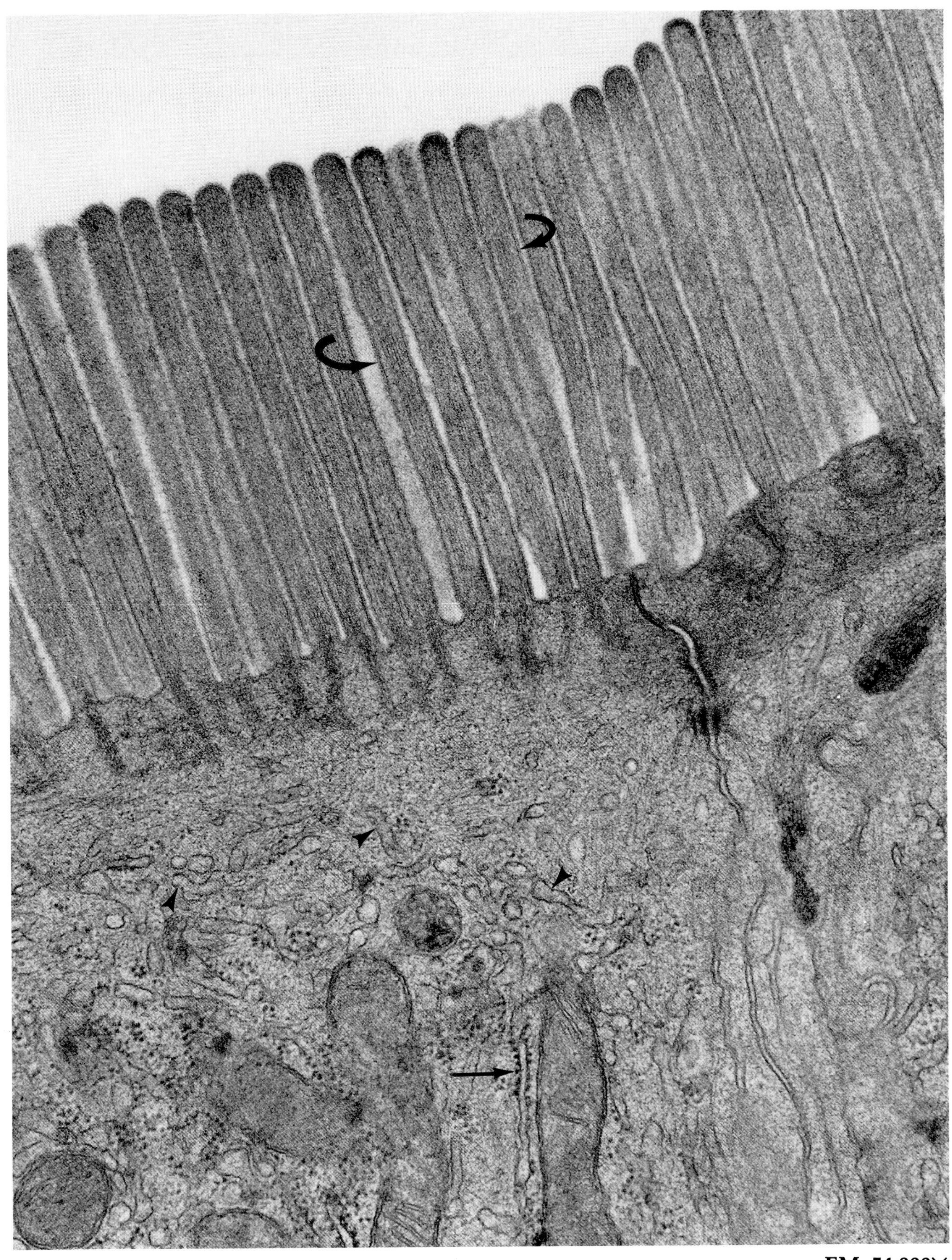

EM: 54,000×

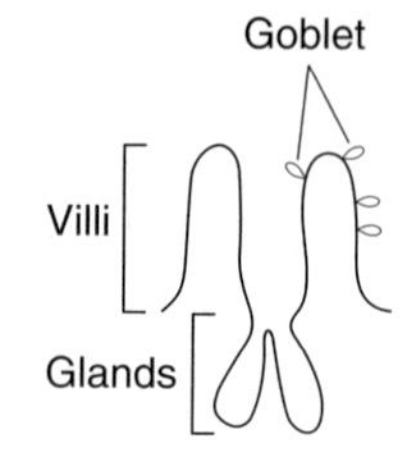

The mucous layer coating the small and large intestine is secreted primarily by **goblet cells.** The micrograph (from small intestine) illustrates a goblet cell interspersed among the enterocytes covering the villi. Each goblet cell is characterized by an apical surface swollen with mucous granules (g, micrograph) and a narrow region of contact (arrows, micrograph) with the lumen. Rough ER (arrowheads, micrograph) and prominent Golgi (G, micrograph) fill the rest of the cytoplasm.

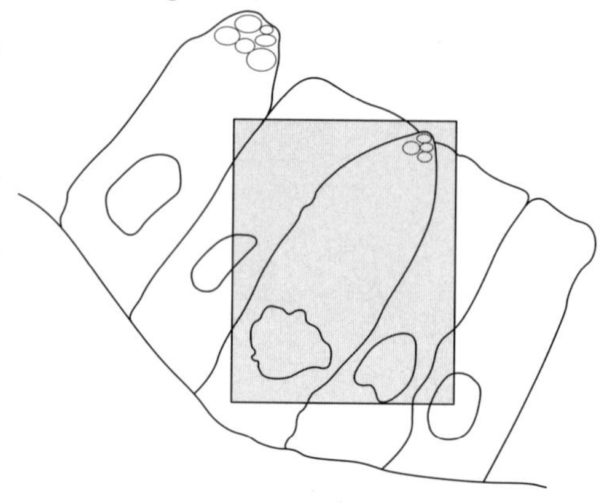

Even though the specific composition of mucus changes from the stomach to the large intestine, the basic structure and path of synthesis and secretion of all mucins are similar. **Mucins** are large glycoproteins that contain greater than 50% (by weight) carbohydrate arranged as chains of oligosaccharides linked to peptide backbones. Subunits consist of large segments of peptide covered by oligosaccharides and small segments of naked peptide. Disulfide bonds connect subunits in either a windmill or chain arrangement to form the mucin polymers.

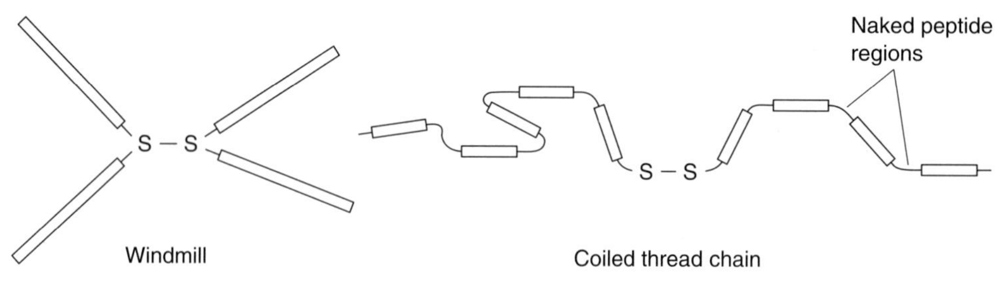

Modified from M. R. Neutra and J. F. Forstner, Gastrointestinal Mucus: Synthesis, Secretion and Function, in *Physiology of the Gastrointestinal Tract,* Vol. 2, Raven Press, New York, 1987.

The synthesis and initial glycosylation of peptide backbones occurs in the rough ER. Following transport to the Golgi, sugars are added as linear or branched chains 2 to 20 residues in length. Subsequent sulfation and subunit polymerization also occurs in the Golgi. Mucin polymers are continually packed into secretory granules on the trans surface of the Golgi and transported to the apical region of the cell. Within seconds after exocytosis, the released mucins are hydrated, resulting in a 600-fold expansion in volume. This addition of water, along with the association of polymers with each other, creates an unstirred layer of **mucous gel** covering the cell surface.

Some of the functions of goblet cell mucus are common to all regions of the gastrointestinal tract, such as (1) trapping sloughed cells and undigested particles and facilitating their clearance and (2) holding IgA that binds to and inactivates microorganisms. Other functions are unique to specific locations. In the colon an important function of mucus is to house ~400 different species of **enteric ("normal") bacteria.** These bacteria perform important dietary functions, such as the formation of Vitamin K; others release enzymes that destroy pathogens or digest mucus. Routine removal of mucus is essential to prevent an abnormal buildup resulting from its continual synthesis.

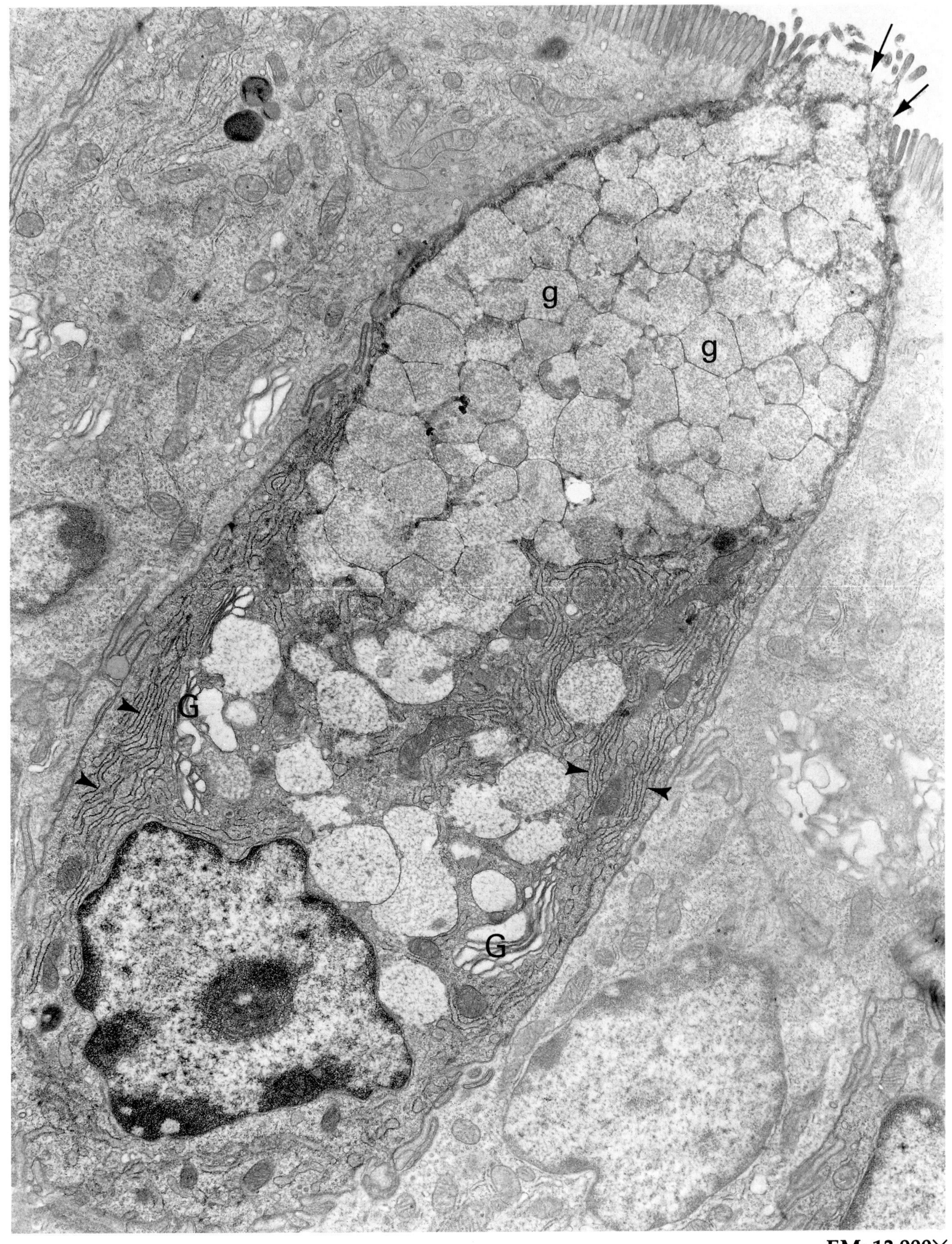

EM: 12,900×

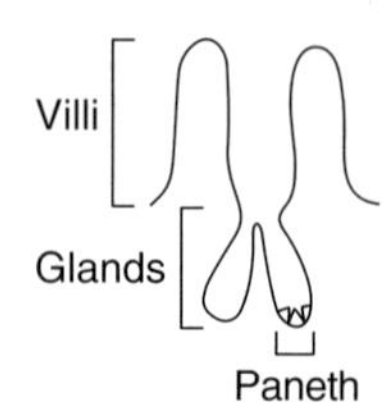

Paneth cells (P, micrograph) are situated, along with stem cells (S, micrograph), in the base of crypts of Lieberkühn in the small intestine. The contrast in morphology of these two cell types reflects their different functions. The free ribosomes in stem cell cytoplasm support synthesis of intracellular proteins needed for mitosis, whereas the rough ER of the Paneth cells reflects their role in the formation of lysosomal and secretory proteins.

Characteristic rough ER (arrowheads) and homogenous dense granules (g) are obvious in the Paneth cells in the micrograph. The cellular morphology, polarization, and location within the gastrointestinal epithelium suggest that, like pancreatic acinar cells, Paneth cells secrete digestive enzymes. Certain digestive enzymes (e.g., trypsin and phospholipase A_2) have been localized by immunohistochemistry to the secretory granules of Paneth cells.

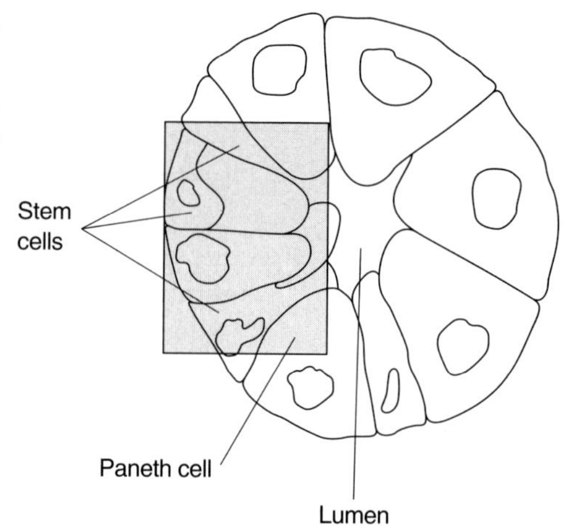

Considerable evidence supports a more significant role for Paneth cells in **defense.** Paneth cells kill foreign organisms outside the cell by releasing enzymes, or within the cell following phagocytosis.

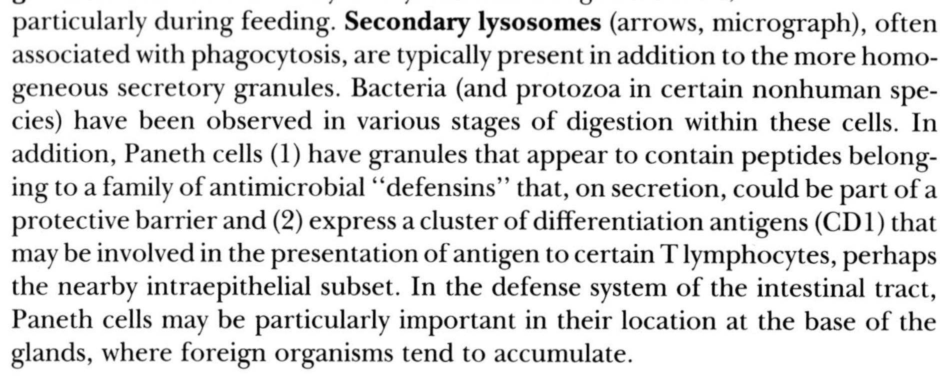

Lysozyme, a cationic protein that digests bacterial cell walls, has been localized in the rough ER, Golgi, and **secretory granules** and is released by exocytosis into the gland lumen, particularly during feeding. **Secondary lysosomes** (arrows, micrograph), often associated with phagocytosis, are typically present in addition to the more homogeneous secretory granules. Bacteria (and protozoa in certain nonhuman species) have been observed in various stages of digestion within these cells. In addition, Paneth cells (1) have granules that appear to contain peptides belonging to a family of antimicrobial "defensins" that, on secretion, could be part of a protective barrier and (2) express a cluster of differentiation antigens (CD1) that may be involved in the presentation of antigen to certain T lymphocytes, perhaps the nearby intraepithelial subset. In the defense system of the intestinal tract, Paneth cells may be particularly important in their location at the base of the glands, where foreign organisms tend to accumulate.

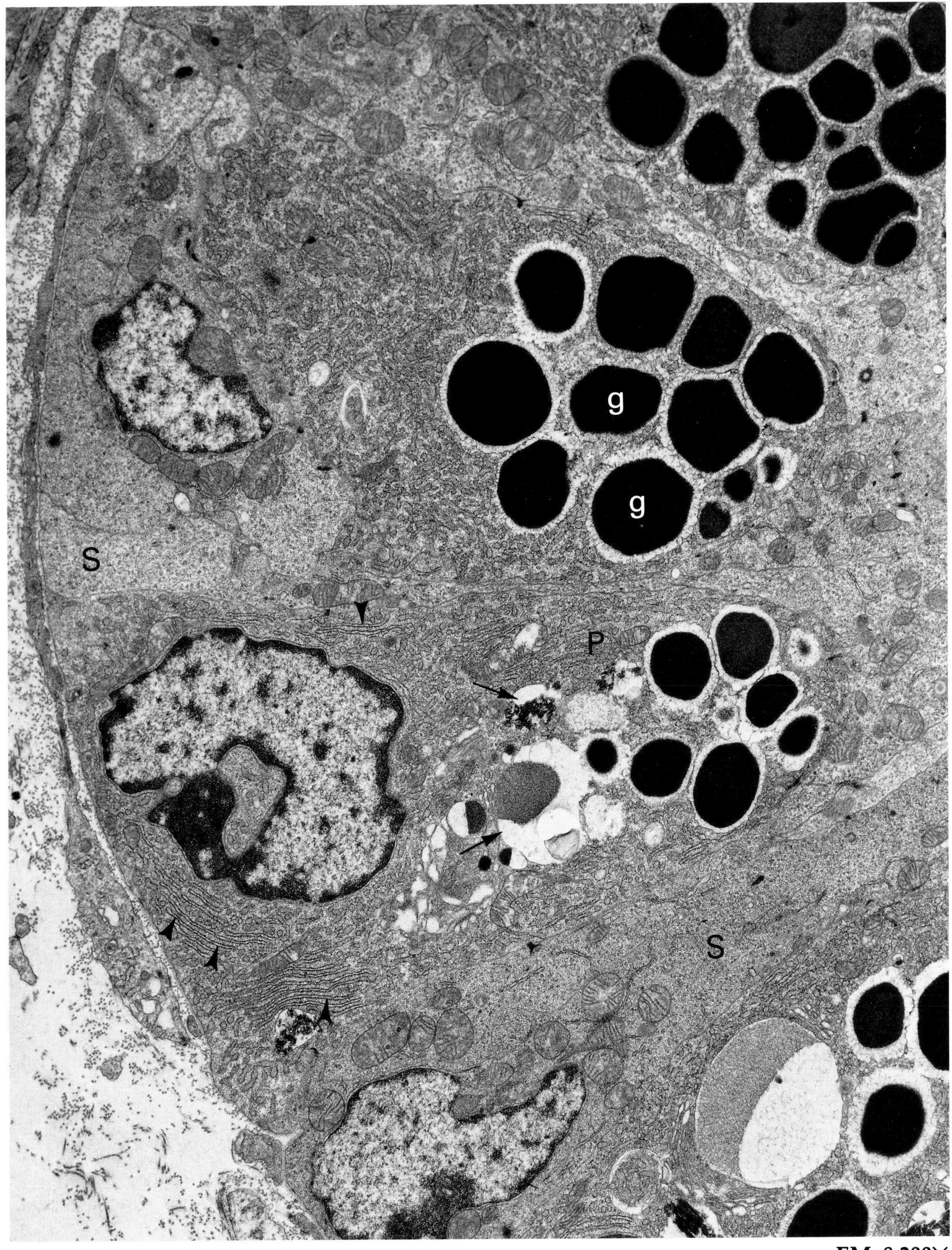

EM: 8,200×

LAMINA PROPRIA

Lamina propria
Villi
Glands

A loose connective tissue layer, the **lamina propria,** underlies the epithelial cells (Ep, micrograph) that line all regions of the gastrointestinal (GI) tract. This layer contains the blood and lymph vessels that collect processed nutrients and many cells that protect this region from foreign invasion from the lumen. **Lymphocytes** (L), an **eosinophil** (E), a **mast cell** (MC), **macrophages** (M), and part of a **plasma cell** (P) are evident in this micrograph. In certain parts of the GI tract, as true here, cells are so tightly packed that collagen, usually prominent in connective tissue, is sparce.

The GI tract is exposed to a continual variety of antigens from the external lumen that contact the body over a large area; the surface area of the small intestine is 60× greater than that of skin. It is not surprising, therefore, that the **immune system** in this organ is extensive. Lymphocytes in the GI tract represent a separate pool of immune cells that migrate mostly within gut-associated lymphoid tissue (GALT) rather than peripheral lymph nodes. Their migration path includes afferent lymphatics in the lamina propria, mesenteric lymph node, efferent lymphatic, thoracic duct, blood vascular system, and return to another lamina propria region.

Isolated lymphocytes, as seen in this micrograph, are initially activated by antigen in organized follicles that are most pronounced in Peyer's patches in the ileum. Germinal centers within these follicles represent a response to viruses, toxins, or bacteria that cross from the lumen. Eighty percent of the plasma cells that result from B cell activation in these areas synthesize IgA, which crosses the epithelial cells and enters the lumen (see Major Exocrine Glands, page 264). Antibodies produced by plasma cells also play an important role in the functions of many of the other defense cells found in the lamina propria. Each cell binds the Fc portion of antibodies, which facilitates subsequent granule release (mast cells, eosinophils), attachment to organisms (eosinophils), or phagocytosis (macrophages).

The vast majority of lymphocytes that are found in the GI tract outside of follicles are T lymphocytes, and those that migrate into the epithelium (arrow, micrograph) are primarily members of the cytotoxic class. Even though these **intraepithelial lymphocytes** are packed between enterocytes, they are not bound to them by junctions and move freely between the epithelial lining and the lamina propria. In their position within the epithelium these lymphocytes are the first immune cells to contact viral-infected cells or foreign antigens entering from the lumen and thus monitor the antigenic character of the luminal environment.

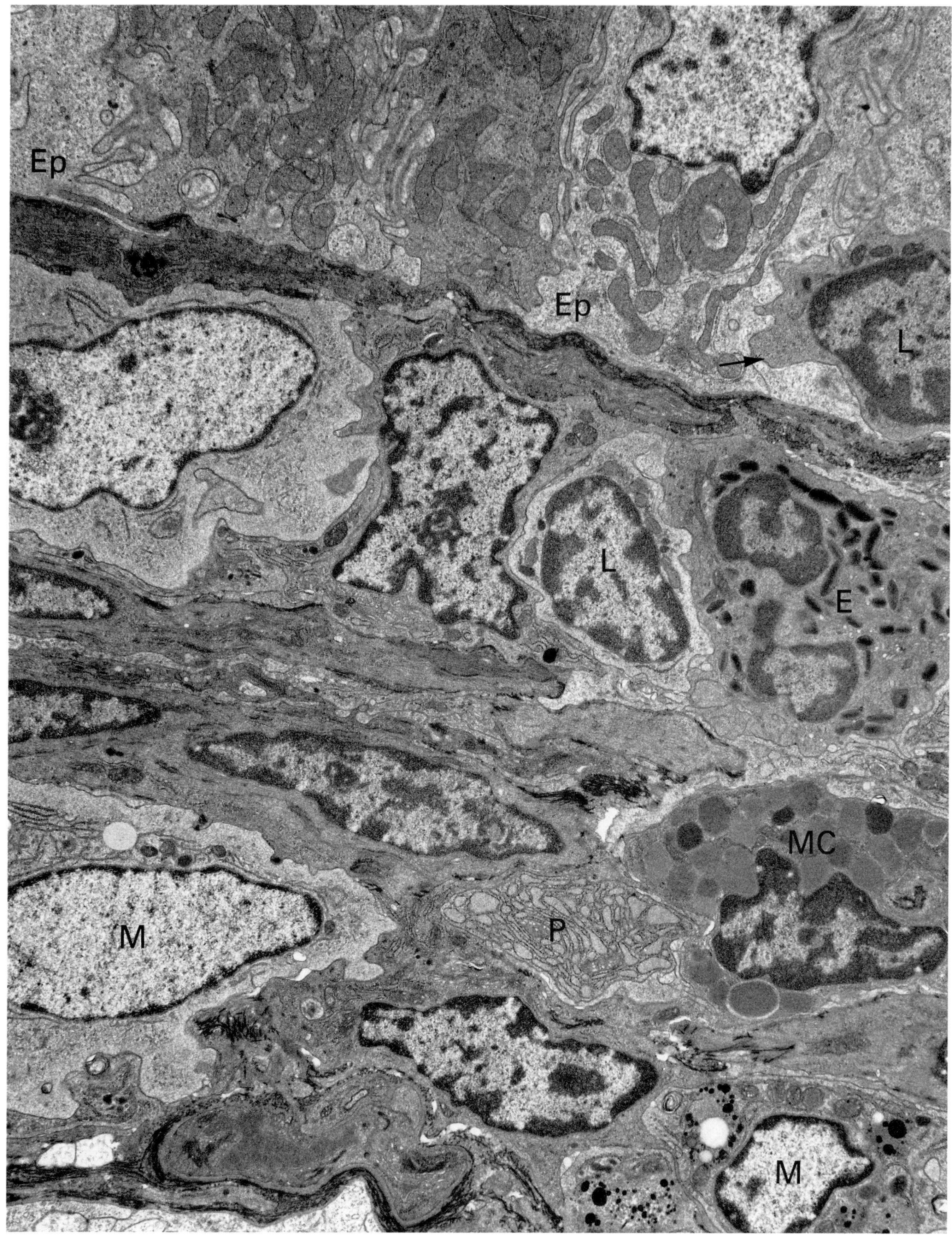

EM: 7,400×

RESPIRATORY SYSTEM

RESPIRATORY MUCOSA: Mucociliary Clearance

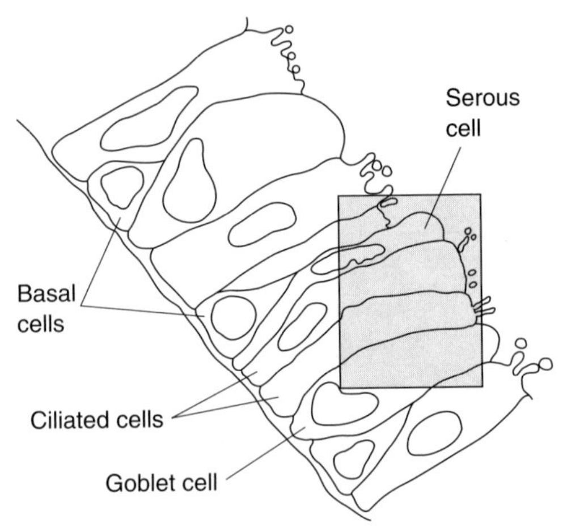

Beginning in the nasal cavity and extending through the larger bronchioles, air passageways are coated with a **viscoelastic sheet** produced by mucous and serous cells located in the surface epithelium and underlying submucosal glands. The surface epithelium (micrograph) is pseudostratified and typically contains mucus-producing **goblet cells** (G, micrograph), **ciliated cells** (C, micrograph), and **basal cells.** In laboratory rodents, serous cells (S, micrograph) are also found in the surface epithelium. In adult humans, however, these cells are found only in the submucosal glands.

The primary structural components of the viscoelastic layer are large glycoproteins, **mucins,** synthesized by surface Goblet cells and mucous cells of the submucosa glands and stored in granules (g, micrograph). Release of mucin is accompanied by hydration, swelling, and gel formation similar to that occurring in the intestinal tract (see Gastrointestinal Tract, page 296). Water and ions are moved into the secretion primarily across the ciliated cells.

In addition to IgA, serous cells, which comprise the majority of cells in submucosal glands, make several other essential contributions to the secretion. The dense granules (arrowheads, micrograph) contain enzymes such as **lysozyme** that destroy bacteria and **antiproteases** that inactivate destructive bacterial enzymes. The antiproteases also destroy proteases released by neutrophils before they cause the secondary damage common to pulmonary disease. Within the serous cell granule the positively charged secretory proteins are bound to negatively charged sulfated **proteoglycans.** Opposite-charge packing, common to many granules, is a mechanism that concentrates contents by reducing osmotic activity to exclude water. In addition to their structural role within the granule, proteoglycans, released with the secretory proteins during exocytosis, are incorporated into the mucin layer and may make important contributions to its physical substructure.

The "blanket" of secretion covering the epithelium is continually moved (3 – 12 mm/min.) toward the pharynx by the coordinated activity of cilia. Cilia project from up to 90% of surface epithelial cells and move as an undulating wave in a low-viscosity fluid layer under the viscoelastic sheet. In the beginning of the effective stroke, each cilium swings upward and attaches to the mucous sheet. As the force is transmitted, cilia movement becomes slower and each cilium detaches from the sheet and completes a recovery stroke beneath the mucous layer. Organisms and particles in solid, liquid, and gaseous forms are trapped within the sticky mucous layer and carried by the **mucociliary elevator** to the pharynx, where they are swallowed or expectorated. This "clearance" of the airways is the primary factor preventing respiratory infection and damage.

In certain major respiratory complications such as chronic bronchitis, asthma, and cystic fibrosis, an early feature is hypersecretion of mucus and hyperplasia of goblet cells at the expense of ciliated cells. Excess mucus obstructs the airway and puts stress on the mucociliary elevator. Bacteria and viruses accumulate and overload the antimicrobial serous cell defense system, resulting in infection.

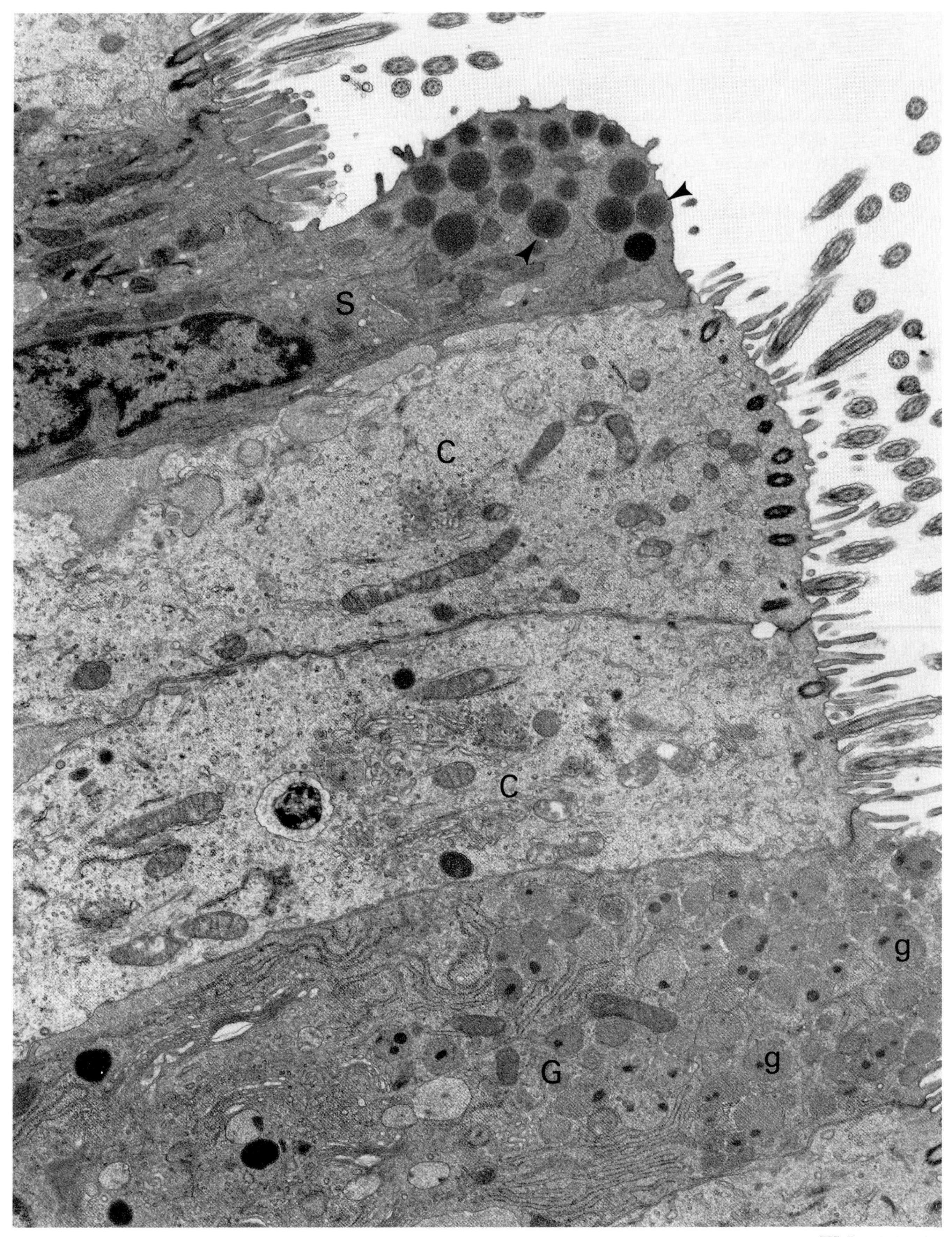

EM: 13,200×

BRONCHIOLE

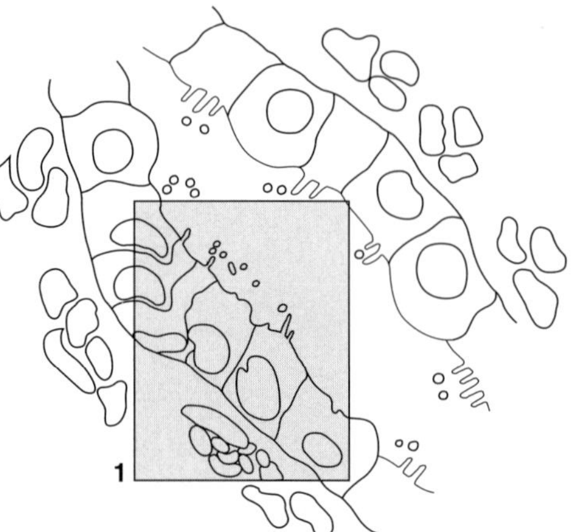

Bronchioles, the final conducting portions of the airways, occupy a strategic position between the larger airways, which contain cartilage and extensive submucosal glands, and the delicate alveoli where gas exchange occurs. They are generally less than 2 mm in diameter, with a simple epithelial lining and a wall containing a relatively high proportion of circularly arranged smooth muscle (M, micrograph 1).

In the terminal bronchioles, the epithelium is cuboidal and consists of ciliated (C, micrograph 1) and nonciliated (NC, micrograph 1) cells. The **ciliated cells** function as all ciliated cells found throughout the conducting portion of the respiratory tree, to move the secretions and trapped airborne particles toward the pharynx. The nonciliated cells, often called **Clara cells,** are unique to bronchioles.

Clara cells cannot be defined by a single morphology. Different types are seen not only in different species, but also at different times in development and even adjacent to each other in the same section. The Clara cells in micrograph 1 (from rabbit) are similar to those most commonly observed in humans, with an apical region packed with irregularly shaped dense granules (arrows) and large mitochondria (m), and a basal region containing the nucleus (N), rough ER, and patches of glycogen (g). Smooth ER (arrowheads, inset) is present in Clara cells of all species but is apparently most abundant in rodents (inset). It may function, in a manner similar to the liver, to detoxify a variety of compounds. The mitochondria in the Clara cells shown have a particularly dense matrix. The matrix of the mitochondria in the inset is so dense that only occasional cristae (arrows) can be distinguished.

A major function attributed to Clara cells is the secretion of the material lining the bronchiolar lumen. The secretion contains proteins important in defense (e.g., lysozyme, antibodies) and in breaking up the mucus produced by the upper airways. The heterogeneous nature of the dense granules suggests that they may not all be secretory. Some may be lysosomal and involved in the recycling of secretions.

Of all the respiratory passageways, the bronchiole is occluded most easily. Because of the small lumen size, factors associated with disease, such as an abnormal production of mucus (e.g., in chronic bronchitis due to smoking) or spasmodic contraction of the smooth muscle in the wall (e.g., asthma), can close bronchioles and reduce airflow enough to be life threatening. The smooth muscle is innervated primarily by parasympathetic nerves that release acetylcholine, a bronchioconstrictor, and is particularly sensitive to mediators, such as leukotrienes from mast cells (see Connective Tissue, page 76), released during the allergic response.

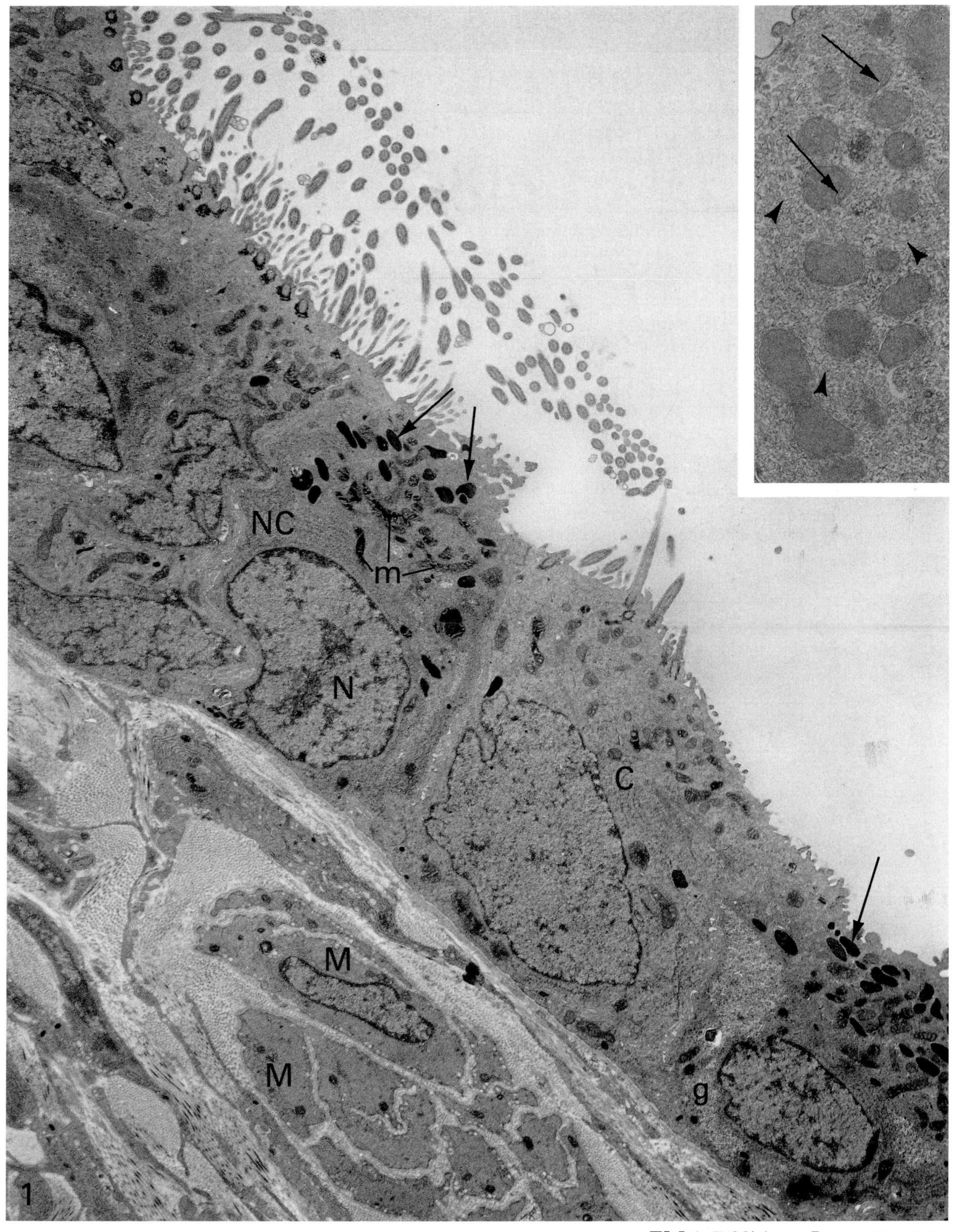

EM 1: 7,00× Inset: 17,500×

ALVEOLUS

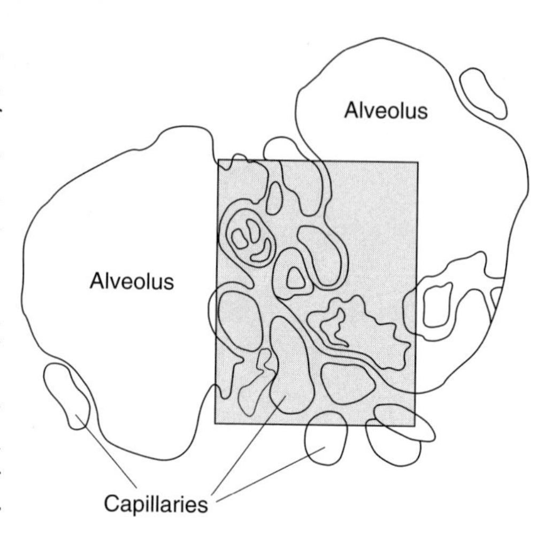

The respiratory passageways that bring air into the lungs terminate in over 300 million bubblelike **alveoli** where gas exchange occurs. These alveoli are the principal structures of the functional unit of the lung, the acinus, which includes the passageways and alveoli that branch out from each terminal bronchiole. The cuboidal epithelium lining terminal bronchioles is continuous with the epithelium lining the alveoli, which consists of simple squamous **(Type I alveolar cells;** AI, micrograph) and intermittent cuboidal secretory cells **(Type II alveolar cells;** AII, micrograph).

A basketlike network of pulmonary **capillaries** (c, micrograph) surrounds each alveolus. The exchange of carbon dioxide and oxygen occurs via diffusion across the barrier (arrows, micrograph) that separates the capillaries from the **air space** (S, micrograph). The total surface area for gas exchange in the lung is approximately 60 m^2.

The activities of cells other than those that make up the barrier are important in maintaining its structure. Macrophages are present within the air space and in the interstitial tissue of the alveolar septa, and here, as in other tissues, provide a critical line of defense. The **macrophage** (M) seen within the air space in the micrograph is walking across the apical surface of a thin cytoplasmic extension of a Type I alveolar cell. When activated by airborne insults, these cells recruit blood cells, such as **lymphocytes** (L, micrograph), to aid in their protective efforts. With progressive disease, blood cells actually enter the air space to join macrophages.

Fibroblasts (F, micrograph), found throughout the interstitium, synthesize proteoglycans and both collagen and elastic fibers. Collagen supports the septum, while elastic fibers accommodate the stretching associated with the respiratory cycle. In the event of injury to the septum, fibroblasts, in an effort to repair the damage, divide and secrete more collagen. Chronic insult leads to overproduction of collagen and scarring, which interferes with gas exchange. The defense and repair mechanisms in the lung must be tightly controlled since what might be a minor response in another location is major in the lung if it interferes with the delicate gas exchange area.

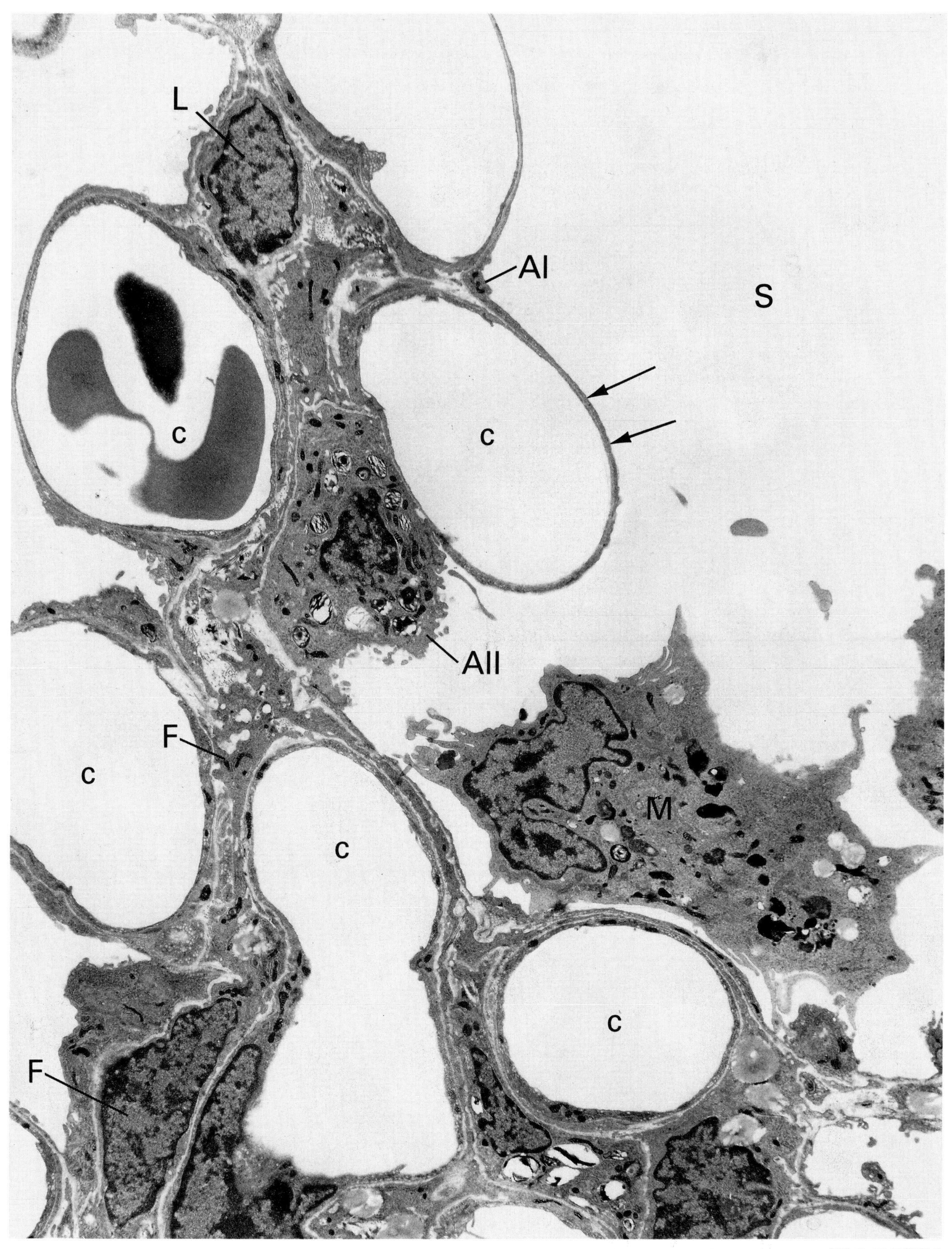

EM: 7,200×

Type I Alveolar Cell

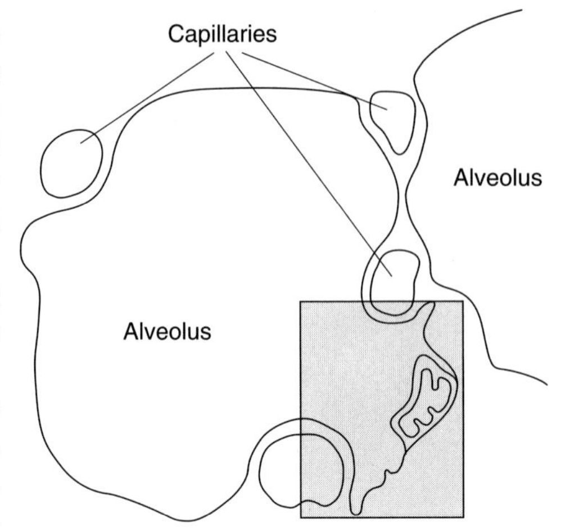

Type I alveolar cells make up only 7% of the lining cells of the alveoli, yet they cover 95% of the surface area. The thinness of this layer (0.15 μm) is an important factor in efficient gas exchange. In the micrograph, the nucleus (N) and cytoplasm (arrows) of a Type I cell are seen lining the alveolar space (S). Two capillaries (c) are closely apposed to extensions of the Type I cell cytoplasm.

Type I cells expend much of their energy maintaining turnover associated with their large volume. The thinness of the cytoplasm makes it difficult to appreciate how active this cell is. However, free and attached ribosomes are packed into the cytoplasm in even the most attenuated areas. Protein synthesis throughout this cell is needed to support membrane-associated events such as ion pump activity, vesicle transport (white spots in the boxed area, micrograph), and the removal of certain airborne invaders. Lysosomes (l, micrograph) play an important role in routine organelle turnover and cleaning the blood-air barrier. Activity along the entire surface of Type I cells is essential to maintain its function as a barrier to most molecules.

The attenuated Type I cells are particularly susceptible to damage and yet are not capable of dividing and replacing themselves following injury. Repair (and routine replacement) is carried out by Type II cells that switch from their secretory activity to divide and differentiate in a period of two days into Type I cells. With chronic injury, Type II cells divide but do not differentiate into Type I cells. As a result, parts of the air spaces become lined by cuboidal Type II cells, thus reducing the area available for gas exchange.

Fluid does not normally enter the air space because of the negative pressure in the interstitium that diverts excess fluid into the nearby lymphatics. When endothelial cells (E, micrograph) are damaged, excess proteins leak into the interstitium, changing this pressure from negative to positive. Even though Type I cells are the least permeable cells in the blood-air barrier, associated by tight junctions with a resistance as high as 2000 $\Omega \cdot cm^2$, once the interstitial pressure becomes positive the Type I lining breaks and fluid and proteins escape into the alveoli.

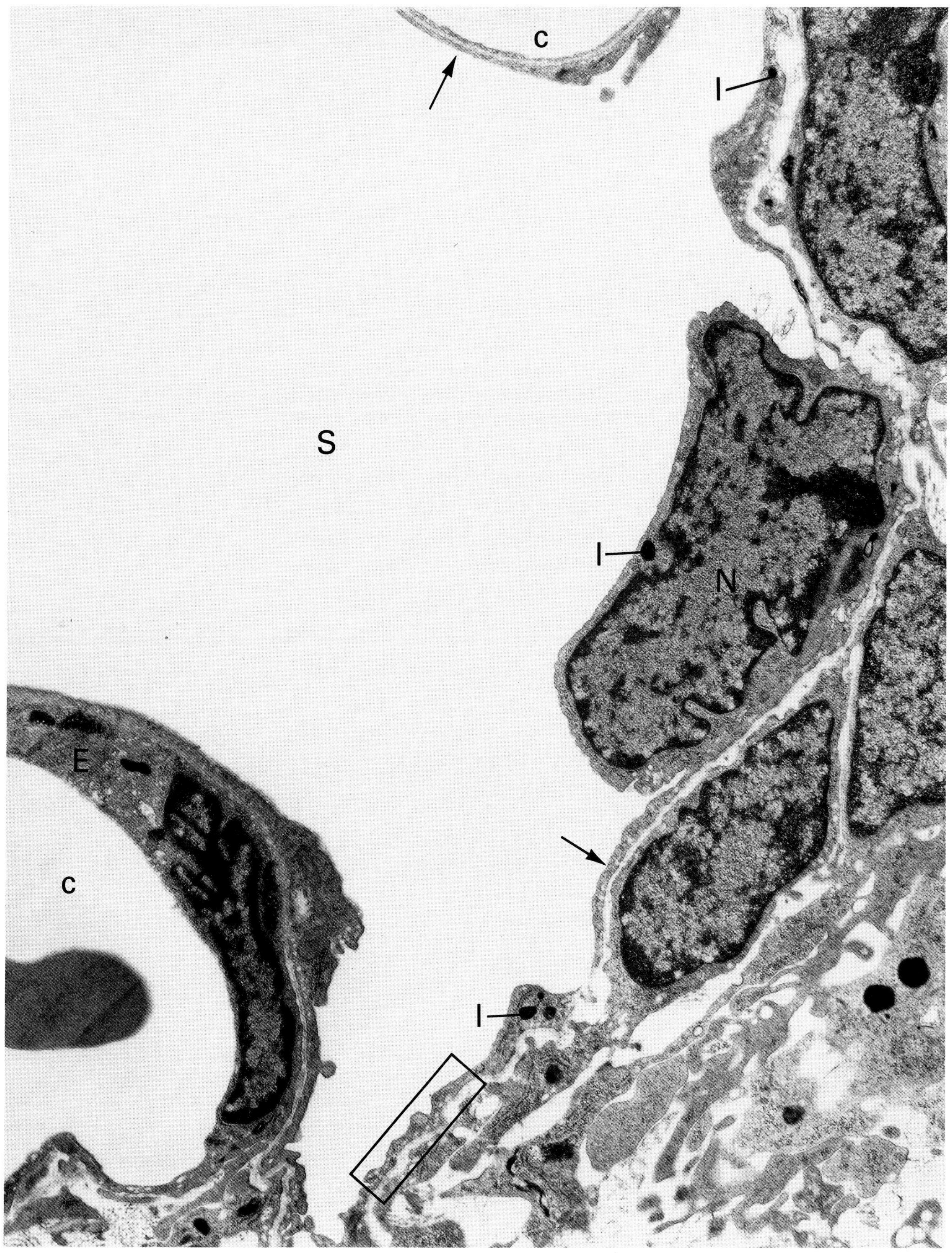

EM: 17,200×

BLOOD-AIR BARRIER

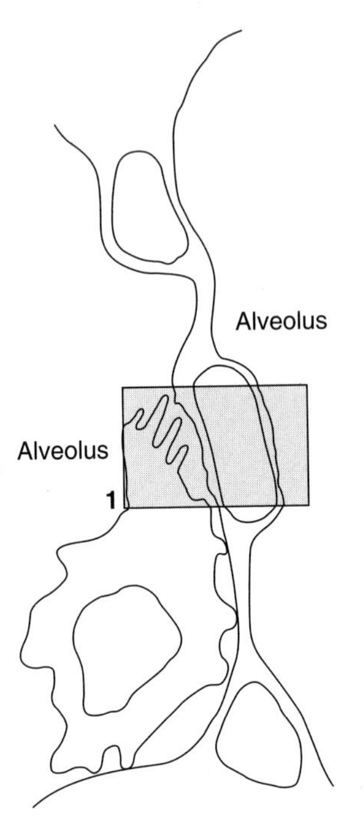

Three components of the blood-air barrier are clearly illustrated in the micrographs: endothelial cells (E), Type I alveolar epithelial cells (AI), and fused basal laminas of both cell types (*). The fourth component, surfactant that covers the apical surface of the Type I cells, is not preserved in routine micrographs. The extreme attenuation of the endothelial and epithelial surfaces (best illustrated in micrograph 1) facilitates efficient exchange of oxygen and carbon dioxide. At the same time that gas exchange is facilitated, the movement of most molecules and water is prevented (see Respiratory System, page 310).

Less attenuated areas (micrograph 2) of some **Type I cells** may contain numerous vesicles of various sizes, some of which are coated (arrowhead). These vesicles are involved in (1) the transport of proteins (serum proteins and immunoglobulins have been localized within vesicles and within the water phase of surfactant) and (2) a limited removal of air borne particles. The removal of excess fluid from the air space primarily involves nonvesicular water movement following active sodium ion transport.

Endothelial cells lining the capillaries in the lung have some unique metabolic functions aside from their role as part of the blood-air barrier. Certain drugs and hormones are modified by these cells as they pass through the lung. Some substances are taken up and modified within the cell, while others bind to the surface of the endothelium and are metabolized there directly. Angiotensin I is converted to angiotensin II, a hormone that regulates salt balance and blood pressure, by an enzyme that is an integral membrane protein evenly distributed along the luminal surface of the lung endothelium.

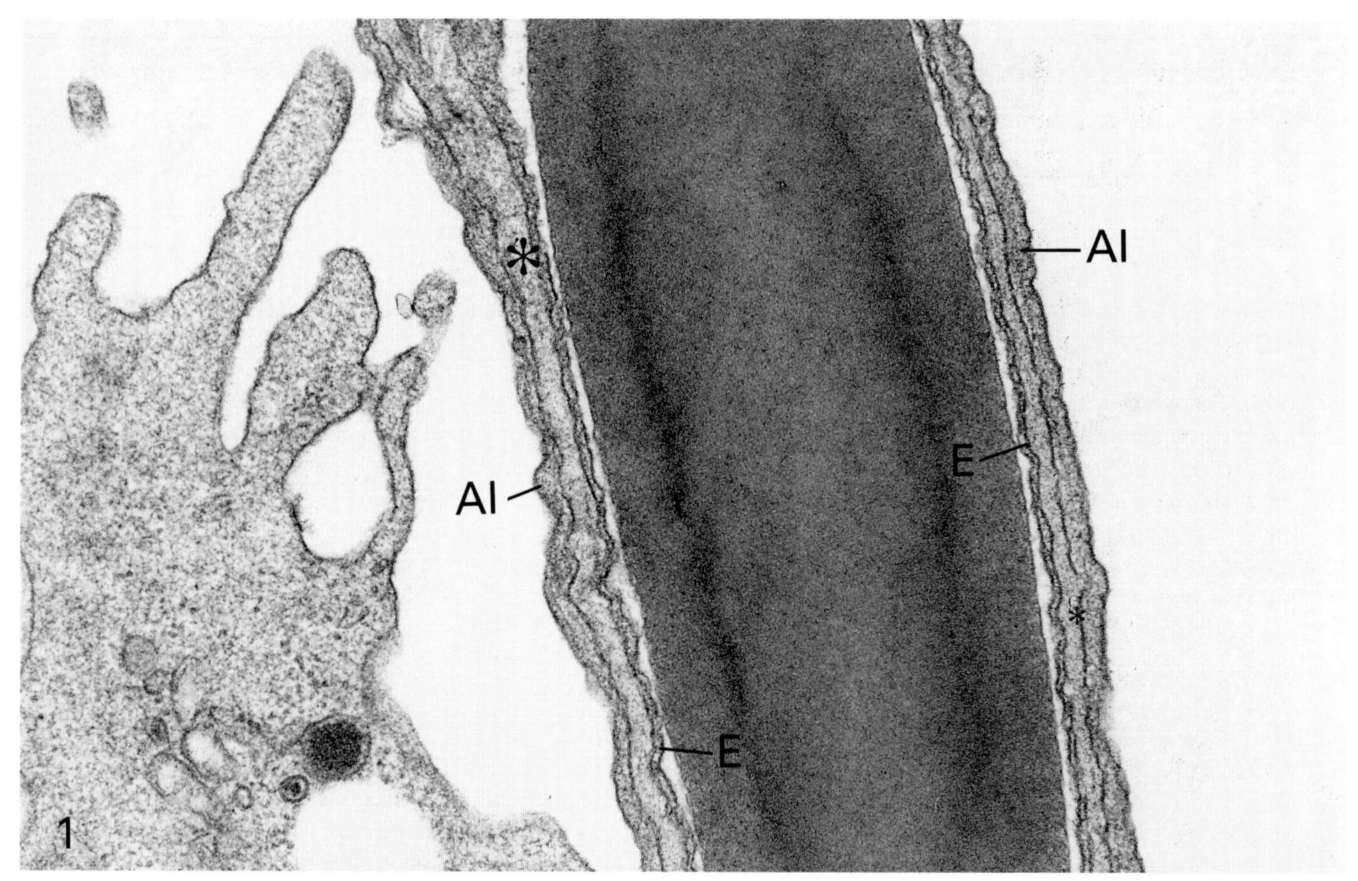

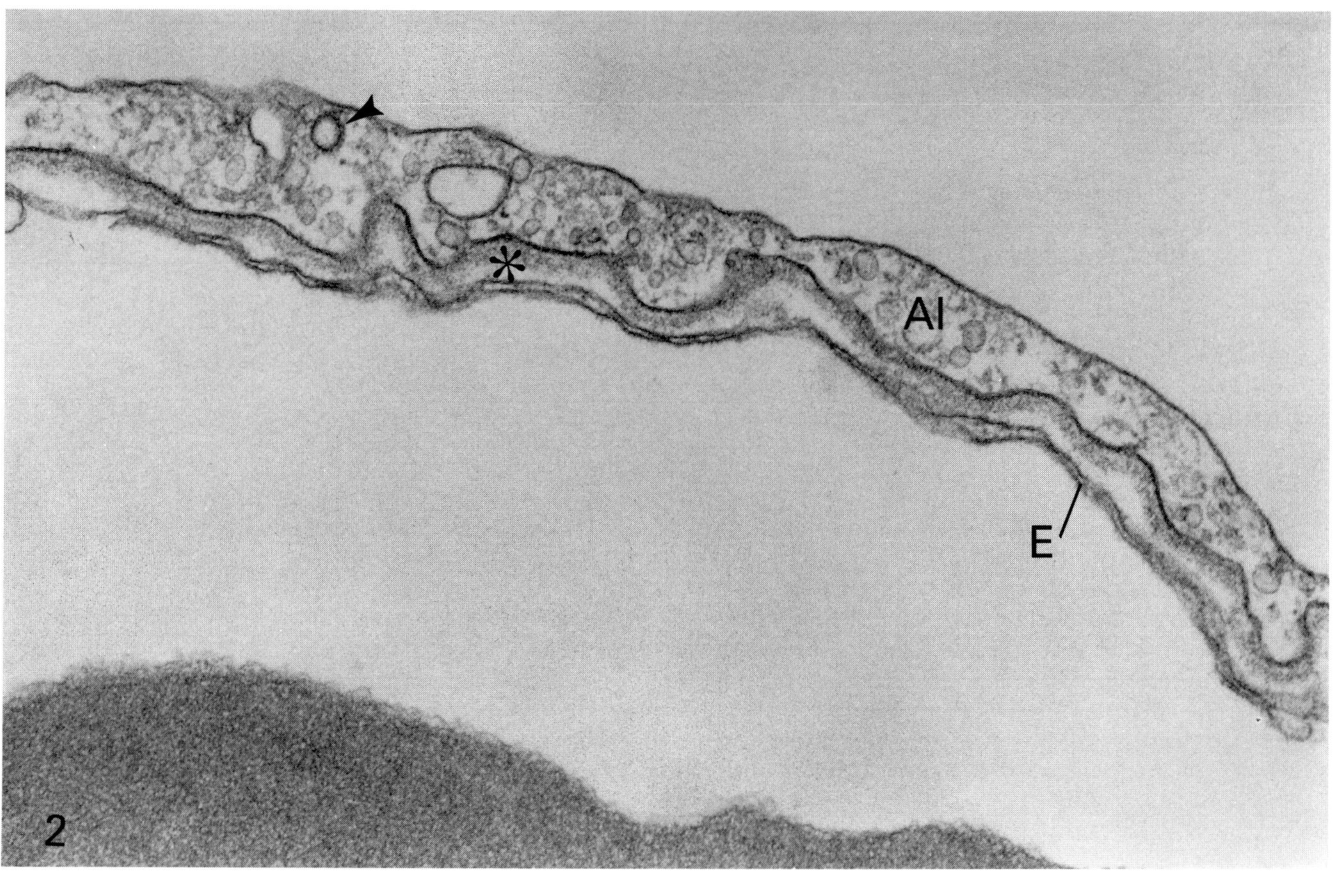

EM 1: 54,000× **EM 2: 62,500×**

Type II Alveolar Cell

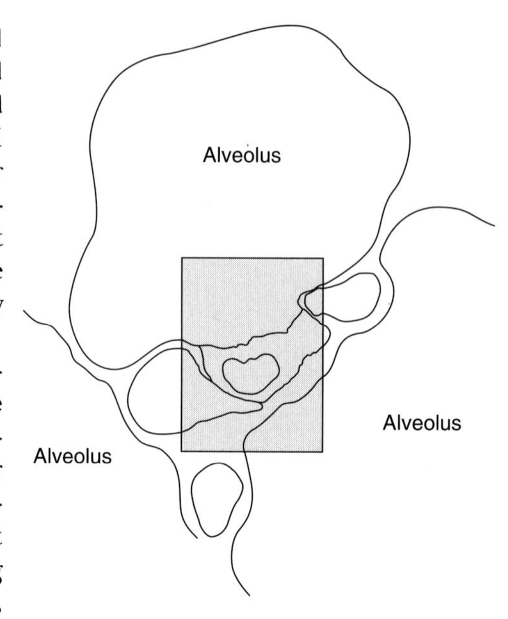

Type II alveolar cells (AII, micrograph) are cuboidal cells that occupy crevices in the alveolar surface, interspersed among the Type I cells (arrowheads, micrograph) and attached to them by junctional complexes (circle, micrograph). Type II cells not only give rise to Type I cells, but in addition their secretion, **surfactant,** coats the entire alveolar surface. Surfactant is a macroaggregate of phospholipids and proteins that reduces surface tension, thus (1) preventing the collapse of the lungs during exhalation and (2) reducing the amount of energy required to inflate the lungs during inhalation.

Ninety percent (by weight) of surfactant is lipid, primarily the phospholipid **dipalmitoylphosphatidylcholine** (DPPC). As phospholipids are synthesized, they are sequestered within granules (arrows, micrograph) in a lipid bilayer lamellar form. Lamellae continue to accumulate as phospholipids are added to the granule. These lamellar bodies, like most secretory granules, release their contents by exocytosis. During this process the lamellae unfold and transform into a thin film, which is the primary surface-active component. It is likely that DPPC lines up as a monolayer, with the nonpolar fatty acid chains projecting into the air space. The polar portion of the resulting monolayer is soluble in the aqueous phase between the monolayer and alveolar surface. Tight packing of the DPPC lowers surface tension by interfering with the attractive forces between the water molecules covering the alveolar surface.

Surfactant secretion is stimulated by a variety of factors, including hormones, growth factors, and simple mechanical stretching. With each inhalation alveolar surface area increases by as much as 80%. The mechanical stretch of the alveolar lining directly causes an increase in the secretion of surfactant necessary to cover the expanded area. During exhalation the excess surfactant is removed to maintain a thin surface-active film.

Type II cells are the major cells responsible for the turnover of surfactant. Certain lamellar bodies and other vesicles seen within Type II cells may actually be a part of an endocytotic recycling pathway rather than the exocytotic secretory pathway. Lysosomal enzymes have been demonstrated within some lamellar bodies in Type II cells.

Ten percent of surfactant is protein that, even though a minor component, is essential to the formation and functioning of surfactant as a whole. Surfactant proteins include a family of large acidic glycoproteins (SP-A) that are essential to the normal turnover of surfactant, and two smaller hydrophobic peptides (SP-B, SP-C) that promote surfactant spreading.

The maturity of the lungs at birth is critical; enough surfactant must be formed to prevent lung collapse at the first breath. Immaturity of Type II cells, with reduced surfactant secretion, results in newborn respiratory distress syndrome, an important cause of perinatal mortality. Current treatment involves glucocorticoids administered at critical development stages, and surfactant replacement therapy, which can be either synthetic, from bovine lung, or from amniotic fluid.

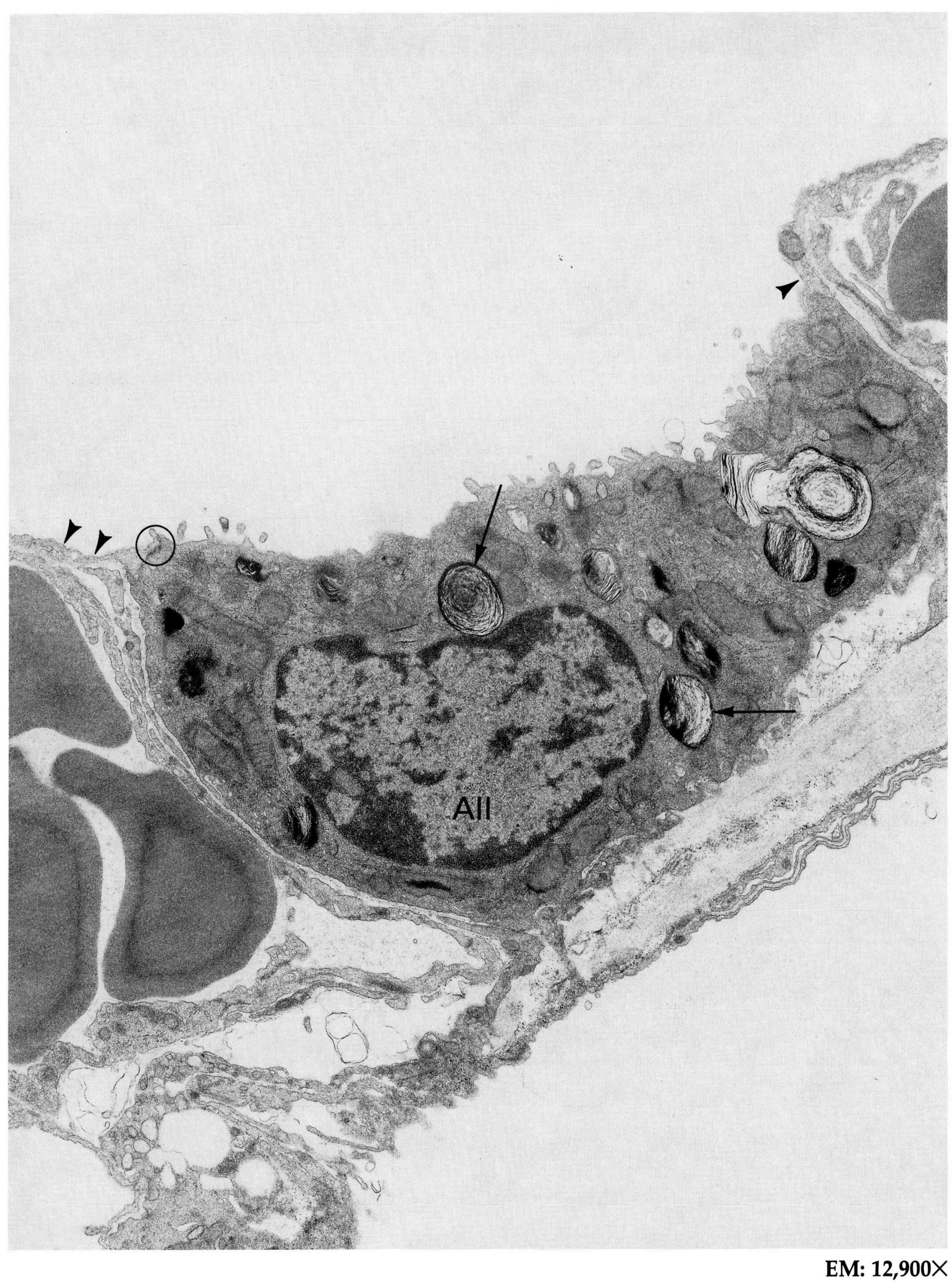

EM: 12,900×

Alveolar Macrophage

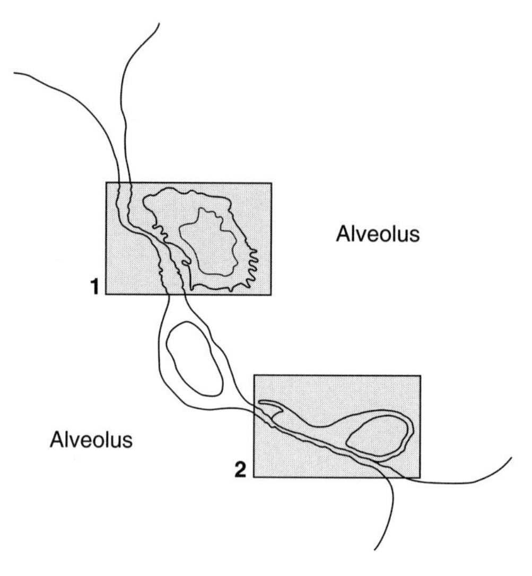

The **alveolar macrophage** is a large, active cell with an irregular surface that reflects the cell's movement, phagocytosis, and secretion. The primary role of the alveolar macrophage, to defend and clear surrounding tissue, is particularly critical in the lung since (1) it protects an extremely delicate layer across which gas exchange occurs, and (2) it is the last line of defense against a vast array of invading organic and inorganic airborne particles.

In their position between air and tissue, alveolar macrophages come into contact with pollutants, viruses, bacteria, and allergens that have been missed by the filtering action of the respiratory airways. Their primary action in defense is to phagocytose and remove these substances. The extensive surface area of the macrophage facilitates binding to these particles, either by the specific receptor-mediated mechanism (e.g., bacteria) or a nonspecific charge interaction (e.g., asbestos). Binding triggers a process of activation that, along with phagocytosis, results in killing (via oxygen radicals), digesting (via acid hydrolases), and recruiting (via chemotactic factors for inflammatory and immune cells).

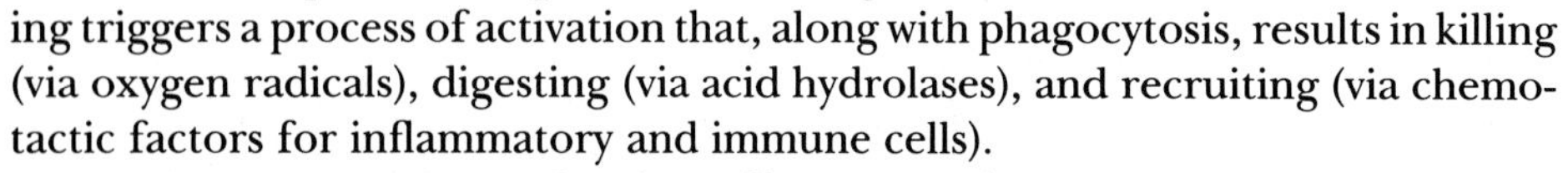

The large number and variety of **lysosomes** (arrows, micrographs 1 and 2) reflect the phagocytotic capacity of these cells. In addition to their role in defense, lysosomes also function in the turnover of cellular components and surfactant. Even though Type II alveolar cells are the main cells responsible for surfactant turnover, alveolar macrophages play a minor role in this activity and, occasionally, lamellar phospholipid material (p, micrograph 1) is seen within a secondary lysosome.

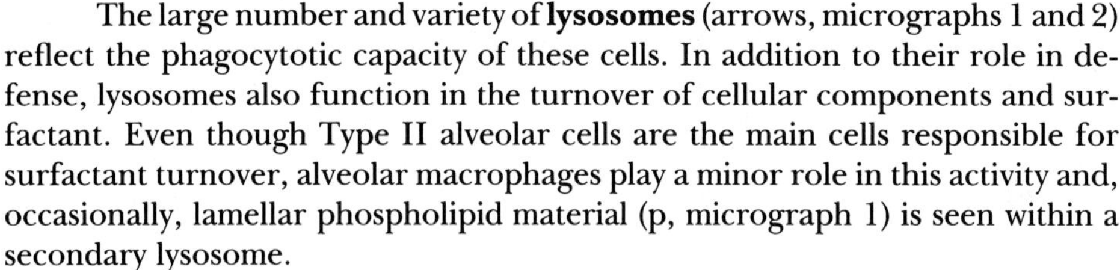

The alveolar macrophages in micrographs 1 and 2 developed from monocytes that left a pulmonary capillary, entered the interstitial area of the alveolar septum, and subsequently passed through a junction between alveolar cells to enter the air space. Such macrophages wander on the surface of the Type I lining under the surfactant layer. They remain in close association with Type I cells, and even make transient junctional contact (circles, micrograph 2) with this epithelium during their journey. After a life span of several months, these cells, full of "dust," either migrate up the respiratory tree to the pharynx, where they are swallowed, or reenter the interstitium and leave the lung in the lymphatics.

If damage to lung tissue occurs, an important aspect of the alveolar macrophage response is the synthesis of growth-regulating proteins that stimulate proliferation of cells critical to repair (Type II alveolar cells and fibroblasts). These proteins, synthesized on the rough ER (arrowheads, micrograph 1), are just a small fraction of the hundreds of secretory products of these cells. As true with all defense cells, overactivity and secretion can, instead of protecting tissues, create further injury.

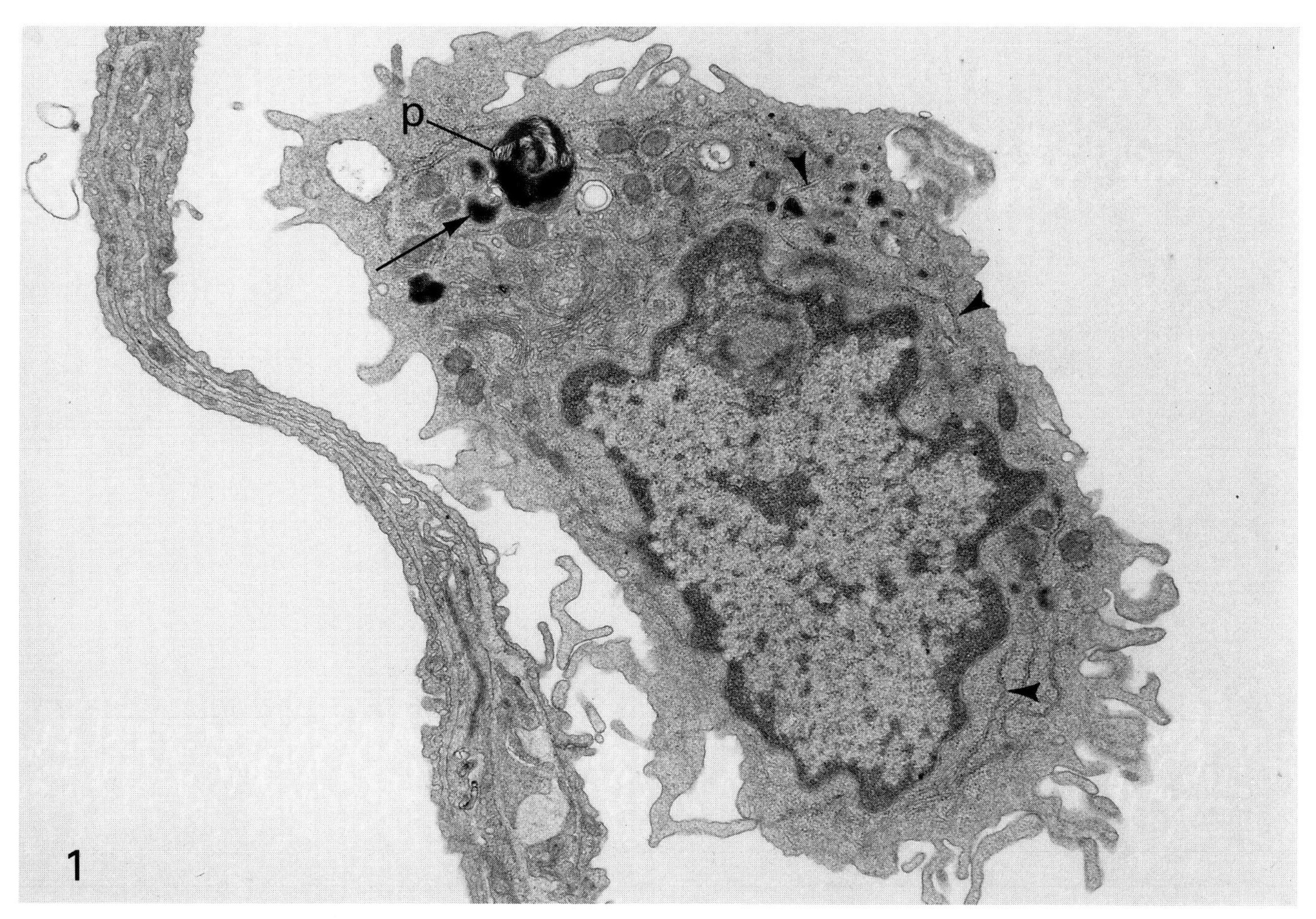

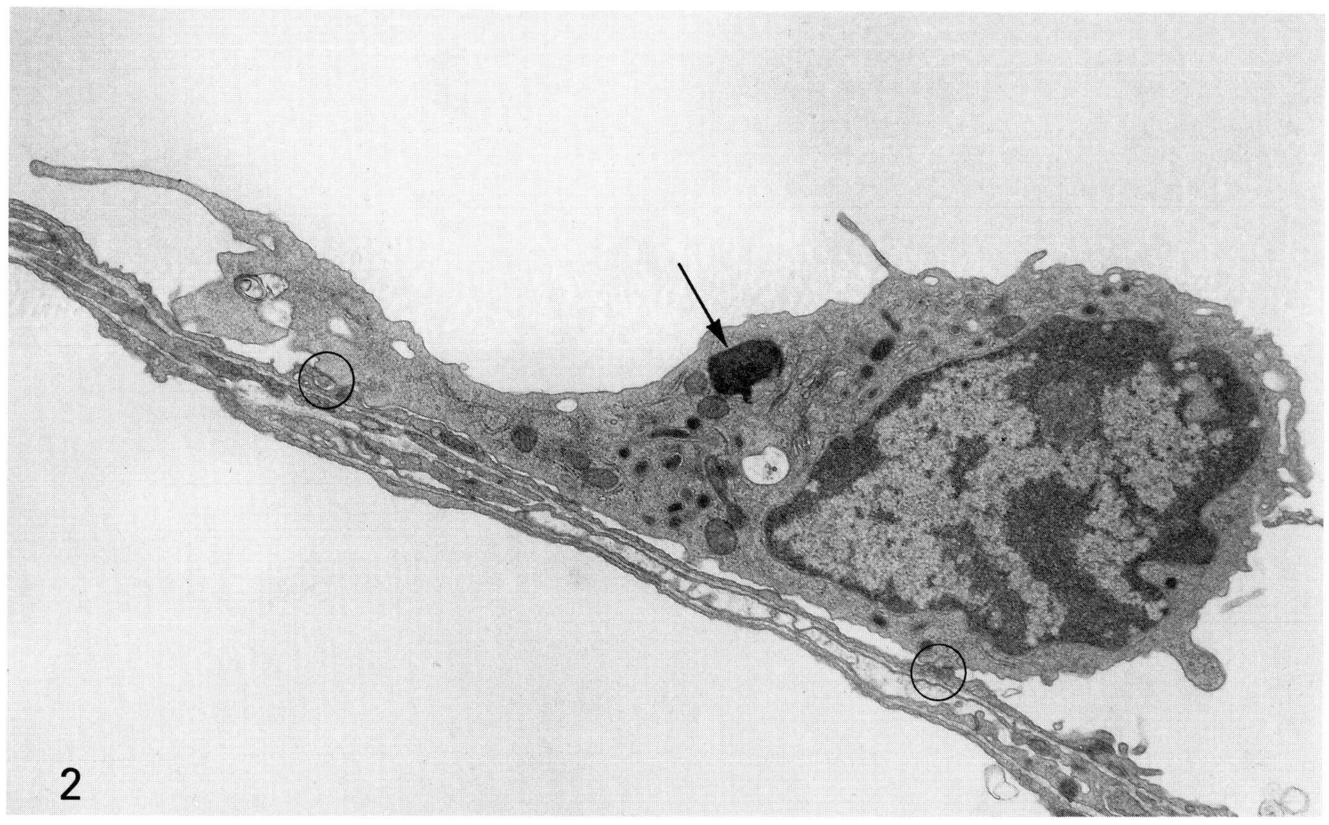

EM 1: 13,500× **EM 2: 12,000×**

Chapter Fourteen

Kidney

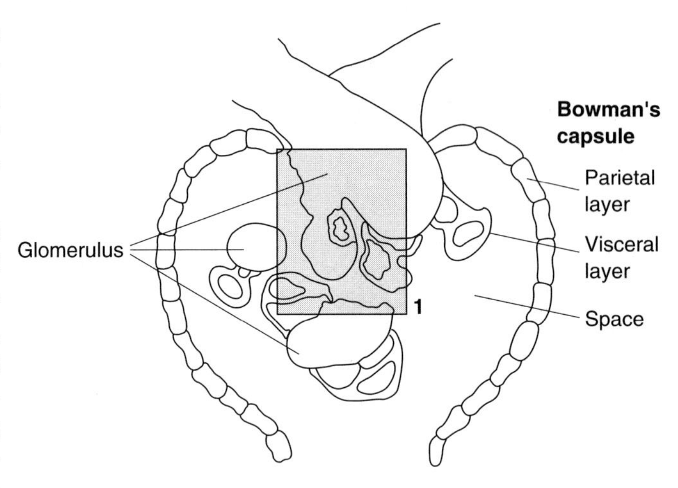

The kidney functions to remove waste products from the blood and to adjust the concentrations of plasma components to maintain homeostasis. It performs these functions by processes of filtration, reabsorption, and secretion.

The initial process of filtration takes place within 1 million **renal corpuscles** scattered throughout the cortex. Within each renal corpuscle blood is delivered to a tuft of capillaries, the **glomerulus,** which comes into close contact with the initial portion of the nephron, **Bowman's capsule.** In the process of filtration most plasma components pass from the glomerular capillaries (c, micrograph 1) to Bowman's space (B, micrograph 1). The relatively high hydrostatic pressure within the glomerular capillaries (45 mm Hg) is the driving force for filtration. Opposing this outward pressure are 10 mm Hg hydrostatic pressure in Bowman's space and 20 mm Hg osmotic pressure within the capillaries, resulting in a net filtration pressure of 15 mm Hg.

The visceral layer of Bowman's capsule consists of **podocytes** (Po, micrograph 1). Podocytes have a central cell body that contains the nucleus (N, micrograph 1) and a hierarchy of cytoplasmic extensions that branch to form many small processes, **pedicels** (p, micrograph 1 and inset), that rest on the glomerular basement membrane. These "foot" processes are of somewhat uniform size and are supported by a core of actin filaments (a, inset). Pedicels extending from one podocyte (long arrows, micrograph 1) interdigitate with pedicels from another podocyte (short arrows, micrograph 1). The 20- to 30-nm space between adjacent pedicels is traversed by a diaphragm (arrows, inset). These thin **filtration slit diaphragms** are the final barrier that plasma encounters after crossing the endothelium (E, micrograph 1 and inset) and glomerular basement membrane (b, micrograph 1 and inset).

The elaborate interdigitation of the podocyte pedicels and their association via filtration slit diaphragms is a result of a complex developmental process that begins with simple epithelial cells associated by apical tight junctions (see diagram below). As the branching processes develop, the tight junctions are replaced by filtration slit diaphragms. In the mature podocyte, the filtration slit diaphragm separates an apical surface covering the cell body and processes adjacent to Bowman's space from a basal surface adjacent to the basal lamina. In nephrotic syndrome, this developmental sequence is reversed; foot processes spread out and podocytes come together, attached by tight junctions. Experimental neutralization of the podocyte surface negative charge has a similar effect. In both cases, once the surface area of the filtration slit is reduced, filtration is compromised. The maintenance of podocyte shape is therefore critical to glomerular filtration.

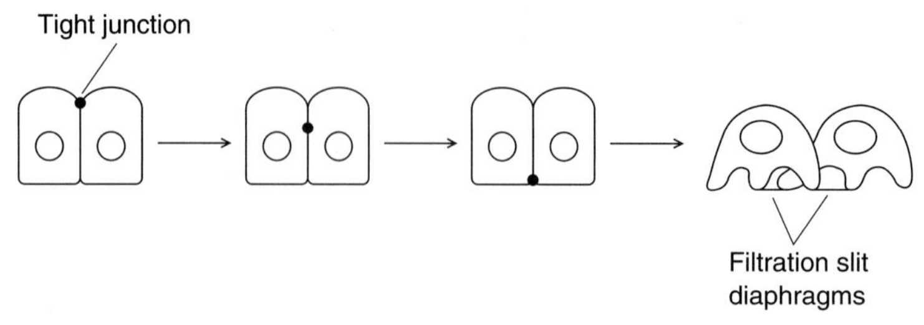

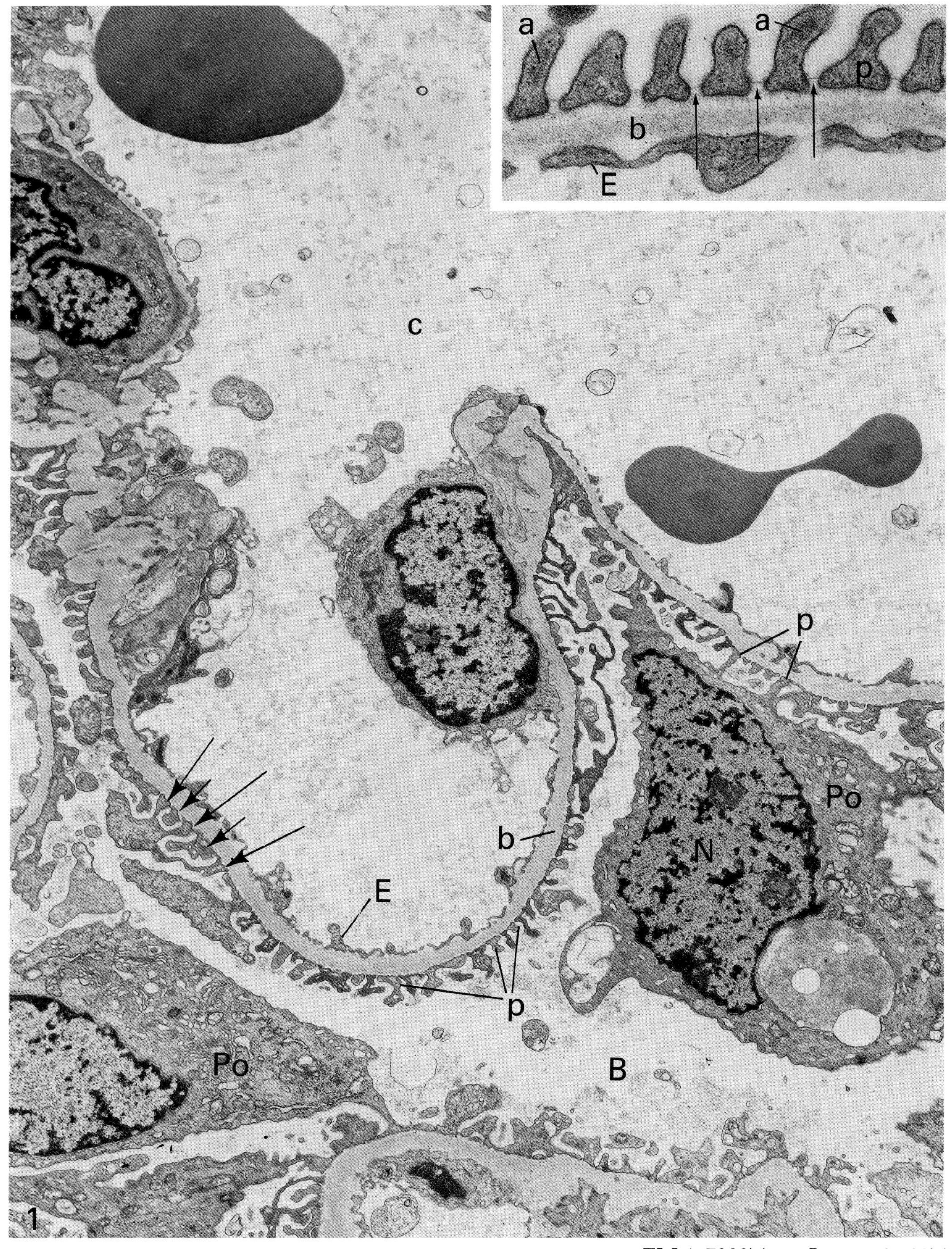

EM 1: 7200× Inset: 49,500×

RENAL CORPUSCLE: Filtration Barrier

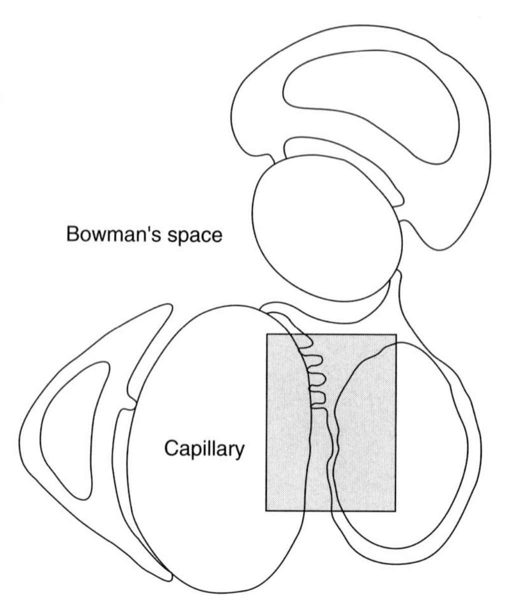

The **filtration barrier** that separates vascular (V, micrograph) from urinary (U, micrograph) space is impermeable to cells and most large molecular weight proteins, but is unusually permeable to water and small solutes.

The major size, shape, and charge barrier is the **glomerular basement membrane** (GBM; brackets, micrograph). This structure, ranging from 250 to 300 nm in diameter, is composed of three substructures: a central lamina densa (includes the fused basal laminas of the podocyte and endothelium) flanked by two lightly stained areas, the lamina rara interna (adjacent to the endothelium) and the lamina rara externa (adjacent to the podocyte). The major components of the GBM are collagen Type IV and heparan sulfate proteoglycan. The collagen network is most concentrated in the lamina densa and is thought to represent the size and shape barrier. The heparan sulfate proteoglycan is concentrated in the lamina rara interna and externa and is thought to represent the major charge barrier.

During the development of the kidney, both the endothelium and podocytes synthesize their respective basal laminas. In the adult kidney, after the fusion of the basal laminas, the podocytes continue to synthesize components to replace those lost with routine turnover. The euchromatic nucleus (N, micrograph) and the rough ER (r, micrograph) within the podocyte cell body reflect activity associated with this protein synthesis.

The role of the endothelium and filtration slit diaphragms in glomerular filtration is much less clear than that of the GBM. In routine electron micrographs the **endothelium** (E, micrograph) is seen to be interrupted by open fenestrations (f, micrograph) of up to 75 nm in diameter. This appearance suggests that this cell layer is capable of serving as a barrier only to whole cells. However, when micrographs are prepared using cationic electron opaque substances, these fenestrations are found to be bridged by an anionic glycocalyx. The glycocalyx may play a role in hindering the passage of anionic macromolecules such as albumin, as well as blood cells.

The **filtration slit diaphragm** (arrows, micrograph), a porous, 4-nm thick extracellular structure connecting pedicels (p, micrograph), appears to slow the rate of filtration of solutes but does not seem to process the filtrate any further than has already occurred. A major role of the diaphragm is to maintain polarity of the podocytes, in a manner similar to the tight junction from which it originates during development. The apical surface membrane above the diaphragm contains podocalyxin, a special negatively charged sialoprotein that is important in the maintenance of the elaborate podocyte shape.

Renal disease can initiate with alterations of the GBM with a resulting loss of the size selectivity of this structure. As more albumin and other proteins of large molecular weight leak out of the glomerular capillaries, plasma colloid osmotic pressure drops and fluid is lost, lowering blood pressure and creating edema. In many instances, a leaky filtration barrier is a result of the deposition of antigen–antibody complexes within or adjacent to the GBM. A subsequent inflammation of the entire glomerulus can lead to the replacement of the GBM by fibrous tissue, resulting in renal failure.

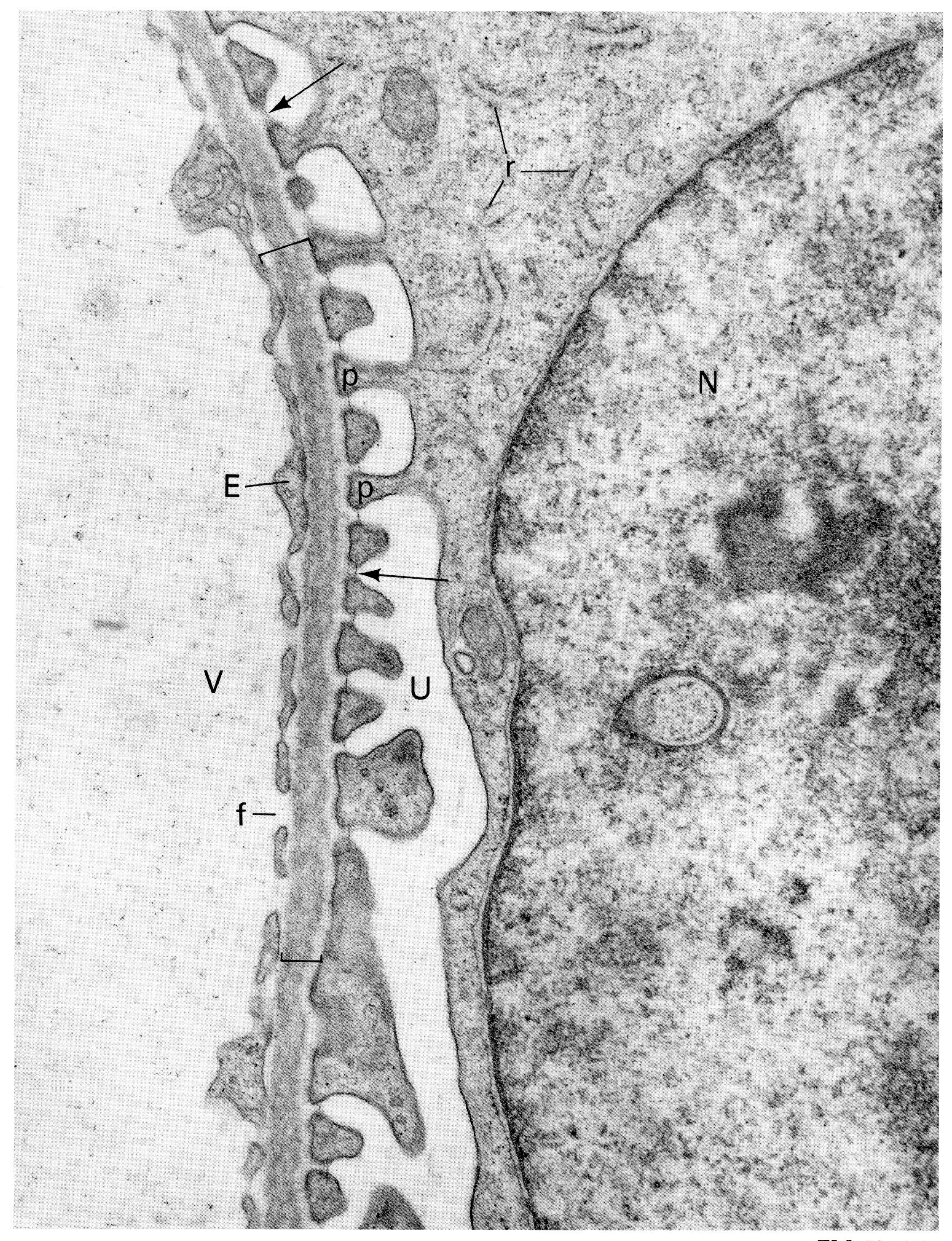

EM: 52,500×

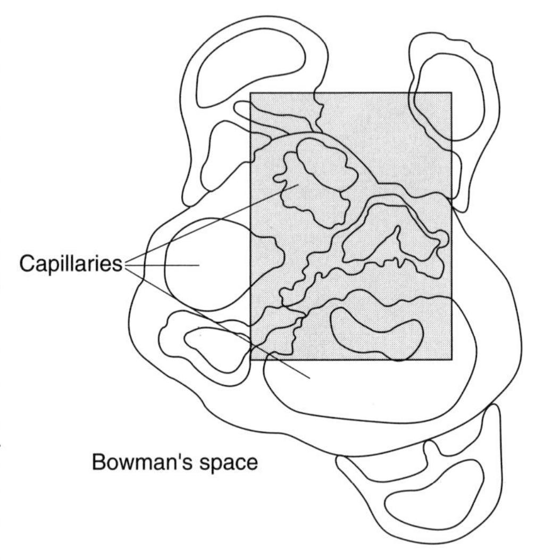

The **mesangium** is the central region of the glomerulus that acts as a supportive framework for the capillaries. It consists of **mesangial cells** embedded in an **amorphous matrix** similar in morphology and composition to the glomerular basement membrane. In the micrograph a mesangial cell (MC) and surrounding matrix (m) occupy the crevice between three capillaries (c). Podocytes (Po) and their associated processes (p) reside within Bowman's space (B) in the upper part of the micrograph.

Mesangial cells have an important function in the maintenance of the integrity of the glomerular basement membrane (GBM) and thus of the filtration barrier. They **phagocytose** aging components of the GBM and maintain the thickness of this structure. In the adult, the basal lamina components are synthesized primarily on the epithelial podocyte side and slowly move to the endothelial mesangial region where phagocytosis and degradation occur. The half-life of the proteoglycan com-

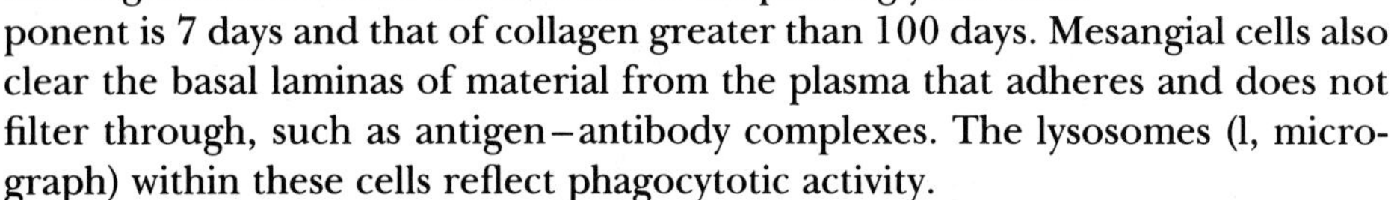

ponent is 7 days and that of collagen greater than 100 days. Mesangial cells also clear the basal laminas of material from the plasma that adheres and does not filter through, such as antigen–antibody complexes. The lysosomes (l, micrograph) within these cells reflect phagocytotic activity.

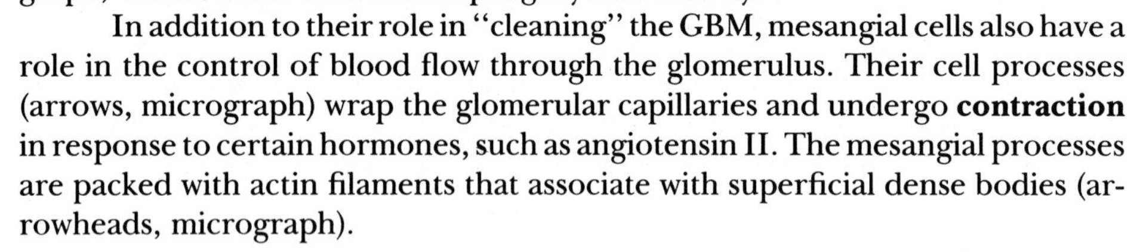

In addition to their role in "cleaning" the GBM, mesangial cells also have a role in the control of blood flow through the glomerulus. Their cell processes (arrows, micrograph) wrap the glomerular capillaries and undergo **contraction** in response to certain hormones, such as angiotensin II. The mesangial processes are packed with actin filaments that associate with superficial dense bodies (arrowheads, micrograph).

Nephrotoxicity is the major side effect of cyclosporin A, a cyclic peptide that is widely used as an immunosuppressant during organ transplants and to treat certain autoimmune diseases. The characteristic reduction in glomerular filtration rate may be linked with a disturbance of the mesangial contraction cycle.

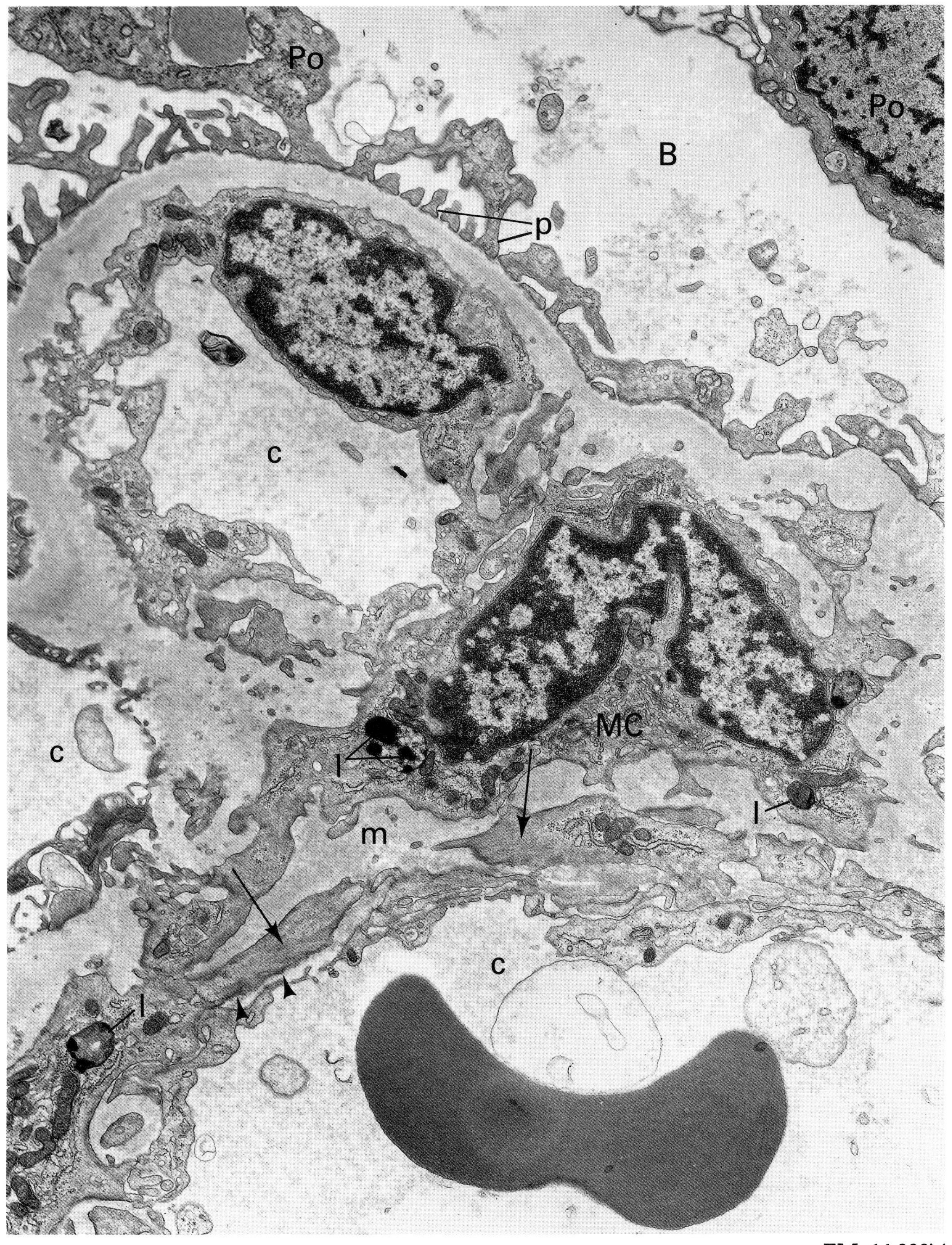

EM: 14,000×

RENAL TUBULES: Overview

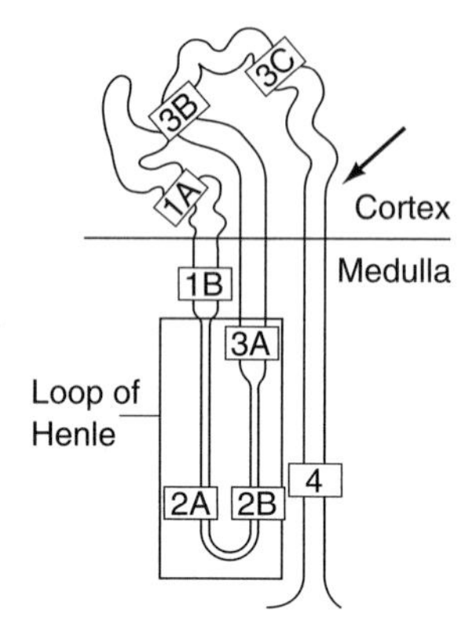

Schematic Diagram

The primary function of the **tubular portion of the kidney** is to selectively reabsorb the metabolically useful components of the glomerular filtrate, while leaving toxic metabolic end products for excretion as urine. The tubules also perform a secretory role, eliminating drugs and secreting ions when they exceed physiological levels. Four types of tubules can be differentiated histologically and are represented by the four micrographs: (1) proximal tubule, (2) thin limb of the loop of Henle, (3) distal tubule, and (4) collecting duct. Each type of tubule occupies a specific region within the kidney, as shown in the schematic diagram. This diagram is representative of the tubules associated with a juxtamedullary renal corpuscle, in which the loop of Henle extends the length of the medulla.

The **proximal tubule** (micrograph 1) is lined with cuboidal cells with an apical brush border (b) of microvilli, basolateral membrane infoldings (arrows) tightly associated with mitochondria (m), and lysosomes (l). Most amino acids, proteins, and glucose are reaborbed here, in addition to 60% or more of the filtered fluid and ions. Transport activity is highest in the proximal convoluted portion (schematic, 1A) in the cortex and diminishes in the proximal straight portion (pars recta) that descends into the medulla (schematic, 1B).

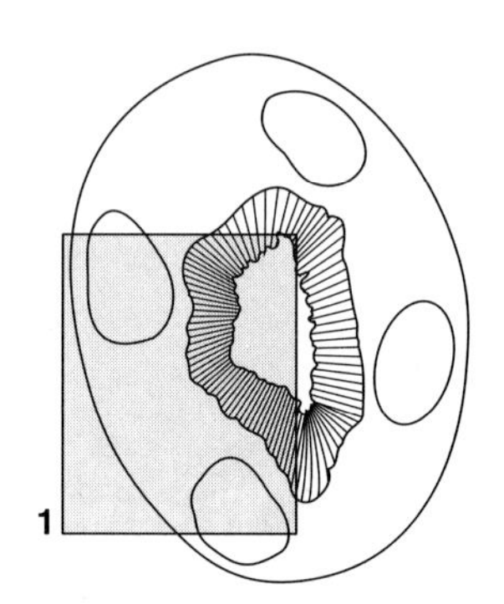

The **thin limb of the loop of Henle** (micrograph 2) is lined with squamous epithelium (E, micrograph 2) with fewer mitochondria and membrane specializations. It consists of a descending portion (schematic, 2A) that is permeable to both ions and water, and an ascending portion (schematic, 2B) that is permeable to ions but not water.

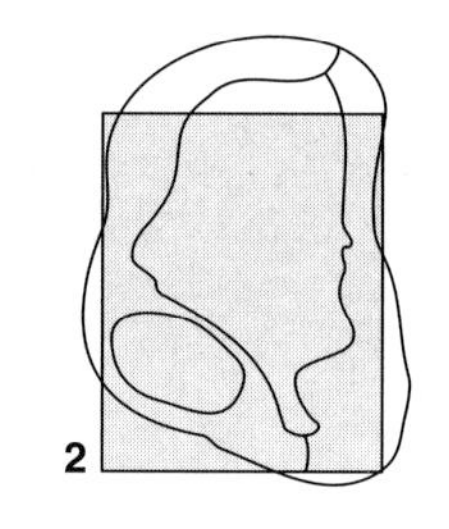

The **distal tubule** (micrograph 3) is lined with cuboidal cells with basolateral infoldings (arrows) and associated mitochondria (m) but no brush border. As the ultrastructure implies, the major function of this tubule is ion reabsorption, driven by a pump concentrated in the basolateral cell membranes. Within this single histological category there are three very different functional classifications.

1. The ascending thick limb (schematic, 3A) of the loop of Henle actively reabsorbs sodium/chloride ions while excluding water, thus concentrating ions in the medullary interstitium.

2. The macula densa (schematic, 3B) makes contact with the vascular pole of the renal corpuscle as the distal tubule enters the cortex. The macula densa appears to function as the salt sensor of the juxtaglomerular apparatus.

3. The distal convoluted tubule (schematic, 3C) actively reaborbs sodium ions in response to aldosterone.

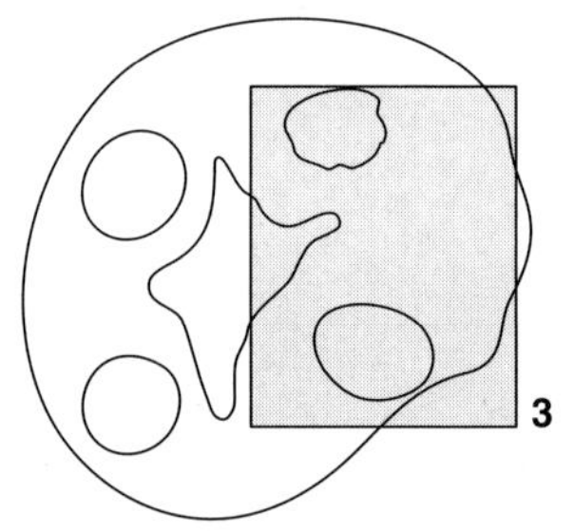

The **collecting duct** (micrograph 4) is lined with neat cuboidal epithelium (E) with well-defined boundaries (arrowheads) and no brush border. The collecting ducts carry the filtrate through its final pass across the hypertonic medulla. Antidiuretic hormone (ADH) increases the permeability of collecting ducts to water, which is drawn osmotically into the interstitium, thus concentrating the urine. ADH also acts on the collecting tubules (arrow, schematic) in the cortex to increase their permeability to water. This effect, in combination with the increased reabsorption of sodium in response to aldosterone, acts to conserve sodium and adjust intravascular fluid volume.

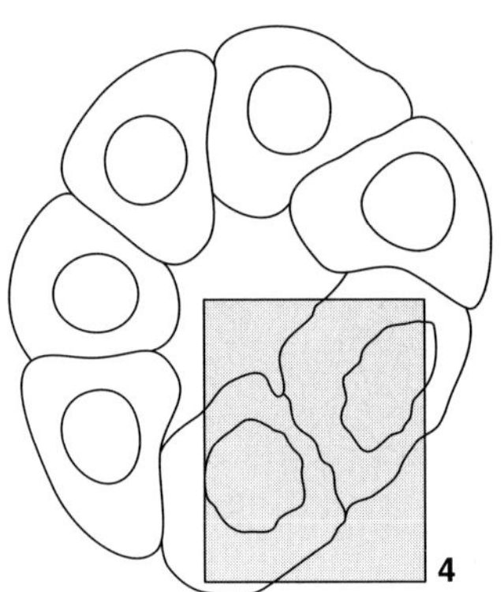

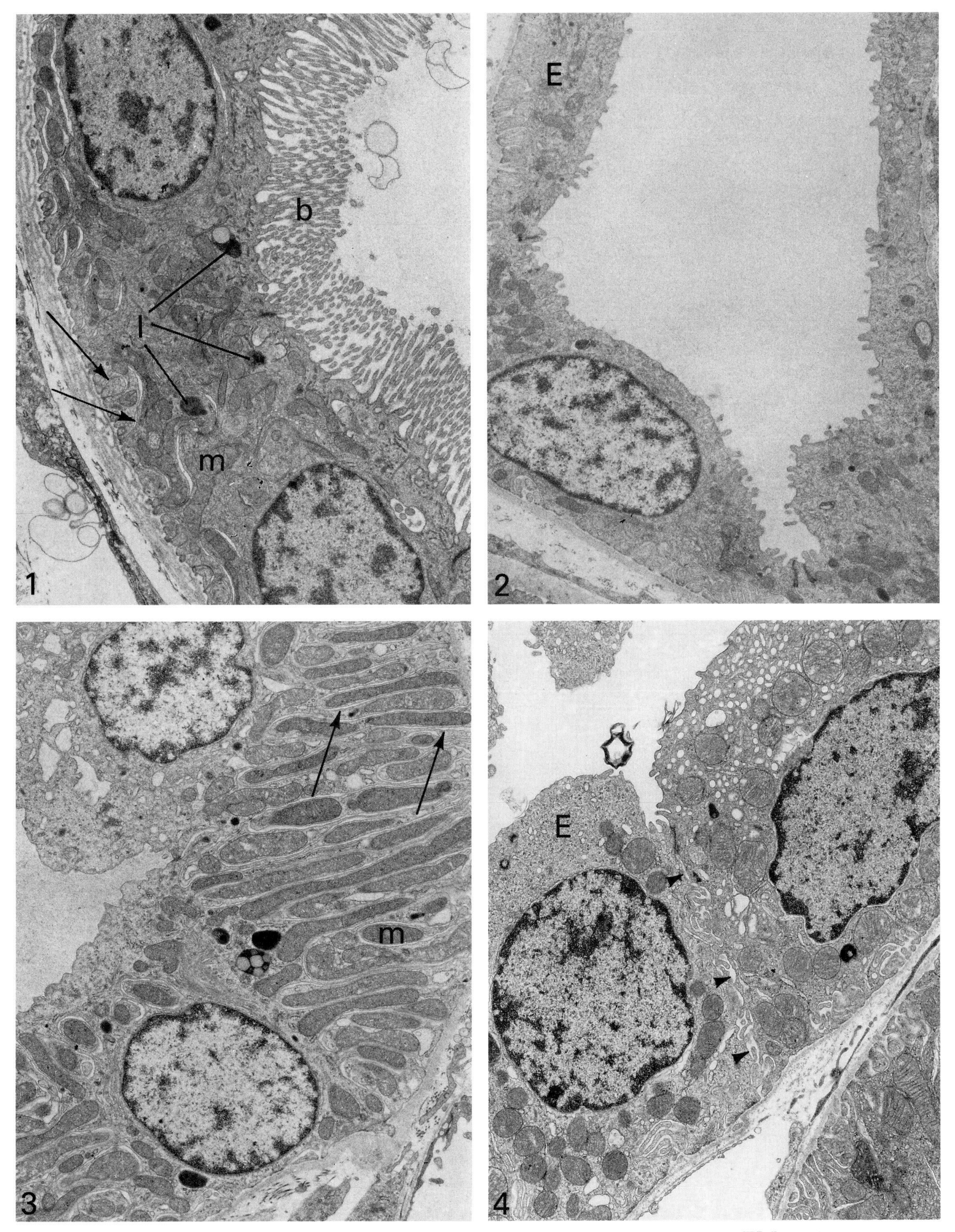

EMs 1, 2, 3, 4: 7500×

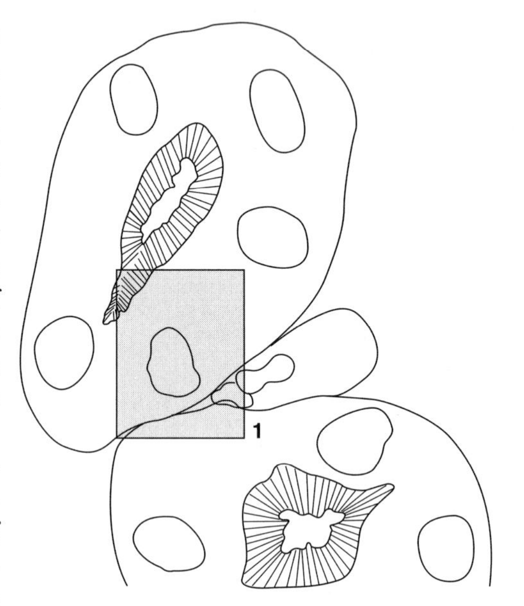

The **proximal tubule** reabsorbs most components of the filtate that are of nutritional significance (e.g., glucose, ions, amino acids). These substances are transported from the lumen across the epithelium to enter capillaries (c, micrograph 1) in the underlying tissue. Both apical and basolateral surfaces have increased surface areas to facilitate this transport. The microvilli (mv, micrograph 1 and inset) comprising the brush border of the apical surface house channels and enzymes involved in reabsorption. On the basolateral surface, the membrane of each cell is thrown into large folds (arrows, micrograph 1) that interdigitate with folds of other cells. Many of the folds that appear to be within the boundaries of the nucleated cell shown in micrograph 1 actually belong to other cells.

The proximal tubule is an excellent example of the significance of polarity and the role of paracellular and transcellular transport (see Epithelium page 50). The **reabsorption of major metabolites** is linked to activity at the extensive basolateral surface. The most significant component of this surface is the **Na^+K^+ATPase enzyme-channel complex.** This complex pumps Na^+ out of the cell across the basolateral surface and is thus a major factor responsible for returning Na^+ to the circulation. Mitochondria (m, micrograph 1), with their long axis parallel to the basolateral infoldings, rest adjacent to the membranes where these pumps are located and provide the ATP for their activities. The transport of Na^+ across the basolateral surface creates an electrochemical gradient across the apical membrane that favors movement of Na^+ from the lumen into the cell. The transport of amino acids and glucose is coupled to this apical movement of Na^+. Thus the Na^+K^+ATPase drives the recovery of several major plasma constituents. Ions that cross the basolateral surface draw water across the cell osmotically, accounting for the large volume of fluid reabsorption in this tubule. Ions and water also flow between the cells through the unusually leaky tight junctions within the junctional complexes (circles, micrograph 1).

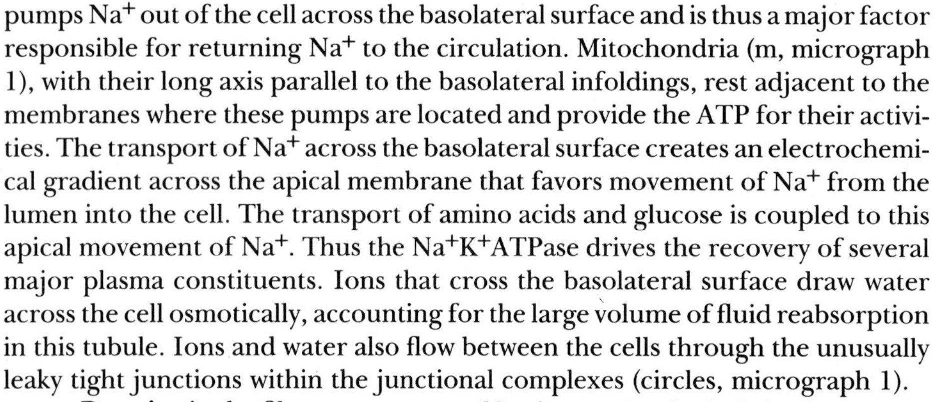

Proteins in the filtrate are removed by the proximal tubule by the process of digestion. Albumin that leaks across the filtration barrier in the renal corpuscle binds to receptors at the base of the microvilli and initially accumulates in elongated canaliculi (arrowheads, micrograph 1 and inset) that are coated on the cytoplasmic surface. Coated vesicles (arrows, inset) originating at the base of these canaliculi lose their coat and become endosomes, which in turn fuse with lysosomes (l, micrograph 1), where the proteins are digested. The resulting amino acids leave the cell across the basolateral membrane to enter the blood.

Oligopeptides (less than 10 amino acid residues) are also removed from the filtrate by digestion, but instead of occurring within lysosomes, digestion takes place on the microvillous border by hydrolytic enzymes that are integral membrane proteins. Many of the filtered oligopeptides are hormones (oxytocin, vasopressin, angiotensin II), and their destruction by the proximal tubule is an important means of regulating their blood levels.

At the same time that the nutritionally important substances are being reabsorbed in the proximal tubule and their concentration in the filtrate is decreasing, the concentration of metabolic waste products (e.g., urea, creatinine, uric acid) that will be excreted is increasing.

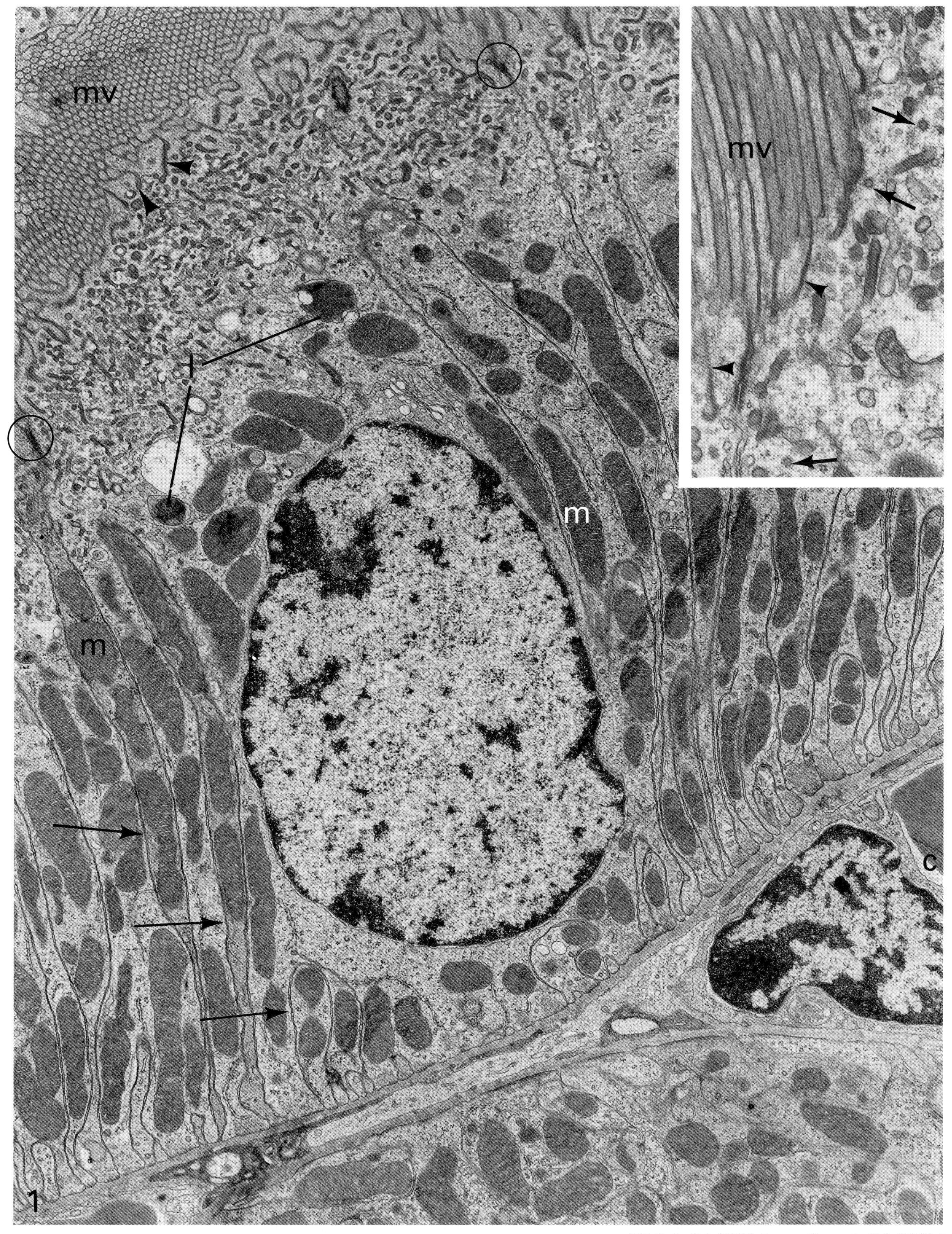

EM 1: 20,300× Inset: 30,000×

MEDULLA: Concentration of Urine

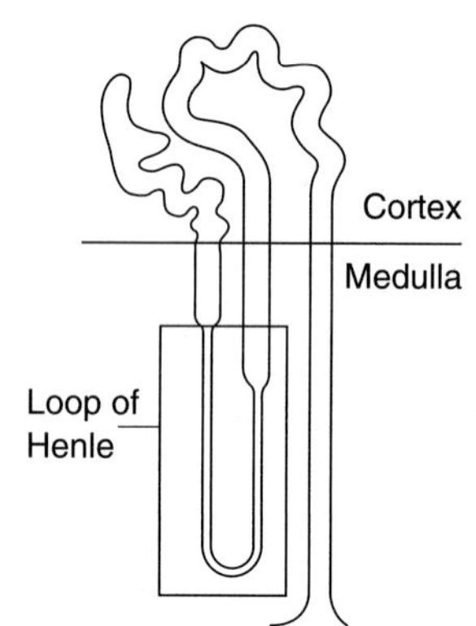

Loops of Henle, vasa recta (medullary blood vessels), and **collecting ducts** each play an important role in the mechanism for concentrating urine. The loops of Henle concentrate ions in the medullary interstitium, vasa recta are structured to maintain this hypertonicity, and the collecting ducts provide the final filtrate pathway in which the concentration of urine occurs.

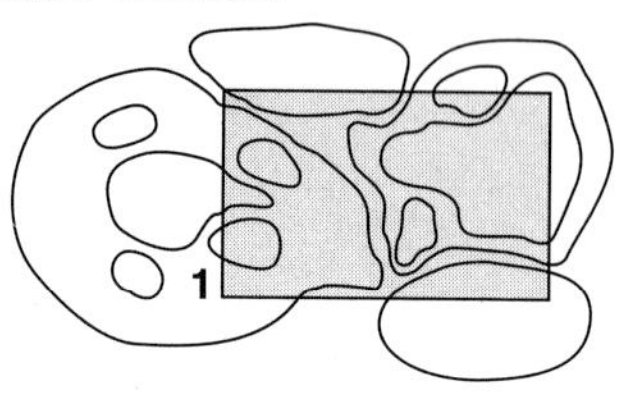

Each loop of Henle consists of a **descending thin limb, ascending thin limb,** and **ascending thick limb.** In micrograph 1 both thick (A) and thin (B) portions of the loop are visible. In the ascending thick portions of the loop of Henle, ions (without water) are transferred actively from the lumen to the interstitium. The resulting hypertonicity of the interstitium is maintained by the loop structure of the vasa recta (V, micrograph 1) that follow the tubules through the medulla. The flattened endothelial lining (arrows, micrograph 1) of these capillaries contrasts with the less attenuated lining of the thin limb. The descending and ascending vasa recta work in a countercurrent fashion to trap solutes in the medullary interstitium. A hypertonic medullary interstitium in the human kidney provides the necessary environment for concentrating urine and conserving body fluid levels.

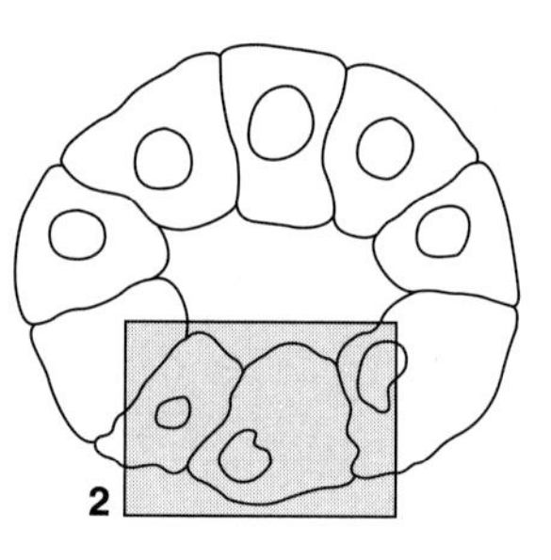

The collecting ducts (micrograph 2) are the gates that control the degree of urine concentration. When the body fluid osmolality rises or volume drops, antidiuretic hormone (vasopressin), released from the posterior pituitary, binds to receptors on the basal surface of the collecting ducts. This binding triggers the opening of water channels on the apical surface. Water is drawn osmotically across the collecting duct cells to the hypertonic interstitium. This water is picked up by the vasa recta and distributed to the body. Some urea, which has become concentrated during passage of the filtrate through the nephron, also crosses these cells and contributes to the hypertonicity of the medulla interstitium.

The final urine passing into the ureter is a concentrated, acidic fluid with high levels of metabolic end products and drugs. The kidney has a limited capacity to concentrate these unwanted substances, and when it is overloaded, pathology results. Patients undergoing chemotherapy for cancer have unusually high levels of uric acid resulting from the breakdown of nucleic acids in the cancerous cells. In rare instances uric acid crystals precipitate in the medulla and interfere with tubular function.

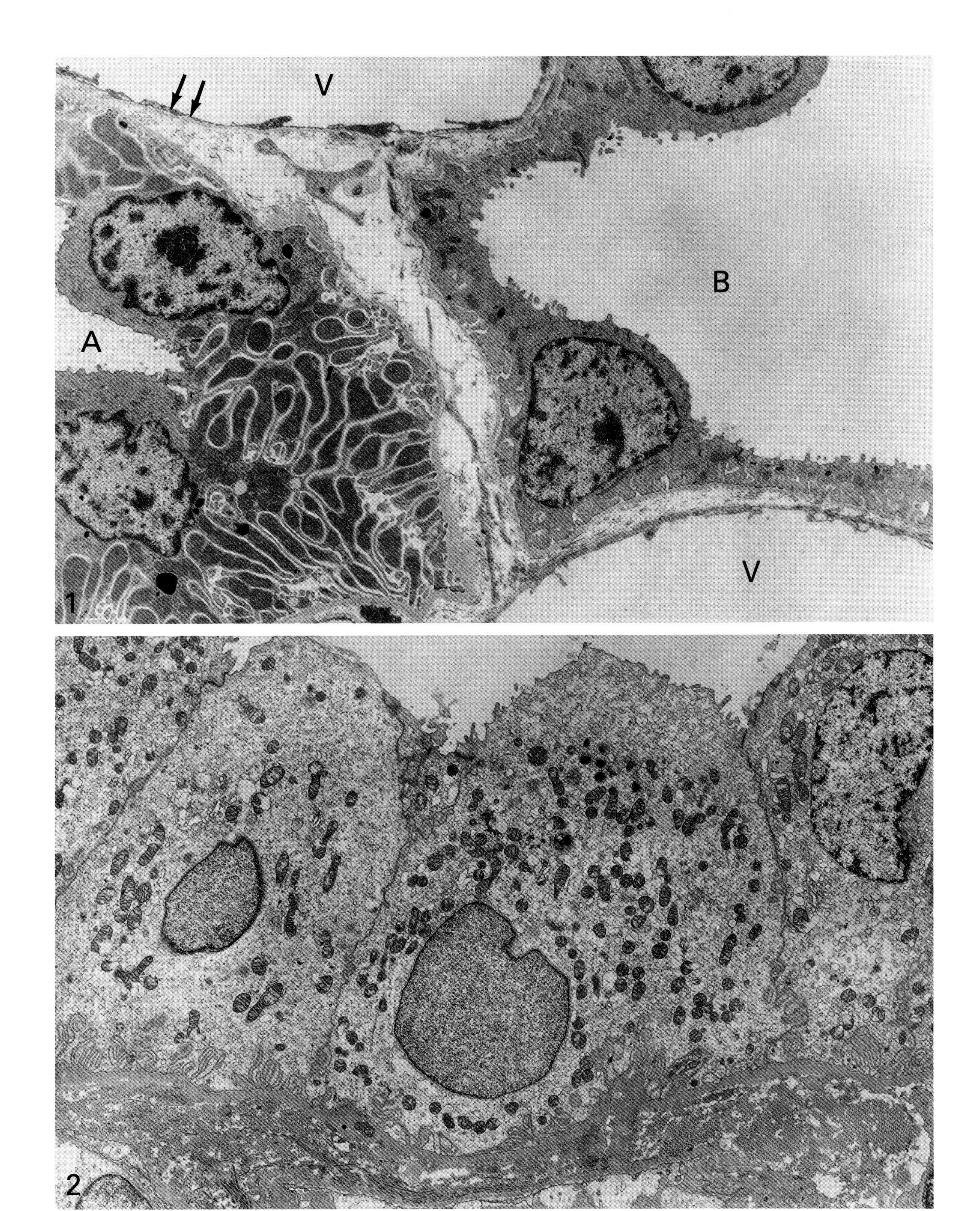

EM 1: 6,400× EM 2: 5,400×

JUXTAGLOMERULAR APPARATUS

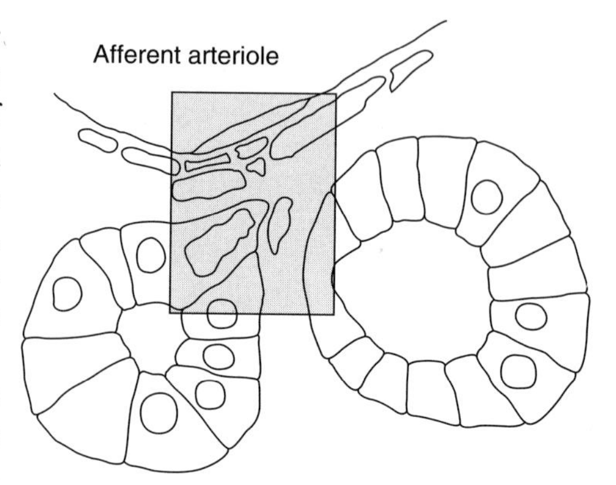

One of the major functions of the kidney is to monitor and control salt balance and associated blood pressure. The **juxtaglomerular apparatus,** situated at the vascular pole of each renal corpuscle, is the region in which a negative feedback loop operates to adjust these physiological parameters. This apparatus consists of three components: juxtaglomerular cells, the macula densa, and extraglomerular mesangial cells. Juxtaglomerular cells (J) and part of the macula densa (MD) can be seen in the micrograph.

The **juxtaglomerular cells** are modified smooth muscle cells in the wall of the afferent arteriole (L: lumen, micrograph) supplying the glomerulus. Even though they possess characteristics of smooth muscle, such as dense bodies (arrowheads,

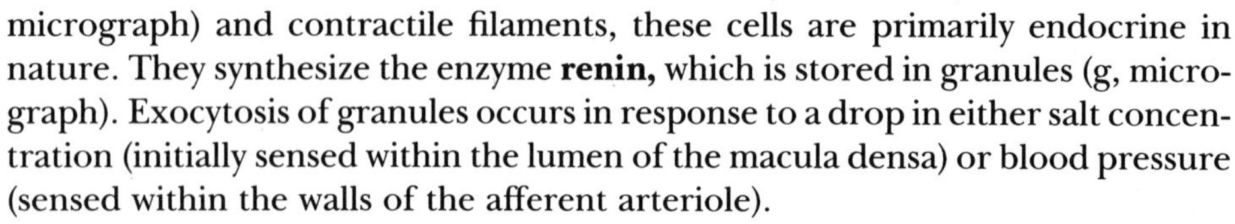

micrograph) and contractile filaments, these cells are primarily endocrine in nature. They synthesize the enzyme **renin,** which is stored in granules (g, micrograph). Exocytosis of granules occurs in response to a drop in either salt concentration (initially sensed within the lumen of the macula densa) or blood pressure (sensed within the walls of the afferent arteriole).

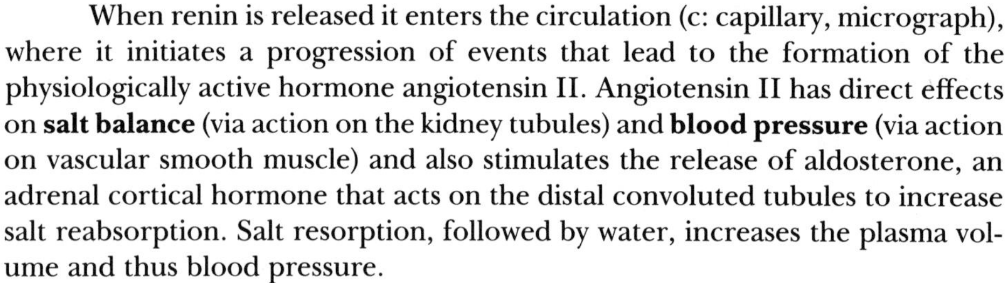

When renin is released it enters the circulation (c: capillary, micrograph), where it initiates a progression of events that lead to the formation of the physiologically active hormone angiotensin II. Angiotensin II has direct effects on **salt balance** (via action on the kidney tubules) and **blood pressure** (via action on vascular smooth muscle) and also stimulates the release of aldosterone, an adrenal cortical hormone that acts on the distal convoluted tubules to increase salt reabsorption. Salt resorption, followed by water, increases the plasma volume and thus blood pressure.

A renin precursor, prorenin is stored in the less dense, immature granules within the juxtaglomerular cell cytoplasm. Normally some prorenin is released along with renin, with little effect. In certain diseases, however, the proportion of prorenin released into the circulation is greatly increased, and in these cases the normal regulating effect of the kidney is disturbed.

The **macula densa** is the region of the distal tubule, adjacent to the vascular pole of the renal corpuscle, in which the nuclei of the cells are packed tightly together. The section through two macula densa cells in the micrograph includes nuclei and characteristic basal membrane infoldings (arrows). Typically, in the macula densa region the mitochondria (m1, micrograph) rest as a group above the infoldings, whereas in other regions mitochondria (m2, micrograph) are squeezed within the infoldings. The mechanism in which information obtained in the macula densa regarding salt concentration is transmitted to the juxtaglomerular cells is not established; however, changes in the distance between components of the juxtaglomerular apparatus may coincide with information transfer.

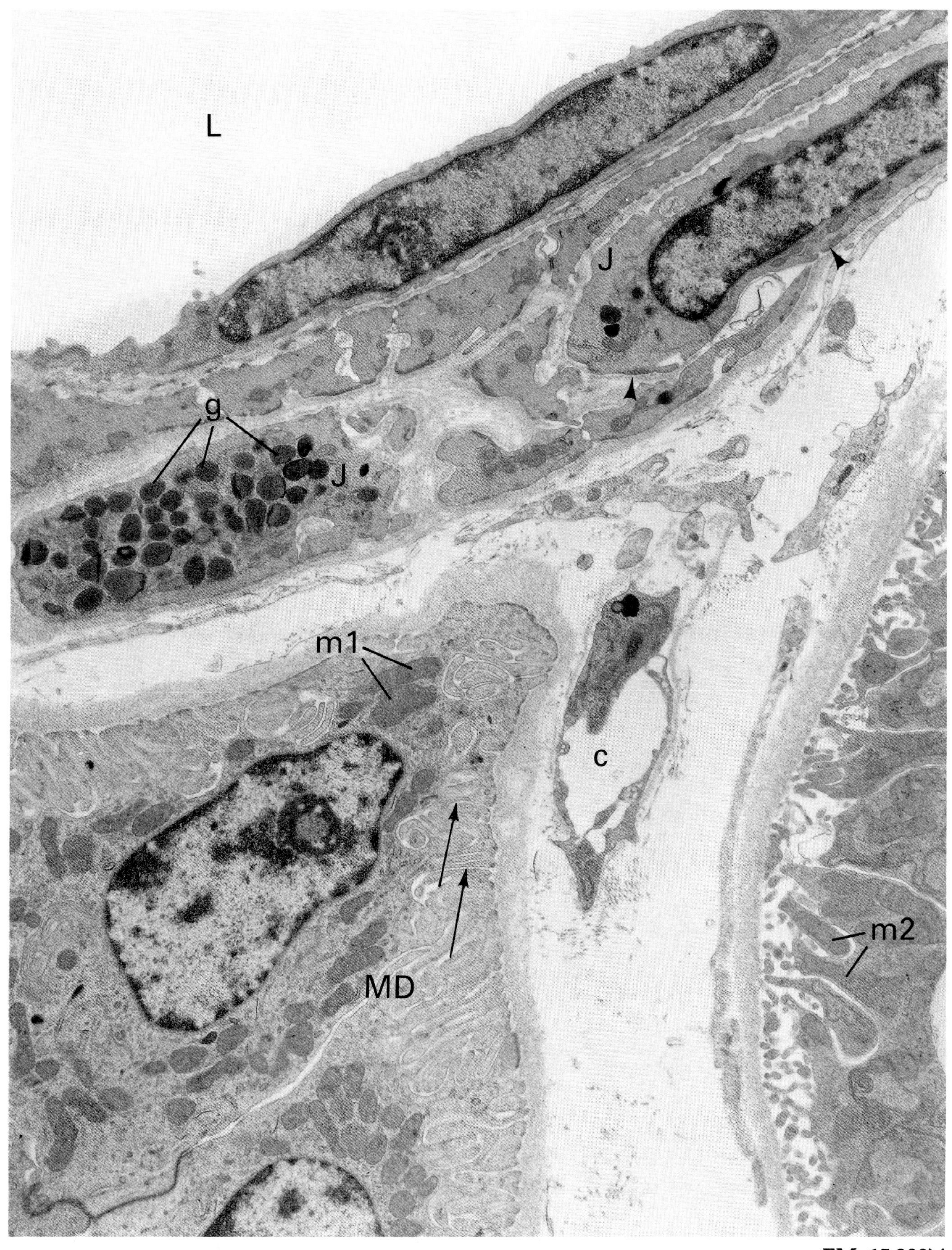

EM: 15,200×

MALE REPRODUCTIVE SYSTEM

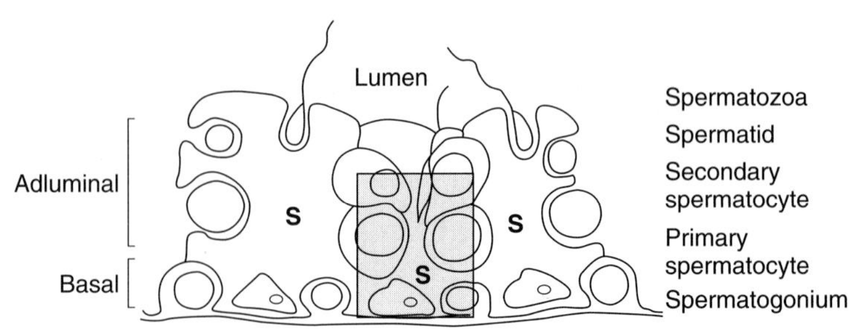

Spermatogenesis (the formation and differentiation of spermatozoa) takes place in the **seminiferous tubules.** Each seminiferous tubule is lined by a single layer of irregularly shaped **Sertoli cells** (S, diagram and micrograph). In the micrograph the nucleus (N) of a Sertoli cell (arrows define boundary of cell) is near the basal surface of the seminiferous tubule. **Spermatogonia,** which rest on the basal lamina of the tubule, give rise to sperm in stages that progress from the basal lamina to the lumen. During this 64-day process, mitosis, meiosis, and complex differentiation increase the numbers, genetic diversity, and specialization of the resulting cell to enable it to perform its role as the carrier of the male genetic complement.

Sertoli cells are associated with one another by a complex of junctions including adherens, gap, and occludens types. As the developing male germ cells move between the Sertoli cells toward the lumen, the Sertoli cell junctions are modified to permit this movement. Zonula occludens junctions between Sertoli cells separate the seminiferous epithelium into two compartments: basal, in communication with underlying tissue, and adluminal, protected and isolated from underlying tissue. Once germ cells have zippered through this junction they are within the blood-testis barrier.

Spermatogenesis begins at puberty. Spermatogonia divide and give rise either to additional undifferentiated stem cells (A type) or to cells that begin the sequence of events leading to sperm formation (B type). Type B cells enter the early stages of the first meiotic prophase and thus become **primary spermatocytes** while they are still situated in the basal compartment. The preleptotene (DNA synthesis), leptotene (chromosome thickening), and zygotene (pairing of chromosome homologs) stages are found closely associated with the basal lamina, along with the spermatogonia. The primary spermatocyte (*) in the basal compartment seen in this micrograph has the chromosome condensation pattern characteristic of one of these early meiotic stages.

The progression into the pachytene stage, in which crossing over occurs between homologous chromosomes, is associated with movement of the spermatocyte into the adluminal compartment. Pachytene is the longest stage of meiosis and therefore most commonly observed in section. In the pachytene (P) stages in the micrograph there is evidence of synaptonemal complexes (circles), indicating crossing over.

Germ cells undergoing the second meiotic division **(secondary spermatocytes)** are rarely observed in section, since this stage is relatively short and does not involve DNA synthesis or chromosome pairing. The resulting haploid **spermatids** then begin the process of differentiation into spermatozoa. Early spermatids (ES, micrograph) are round, like the earlier stages, but they are easily distinguished by the acrosome (a, micrograph) that is beginning to form. As differentiation proceeds and remodeling occurs, spermatids take on the elongated shape characteristic of mature sperm. The heads of two late spermatids (LS, micrograph) with condensed chromatin are seen embedded within Sertoli cytoplasm.

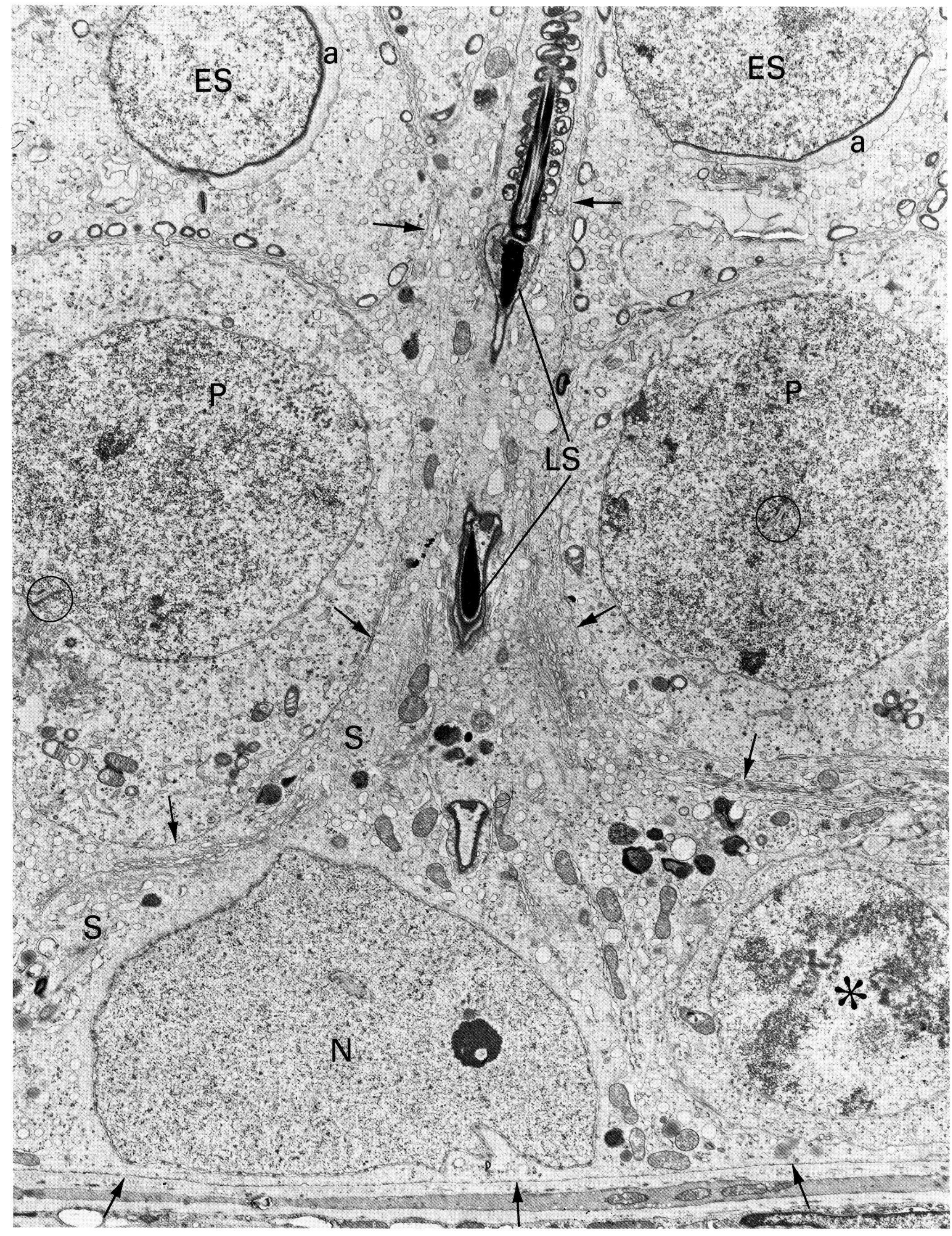

EM: 8200×

Sertoli–Sertoli Junctions

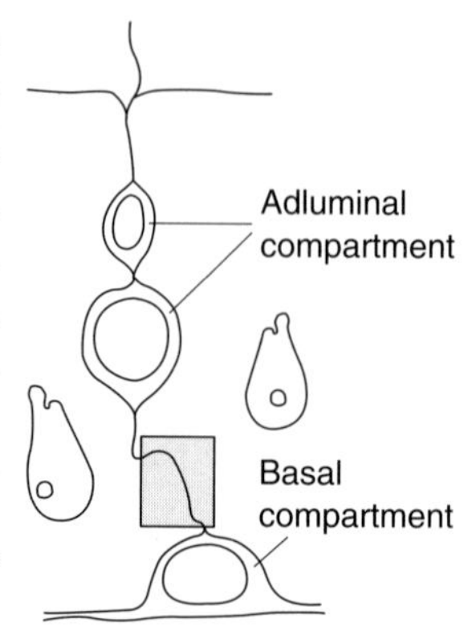

The **junctional specializations** that bind **Sertoli cells** to one another are similar to the classical junctions that are found between other epithelial cells: gap junctions provide communication, adherens junctions bind cells together, and tight junctions control permeability. The two Sertoli cells (S) in the micrograph are tightly apposed through most of the junctional region. A close examination permits a distinction between tight junctions (A, micrograph), with a central dense line parallel to the cell surface, and gap junctions (B, micrograph), with dense lines (connexons) perpendicular to the cell surface.

Tight junction components between Sertoli cells are localized to regions separating the basal and adluminal compartments. These junctions prevent the passage of proteins between the two compartments and thus define a type of **blood-testis barrier.** This barrier maintains the sperm in a unique environment largely controlled by the Sertoli cells. Within this barrier sperm are isolated from recognition and subsequent destruction by the immune system. The proteins synthesized and inserted into sperm membranes during development are "foreign," since sperm develop after the immune system has completed its registry of self proteins. Many of the new sperm proteins are synthesized directly after the spermatocytes have moved into the protected adluminal compartment. In cases in which this barrier is disturbed, a reduction or loss of fertility occurs.

The tight junctions are not the only factors responsible for protection of developing sperm from immune destruction. The testis itself is an immunologically privileged site, and even though some autoantigens appear before the germ cells are enclosed within the blood-testis barrier, an immune reaction does not occur. The immune system within the testis seems to be locally suppressed, and both steroidal and nonsteroidal products of Leydig cells may mediate this suppression.

Sertoli cell junctions have a unique association with actin and the endoplasmic reticulum. Ordered bundles of actin filaments (arrowheads, micrograph) appear regularly between the junctions and a flattened cisterna of endoplasmic reticulum (arrows, micrograph). While the ER membrane adjacent to the junction is typically smooth (s, micrograph), the distal ER membrane is frequently studded with ribosomes (r, micrograph). A variety of proteins have been localized that suggest that the ER, actin, and junctions are all cross-linked. The actin associated with Sertoli cell junctions is found adjacent to adherens, tight, and gap junctions, and it is more abundant and highly ordered than in other cell types. The major role of this ER–actin–junction complex in these regions may be to facilitate the synthetic and architectural changes needed to accommodate junctional plasticity: as germ cells move toward the lumen, junctions must continually be broken and remade.

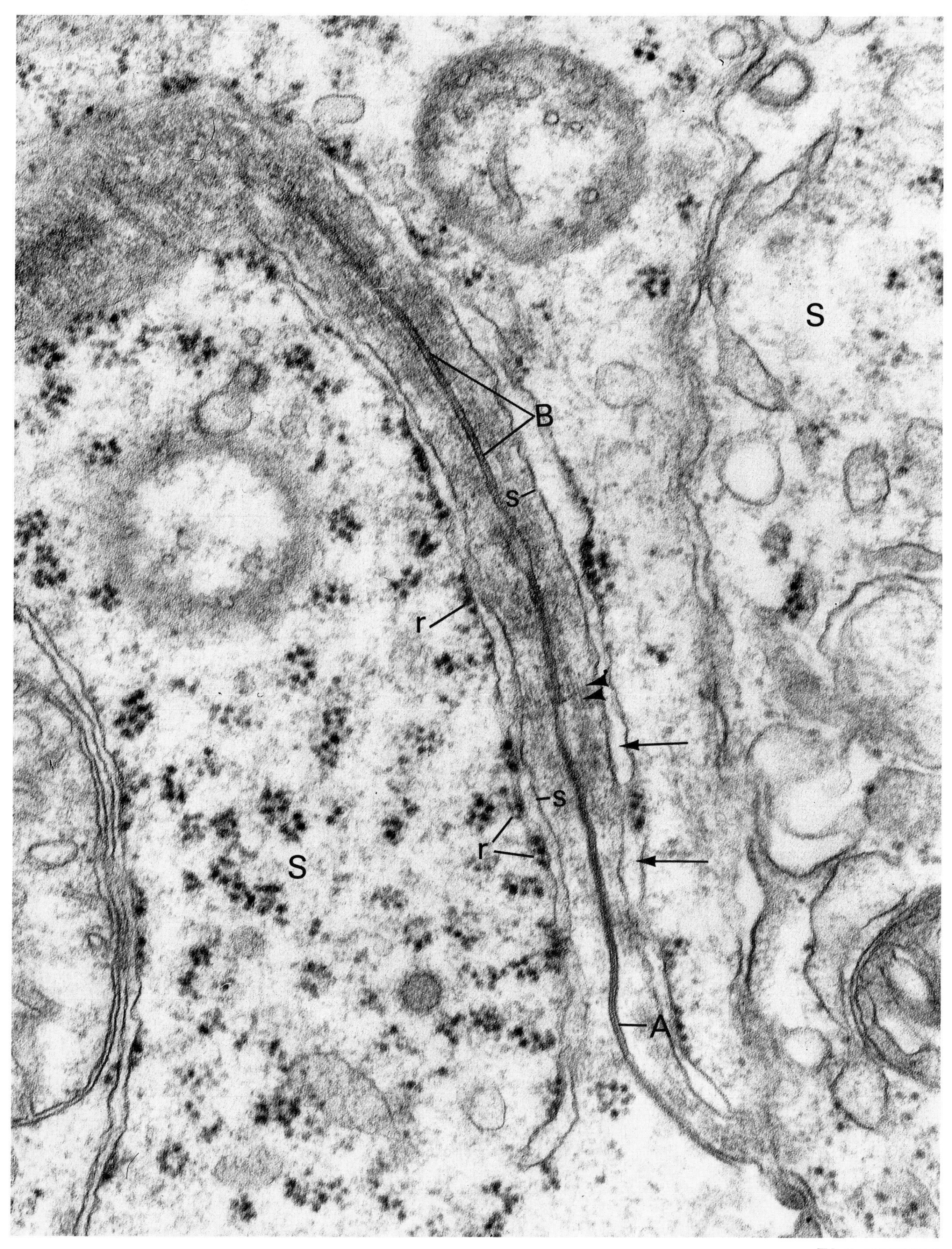

EM: 88,500×

SERTOLI CELL

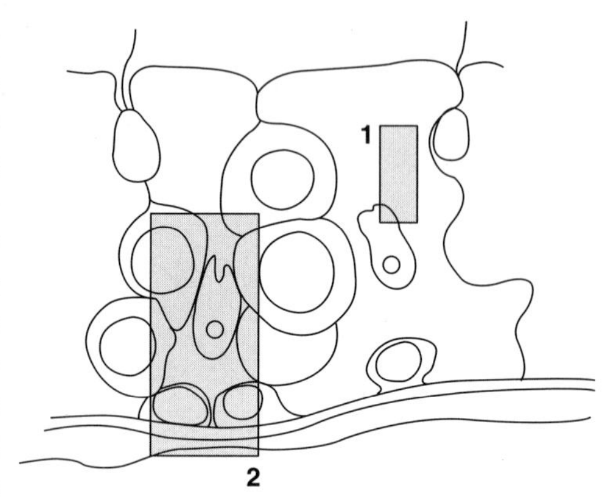

The **Sertoli cell** is one of the most active and functionally complex cells in the body. It responds to growth factors, hormones (particularly FSH), and the developing male gametes to provide the structural and metabolic support needed for sperm development. The euchromatic nucleus (N) and large nucleolus (arrow) evident in micrograph 2 indicate the intense activity of this cell. Over 100 different proteins are synthesized on the ribosomes (arrows, micrograph 1). Some of the **filamentous proteins,** such as intermediate filaments (arrowheads, micrograph 1) and microtubules (t, micrograph 1), are needed to maintain the integrity of this large cell that supports and "carries" the developing gametes from the basal lamina to the lumen. Other **enzymatic proteins** are concentrated in lysosomes (l, micrograph 1), which function in the continual turnover of junctions as they are formed and reformed to accommodate the movement of the germ cells. Lysosomes are also important sites of digestion following the phagocytosis of excess spermatid cytoplasm in the final stages of development.

A significant portion of the synthetic activities of the Sertoli cell are directed toward secretion. The Sertoli cell secretes the fluid that is the vehicle for transport of the sperm in the excurrent duct system. This fluid provides a unique environment for sperm and contains molecules essential to sperm survival such as androgen-binding protein, which concentrates testosterone.

Sertoli cell ultrastructure changes in relationship to the developmental stage of associated germ cells. In certain associations (micrograph 2) the Sertoli cell nucleus is elongated and more apically situated, and the predominant organelle in the basal cytoplasm is the lysosome (l). In other associations (see Male Reproductive System, page 336) the nucleus moves to a basal location and smooth ER is the predominant organelle. The stage represented in micrograph 2 is associated with the highest secretion rate for several specific proteins, including androgen-binding protein and a meiosis-inducing substance.

The developing gametes have pronounced effects on Sertoli function and are most likely the major factors that control changes in the structure and function of the Sertoli cell. It has been demonstrated that the synthesis and release of the iron-binding protein transferrin by Sertoli cells are stimulated by the presence of germ cells and that the amount of stimulation varies with the stage of germ cell development. Male germ cells thus regulate the availability of iron and, as a result, regulate their own development. The reciprocal interchange between the Sertoli cell and germ cells is one of the most significant and interesting events in spermatogenesis.

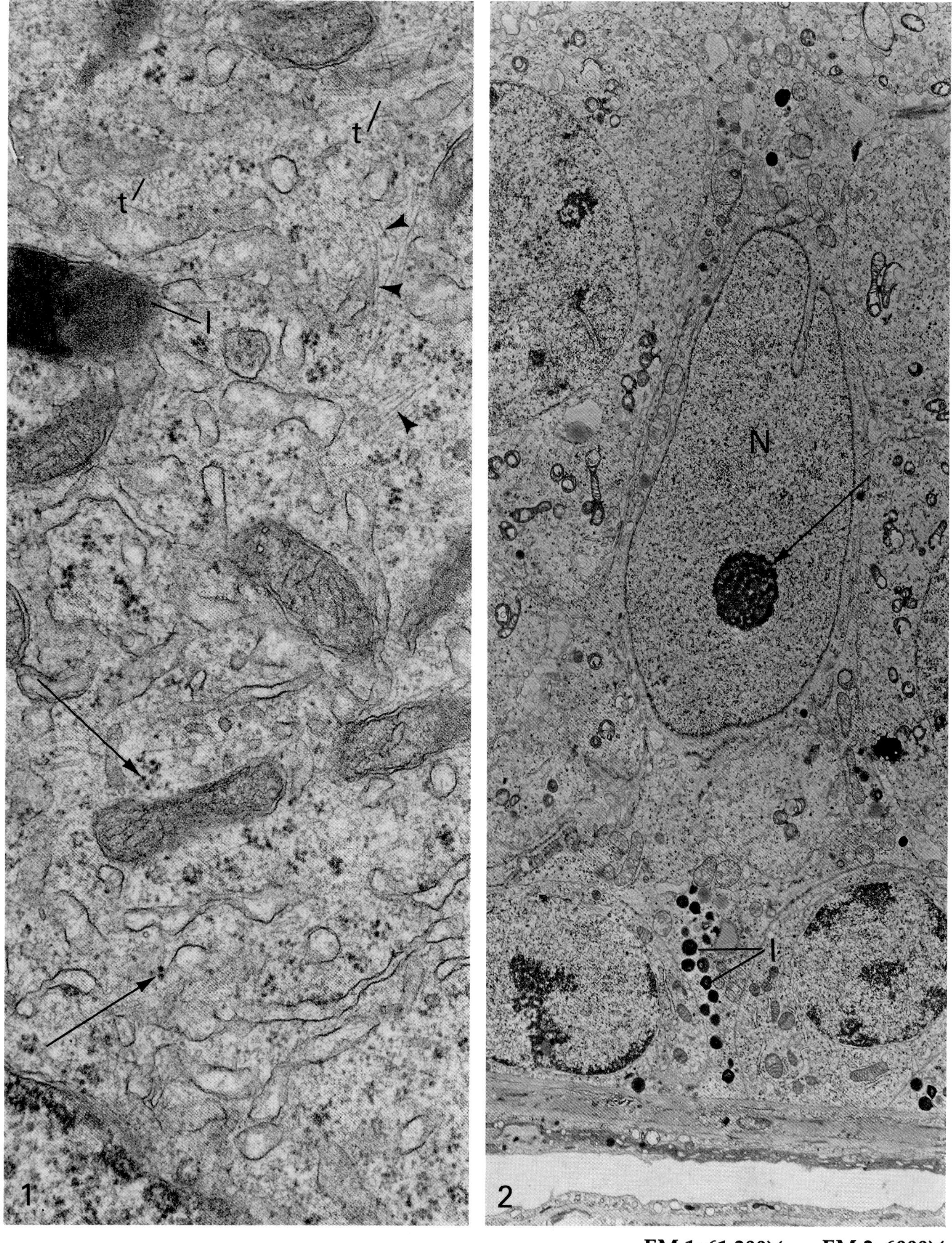

EM 1: 61,200× EM 2: 6000×

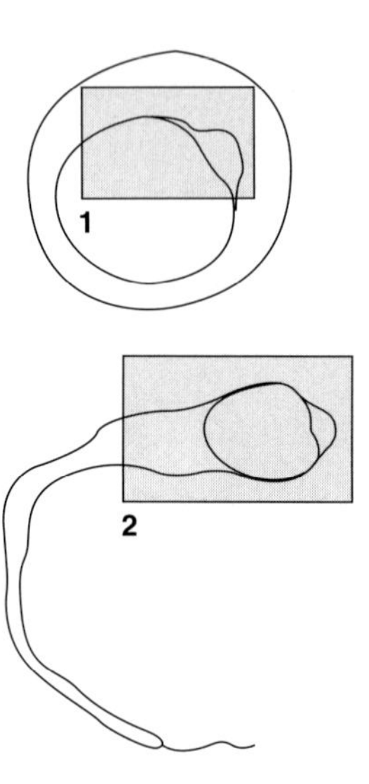

After completion of meiosis, the haploid male gamete, now called a spermatid, undergoes **spermiogenesis,** a dramatic series of events that alters both shape and activity in preparation for the journey to the site of fertilization. One of the initial steps in this process is the formation of the acrosome.

The **acrosome** (arrows, micrographs 1 and 2) is a vesicle that covers the anterior pole of the spermatid nucleus and acts as a storage vesicle for enzymes that are important in the process of fertilization. The vesicle originates from the Golgi (G, micrograph 1), where the acrosomal hydrolytic enzymes are processed and packaged. Even after the acrosomal vesicle has spread over the nucleus to form a cap (micrograph 1), coated vesicles (arrowhead, micrograph 1) fuse with the acrosome, adding newly processed enzymes and concentrating the contents to form an electron-dense mass (micrograph 2). It is not until the sperm binds to the zona pellucida of the egg that the enzymes are released during the "acrosome reaction."

In **early spermatids** (micrographs 1 and 2), the haploid genome within the nucleus (N) is active before its condensation in the second half of spermiogenesis. The genome directs the synthesis of the acrosomal enzymes that have not already been synthesized during the spermatocyte stage. As nuclear activity begins to decrease, a dense peripheral nuclear band (b, micrograph 2) can be seen directly adjacent to the inner acrosomal membrane. Nuclear pores are eliminated in this region, thus decreasing communication between the nucleus and cytoplasm.

> As the polarity of the sperm develops during this early spermatid stage, certain membranes acquire distinct compositions that will become particularly significant at the time of fertilization. The plasma membrane acquires a protein on its anterior surface that binds to a zona pellucida receptor and one on its lateral surface that binds the oocyte plasma membrane. The inner acrosomal membrane (a, micrograph 1), which becomes contiguous with the plasma membrane after the acrosome reaction, acquires highly specialized enzymes. Acrosin, one of the enzymes important in zona pellucida penetration at fertilization, is inserted at this time as an integral membrane protein of the inner acrosomal membrane.

At the same time that acrosome formation is taking place, the shape of the entire sperm is changing, and the differentiation of the sperm tail (not observed in micrographs 1 and 2) is occurring. Even though the plane of the section of the micrographs does not pass through the region of differentiation of the tail, the micrographs do illustrate a shape change. Cytoplasm, originally present anterior (C, micrograph 1) to the acrosome shifts posteriorly (C, micrograph 2) away from the developing head such that the plasma membrane (arrowheads, micrograph 2) comes to lie tightly apposed to the acrosomal membrane. Organelles such as the lysosomes (l) and endoplasmic reticulum (e) adjacent to the acrosome in micrograph 2 are actually within Sertoli cell cytoplasm.

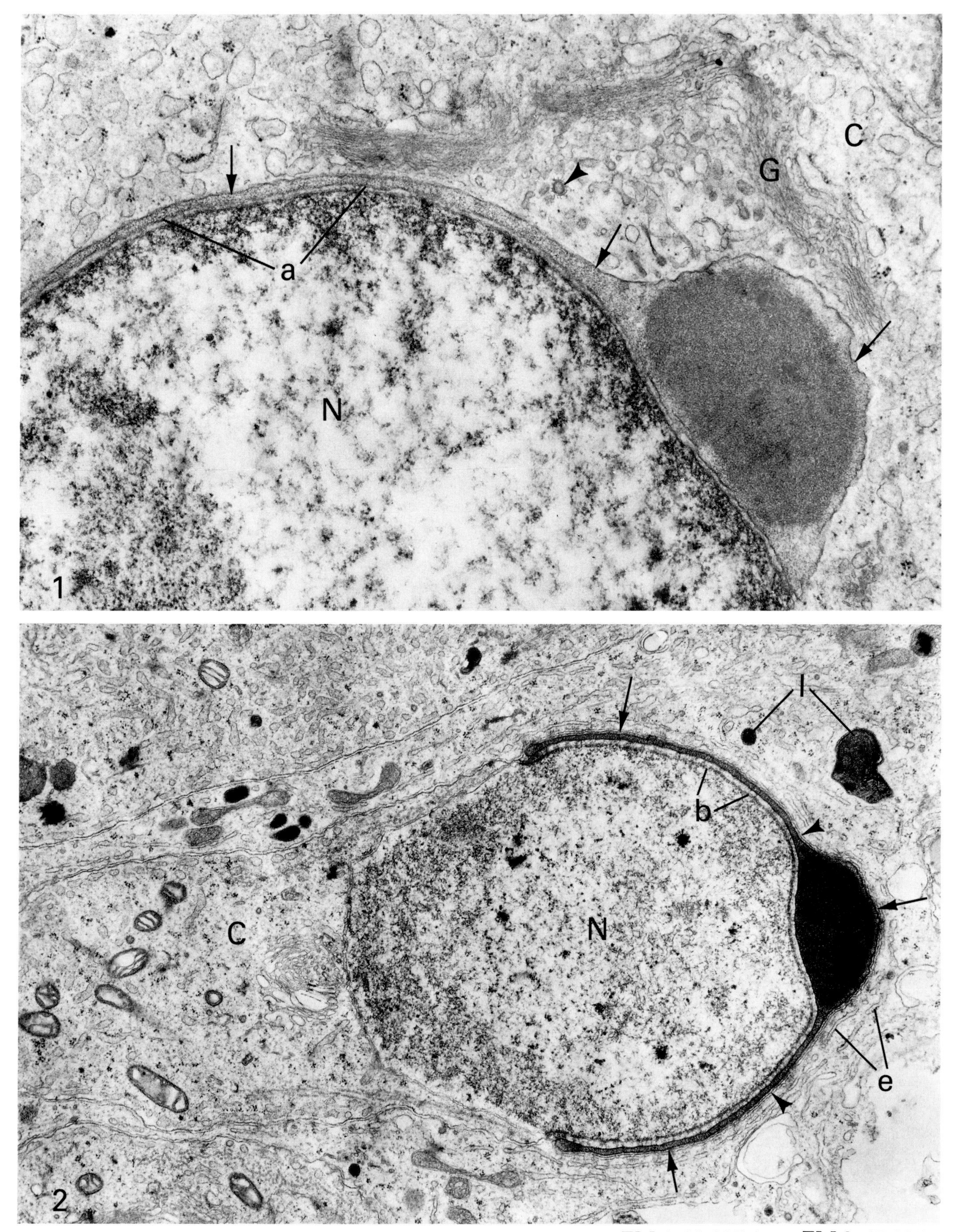

EM 1: 28,800× EM 2: 17,000×

SPERMATID: Late Development

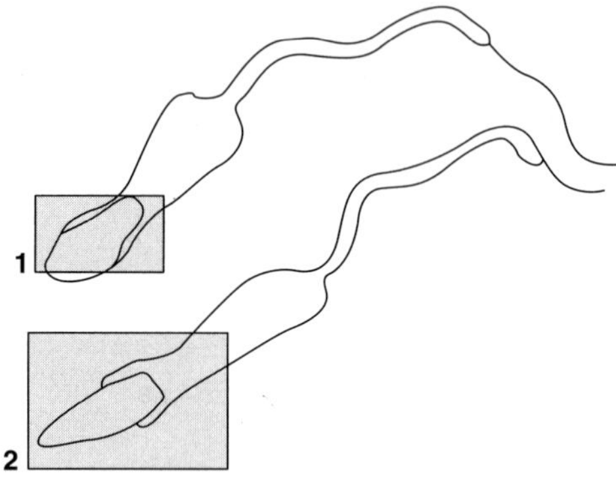

In the **late spermatid stage,** the nucleus elongates and the chromatin condenses (illustrated in the progression from micrograph 1 to micrograph 2). The elongation process is typically associated with the appearance of a ring of microtubules, the **manchette** (arrowheads, micrographs 1 and 2), that extends posteriorly from the base of the acrosome parallel to the axis of elongation. Cross-links between these microtubules suggest that they act together; they may move organelles or effect shape changes at this critical time in development.

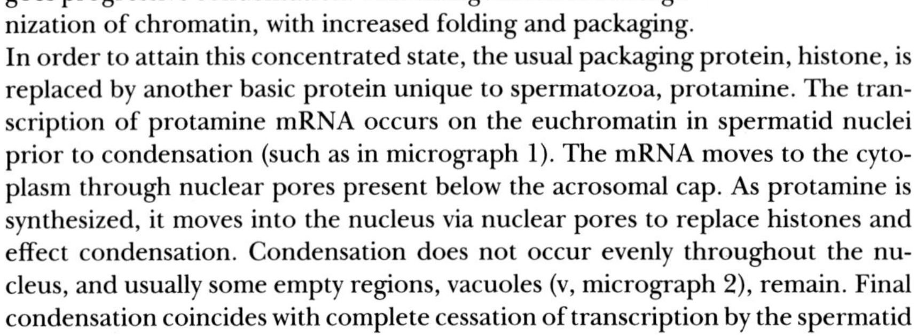

As spermiogenesis comes to an end, the chromatin undergoes progressive condensation. This change involves a reorganization of chromatin, with increased folding and packaging. In order to attain this concentrated state, the usual packaging protein, histone, is replaced by another basic protein unique to spermatozoa, protamine. The transcription of protamine mRNA occurs on the euchromatin in spermatid nuclei prior to condensation (such as in micrograph 1). The mRNA moves to the cytoplasm through nuclear pores present below the acrosomal cap. As protamine is synthesized, it moves into the nucleus via nuclear pores to replace histones and effect condensation. Condensation does not occur evenly throughout the nucleus, and usually some empty regions, vacuoles (v, micrograph 2), remain. Final condensation coincides with complete cessation of transcription by the spermatid nucleus.

Throughout spermatogenesis, the gametes remain intimately associated via junctions with a Sertoli cell. Desmosomes and gap junctions found in the early stages (spermatogonia and spermatocytes) are replaced in the spermatid stages by unique, specialized junctions that bind the spermatid head region directly over the acrosome to the opposing Sertoli cell. The Sertoli contribution to this junction is clearly illustrated in micrograph 1 as dense groups of actin filaments (short arrows) situated between the Sertoli plasma membrane and Sertoli endoplasmic reticulum (long arrows). The Sertoli aspect of this junction is similar to the Sertoli–Sertoli junction that defines the blood-testis barrier; however, there are no tight-junction components in the Sertoli-germ cell type. These junctions are a major final association between the Sertoli cell and the spermatid. The disruption of this junction is important in **spermiation,** the process of release of the spermatids into the seminiferous tubule lumen.

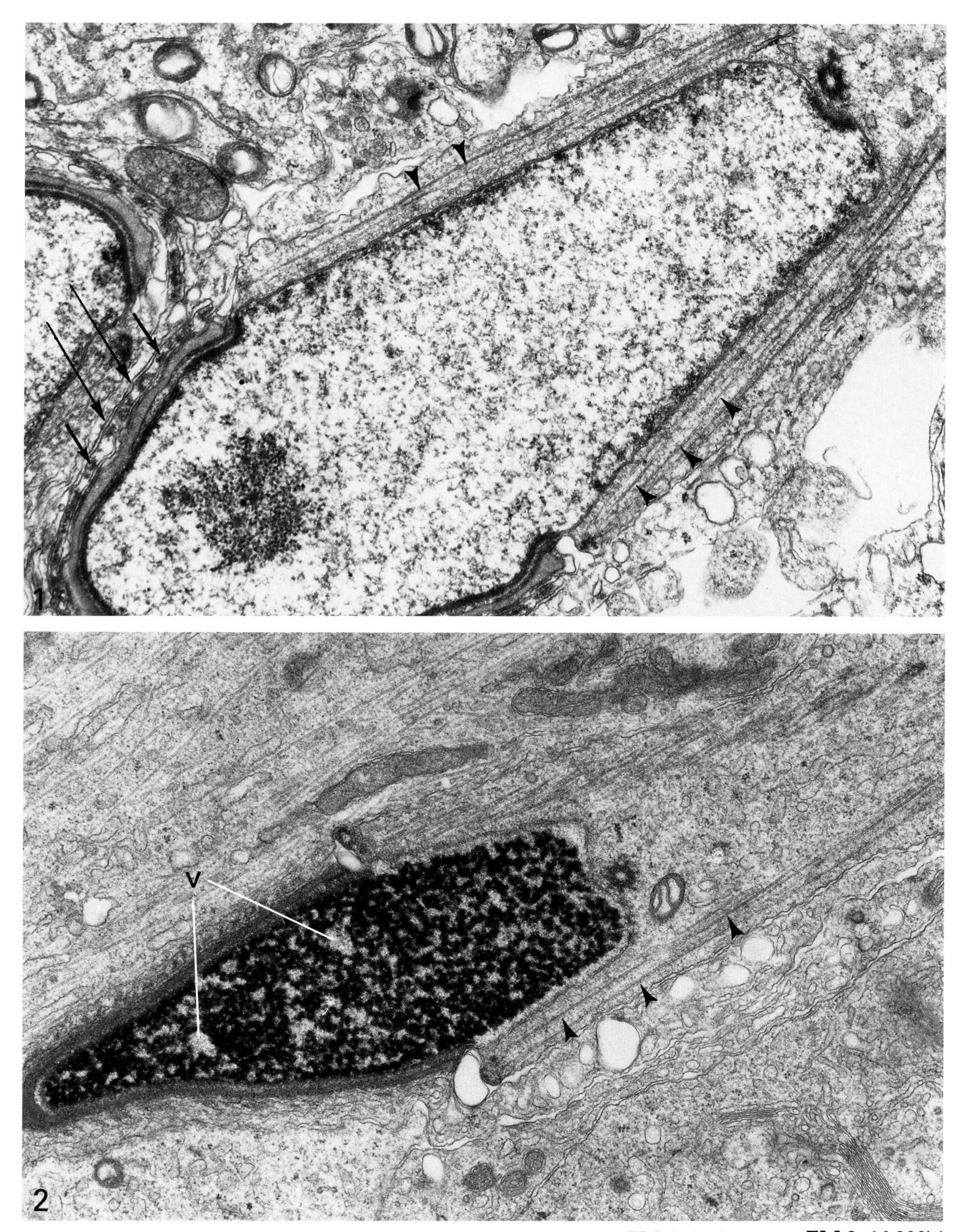

EM 1: 23,000× EM 2: 16,800×

SPERMATID: Intercellular Bridges

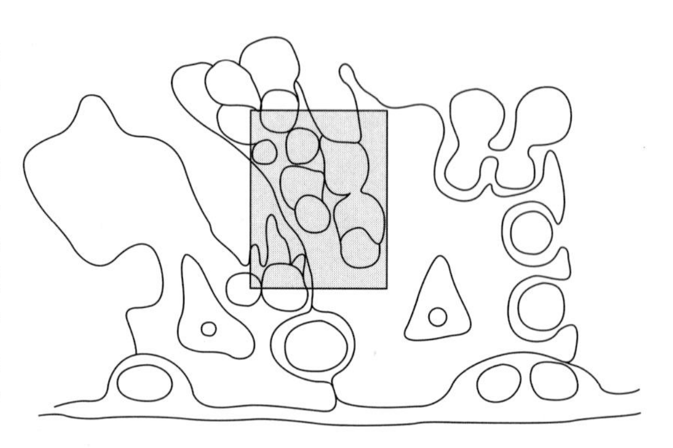

From the time spermatogonia are committed to becoming spermatozoa, their mitotic and meiotic nuclear divisions proceed without complete separation of the cytoplasm. As a result, a large number of progeny from a single B spermatogonium remain connected by **intercellular bridges** up until their release as spermatozoa into the lumen of the seminiferous tubule. Even though all gametes are connected by these bridges, it is relatively rare that a section passes through one. In the micrograph such a bridge (arrowhead) is seen connecting two early spermatids (S).

The membranes defining intercellular bridges are unique, possessing fewer proteins than cell membranes in other areas. The electron density typically associated with this part of the membrane reflects tightly packed actin filaments. Since all gametes that are connected develop originally from one cell, their development is synchronous. Gametes that are joined together, even though frequently referred to as a clone, are genetically distinct following meiosis. Nevertheless, they share the products of their activity via the intercellular bridges. Messenger RNAs from one spermatid cross to other spermatids and thus establish a common translational machinery, even in light of genetic diversity.

At the time of spermiation (when spermatids are released from their Sertoli cell association and enter the lumen of the seminiferous tubule) spermatids separate from one another. This separation does not occur at the intercellular bridge region. Instead it occurs by removal of excess cytoplasm and the formation of **residual bodies.** Residual bodies can be found associated with one another via intercellular bridges.

Modified from D. M. de Kretzer and J. B. Kerr, The Cytology of the Testis. In *The Physiology of Reproduction,* Vol. 1, Raven Press, New York, 1988.

During the differentiation of the tail considerable remodeling will occur. Spermatid mitochondria (arrows, micrograph) will localize and wrap around the axonemes. Many of the organelles that will not become part of the mature spermatozoa, such as the Golgi (G, micrograph) and extensive tubules found throughout, will be destroyed by autophagy within local lysosomes (l, micrograph). Cytoplasm that is lost as the residual body will be either phagocytosed by Sertoli cells (SC, micrograph) and destroyed within their lysosomes, or carried with the mature sperm along the sperm channels to the epididymis and eventually removed by the epididymal lining.

Remodeling in the spermatids shown in the micrograph has occurred to the extent that the spermatid cytoplasm has shifted from the head region to a region posterior to the acrosomal cap. The unique, often dumbbell shape of the Sertoli cell mitochondria (m, micrograph) aids in the distinction between Sertoli and spermatid boundaries.

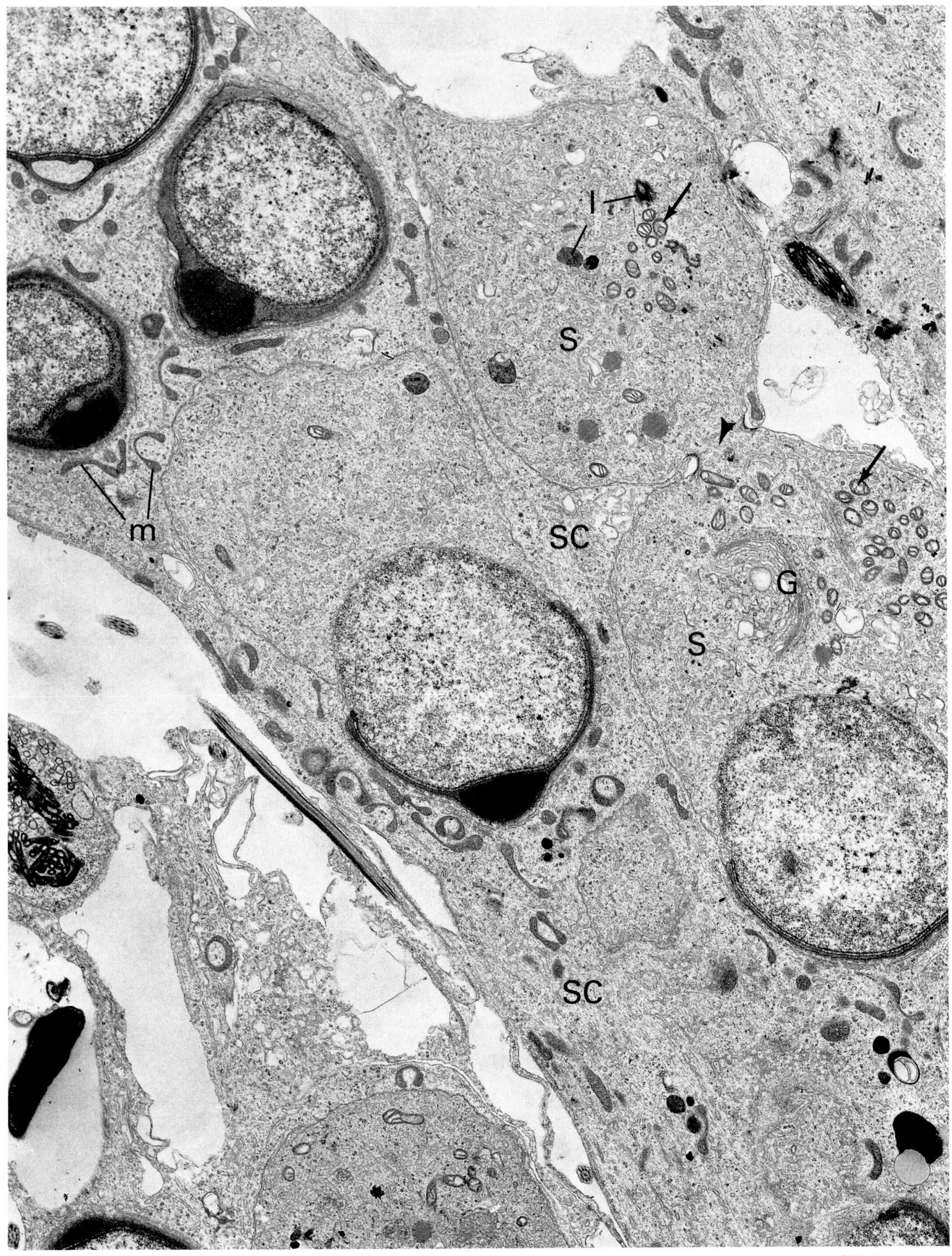

EM: 7650×

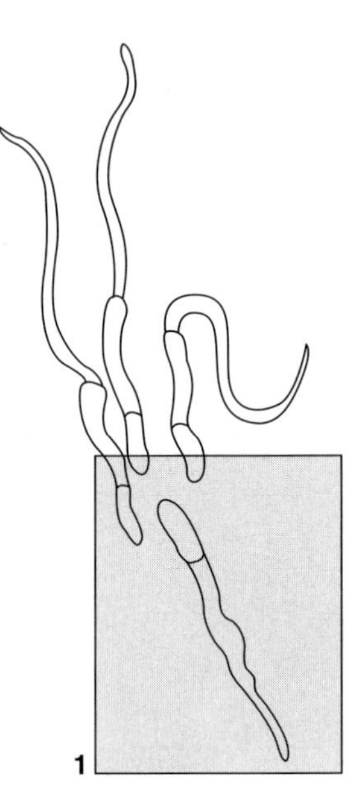

Spermatozoa are highly specialized cells that are designed to deliver the haploid male genome to the egg. The **head** of the sperm contains the genome, protected in the nucleus (N, micrograph 1) in a specialized package of unique basic protein, protamine. It is carried to the egg surface by the vigorous movement of the tail. Once at the egg surface, movement of the tail and the release of acrosomal enzymes from the acrosome (a, micrograph 1) assure penetration and fertilization of the egg.

The **tail** is divided into three regions: the middle piece (inset A), principal piece (inset B), and the end piece (inset C). In the longitudinal section of the sperm in the center of the micrograph, only the middle piece (mp) and principal piece (p) were caught in the plane of section. A central axoneme extends through all of these regions and is most easily observed in the cross sections in the inset. Even though sperm are capable of movement within the male tract, they do not undergo autonomous movement until after ejaculation. In the male tract they are carried by fluid movement and muscular contractions of the wall of the excurrent ducts. Mitochondria (m, micrograph 1 and inset A) wrapped around the midpiece provide the energy for flagellar movement following ejaculation.

Much of the electron density in the tail region represents cytoskeletal proteins that are arranged in two specific patterns, longitudinally and circumferentially. The **outer dense fibers** that extend through the middle piece and principal piece are the major longitudinal cytoskeletal components. These fibers form bands that can be seen longitudinally in micrograph 1 (arrow) and in cross section in inset A (arrow) as nine distinct densities between the mitochondria and axoneme. The **fibrous sheath,** found in the principal piece, comprises the major circumferential components. Circular ribs, in glancing and cross section (arrowheads, micrograph 1), of this sheath wrap around the tail directly under the plasma membrane. In cross section of the tail (inset B) each rib is seen to be stabilized by connection to two longitudinal columns (c).

The outer dense fibers and fibrous sheath are composed in part of intermediate filaments unique to sperm. These cytoskeletal elements do not play an active part in flagellar movement but directly influence the character of the movement. The combination of stiffness imparted by the protein and the flexibility of many cross-linking disulfide bonds provide passive elasticity that affects the amplitude of the bending waves. The symmetry and arrangement of the fibers and sheath may also affect the plane of flagellar movement.

The sperm plasma membrane is highly polarized into at least five specific domains from the head to the end piece of the tail. The integrity of each domain, maintained in the absence of junctions, is essential to the functioning of the underlying region. As sperm travel through extremely different environments in both male and female tracts, the domains are altered and adjusted to modify function in a very specific manner. Sperm maturation and storage in the male tract and capacitation in the female tract are initiated by environmental factors that change the character of one or more of the plasma membrane domains.

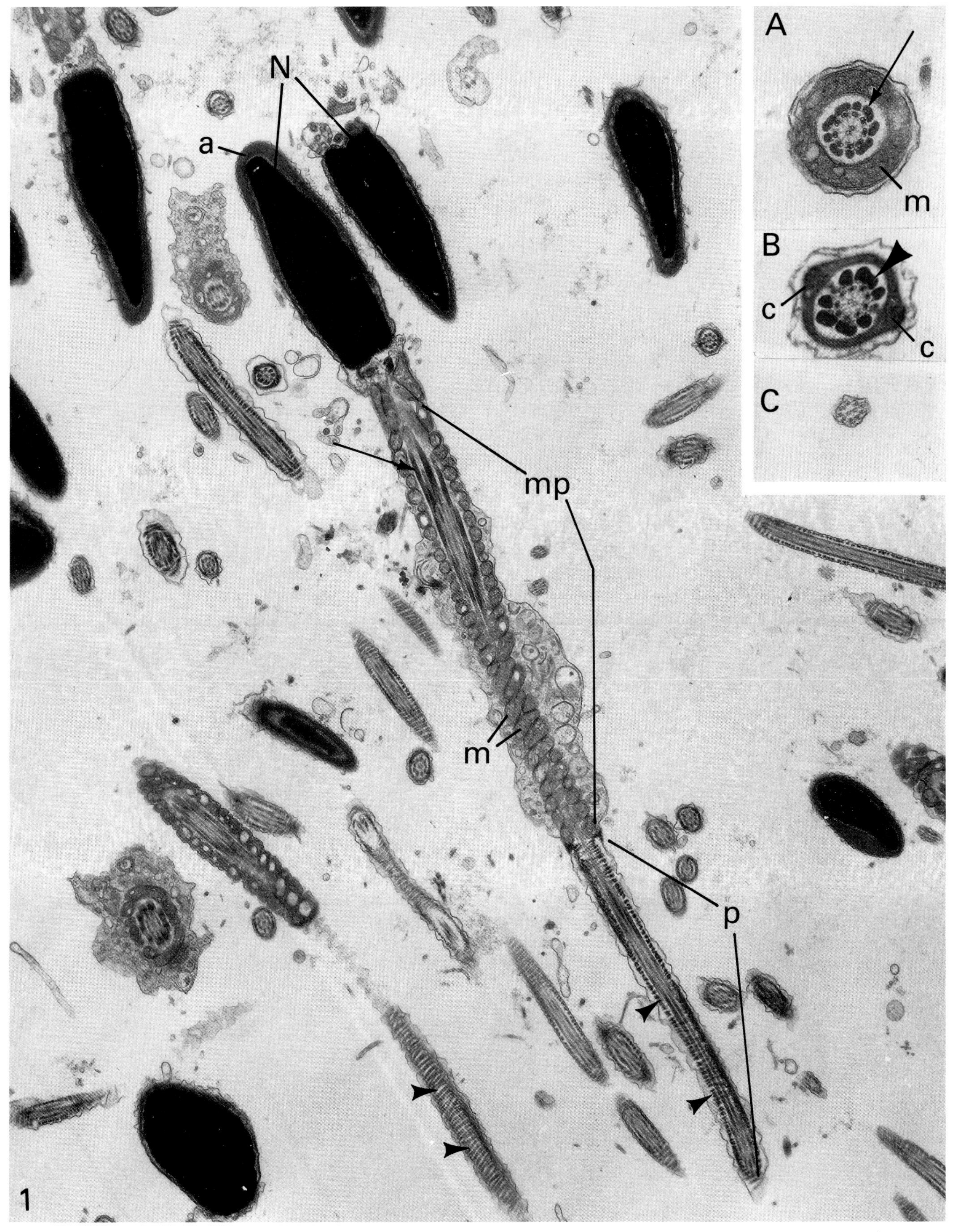

EM 1: 14,400× Insets: 38,400×

Leydig Cell

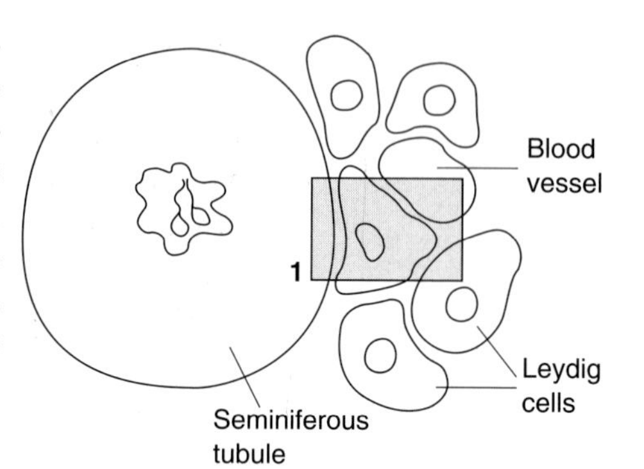

The formation of sperm and the production of androgens are the major functions of the testis. Androgens are essential not only within the testes, to stimulate and maintain spermatogenesis, but also outside of the testis to stimulate the development and maintenance of the male reproductive ducts, glands, and genitalia and to initiate the appearance of the secondary sex characteristics. Androgens, primarily **testosterone,** are produced by the **Leydig cells** (L, micrograph 1) that reside in the interstitial tissue between the seminiferous tubules.

Testosterone is derived from the precursor cholesterol, which is obtained either directly from synthesis on the smooth ER (arrows, micrograph 2) or from low-density-lipoprotein particles that arrive in local vasculature. Frequently cholesterol is stored as esters in lipid droplets (l, micrograph 1) within the cytoplasm. These droplets are depleted during active testosterone synthesis, primarily in response to luteinizing hormone (LH) from the anterior pituitary. LH stimulates the mobilization of cholesterol from storage droplets and its transport to the tubular cristae of the mitochondria (m, micrograph 2). The increase in surface area of the inner mitochondrial membrane provides space for the enzymes that convert cholesterol to pregnenolone. Pregnenolone is then transported from the mitochondrial tubular cristae to the smooth ER, where it is converted to testosterone. The extensive surface area of the smooth ER within each Leydig cell accommodates the enzymes necessary for steps in testosterone synthesis.

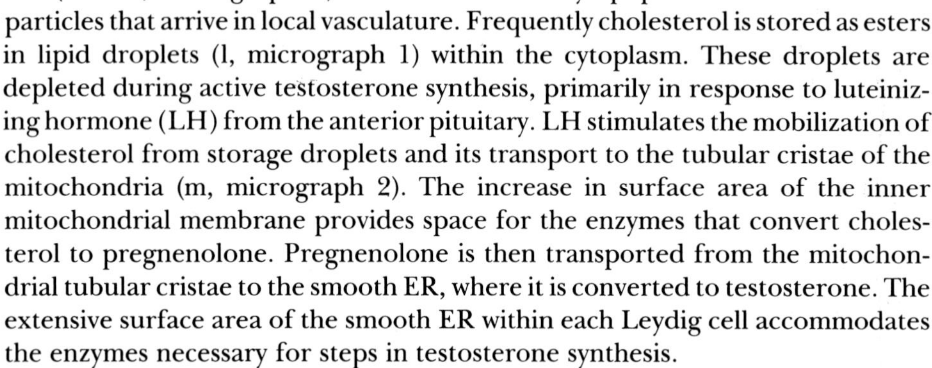

Most of the testosterone secreted from the Leydig cells enters the nearby vessels (v, micrograph 1). One of the major targets is the excurrent duct system and accessory sex glands. In addition, a critical amount of testosterone diffuses through the interstitial tissue fluid to enter Sertoli cells within the seminiferous tubule (ST, micrograph 1). Sertoli cells and, in turn, developing sperm depend upon local testosterone.

The rough ER (arrowheads, micrograph 2), in addition to providing proteins needed for plasma membrane and organelle turnover, synthesizes proteins such as pro-opiomelanocortin (POMC), whose components, MSH, ACTH, and β-endorphin, may have local regulatory roles.

Even though LH is the primary factor controlling the activity of the Leydig cells, local factors are also important. Testosterone production is affected by catecholamines released from local nerve endings and factors (e.g., transforming growth factor-β and estradiol-17β) secreted by Sertoli cells. Leydig cells next to seminiferous tubules in different stages have different morphologies. Therefore, as the Sertoli cells respond to particular germ cell stages, the Leydig cells seem to respond to Sertoli cells in turn.

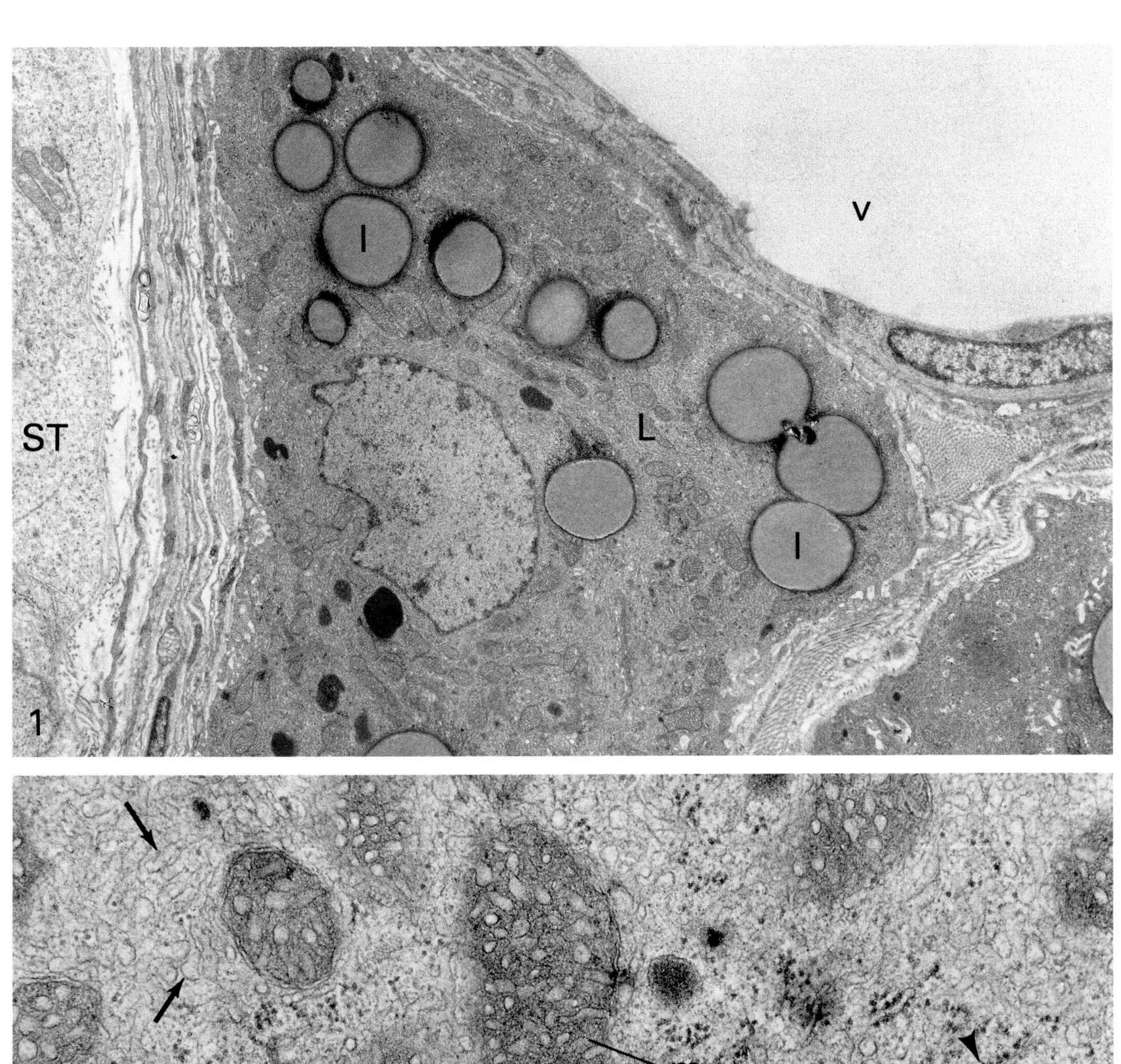

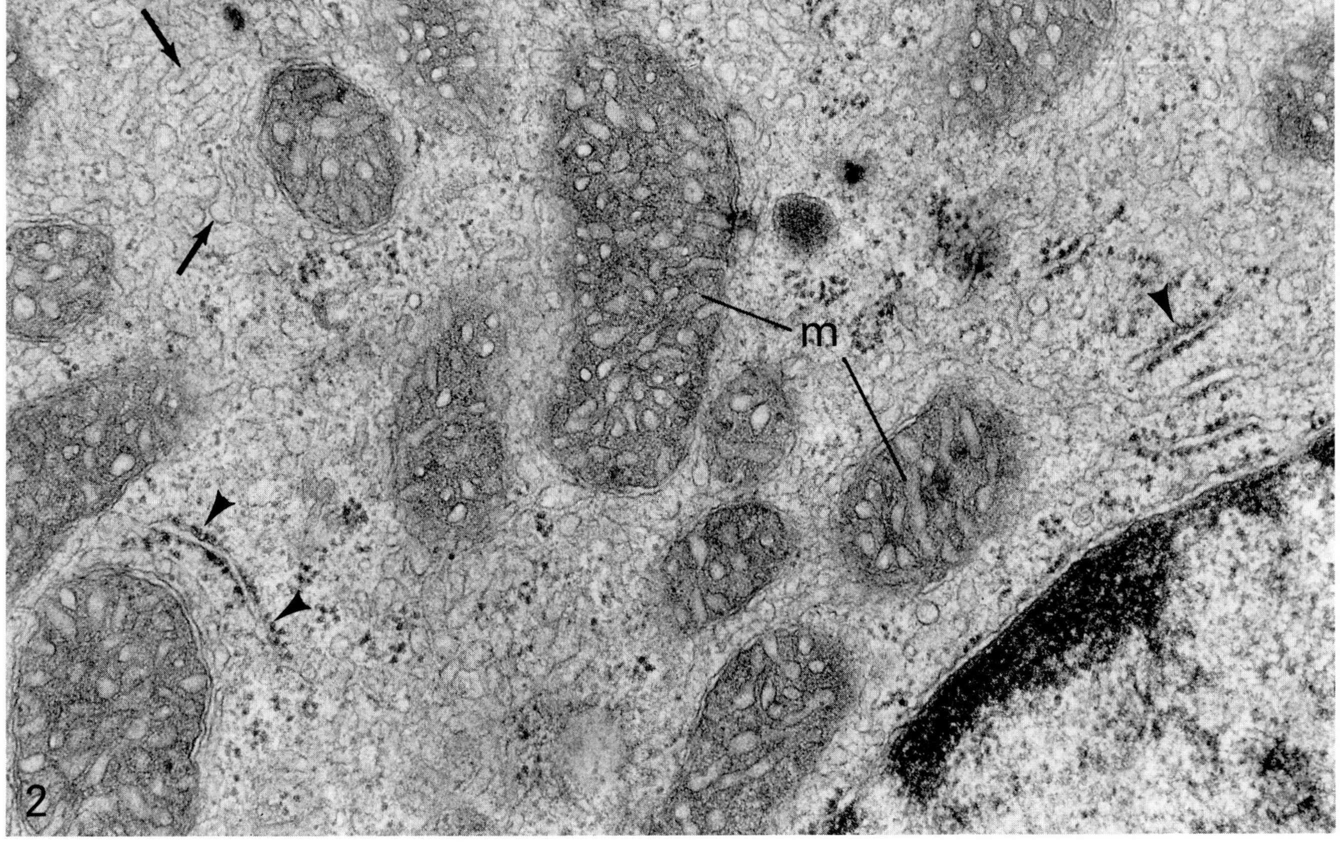

EM 1: 7500× **EM 2: 54,000×**

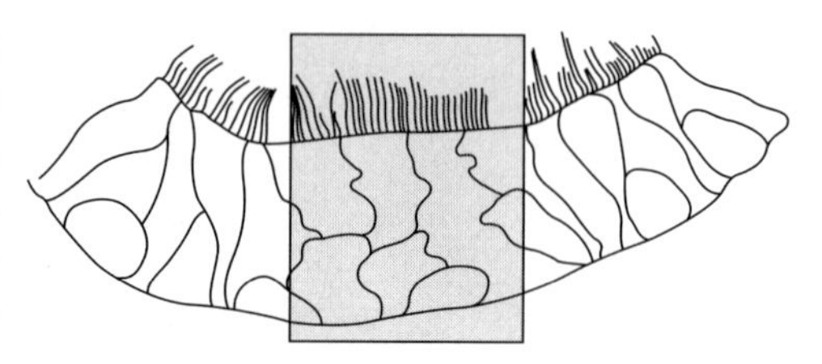

The **epididymis** is the part of the excurrent duct system in which critical stages of sperm maturation take place. As sperm travel through the epididymis from the caput through the corpus to the caudal region, many aspects of differentiation that began in the testis are completed. The plasma membrane of the sperm changes significantly, rendering the sperm capable of motility and fertilization. In addition, the chromatin undergoes further condensation, and the cytoskeletal proteins within the tail continue to cross-link and assume their final physical characteristics. At the same time, the overall form of the sperm is streamlined as superfluous cytoplasm of the midpiece is removed.

Some of the changes that take place in the sperm within the epididymis can be correlated specifically with the development of a function. For example, from caput to caudal epididymis, the capacity for sperm to bind to the zona pellucida increases along with their ability to fertilize the oocyte. In addition, as sperm move down the epididymis, the phosphorylation of sperm membrane proteins increases along with sperm motility potential. Many changes associated with maturation are ascribed to **secretory activity** of the epididymal lining cells (E, micrograph), since the sperm is not an active synthetic cell. The lining cells are known to synthesize several different glycoproteins and enzymes, and the necessary synthetic and secretory machinery are present (rough ER, Golgi); however, the mechanism of action of these synthetic products has not been ellucidated.

The active transport of ions across the lining cells from the lumen to the interstitial fluid draws water across the epithelium. A current is created by the absorption, pulling fluid produced by Sertoli cells to the epididymis. Sperm are carried passively via this current from the seminiferous tubules to the epididymis. The movement of ions and water across the epididymal epithelium is facilitated by the increased surface area of stereocilia (arrows, micrograph) on the apical surface and membrane folds (arrowheads, micrograph) on the basolateral surface. Different ionic concentrations are maintained by tight junctions that are a part of the junctional complexes (circles, micrograph) between these cells.

Endocytosis is an important event at the apical surface of the epididymal epithelium. Endocytosis includes (1) receptor mediated, which carries specific proteins such as androgen-binding protein and transferrin into the cells via coated vesicles, (2) nonspecific vesicle transport of packets of luminal fluid (large vesicles, v, micrograph, are endosome transport packages that form following both specific and nonspecific endocytosis), and (3) phagocytosis. Phagocytosis is an important activity of the lining cell in the removal of excess sperm cytoplasm during normal maturation, and in the removal of whole sperm during chemical or mechanical insult such as occurs following vasectomy. The lysosomes (l, micrograph) are the final station for small endosomes and phagosomes and play an essential role in the processing of ingested material.

Most functions of the epididymis depend upon androgens. Testosterone reaches this duct both on the apical side in the lumen, bound to androgen-binding protein, and on the basolateral side from the underlying vessels. It is converted within the lining cells to the active hormone dihydrotestosterone. The euchromatin within the nuclei (N, micrograph) contains the sites of transcription essential for the formation of the enzyme involved in this critical conversion.

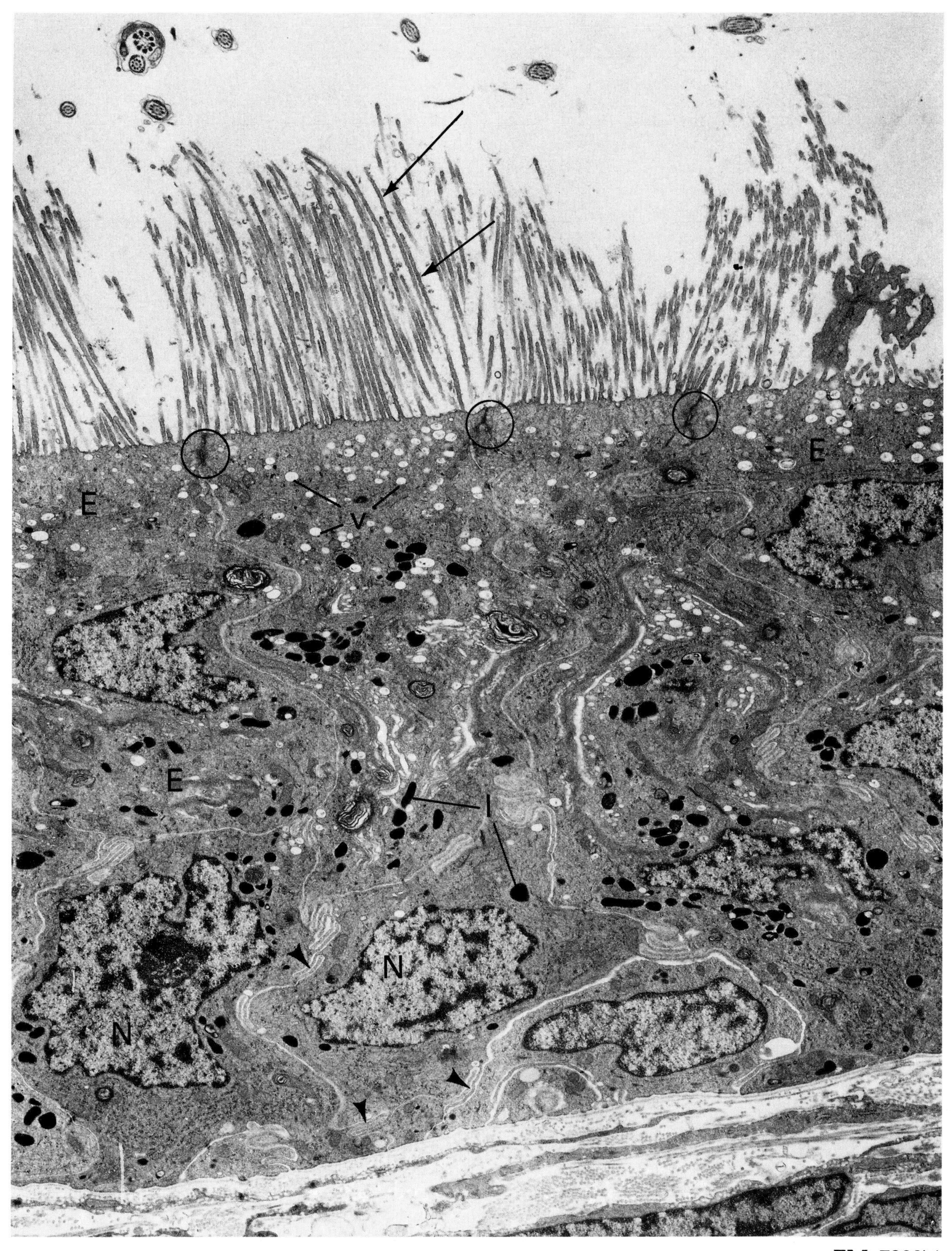

EM: 7200×

Accessory Sex Glands

Seminal plasma consists of the secretions of the excurrent duct system and the accessory sex glands. These secretions, released in sequence during ejaculation, carry spermatozoa (100 million sperm/ml) into the female reproductive tract. **Seminal vesicles** (micrograph 1) and the **prostate** (micrograph 2) account for over 70% of the seminal plasma. The epithelial ultrastructure of these glands clearly reflects their major activities, protein synthesis and secretion. The differences in secretory vesicles distinguishes the seminal vesicles (dense eccentric core) from the prostate gland (central, less dense core).

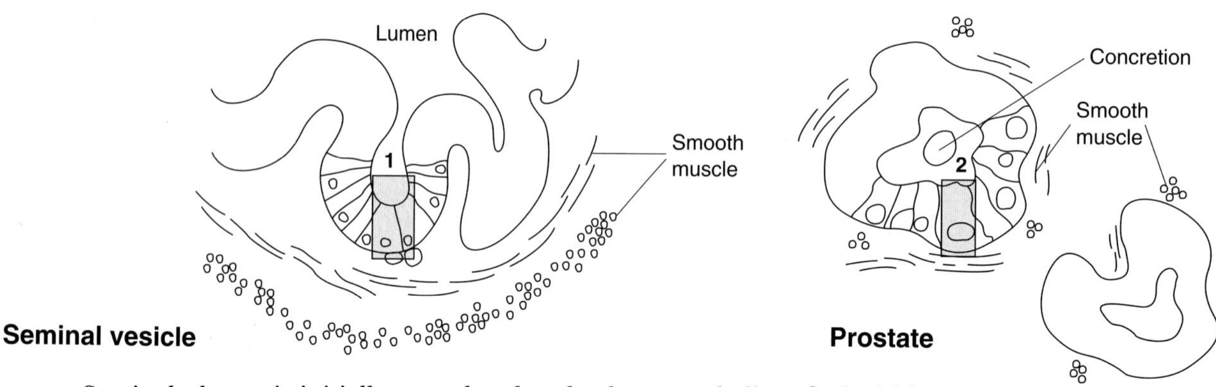

Seminal plasma is initially coagulated and subsequently liquefied within the female reproductive tract. These changes play a significant role in the initial protection of sperm and its subsequent capacitation, which finalizes its ability to fertilize the oocyte. Proteins synthesized on the rough ER (arrowheads, micrograph 1) of the seminal vesicle include proteins (semenogelin and fibronectin) that provide the structural component of the coagulum, a semisolid gel. Proteins synthesized on the rough ER (arrowheads, micrograph 2) of the prostate include the enzymes that induce the polymerizing reaction between these structural proteins. This coagulum, which forms following ejaculation, is similar to that which forms between fibronectin, fibrin, and platelets during blood clotting.

The subsequent dissolution of the coagulum is effected by the activation of proteolytic enzymes (plasminogen activators, seminin, prostate-specific antigen) and possibly other enzymes (e.g., acid phosphatase) secreted by the prostate. Prostate-specific antigen (PSA) has replaced acid phosphatase as the major indicator of abnormal prostatic growth (benign and cancerous).

A wide variety of other substances are also produced by these glands. For example, seminal vesicles secrete fructose, the major energy source of the sperm; the prostate secretes prostaglandins, which effect contractions of the female reproductive tract, and zinc, which is bactericidal. With age, prostatic secretion becomes calcified forming masses referred to as concretions.

The height and activity of the seminal vesicle and prostate epithelium depend upon testosterone levels, most of which reaches the glands in the capillaries (c, micrograph 1) that rest in the underlying connective tissue. With adequate testosterone levels, secretory granules fill the apical cytoplasm and secretion is released into the lumen of the glands, where it accumulates. Sympathetic activity at ejaculation causes the contraction of smooth muscle (organized bands in the wall of the seminal vesicles and scattered bundles in the prostate), forcing the secretions into the excurrent ducts. Individual smooth muscle fibers (M, micrograph 2) rest very close to the epithelium in the prostate.

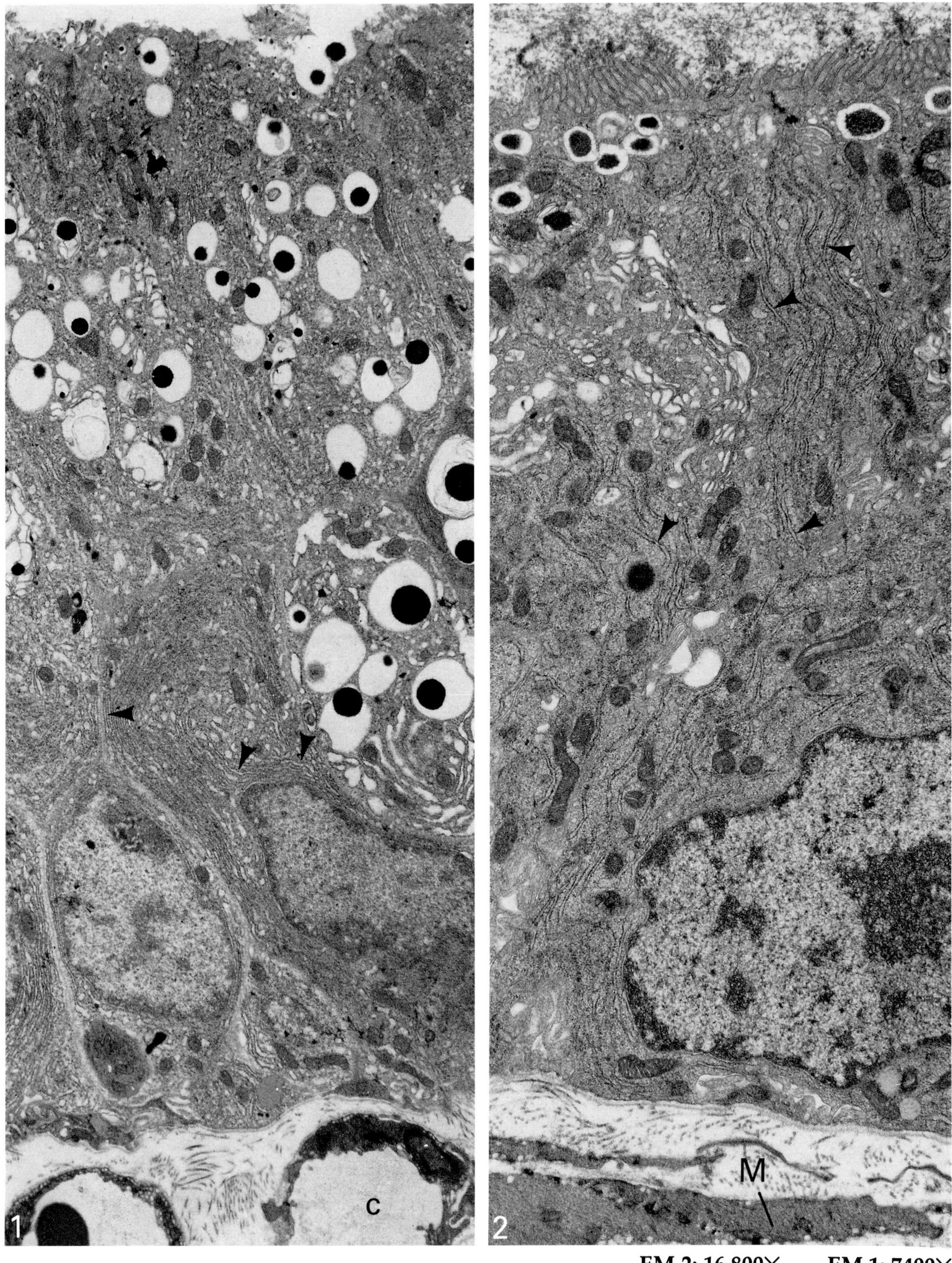

EM 2: 16,800× EM 1: 7400×

Chapter Sixteen

Female Reproductive System

OVARY: Primordial Follicle

The **primordial follicle,** the smallest follicle found in the ovary after birth, consists of an oocyte (O, micrograph) surrounded by a single layer of **squamous granulosa cells** (GC, micrograph). At this stage of development the basal lamina (arrows, micrograph) of the granulosa cells defines the outer boundary of the follicle. The entire follicle is arrested in development. Even though the oocyte is not growing in size, mitochondria (m), Golgi (G), polyribosomes (p), and other organelles required for turnover and maintenance of cellular structure are evident (micrograph).

At birth there are approximately 2 million primordial follicles resting in the outer cortex of the ovary. On a continual basis, from birth until the pool is exhausted at menopause, a certain number of these follicles leave this pool to develop further. Only 500 out of the original 2 million survive to undergo ovulation. The remainder degenerate (i.e., undergo atresia) at some stage of development. Before puberty and the first ovulation, all primordial follicles that leave the pool become atretic.

The large euchromatic nucleus or germinal vesicle (V, micrograph) of the oocyte appears to be in an interphase stage. Actually it is arrested in a unique prophase stage of the first meiotic division, the **dictyate stage.** Prior to this stage, before birth, oogonia complete mitosis, proceed through the early prophase stages of the first meiotic division, and enter diplotene. Instead of condensing further and entering metaphase, as occurs during gametogenesis in the male, the duplicated chromosomes extend and nucleoli reappear. The nucleus will remain in this arrested stage until directly before ovulation at the time of the luteinizing hormone (LH) surge.

Oocytes ovulated at the onset of puberty (average age 12) are at least 38 years younger than oocytes ovulated at menopause (average age 50), since no new oocytes are produced after birth. It is known that the incidence of aneuploidy in births increases with maternal age, and further, that the incidence of nondisjunction at metaphase I increases with maternal age. Maternal nondisjunction during the first meiotic division is the most common underlying mechanism in the trisomy 21 Down syndrome, a common congenital disorder. It remains to be determined whether the increased incidence of nondisjunction is a result of aging of the oocyte per se, aging of other ovarian parameters, or a combination of these two factors.

Granulosa cells are attached to the oocyte by well-defined junctions (circles, micrograph). These cells will remain attached throughout follicle development in the ovary and through ovulation. This intimate association is a statement of the interdependence of these two cell types.

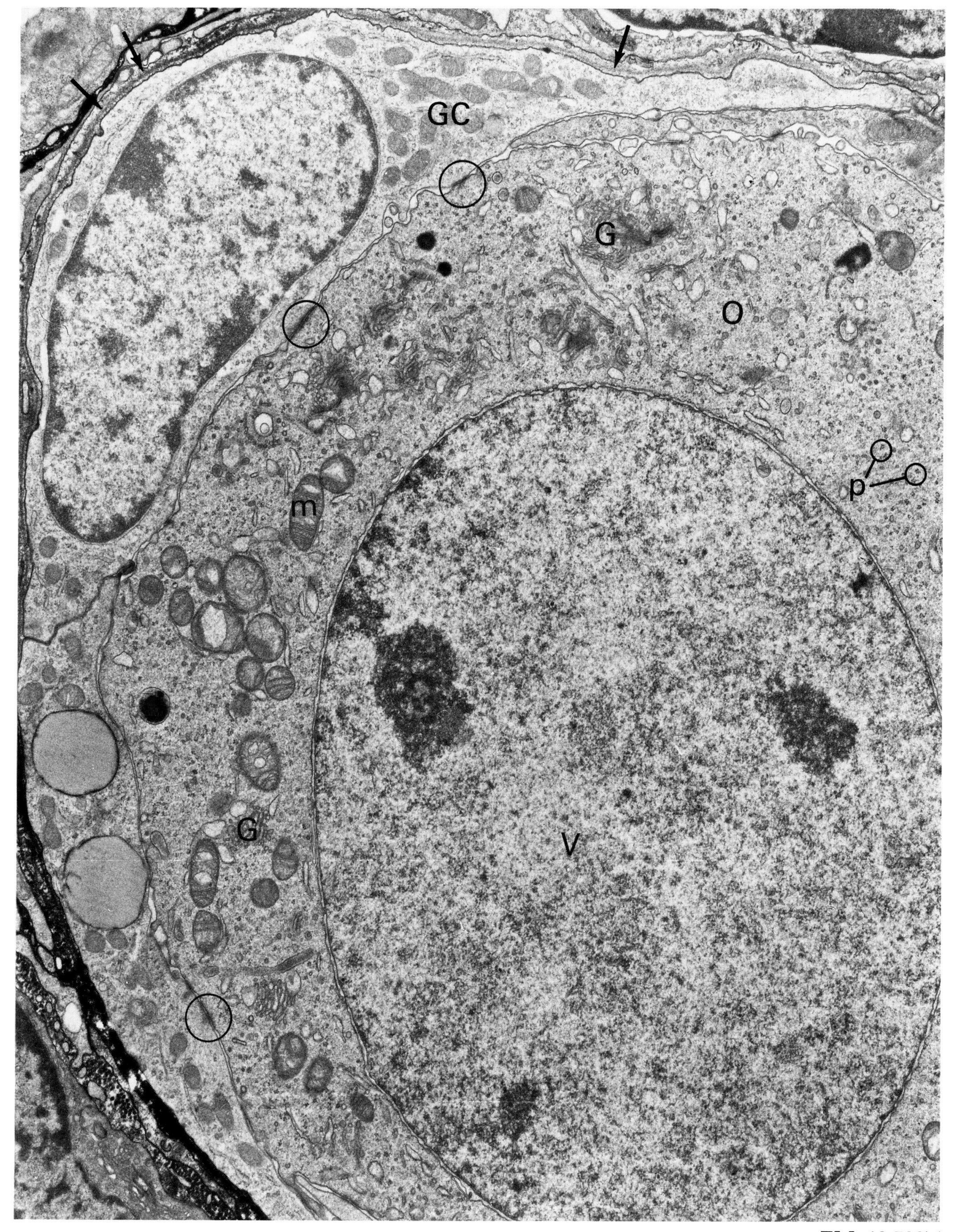

EM: 13,500×

As primordial follicles leave the arrested pool, they develop into **primary follicles.** Oocytes (O, micrographs 1 and 2) complete most of their growth during this stage, and the granulosa cells (GC, micrographs 1 and 2) take on a cuboidal shape as they acquire more organelles and molecules for synthetic activities. Micrograph 1 illustrates a very early primary follicle in which there is only a single granulosa layer. Some of these cells will divide, giving rise to several **granulosa cell layers** as primary follicle development proceeds. A fully grown oocyte is typically surrounded by 5 to 6 layers of granulosa cells. Outside of the granulosa cell layer, a new follicular layer, the **theca** (t, micrograph 1), is being defined. Primary follicles are apparently dependent upon a certain level of gonadotropins for maintenance but do not undergo changes with the cycle as do secondary follicles.

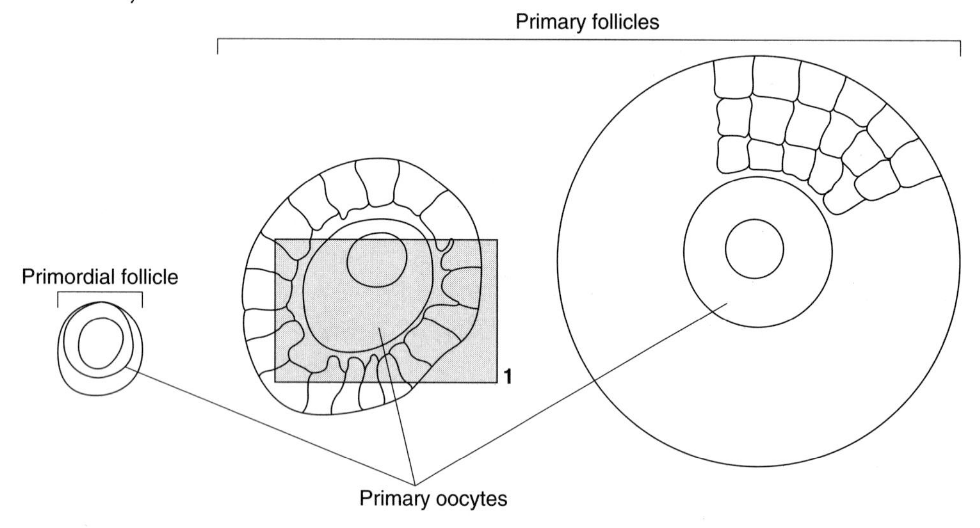

During the **growth of the oocyte** the nucleus remains at the dictyate stage of the first meiotic prophase. The activity of the primary oocyte nucleus, with its characteristic nucleolus (arrow, micrograph 1), is directed toward (1) an increase in cell size to reach the critical mass from which the embryo will form, (2) the accumulation of specific proteins and nucleic acids that will be needed for fertilization and subsequent development, and (3) the synthesis of the **zona pellucida** (Z, micrographs 1 and 2), a thick, filamentous extracellular coat that surrounds the oocyte up to implantation. Zona glycoproteins are processed and packaged in the numerous Golgi (G, micrograph 2) at the surface of the oocyte. The porous zona does not facilitate or prevent the movement of viruses and macromolecules. It does, however, exert fine control over sperm movement during fertilization. Initially the zona aids sperm penetration (certain glycoproteins are sperm receptors); later it prevents penetration by more than one sperm.

As the zona pellucida is synthesized, points of attachment (large circles, micrograph 2) between oocyte and granulosa cell processes (arrowheads) remain. Even though only adherens junctions are observed in this micrograph, gap junctions are also common. Oocyte growth is dependent upon substances (e.g., ribonucleosides, ribonucleotides, and certain amino acids) provided by the granulosa cells via these junctions. Granulosa cells are extremely active during the primary follicle stage, as evidenced by a cytoplasm packed with polyribosomes (p, micrograph 2) and mitochondria (m, micrograph 2).

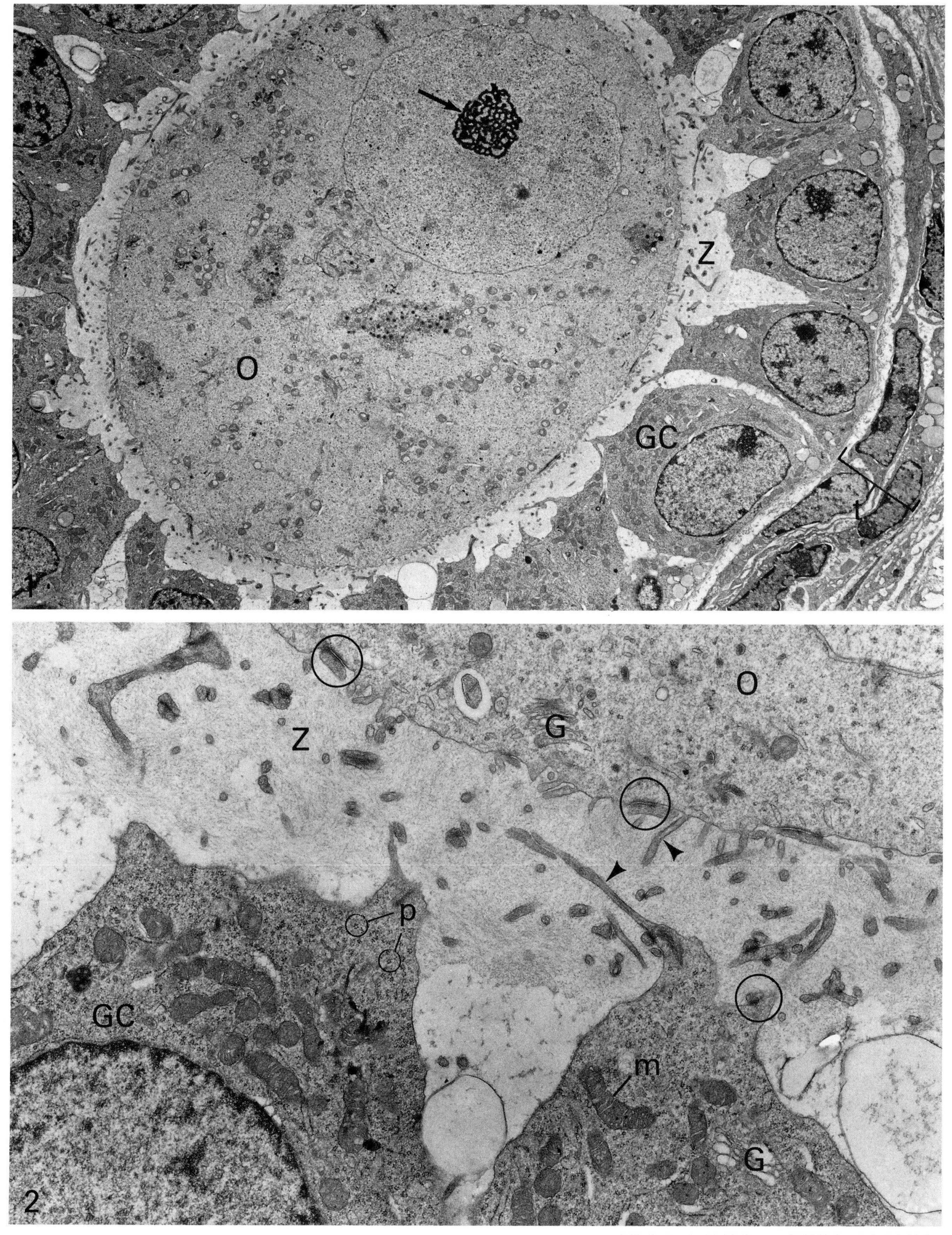

EM 1: 3,600× EM 2: 16,000×

OVARY: Secondary Follicle, Peripheral Granulosa Cells

Beginning at puberty, **secondary follicles** develop from primary follicles in response to increasing levels of follicle-stimulating hormone (FSH). Secondary follicles are large follicles in which a cavity or **antrum** has developed between the granulosa cells. During the first half of each cycle, when FSH is the dominant gonadotropin, secondary follicles grow in size and number. As the antral cavity enlarges, the granulosa cells are divided into **peripheral cells** that line the inner follicle wall and **cumulus cells** that surround the oocyte. Shortly before ovulation, the luteinizing hormone (LH) surge acts on a single large secondary follicle (Graafian follicle) to stimulate events leading to ovulation.

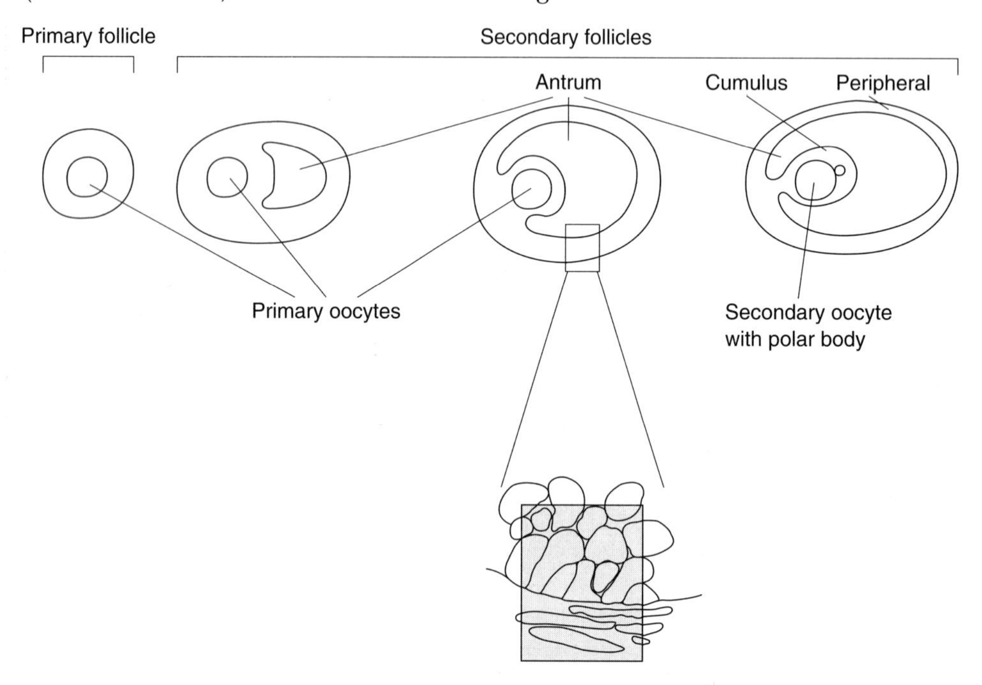

Secondary follicles are the source of estrogen, the major steroid synthesized during the first half of the menstrual cycle. Estrogen exerts effects on the oviduct, uterus, and vagina essential for receptivity of the female reproductive tract to the male gamete. The cells that make up the wall of the secondary follicle, the **theca interna cells** (T, micrograph) and **peripheral granulosa cells** (G, micrograph), cooperate in the synthesis of estrogen. Lipid droplets (l, micrograph) packed in the theca interna cells are accumulations of cholesterol, the steroid hormone precursor. Cholesterol is converted within the theca interna to androgens (in response to LH), which diffuse across the basal lamina (arrowheads, micrograph) to the granulosa cells, where they are converted to estrogens (in response to FSH).

Cholesterol is obtained primarily from low-density lipoprotein particles (LDL) arriving in capillaries (c, micrograph) that permeate the theca layers outside of the basal lamina. The basal lamina of primary and early secondary follicles is relatively impermeable to these lipoprotein particles; however as secondary follicles develop and the basal lamina increases in permeability, some LDLs diffuse to the granulosa cells and they too accumulate a cholesterol store (micrograph). The cholesterol within the granulosa cells will be used primarily for the synthesis of progesterone after ovulation.

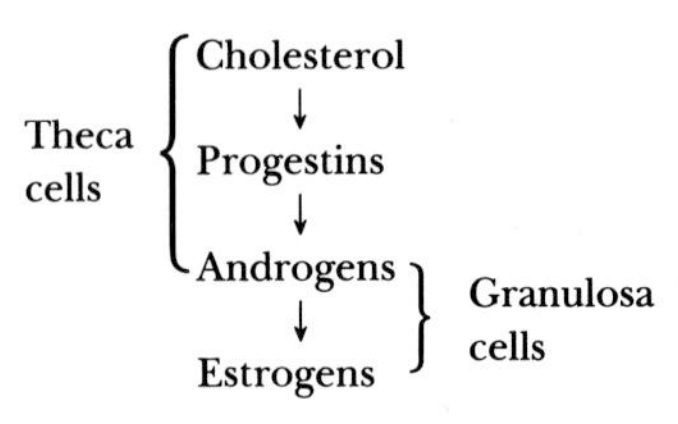

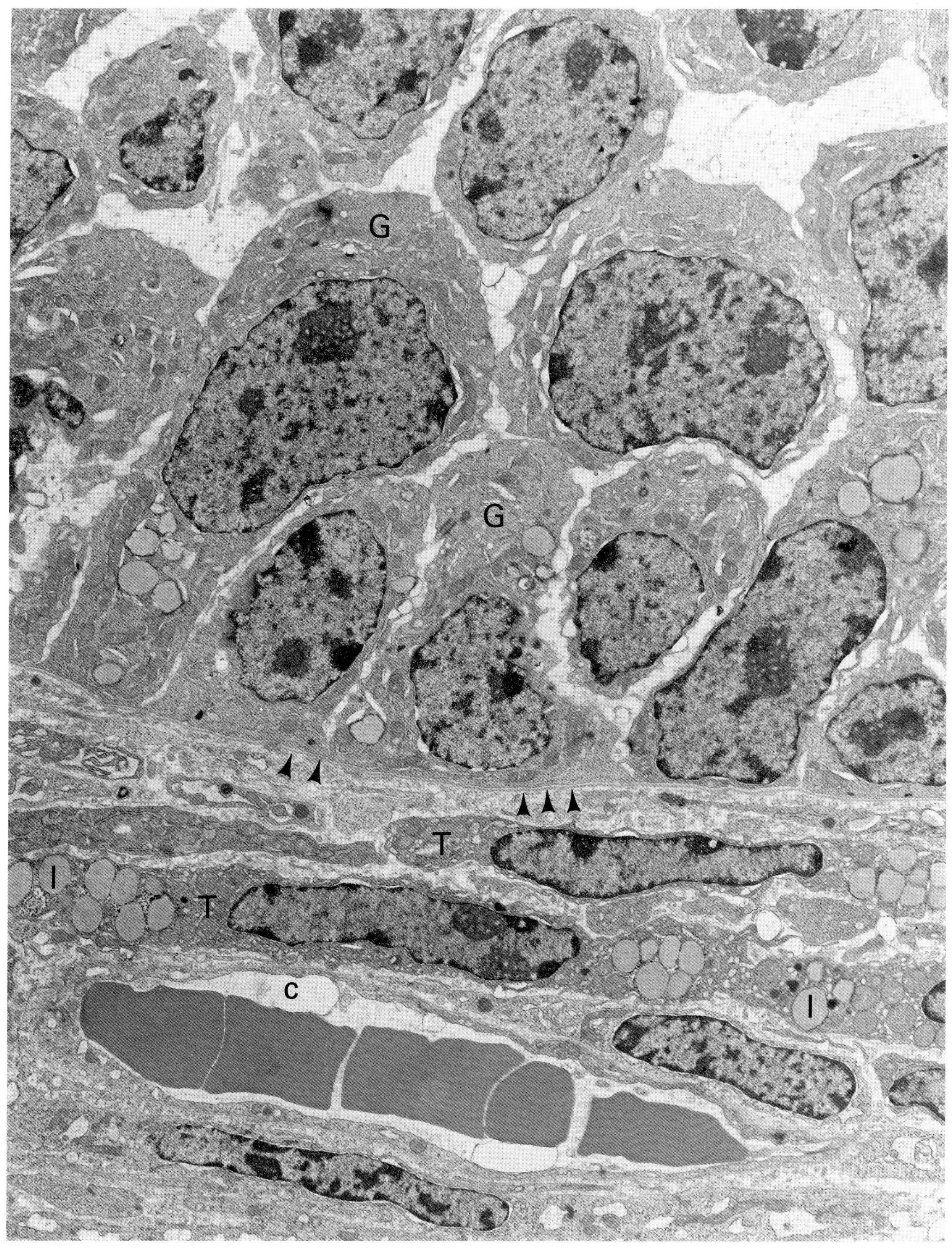

EM: 7,200×

OVARY: Secondary Follicle, Cumulus and Antrum

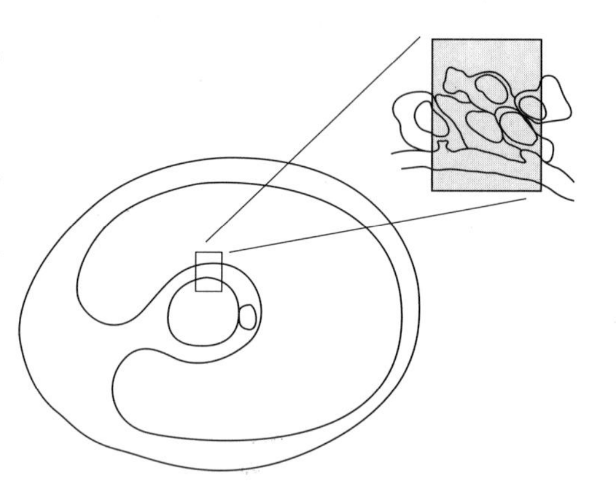

A mid-cycle LH peak induces the **resumption of meiosis** in the oocyte within the preovulatory follicle. Between the LH peak and ovulation the oocyte completes the first meiotic division to form the first polar body and continues in the second meiotic division to metaphase. During this process of maturation, the cumulus-oocyte mass maintains interconnections via adherens and gap junctions. Processes of **cumulus cells** (C, micrograph) extend through the zona pellucida (Z, micrograph) to attach to the maturing oocyte (O, micrograph). If the cumulus cells, including the inner layer, the **corona radiata,** are experimentally removed and the oocytes cultured in vitro, meiosis is resumed and the first polar body forms. However, the fertilizability of such apparently normal denuded oocytes is compromised. Therefore, the cumulus cells appear to be necessary for normal oocyte development, not only during the growth phase (see Female Reproductive System, page 360) but also during maturation directly prior to ovulation.

The **antrum** (A, micrograph) of secondary follicles begins as an accumulation of a viscous material between the granulosa cells. Initially, this material is predominantly proteoglycans synthesized by the adjacent granulosa cells. As the antral cavity increases in size, its contents become less viscous as a result of the addition of fluid via transudation from the capillaries in the theca interna. The composition of the antral fluid is similar to that of plasma, with essential contribution from the granulosa and theca interna cells, such as steroids and peptide hormones (e.g., inhibin) that act at local and distant sites.

The LH peak induces (in the single follicle that will undergo ovulation) (1) the synthesis of enzymes that weaken the basal lamina and the connective tissue wall, particularly at the site of future follicle rupture; (2) an increase in permeability of theca interna vessels and their subsequent growth across the basal lamina into the granulosa cell layers; and (3) changes in the steroidogenic potential of the granulosa cells as they transform into progesterone-producing luteal cells.

The follicular fluid of normal secondary follicles contains a higher concentration of estrogen than do atretic follicles. Oocytes obtained from follicles with a high estrogen : androgen ratio undergo fertilization in vitro at a higher rate than do oocytes obtained from follicles with a low estrogen : androgen ratio.

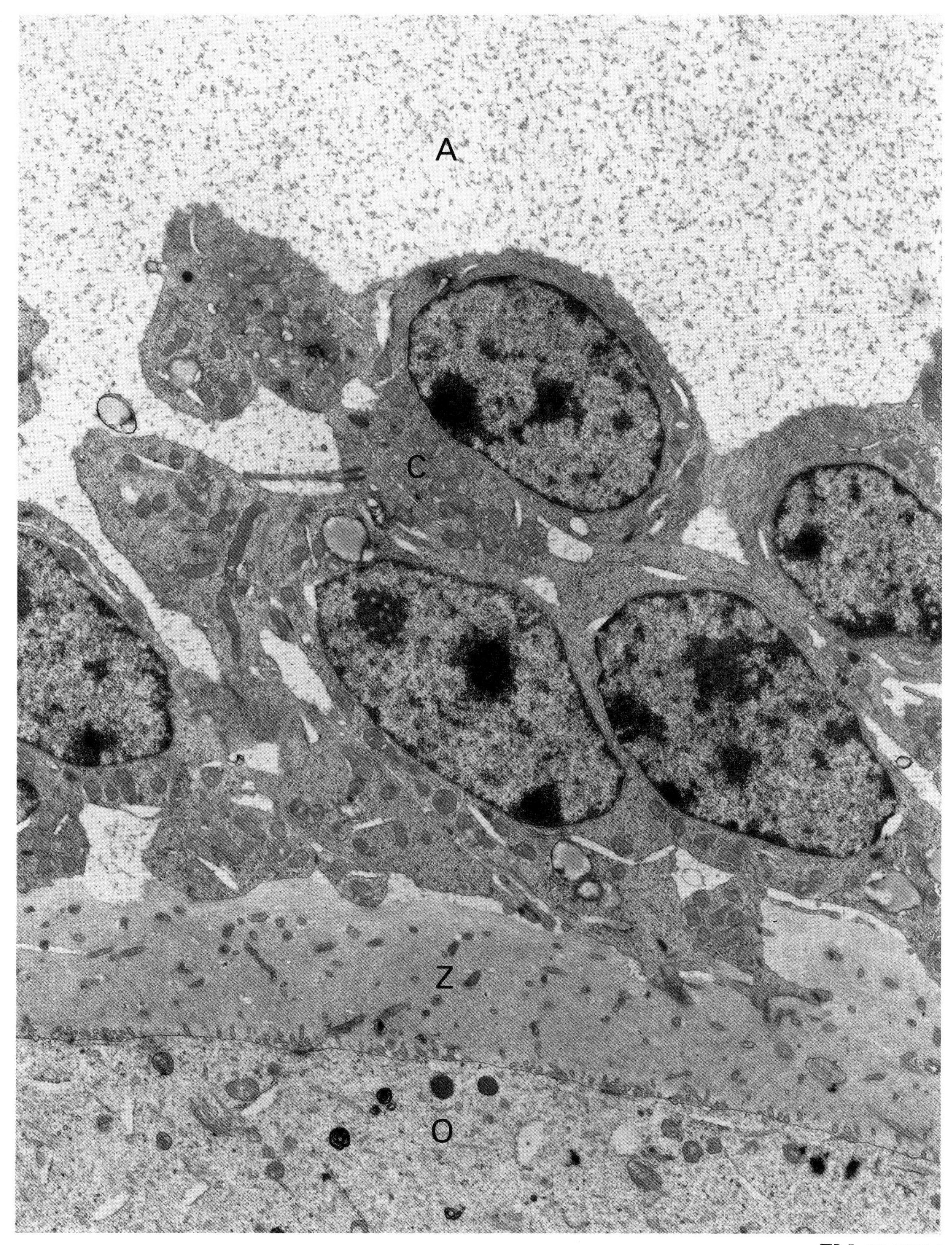

EM: 10,500×

OVARY: Corpus Luteum

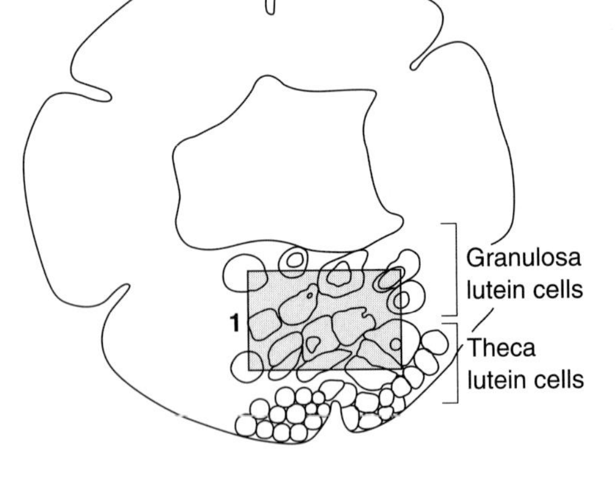

The wall of an ovulated follicle develops into a major endocrine gland, the **corpus luteum.** The basement membrane of the follicle breaks down and blood vessels (v, micrograph 1) invade the granulosa layer. This previously avascular tissue becomes a gland in which blood vessels occupy 20% of the volume of the gland and 60% of each cell surface.

Both theca interna and granulosa cells undergo transformation to become corpus luteum cells. Changes are most pronounced in the **granulosa lutein cells** (G, micrographs 1 and 2), which begin to synthesize and secrete large amounts of **progesterone.** Cholesterol, stored in **lipid droplets** (l, micrograph 2), is shuttled to **mitochondria** (m, micrograph 2), where the first and rate-limiting step in progesterone synthesis, the side-chain cleavage of cholesterol, is carried out. Mitochondria proliferate and the number of tubular cristae increases to accommodate the newly synthesized enzyme that catalyzes the cleavage. The classical picture of a steroid-synthesizing cell is completed by the presence of **smooth ER** (arrowheads, micrograph 2), where some of the subsequent steps in steroid synthesis occur. Progesterone, synthesized in the second half of the cycle, prepares the uterus for a possible implantation. At precisely the time that progesterone becomes the dominant steroid, five days after ovulation, the embryo enters the uterus.

Blood flow per gram of tissue to the corpus luteum is greater than to any other major gland, and represents 90% of ovarian blood flow during peak luteal activity. New capillaries bring low-density lipoprotein particles (LDL) containing cholesterol directly to the granulosa cells. The increased availability of LDL, along with newly acquired receptors for LDL, accounts for the cell's ability to take up and store large amounts of this progesterone precursor.

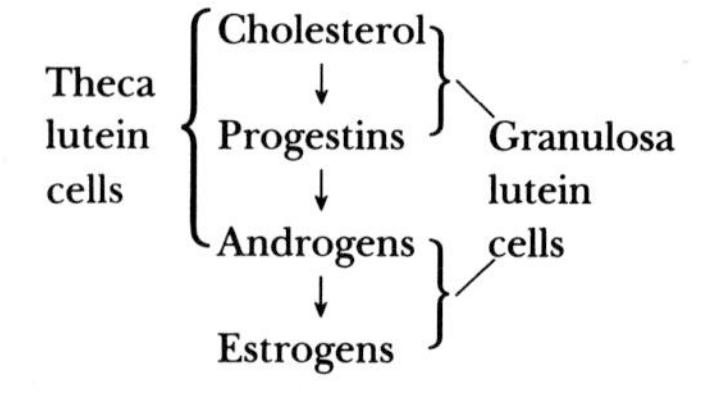

Estrogen is also synthesized in large quantities by the corpus luteum and is particularly important during pregnancy in the development and maintenance of the placenta and mammary gland. Because granulosa lutein cells still lack the enzyme required to convert progestins to androgen, theca lutein cells appear to cooperate with the granulosa lutein cells in the synthesis of estrogen by providing androgen precursors in a manner similar to that which occurs prior to ovulation. Theca lutein cells are smaller than granulosa lutein cells and form the thin outer layer of the corpus luteum.

During the regular cycle, the corpus luteum begins to degenerate on approximately day 23 (4 days after its peak activity), whereas a corpus luteum of pregnancy is maintained for several months, initially by pituitary LH and subsequently by placental human chorionic gonadotropin. At the end of eight weeks the placenta begins to take over the role of synthesizing steroids necessary for pregnancy and the preparation for lactation. The corpus luteum of the menstrual cycle and of pregnancy degenerates to form a scar, the corpus albicans. Some of these remain a lifetime.

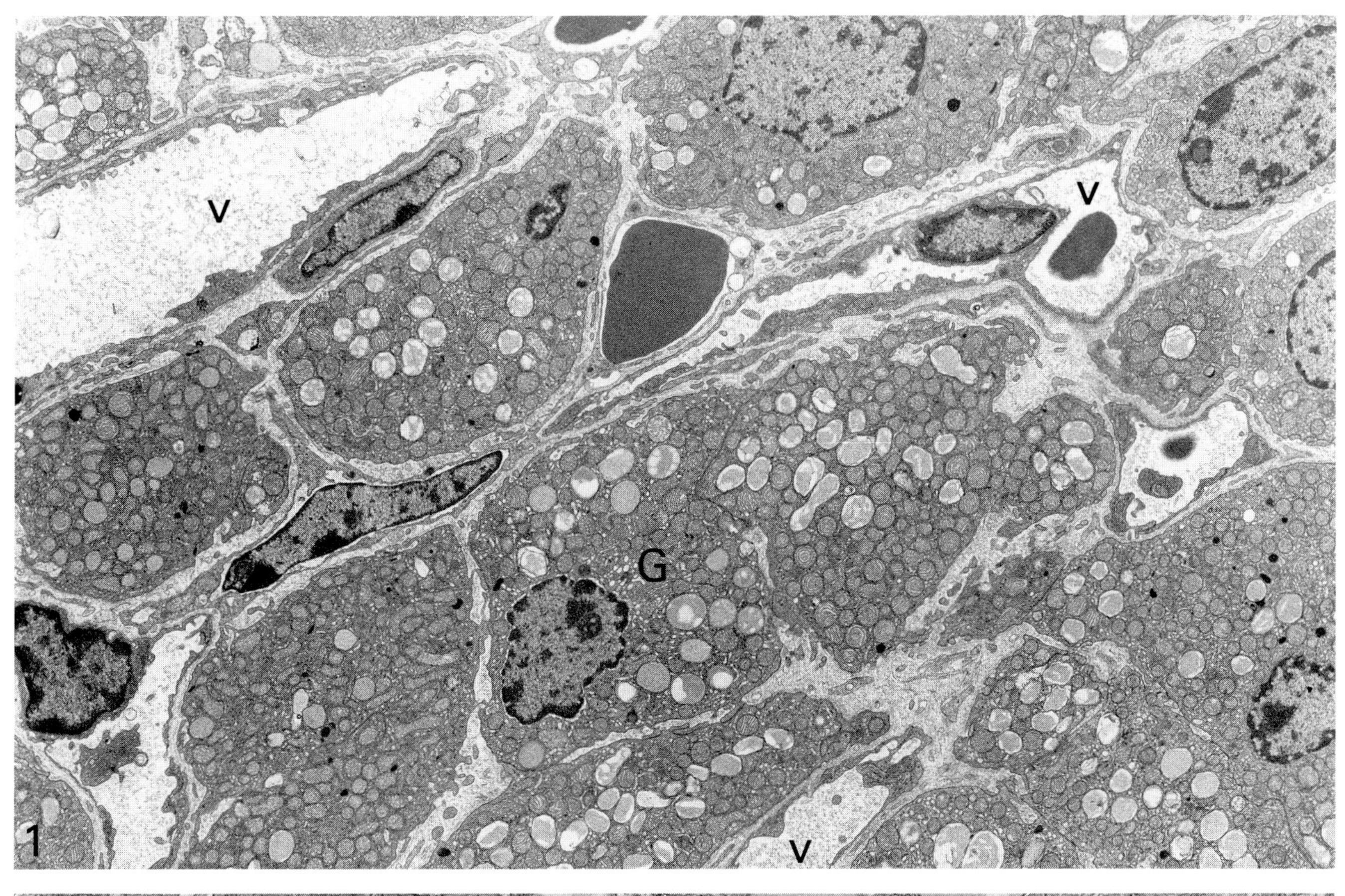

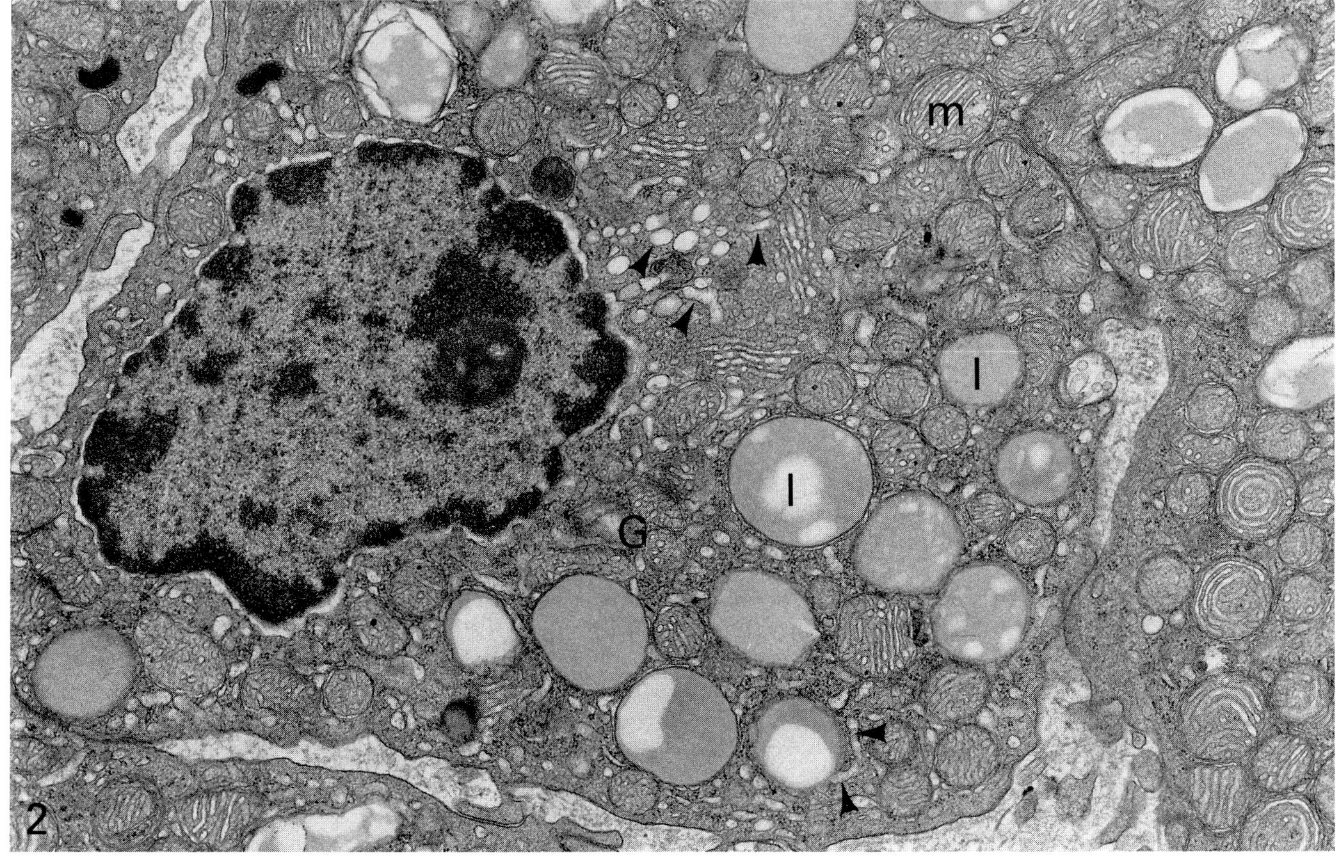

EM 1: 4,800× **EM 2: 18,900×**

Oviduct

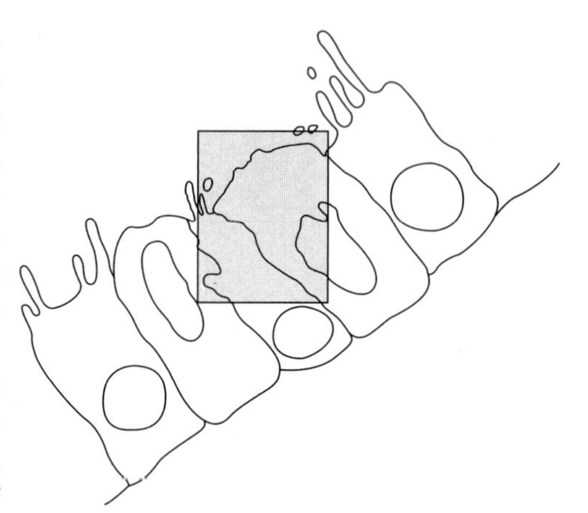

The **oviduct** (i.e., Fallopian tube), the muscular passageway between the ovary and uterus, is lined by a simple epithelium of **ciliated** (C, micrograph) and **secretory** (S, micrograph) **cells.** These two cell types are critical to survival of the gametes, fertilization, and development of the early embryo. This micrograph was obtained during mid-cycle, when oviduct epithelial cells are at their peak activity. Large numbers of polyribosomes (arrows, micrograph) reflect high levels of protein synthesis characteristic of this time in the menstrual cycle.

The **unattached polyribosomes** of the ciliated cells direct the synthesis of proteins used within the cell. Many of these are structural proteins and enzymes necessary to the formation and functioning of the cilia. In contrast to cilia in other parts of the body, the number, height, and activity of cilia in the female reproductive tract undergo extreme changes with changing steroid levels during the cycle. At the time of ovulation and estrogen dominance (micrograph),well-developed cilia mix and move oviduct secretions to optimize successful fertilization. Activity of the muscular wall, however, is apparently as important, if not more important, to gamete and early embryo transport. Most women with Kartagener's syndrome, who lack functional cilia, remain fertile.

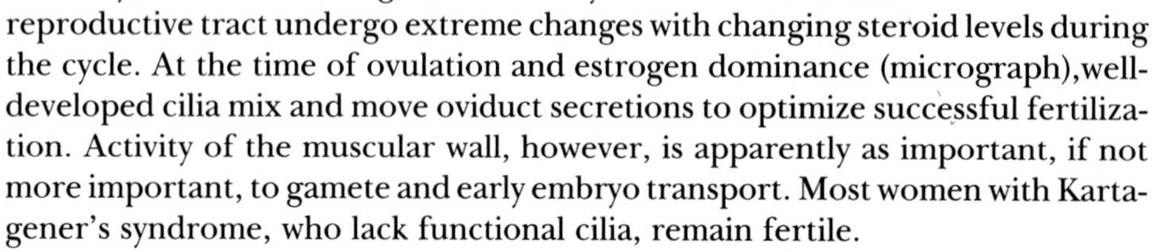

The **attached polyribosomes** of the secretory cells direct the synthesis of proteins destined for secretion. The synthesis of glycoproteins increases as estrogen levels rise prior to ovulation. These glycoproteins accumulate within secretory vesicles (v, micrograph). After ovulation, rising levels of progesterone appear to stimulate exocytosis.

The glycoproteins released at this time may be those that are incorporated into the zona pellucida following sperm penetration. These proteins may represent an important oviduct contribution to development. The difference in success rate between the low rate of in vitro fertilization in the absence of oviduct tissue (18% pregnancy rate), and the improved rate of success when recovered oocytes and sperm are placed in the oviduct (33% pregnancy rate), suggest that the oviduct makes a unique contribution to normal development.

Glycoproteins synthesized and secreted by the oviduct-lining cells are just one of many components of the oviduct fluid. In fact, the major source of secretion is the plasma in underlying capillaries. Metabolites such as pyruvate and glucose, proteins such as albumin and immunoglobulins, ions, and water move (or are moved) across the oviduct epithelium. While many of these substances, such as IgA and IgG, are moved through the secretory cells themselves, others, such as ions and water, pass to the lumen through the unusually leaky tight junctions (circle, micrograph) between epithelial cells. The permeability of the blood vessels and the contribution of plasma to oviduct secretion is maximum at mid-cycle. Precise determination of the components of oviduct fluid is difficult since fluids from the ovulated follicle, peritoneal cavity, and uterus all mix with oviduct secretions to varying degrees.

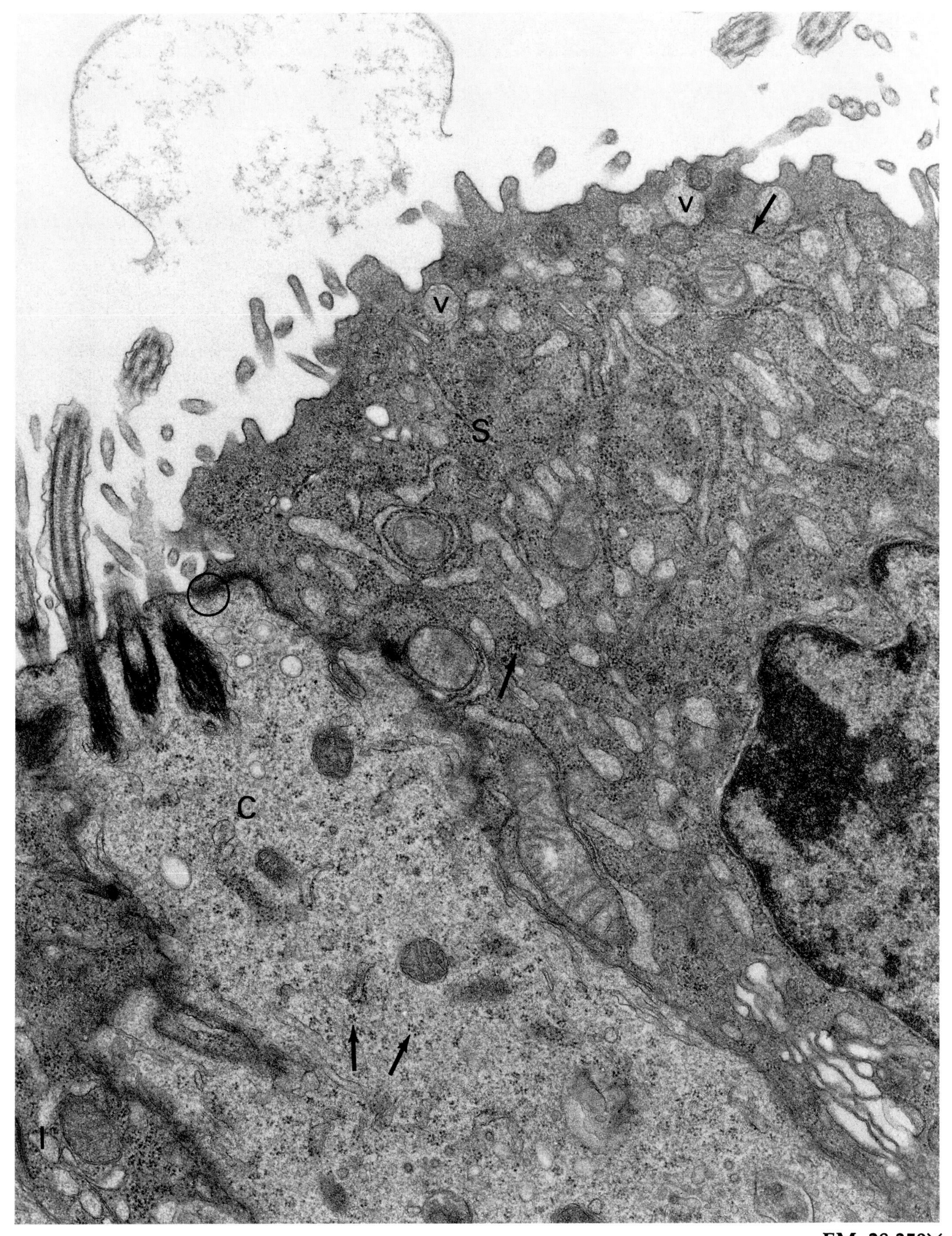

EM: 28,350×

UTERUS: Endometrial Epithelium

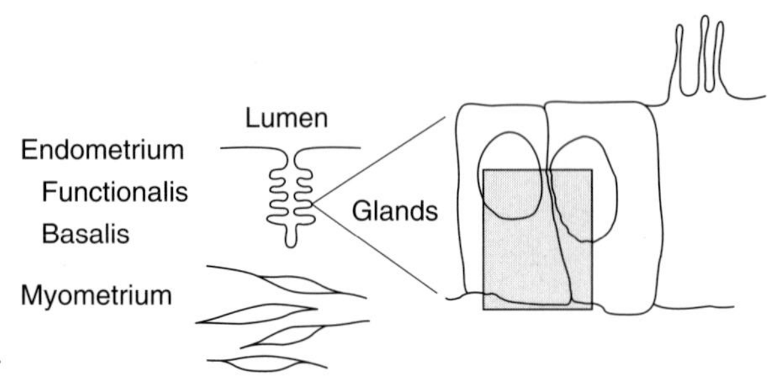

The function of the endometrium is to support implantation and development of the embryo. During each menstrual cycle the most superficial layer, the **functionalis,** undergoes dramatic changes in preparation for these events. If fertilization and implantation do not occur, the functionalis degenerates. Degeneration is a normal, common event in many parts of the female reproductive system (e.g., atresia in the ovary). However, what is striking in the uterus is the degree of tissue loss. The entire functionalis layer is lost every 28 days with menstruation.

Endometrial changes, like other changes within the female reproductive tract, are brought about by variations in the secretion of ovarian estrogen and progesterone. During the first half of the cycle, the proliferative phase, rising estrogen levels stimulate the division of epithelial and stromal cells in the functionalis. The uterine lining is reconstituted by the time of ovulation (day 14). During the second half of the cycle, the secretory phase, the endometrial cells undergo a characteristic differentiation in response to rising levels of progesterone.

Endometrial changes following ovulation occur initially in the **glandular epithelium** (micrograph, courtesy of Dr. Luciano Zamboni) and subsequently in the stroma. Beginning as early as day 15, **glycogen** (arrows, micrograph) appears in the basal region of epithelial cells and displaces the nuclei apically. This appearance of glycogen in uterine biopsies is used as an important diagnostic tool in fertility studies to signify that ovulation has occurred. By day 18, the glycogen has dispersed, nuclei have returned to a basal position, apical Golgi are prominent, and secretion is maximal. Early in the secretory phase, the nuclear envelope indents to form a **channel system** (arrowheads, micrograph) that is typically associated with the nucleolus. This system may facilitate a rapid transfer of ribosomal components between the cytoplasm and nucleus. Uterine secretions during the early secretory phase contain significant amounts of glucose and certain specific glycoproteins secreted by the epithelium, for example PP14, that may participate in local immunosuppression in preparation for contact with a "foreign" embryo.

Oxytocin is secreted locally by endometrial epithelial cells and seems a likely candidate for the induction of contraction by myometrial smooth muscle during labor. The existence of such a local mechanism could account for the paradoxical low levels of circulating oxytocin during the onset of labor.

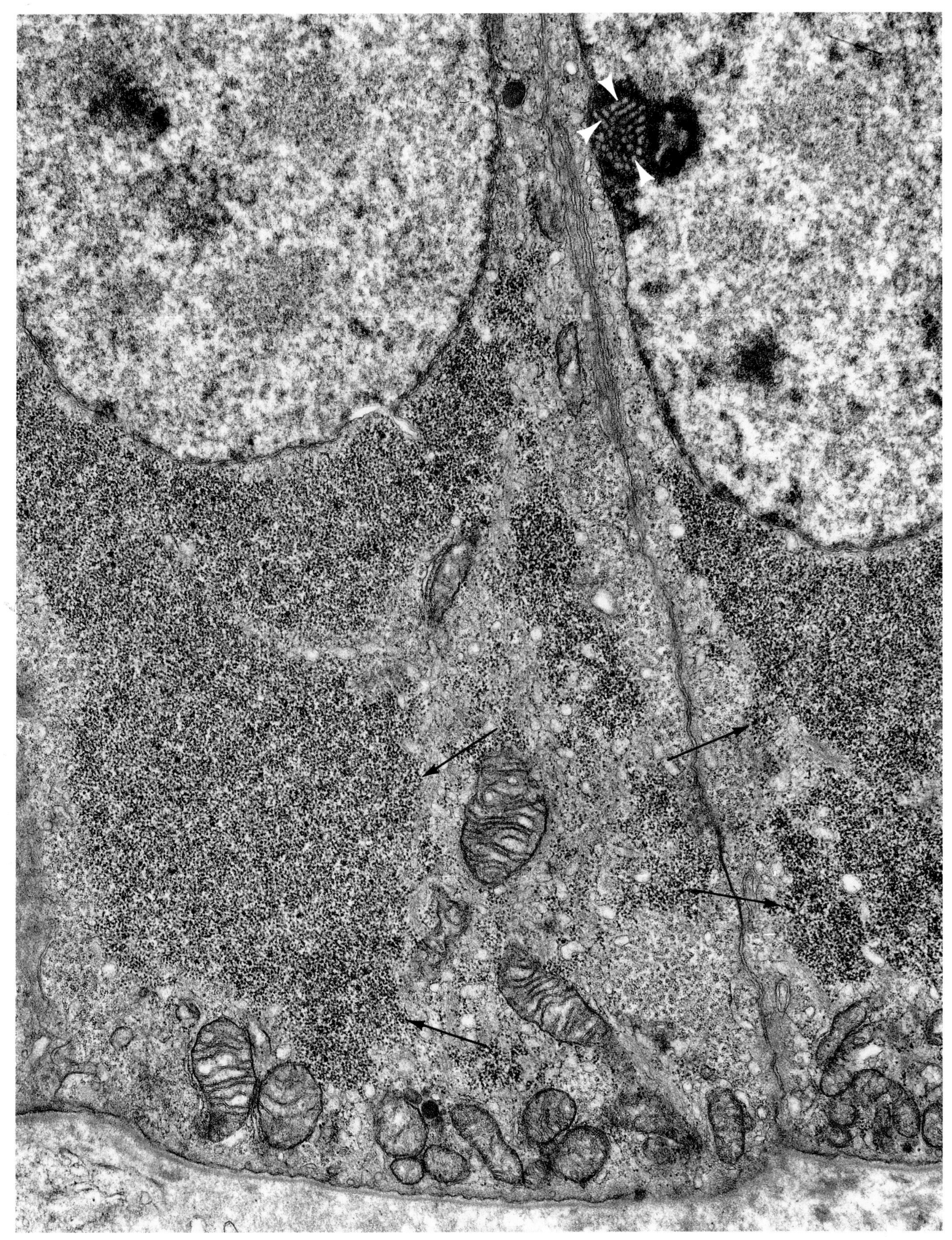

EM: 15,000× (estimated)

In most species implantation induces a **decidual response** that is characterized by pronounced changes in the endometrial stroma (micrograph, from rodent). Fibroblastlike cells transform into large, active **decidual cells** (D, micrograph) that become an important component of the maternal portion of the placenta. In humans, **predecidual cells,** considered to be forerunners of the decidual cells, appear in the stroma during every fourth week of the menstrual cycle. As they develop, predecidual cells form a cuff around small vessels (v, micrograph) in the stroma. The vessels become more permeable as menstruation or placental development approaches.

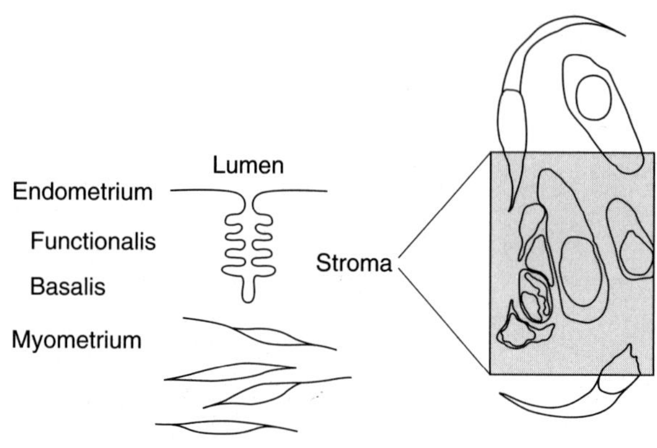

Predecidual cells appear to limit embryo invasion, play a role in embryo nutrition, and protect fetal tissue from rejection. These cells produce prolactin (and possibly relaxin), secrete prostaglandins, and have receptors for both estrogen and progesterone. It is known that the effects of estrogen and progesterone on the endometrium, both during the cycle and following implantation, are complemented (and implemented) by a variety of growth factors. Insulinlike growth factors (IGFs) have a major role in the stimulation of endometrial cell division. With rising levels of progesterone after ovulation, IGF-binding proteins, including the placental protein PP12 synthesized by the predecidual cells, are secreted. IGF-binding proteins reduce the availability of IGFs and thus play a role in the shift from a proliferative to secretory endometrium at this time.

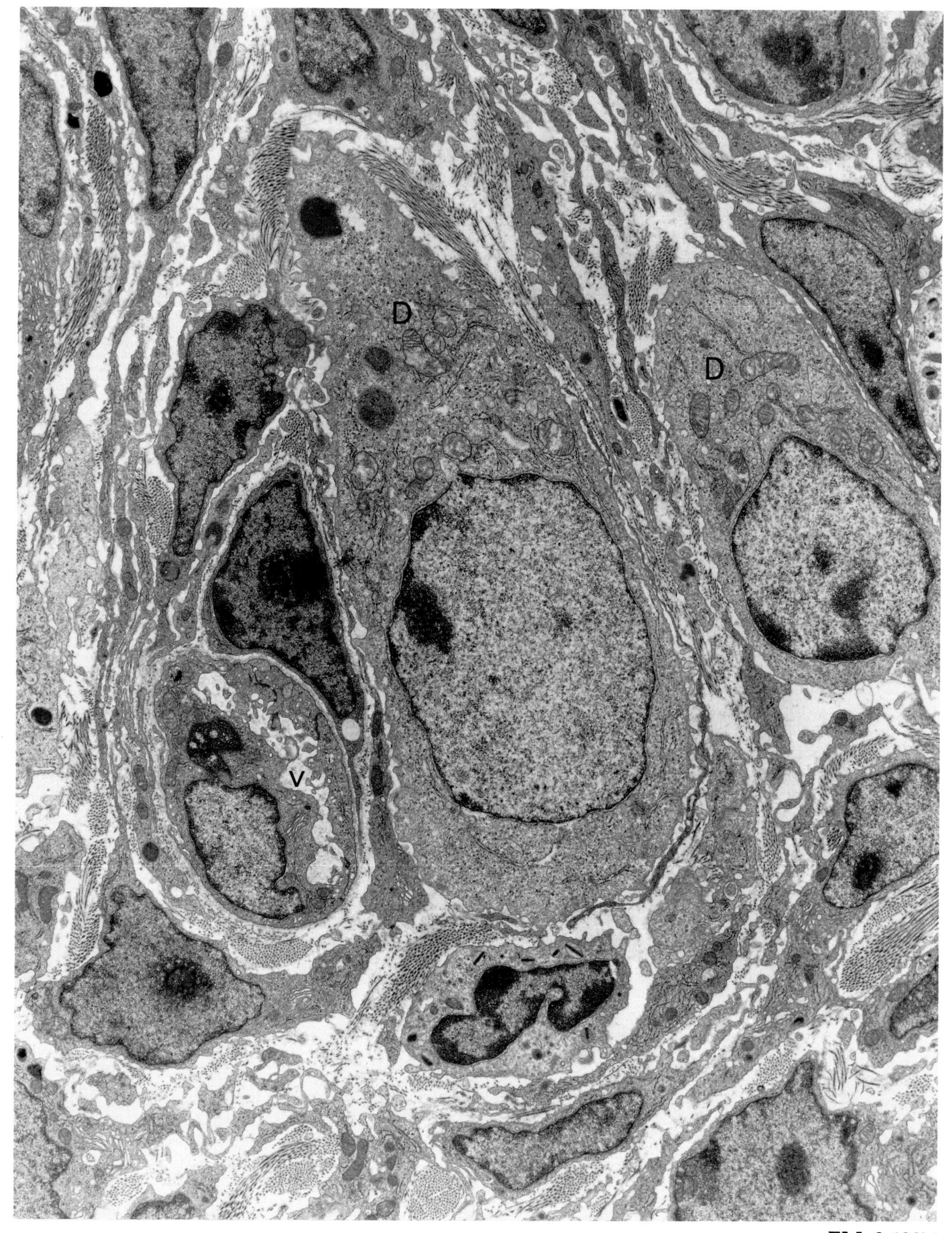

EM: 8,400×

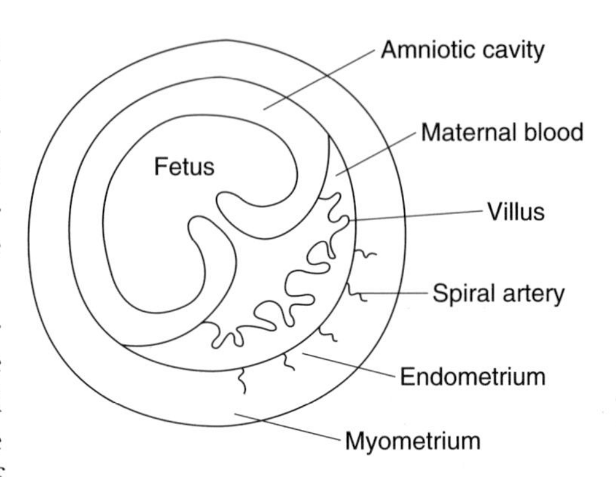

The **placenta** is an organ made up of **maternal** and **fetal tissues** that work together for the survival of the fetus. The functional units of the placenta are the villi, which are finger-like projections of fetal tissue that extend into a lacuna (space) within the endometrium. Maternal blood released under pressure from endometrial spiral arteries fills the lacuna. As villi are formed, fetal blood vessels develop within their core.

In the micrograph, **maternal blood** (M) is shown adjacent to the surface of a **villus.** The fetal villus at this stage consists of an outer epithelial lining (note the thick epithelial basal lamina, b, micrograph) covering underlying mesenchyme that contains cells (C), collagen (c), and fetal vessels (F). One of the fetal vessels contains several mature fetal red blood cells (R) and one reticulocyte (Re).

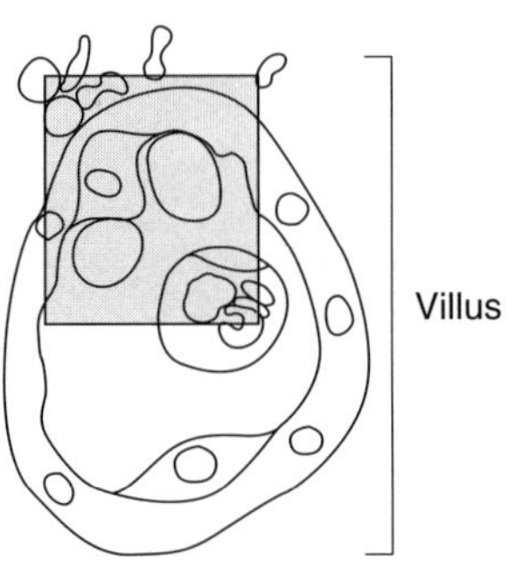

The villous epithelium consists of two cell layers, the **syncytiotrophoblast** (S, micrograph) adjacent to the maternal circulation, and a **cytotrophoblast** (Cy, micrograph) of underlying individual cells. During pregnancy, the surface area of the villi increases dramatically as a result of division of cytotrophoblast cells and their differentiation into syncytiotrophoblast. In this late placenta, most of the cytotrophoblast has become syncytiotrophoblast, leaving only a few individual cytotrophoblast cells that form a discontinuous basal layer.

Exchange of nutrients and wastes between maternal and fetal circulation takes place across two cell layers, the syncytiotrophoblast and the endothelium (E, micrograph) of the fetal blood vessels. The complexity of this bidirectional transport is a reflection of the function of these layers as the equivalent of at least three organ systems: respiratory, gastrointestinal, and urinary. The mechanism of transport is extremely varied, ranging from the simple diffusion of gases to many types of receptor-mediated transport. Receptors within the apical microvilli (arrows, micrograph) facilitate the transport of glucose, the active transport of amino acids, and the special shuttle that carries IgG directly from maternal blood to fetal blood.

Villi express foreign paternal antigens and rest directly adjacent to maternal blood. This fetal tissue, as an allograft, is not typically rejected in this environment, even though a maternal immune response occurs. The type of fetal expression (e.g., MHC I, but not MHC II; low antigen density) and maternal response (suppressor cells and molecules) all contribute to the complexity of this unique tolerance. The absence of MHC II may be particularly significant, since these antigens have been implicated in the rejection of other organ allografts.

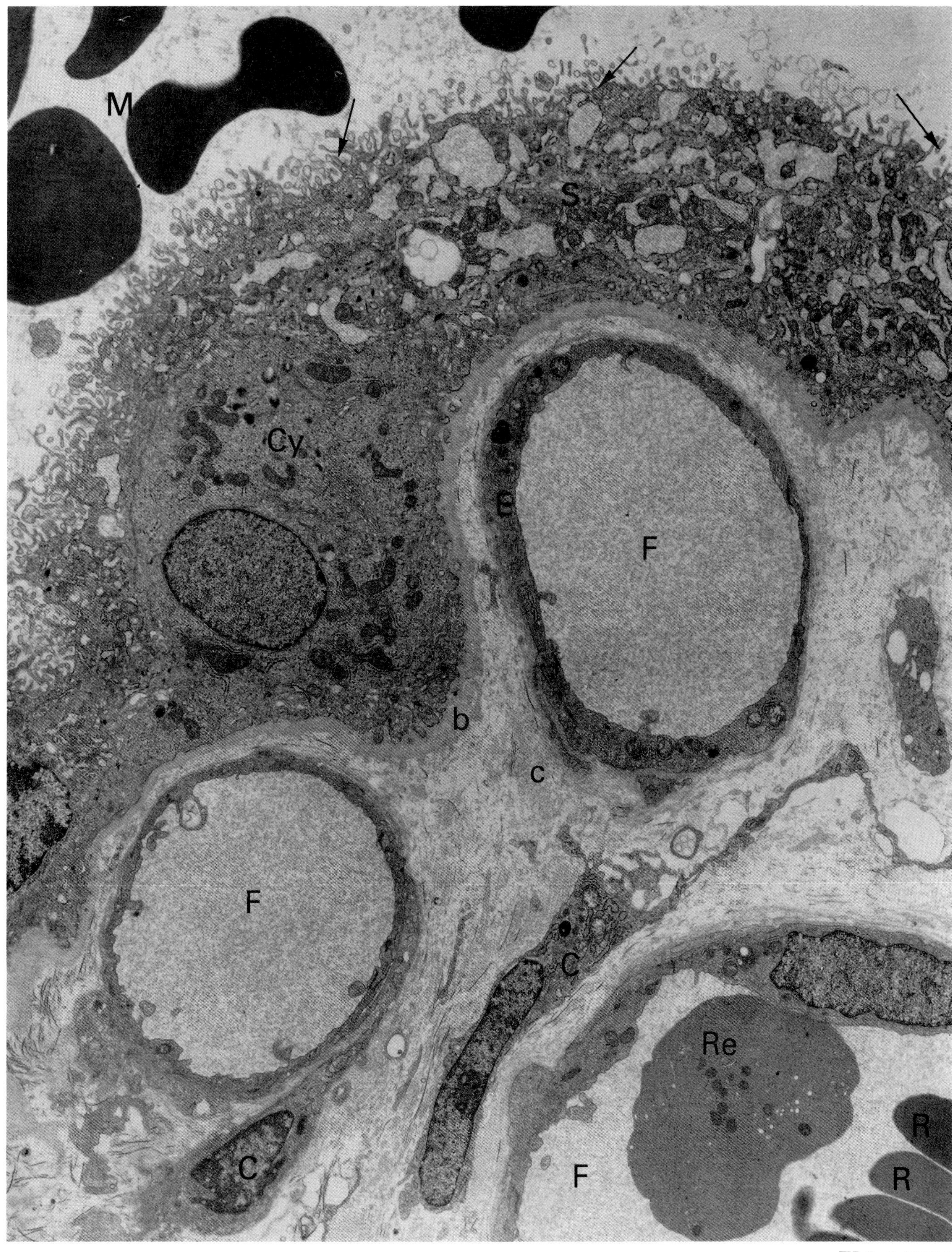

EM: 6,600×

PLACENTA: Hormone Production

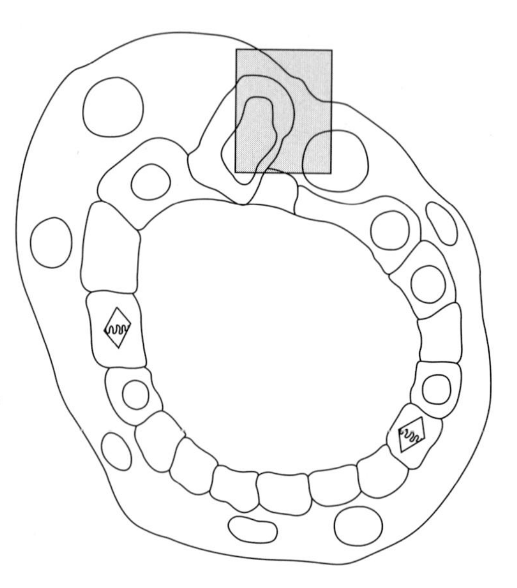

From as early as one day after implantation, the fetal villi begin to assert control over maternal physiology to create an optimal environment for fetal development. This control is effected by hormones synthesized primarily by the **syncytiotrophoblast** (S, micrograph). Both **protein and steroid hormones** are synthesized and released at the apical surface of this cell into maternal blood (M, micrograph). These hormones act to maintain both the quantity and nutritional content of the maternal blood supply.

The syncytiotrophoblast synthesizes **human chorionic gonadotropin** (HCG) immediately after implantation. This glycoprotein hormone, which has a subunit identical to that of luteinizing hormone and follicle-stimulating hormone, acts on the corpus luteum to stimulate estrogen and progesterone synthesis. In the micrograph, obtained during the first trimester, the rough ER (arrows), where HCG is synthesized, is predominant.

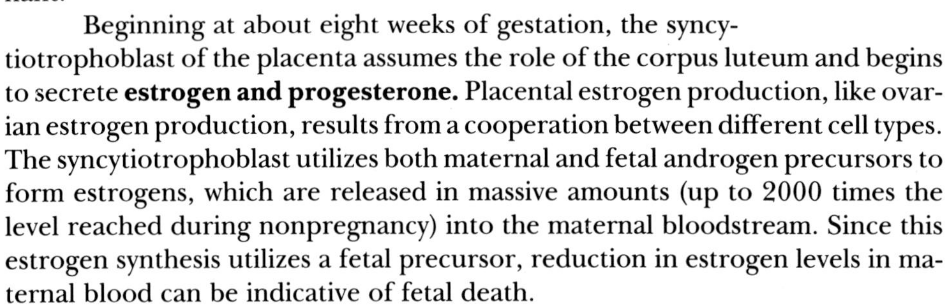

Beginning at about eight weeks of gestation, the syncytiotrophoblast of the placenta assumes the role of the corpus luteum and begins to secrete **estrogen and progesterone.** Placental estrogen production, like ovarian estrogen production, results from a cooperation between different cell types. The syncytiotrophoblast utilizes both maternal and fetal androgen precursors to form estrogens, which are released in massive amounts (up to 2000 times the level reached during nonpregnancy) into the maternal bloodstream. Since this estrogen synthesis utilizes a fetal precursor, reduction in estrogen levels in maternal blood can be indicative of fetal death.

Placental progesterone, like corpus luteum progesterone, is synthesized from cholesterol obtained primarily from circulating low-density lipoprotein (LDL). Membranes of the microvilli (arrowheads, micrograph) provide surface area for receptors that bind LDL. LDL is initially shuttled to lysosomes and broken down by acid hydrolases. Freed cholesterol is then transported to mitochondria (m, micrograph), where it is initially acted upon by enzyme complexes localized within the characteristic tubular cristae. In contrast to cholesterol, which is used within the cell, other molecules released within the lysosomes, such as essential fatty acids and amino acids, are transported to the capillaries of the villi and provide critical nutrition for the fetus.

In addition to HCG, another protein hormone, **human chorionic somatomammotropin** (HCS), has been localized to the rough ER of the syncytiotrophoblast and increases progressively during pregnancy. HCS is very similar to growth hormone and has effects on carbohydrate, fat, and protein metabolism of the mother. It increases the utilization of fatty acids by the mother, leaving available glucose for the fetus. In addition, HCS has a major effect, in conjunction with prolactin, on development of the mammary gland.

The cytotrophoblast (C, micrograph) is tightly associated with the syncytiotrophoblast by desmosomes (circle, micrograph) and exerts control over many hormonal activities of the syncytiotrophoblast. This interaction is particularly interesting since the cytotrophoblast is the stem cell of the syncytiotrophoblast.

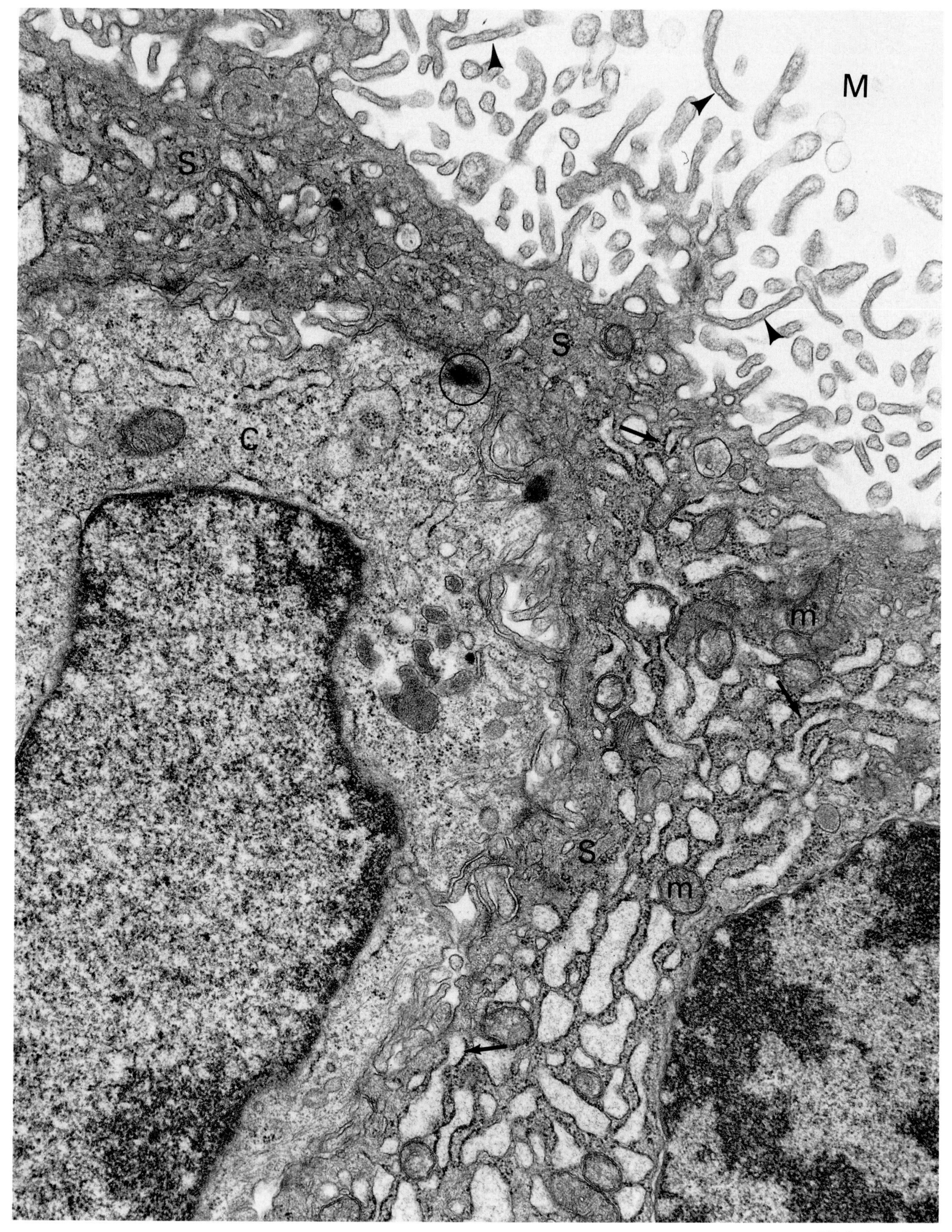

EM: 24,500×

Mammary Gland

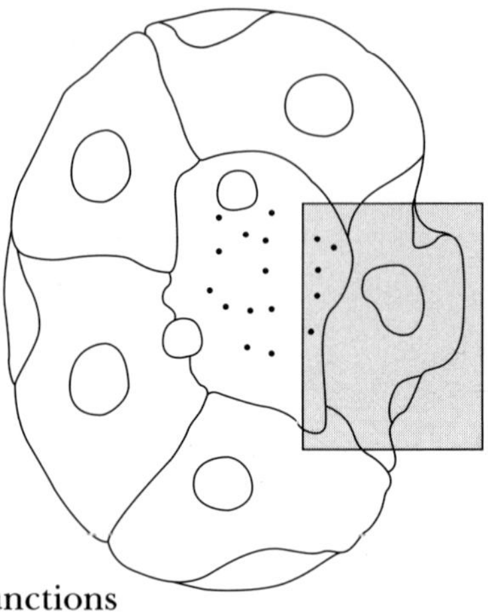

The **mammary gland** is an exocrine gland highly specialized for the production of milk. Milk is synthesized and secreted by a single layer of epithelial cells arranged as alveoli at the termination of the duct system leading to the nipple. Synthesis begins in the third trimester of pregnancy, and this early secretion, colostrum, accumulates in the lumen of the alveoli. Colostrum is the first milk released after birth, and its high concentration of IgA provides the newborn with an important antiseptic coat for the gastrointestinal tract. IgA, together with IgG absorbed across the placenta from maternal blood, provides the newborn with a passive immunity that affords protection until the newborn's immune system matures.

Individual **alveolar cells** (A, micrograph 1) are attached to one another by junctional complexes (circles, micrograph 1). At birth, with the dramatic drop in progesterone levels, the permeability of tight junctions decreases, a change in cellular metabolism occurs, and different proportions and types of milk components are released from the alveolar cells as true milk synthesis begins.

1

Ninety percent of milk **lipids** are triglycerides, the primary energy source of the newborn. Triglycerides accumulate as droplets (l, micrograph 1 and inset) that can become as large as a nucleus. Droplets move to the surface of the cell and are released into the secretion with a covering of cell membrane (arrowheads, inset). The glycocalyx coat of this cell membrane may act as a site of attachment to the newborn intestine to facilitate digestion and absorption. The type of fatty acid incorporated into the droplet is a significant variable in newborn nutrition. The type not only is different between different sources (uptake from maternal circulation or de novo synthesis by alveolar cells) but also is widely variable with differences in maternal diet.

At birth, the **protein** content of milk shifts from proteins associated with defense (IgA, lactoferrin) to those high in nutritional value (caseins, α-lactalbumin). Not only are caseins a source of many essential amino acids, but they also bind and carry phosphorus and calcium into the secretion. Caseins are synthesized on rough ER (arrows, micrograph) and processed in the Golgi (G, micrograph). A defined calcium arrangement with caseins forms densely packed globular micelles (m, micrograph 1 and inset) that develop in the secretory vesicles and, following exocytosis, maintain their integrity.

Lactose, the principal milk **carbohydrate,** is synthesized by an enzyme complex that is a part of the trans Golgi and secretory vesicle membranes. Alpha-lactalbumin, important as a nutritional protein, is also a critical factor in this enzyme complex. As lactose is synthesized, it remains dissolved within the secretory vesicle and creates an osmotic gradient that draws water into the vesicle. As secretory vesicles move to the cell surface, they become larger as they accumulate fluid. Lactose is dissolved in the solution surrounding the casein micelles and is released along with casein upon exocytosis. The galactose component of lactose is an important component of myelin and necessary for the development of the nervous system of the newborn.

Milk, synthesized and secreted primarily in response to prolactin from the anterior pituitary, is pushed out of the ducts only when **myoepithelial cells** (M, micrograph) stimulated by oxytocin from the posterior pituitary contract. A suckling reflex acts to increase both prolactin and oxytocin release.

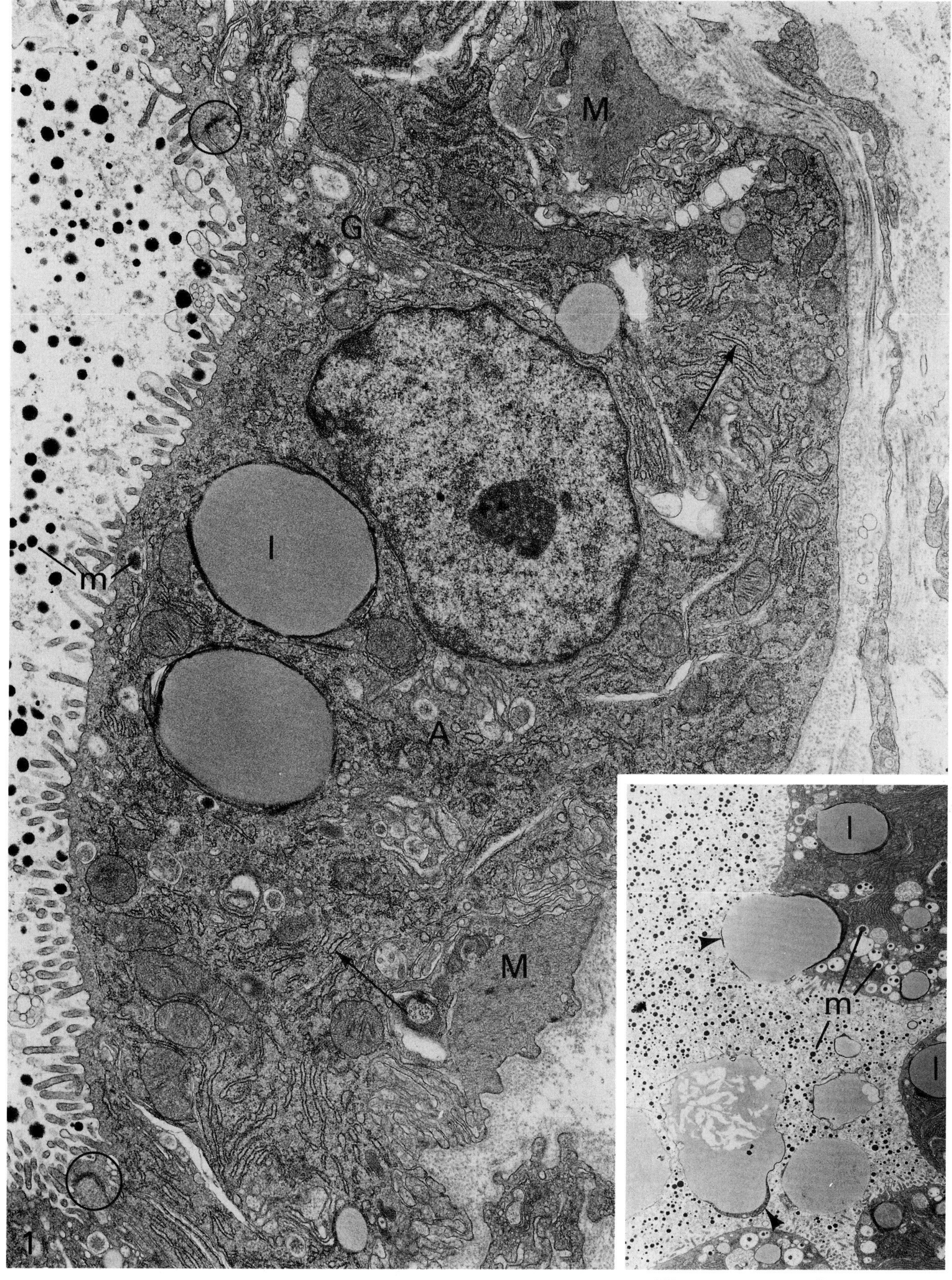

EM 1: 13,050 Inset: 3,900×

SENSORY REGIONS

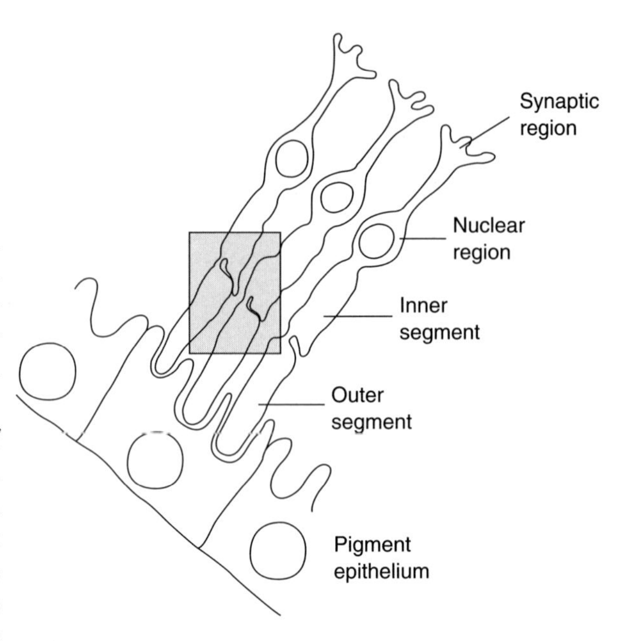

Photoreceptor cells, **rods and cones,** are concentrated in specific regions of the neural retina. Rods occupy the periphery and function in peripheral and night vision; cones are concentrated in the central fovea and function in the perception of detail and color. Both are elongated cells with a highly specialized polarity that includes, from basal to apical, a synaptic region, nuclear region, inner segment, and outer segment. Outer segments (closer to the outer surface of the eye) are connected to inner segments by a thin process containing a modified cilium (c, micrograph).

The **outer segments** are packed with layers of flattened membrane discs that are the sites of visual excitation. In rods (micrograph) the membranes form separate disclike sacs (arrowheads, micrograph) that are not connected to the outside plasma membrane except at the base near the cilium. Components of the disc membranes are synthesized in the inner segment

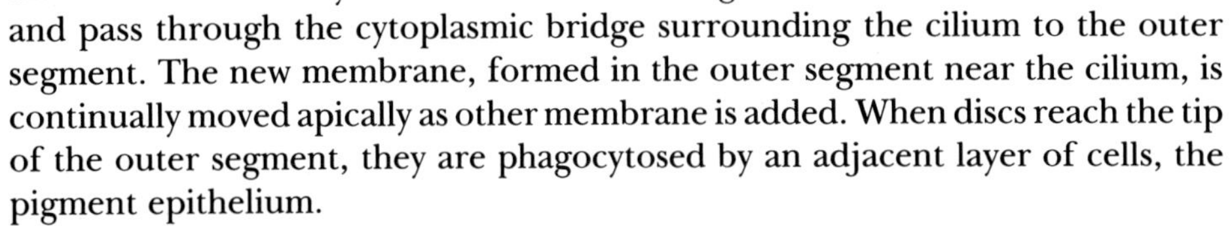
and pass through the cytoplasmic bridge surrounding the cilium to the outer segment. The new membrane, formed in the outer segment near the cilium, is continually moved apically as other membrane is added. When discs reach the tip of the outer segment, they are phagocytosed by an adjacent layer of cells, the pigment epithelium.

Visual pigments are concentrated in the membranes of the outer segment. In both rods and cones, 11-*cis* retinal, a derivative of vitamin A, is the part of the pigment that is light sensitive. The 11-*cis* retinal is attached to a protein or opsin that is altered as a part of the cascade of visual excitation. All rods contain the visual pigment rhodopsin and respond to low-intensity light of several wavelengths. In contrast, there are three types of cones, each with a different opsin that responds preferentially to light of red, blue, or green wavelengths. At least 80% of rod and cone disc membrane proteins are opsins.

In the dark, rhodopsin is magenta. Light bleaches rhodopsin by changing the configuration of 11-*cis* retinal to all-*trans* retinal. During this conversion opsin undergoes a conformation change, which leads to an amplified cascade involving an intermediate protein, transducin, and cGMP phosphodiesterase. The cascade ends with a reduction in the level of cGMP and a subsequent closure of Na^+ channels in the outer plasma membrane. The resulting hyperpolarization reduces the rate of release of neurotransmitter in the synaptic terminal part of the rod. This activates bipolar neurons in an adjacent layer and thus carries the excitation to the next level in the visual pathway. Energy requirements of the photoreceptor cells are met by mitochondria (m, micrograph) concentrated in the inner segment.

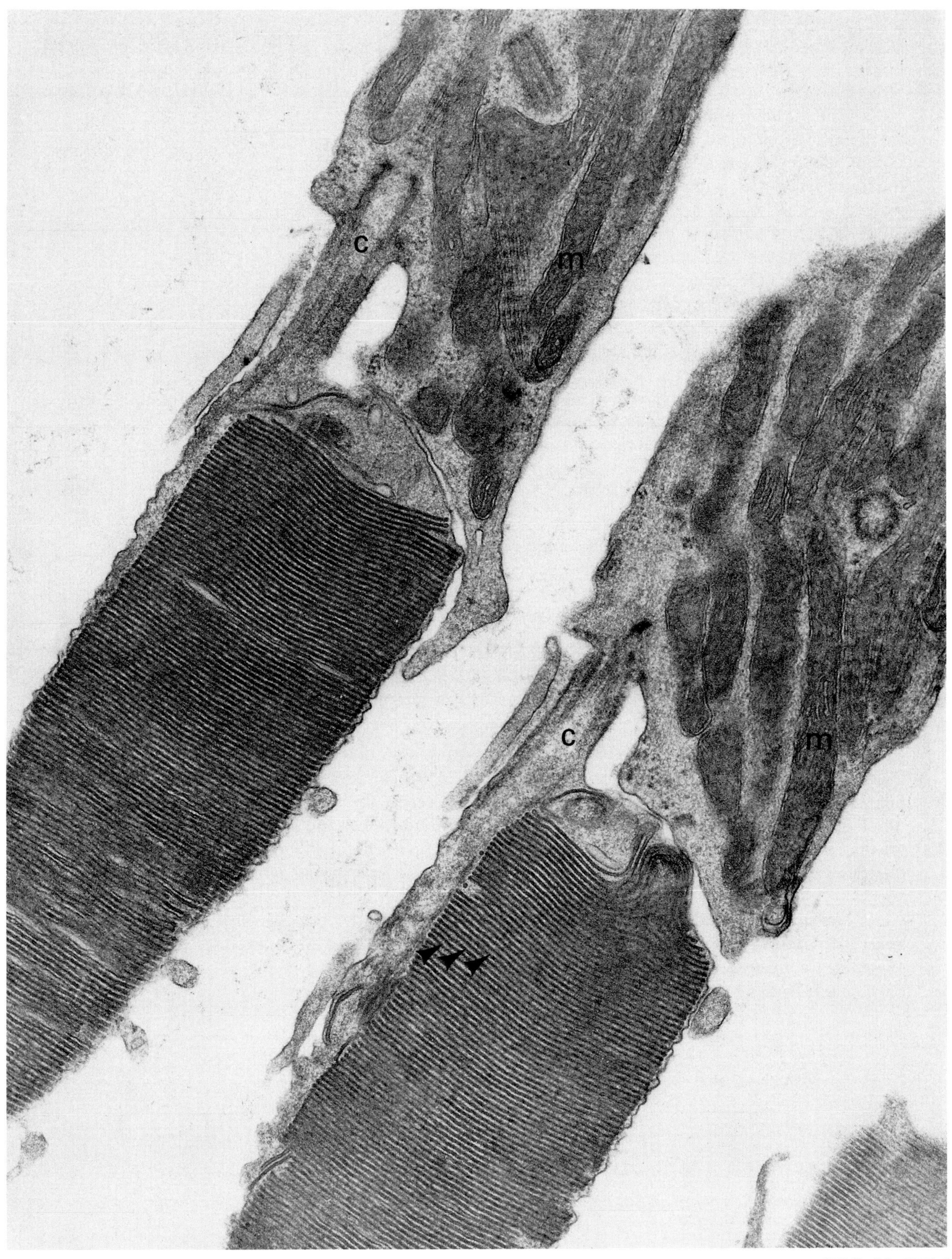

EM: 32,400×

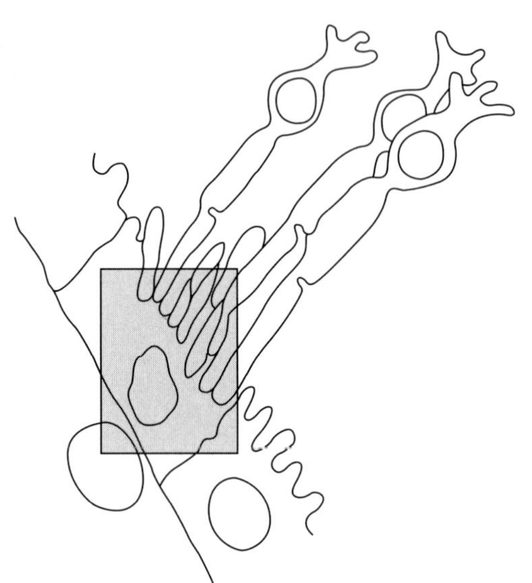

The **pigment epithelium** (PE, micrograph) consists of a single layer of cuboidal cells that extend narrow apical cytoplasmic processes (arrows, micrograph) around and between the outer segments (os, micrograph) of the photoreceptors. A major function of this epithelium is the phagocytosis of the outer segments. Phagosomes containing the outer segments fuse with lysosomes (ly, micrograph), where the process of digestion occurs. Occasionally lysosomes (*, micrograph) are found within the fine pigment epithelium processes directly adjacent to the outer segments. Outer segments, digested in secondary lysosomes, leave residual lamellar structures (rl, micrograph). Phagocytosis of rod and cone outer segments varies diurnally, with rod phagocytosis occurring one half hour after the onset of light and cone phagocytosis occurring immediately after dark. It has been estimated that each pigment epithelium cell phagocytoses 2000–4000 discs daily.

The pigment epithelium is essential to the regeneration of the visual pigment and thus plays a major role in the completion of the visual cycle. The all-*trans* chromophore that results from photoisomerization in the outer segment membranes is continuously delivered to the pigment epithelium and converted back to the 11-*cis* form by an isomerase located within the membranes (most likely the extensive smooth endoplasmic reticulum, arrowheads, micrograph) filling the pigment epithelium cytoplasm. 11-*cis* retinal is then recycled back to the rods to be combined with opsin in newly formed membrane. The pigment epithelium also converts dietary vitamin A (all-*trans* retinol) to the 11-*cis* form that is delivered to the photoreceptors. Night blindness is one outcome of vitamin A deficiency.

The retina has a **dual blood supply.** Capillaries (c, micrograph) of the choroid layer underlying the pigment epithelium supply the outer one third, including the rod and cone cells. Branches of the retinal artery course over the interior surface and supply the inner two thirds of the retina. In instances in which the neural retina detaches and loses the choroid blood supply, the retinal vessels become essential to survival of the photoreceptor cells. Retinal detachment in the fovea, where retinal arteries are absent, results in irreparable damage.

Melanin granules are an important component of pigment epithelium and act to absorb excess light to reduce interference and reflection. Melanin granules are present in the micrograph on page 387 but not in the facing micrograph from an albino rat.

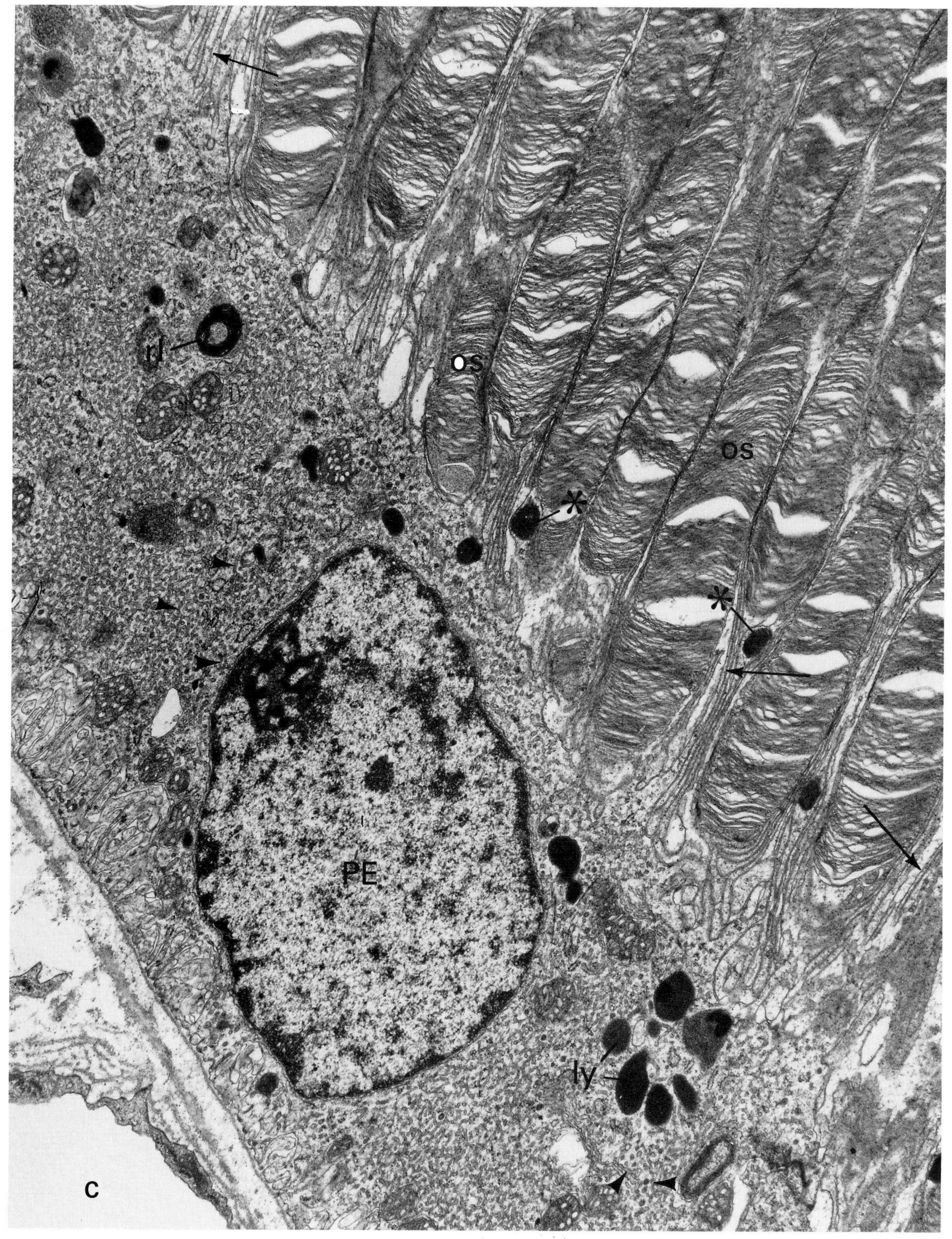

EM: 13,600×

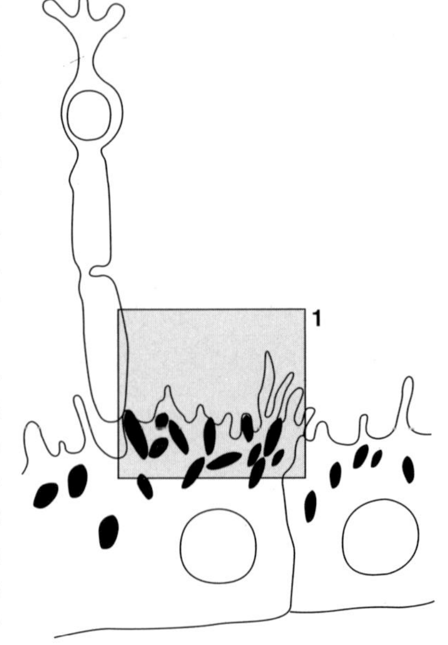

Micrograph 1 depicts a **pigment epithelium** cell in which **melanin granules** (g) pack the apical cytoplasm and the fine extensions (arrows) that normally surround the rod outer segments. In this preparation only a single rod outer segment (R) is found in the vicinity of the epithelium. Photoreceptor outer segments and pigment epithelium frequently separate during tissue preparation. Such separation is dramatically illustrated in scanning electron micrographs 2 and 3, in which the outer segments were left dangling free (micrograph 2) and pigment epithelium processes extend out, released from their attachment to the outer segments (micrograph 3).

In vivo, a tight association between the pigment epithelium and photoreceptor cells is essential for vision to occur; rhodopsin is not regenerated if close contact is not maintained. This association, as necessary as it is, is the most tenuous in the eye. During the development of the eye the pigment epithelium and the neural layers of the retina originate from separate regions of the optic vesicle. Even though the apical regions of the two layers come together as the optic cup forms, they do not appear to attach via junctions; they are held together by the pigment epithelium cellular processes that project out and wrap around the outer segments.

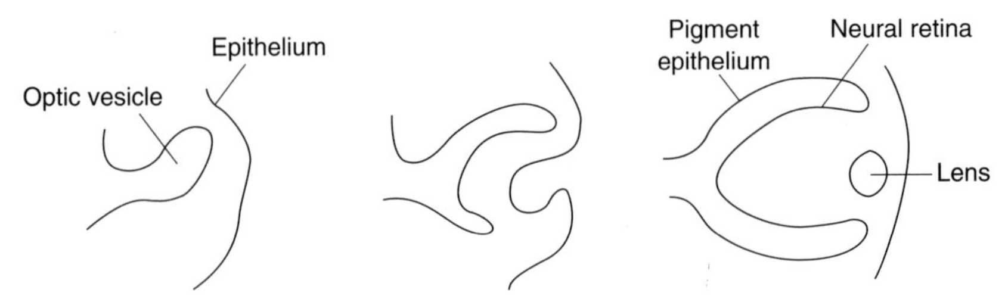

Retinal detachment occurs in 1 out of 15,000 persons as a result of aging, metabolic disorders, trauma, or vascular disease. In certain instances operations can be performed to approximate the two layers, and, using laser surgery, weld the tissues together.

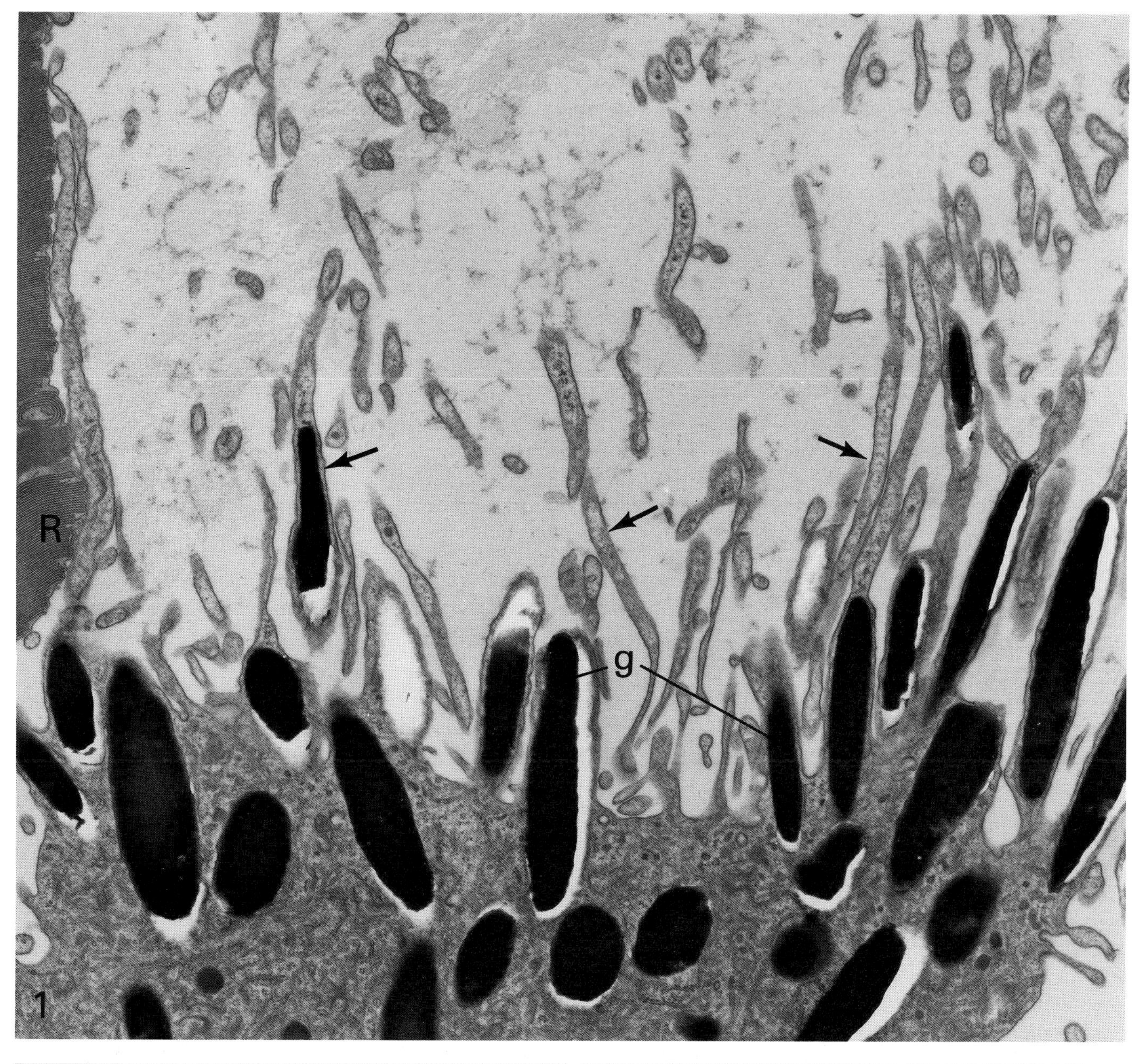

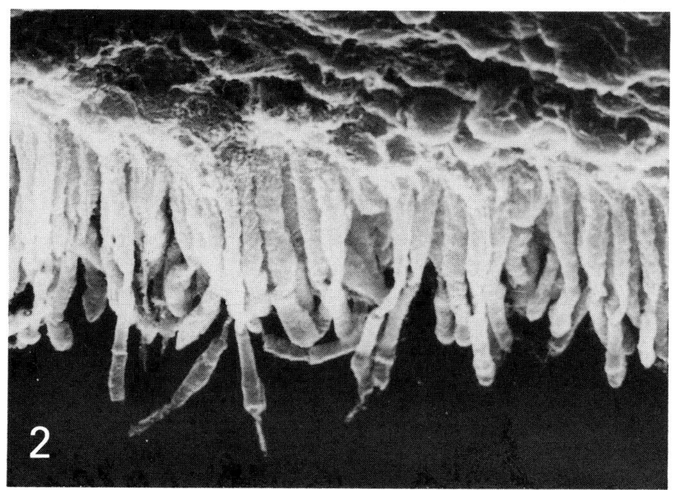

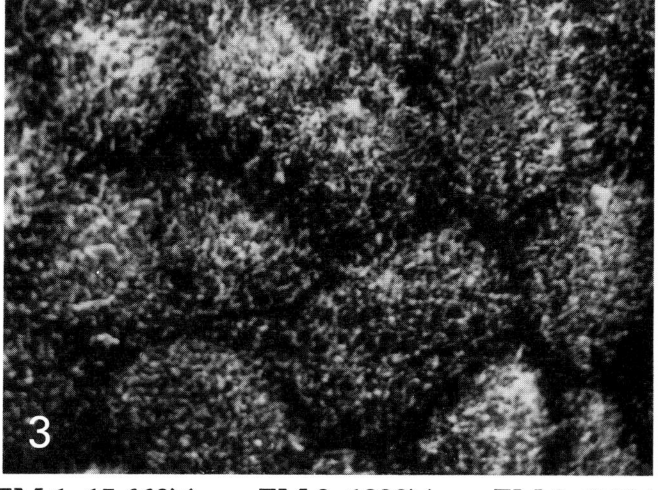

EM 1: 15,660× **EM 2: 1320×** **EM 3: 970×**

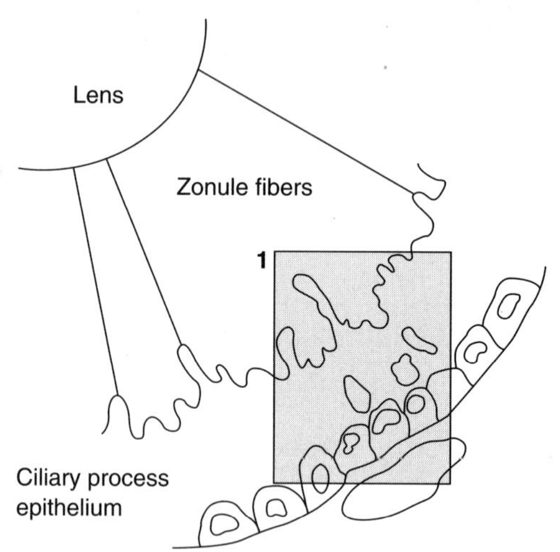

Ciliary processes are ridgelike extensions that project from the ciliary body toward the lens. The processes are covered by a two-layered epithelium (E1 and E2, micrograph 1) whose apical surfaces face one another. The inner layer of this epithelium is continuous with the neural retina in the back of the eye, the outer layer with the nonneural pigment epithelium. Normally the outer layer (E2, micrograph 1) of the ciliary epithelium is pigmented, as in the pigment epithelium of the retina. Micrograph 1, however, was taken from an albino rat.

A primary function of the ciliary epithelium is the secretion of the **aqueous humor,** a fluid that fills the posterior and anterior chambers of the eye. Aqueous humor originates from the plasma circulating in the underlying stromal capillaries (c, micrograph 1). Components are selectively moved across the epithelium from the stroma to the posterior chamber. The selectivity is maintained to a large degree by tight junctions that are a part of apical junctional complexes (arrows, micrograph 1) between cells of the inner epithelial layer (E1, micrograph 1). The ultrastructure of the epithelium reflects its role in active transport. Mitochondria provide the energy, and extensive infoldings (arrowheads, micrograph 1) on both basal surfaces provide the increased surface area needed to effect transport. The two layers of epithelial cells cooperate to (1) transport glucose, the energy source of the avascular lens and cornea, (2) transport and concentrate ascorbic acid, an important detoxicant for radicals formed during UV damage, and (3) exclude most proteins, an important part of the mechanism of protection from foreign antigens. Proteins that leave the underlying fenestrated capillaries with relative ease are blocked in their transit at the tight junctions.

Aqueous humor is a circulating fluid. The rate of production and the removal are balanced to maintain an intraocular pressure that is essential to maintaining the mechanical stability and optical properties of the eye. The fluid produced by the ciliary processes passes to the posterior and then anterior chambers and is finally absorbed into the canal of Schlemm and scleral veins in the limbic region (where the sclera and cornea meet). Any disturbance in this flow that causes increased pressure (glaucoma) can reduce and even irreversibly affect sight.

The ciliary processes also play an important mechanical role as the site of attachment of the lens to the ciliary body. **Zonule fibers** (z, micrograph 1) extend from the basal lamina of the inner epithelial layer to the lens capsule. The association of zonule fibers with this basal lamina is illustrated particularly well in the inset (arrowheads). The zonule fibers, similar in composition to elastic fibers, along with the smooth muscle of the ciliary body, control changes in the shape of the lens and thus its focal length. Zonule fibers are taut and the epithelial layers are stretched toward the lens (micrograph 1) when the smooth muscle is relaxed. The tension pulls the lens into an ellipsoid shape for far vision. When the ciliary muscles contract, the ciliary processes move closer to the lens and the tension on the epithelial cells, zonule fibers, and lens is reduced. The lens takes on the more spherical shape necessary for near vision. The cytoplasm of the inner epithelial layer is packed with tonofilaments (t, inset) that maintain the integrity of these cells as they relax and stretch with accommodation.

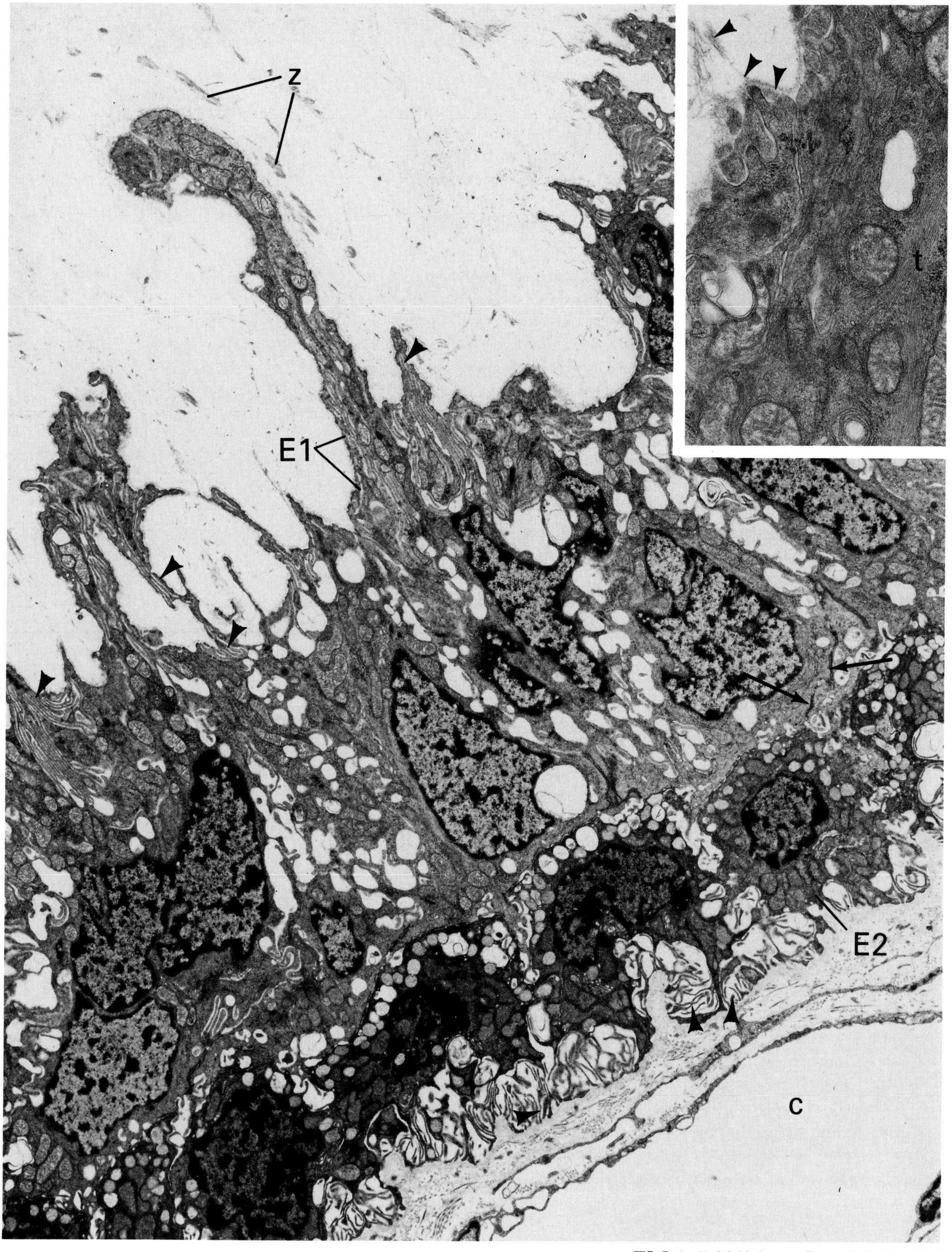

EM 1: 7,800× Inset: 20,300×

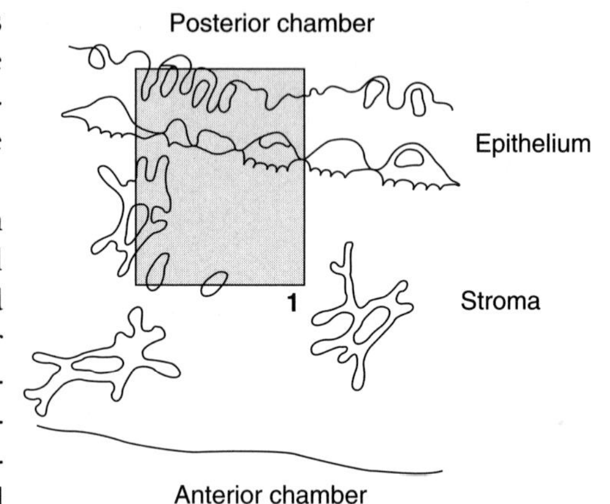

The **iris** rests on the anterior lens surface and separates the posterior from the anterior chamber of the eye. The surface of the iris facing the anterior chamber is lined by a discontinuous layer of fibroblastlike cells while the surface facing the posterior chamber is lined by two layers of epithelial cells (E1 and E2, micrograph 1) that are continuous with the epithelium of the ciliary process. As in the ciliary process, the epithelial cells in the iris are positioned with their apical surfaces attached and one basal surface adjacent to the aqueous humor, the other adjacent to a connective tissue stroma. Unlike the ciliary process, the epithelial cells of the iris are not involved in the production of aqueous humor and do not possess the characteristic basal infoldings associated with ion movement and secretion.

The layer of epithelial cells adjacent to the stroma (E2, micrograph 1) consists of myoepithelial cells that comprise the **dilator muscle** of the iris. These cells have a unique layered organization evident in both micrograph 1 and the inset: (1) a region containing the nucleus (N) and surrounding mitochondria (m) that controls the synthetic and metabolic activities; (2) an underlying band (b) of highly organized filaments that effect contraction; (3) basal interdigitations (arrows) with the stoma that increase the surface area of attachment so critical during the frequent contraction/relaxation cycles. Contraction of the radially oriented filaments opens the iris diaphragm, increasing pupil size. Lateral interdigitations (arrowheads, micrograph 1) between cells accommodate the shape changes associated with this movement.

The dilator myoepithelial cells attach to the stromal sphincter muscle that circles the pupillary margin. Contraction of the dilator and sphincter is controlled primarily by autonomic nervous innervation. A sympathetic release in response to a number of factors, including changes in the amount of light, fear, or pain, results in dilator contraction and pupil enlargement.

Pupil size is adjusted frequently to ensure the greatest degree of visual acuity and depth of focus. Other structures within the iris also function to improve visual acuity. Typically, pigment in both epithelial layers and stromal melanocytes absorbs stray light that would otherwise create aberrations. Vision is poor in the absence of pigment, as in the albino condition (micrograph 1).

The stroma of the iris contains fibroblasts (F, micrograph 1) that have numerous fine processes. The vacuolelike structures (v, micrograph 1) that appear to be within these cells actually represent extracellular regions between these processes. Capillaries (c, micrograph 1) sitting among the fibroblasts are lined by nonfenestrated endothelium. The endothelium acts as a barrier in one direction, preventing the movement of most molecules from blood to stroma, while facilitating movement in the opposite direction. These permeability characteristics enable these vessels to act as "sinks" to remove large molecules that reach the aqueous humor and could interfere with visual acuity. The structure of stromal iris capillaries contrasts with the highly permeable fenestrated lining of the vessels in the ciliary processes.

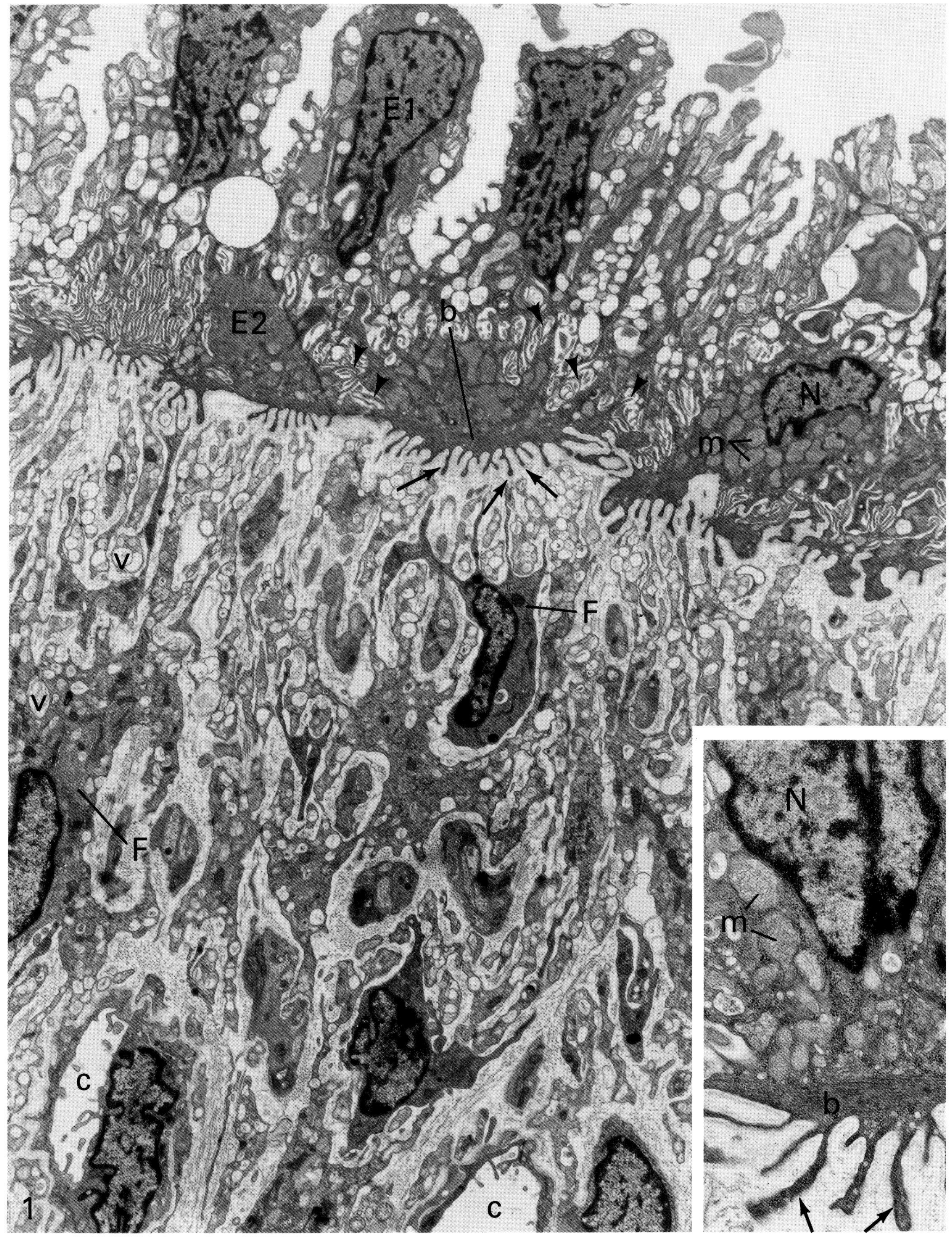

EM 1: 6,600× Inset: 17,500×

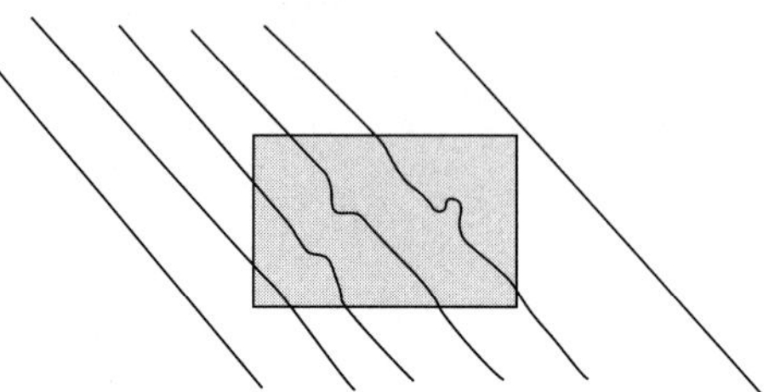

The lens and the cornea are the major refractile organs in the eye. The lens has less refractile power than the cornea, but it has the unique ability to change shape and thus change focal length. The lens is entirely cellular and its refractibility and transparency are characteristics of the special composition and arrangement of these cells.

The **lens** consists of a single type of epithelial cell in different stages of differentiation. The epithelial cells continue to grow and develop throughout life. Stem cells form a single cuboidal layer over the anterior lens surface. At the lateral periphery of the lens they divide and begin to elongate and differentiate into **lens fibers** as they move to the lens interior. The fibrous nature of the lens is evident in low-magnification scanning electron micrograph 1, in which the elastic capsule (c) of the lens has peeled back, revealing the fibers (f). Lens fibers (micrograph 2) are mature cells that have lost their organelles, including nuclei, and are packed with soluble structural proteins called crystallins. The age-related decrease in the ability of the lens to accommodate for near vision is, in part, related to the accumulation of more lens fibers, but it is due primarily to decreased elasticity of the capsule.

Mature lens fibers are tightly packed and join with one another via knob- and socketlike associations (k, micrograph 2). These elaborate cell interdigitations maintain the lens organization during shape changes associated with accommodation. In addition, close packing of cells prevents excess light scattering and facilitates communication between adjacent cells.

Aqueous humor bathes the avascular lens, providing the energy source, glucose, and collecting ions and water removed from the lens. **Gap junctions** (arrowheads, micrograph 2), which occupy an extensive portion of the lens fiber surface, provide the only means for transport of glucose and ions from one cell to the next to maintain the functioning of cells isolated in the lens interior.

An important role of glucose is to provide energy to maintain crystallins in a reduced form, thus preventing aggregation. Glucose also provides energy for the operation of an $Na^{+}K^{+}$ATPase, a major cell membrane protein. This ion pump maintains a net ion concentration that draws water out of the cells, thus preventing osmotic swelling within the lens. Since crystallin aggregation and osmotic swelling are the two major causes of cataracts (lens opacity), it is clear how the communication via gap junctions is critical to vision. It has been shown that gap junctions in the lens change with age, as does the incidence of cataracts.

Embryonic lens

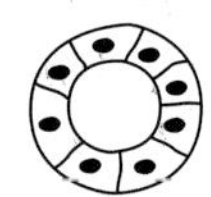

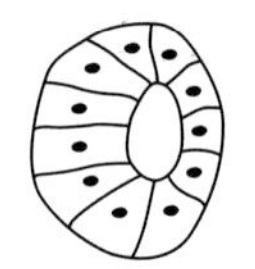

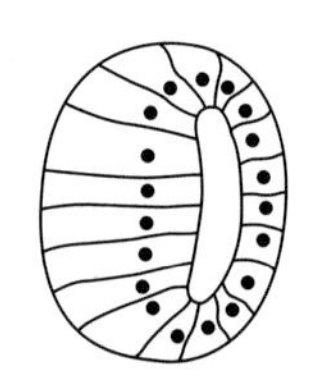

Adult lens

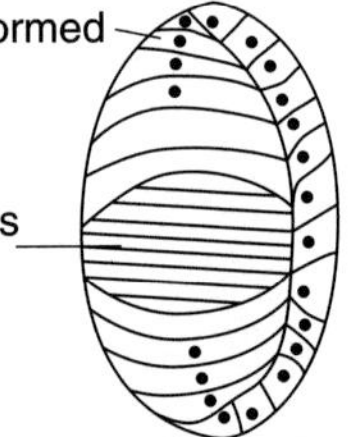

Modified from B. Alberts et al., *Molecular Biology of the Cell,* Garland, New York, 1989.

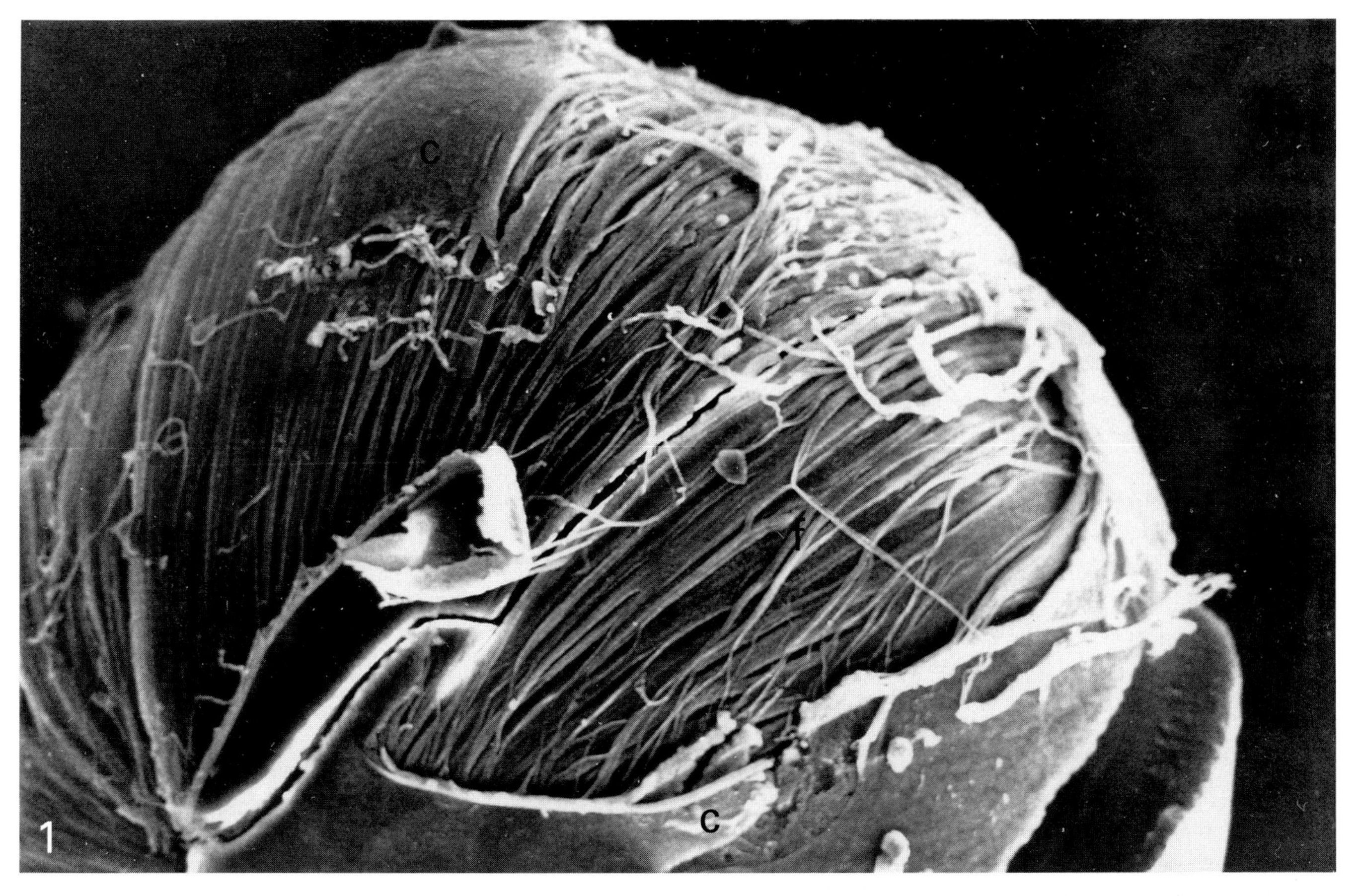

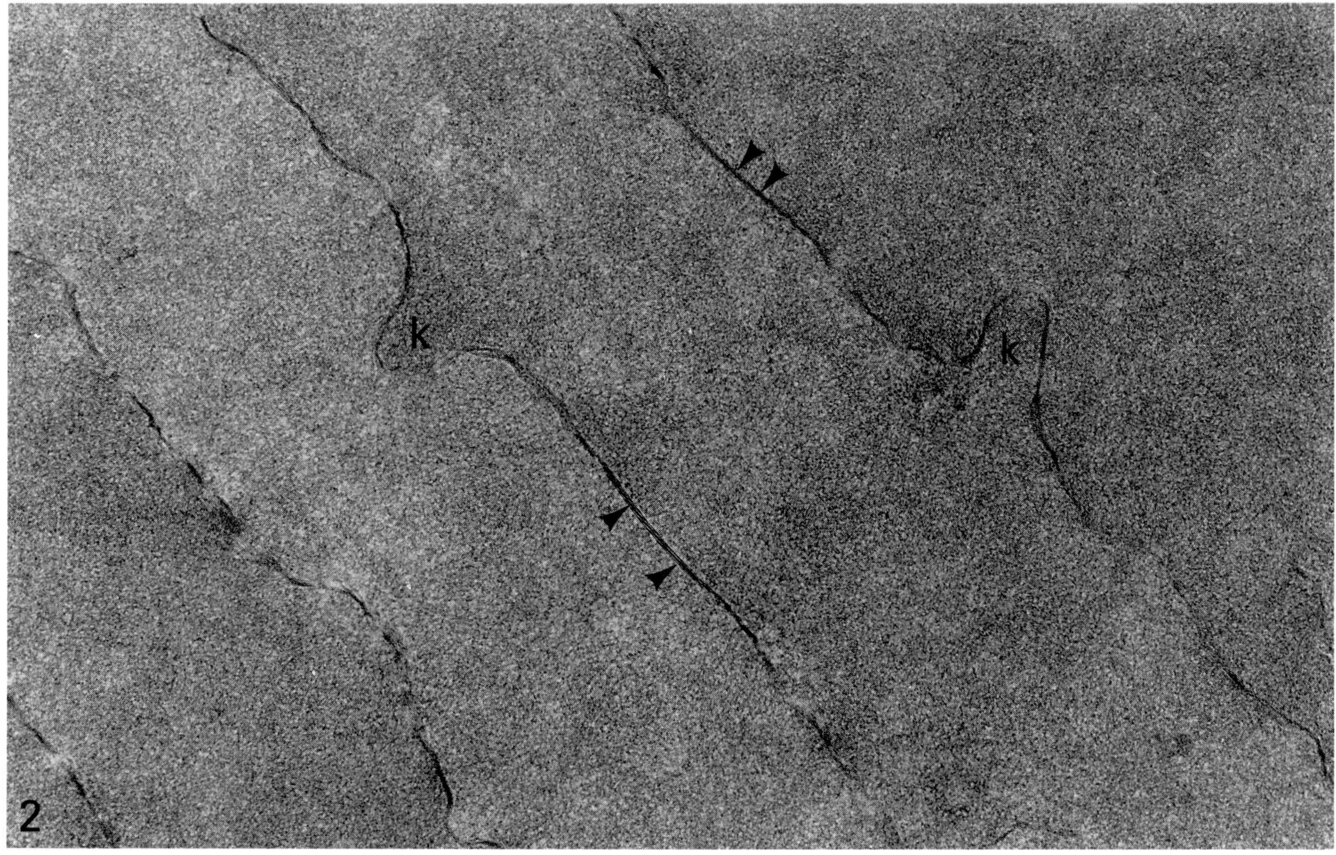

EM 1: 150× **EM 2: 48,000×**

EYE: Cornea

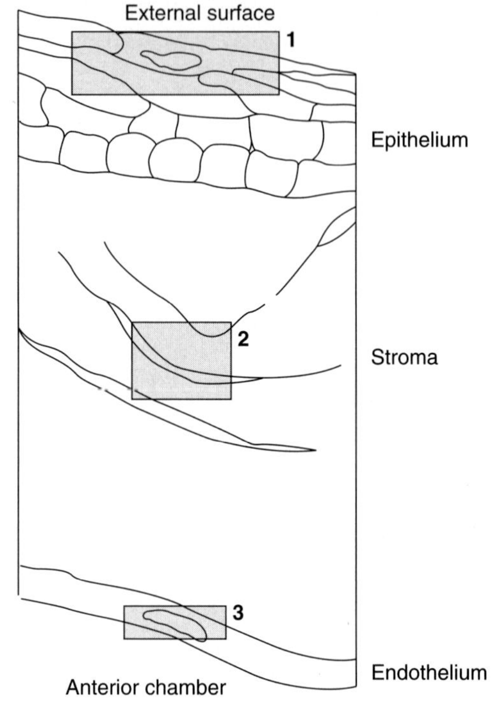

The **cornea** is a fixed-focal-length lens and, as such, its most important characteristics are transparency and refractive ability. The overall smooth, spherical shape of the cornea is maintained in large part by the pressure exerted by the aqueous humor on the inner surface. The finer aspects of maintaining shape and transparency are dependent upon the specific organization within each corneal region. The cornea is similar to the lens in being avascular, but, unlike the lens, it is rich in extracellular matrix.

The **stratified squamous epithelium** (micrograph 1) on the outer corneal surface is not keratinized (conserving image quality) and yet is exposed to many different types of insults. The epithelial cells have a short turnover time of seven days, and are covered by a protective tear film (t, micrograph 1) that can be as thick as 7 microns. This film, secreted by the epithelium itself, lacrimal glands, and goblet cells in the conjunctiva, is held in place by microvilli (arrowheads, micrograph 1) on the apical surface of the upper cell layer. The epithelium is richly innervated with free nerve endings (N, micrograph 1), which function in the blinking reflex that cleanses and spreads the fluid layer evenly across the corneal surface.

The **simple squamous endothelium** (micrograph 3) lining the inner corneal surface is the site of transport between the aqueous humor and corneal stroma. Ions and water are moved across this layer out of the cornea, preventing edema, and glucose, the energy source for all corneal cells, is moved into the cornea. The rough endoplasmic reticulum (arrows, micrograph 3) synthesizes the membrane pumps and channels important to this movement, and is also responsible for the synthesis of the unique endothelial basement membrane, **Descemet's membrane** (D, micrograph 3). The characteristic ladderlike structures (circles, micrograph 3) are cross-sectional views of a latticelike network of Type VIII collagen. This thick basement membrane is one of the few regions that contains a high concentration of this relatively rare type of collagen.

A dense collagen **stroma** (micrograph 2) that occupies 90% of the cornea is sandwiched between the surface epithelum and the inner endothelial layer. Fibroblasts (F, micrograph 2) extend processes to distant neighboring cells to form an interconnected network. The Type I collagen fibrils (c, micrograph 2), synthesized by the fibroblasts, have a small, uniform diameter and are tightly packed and evenly spaced. A fibril diameter less than one-half the wavelength of light and the similarity between the refractive index of the fibrils and the intervening ground substance are the most critical aspects of structure needed to maintain transparency. The larger aspect of organization, the alternating layers of collagen, which in certain regions appear to be at right angles, functions in providing the strength needed to maintain shape.

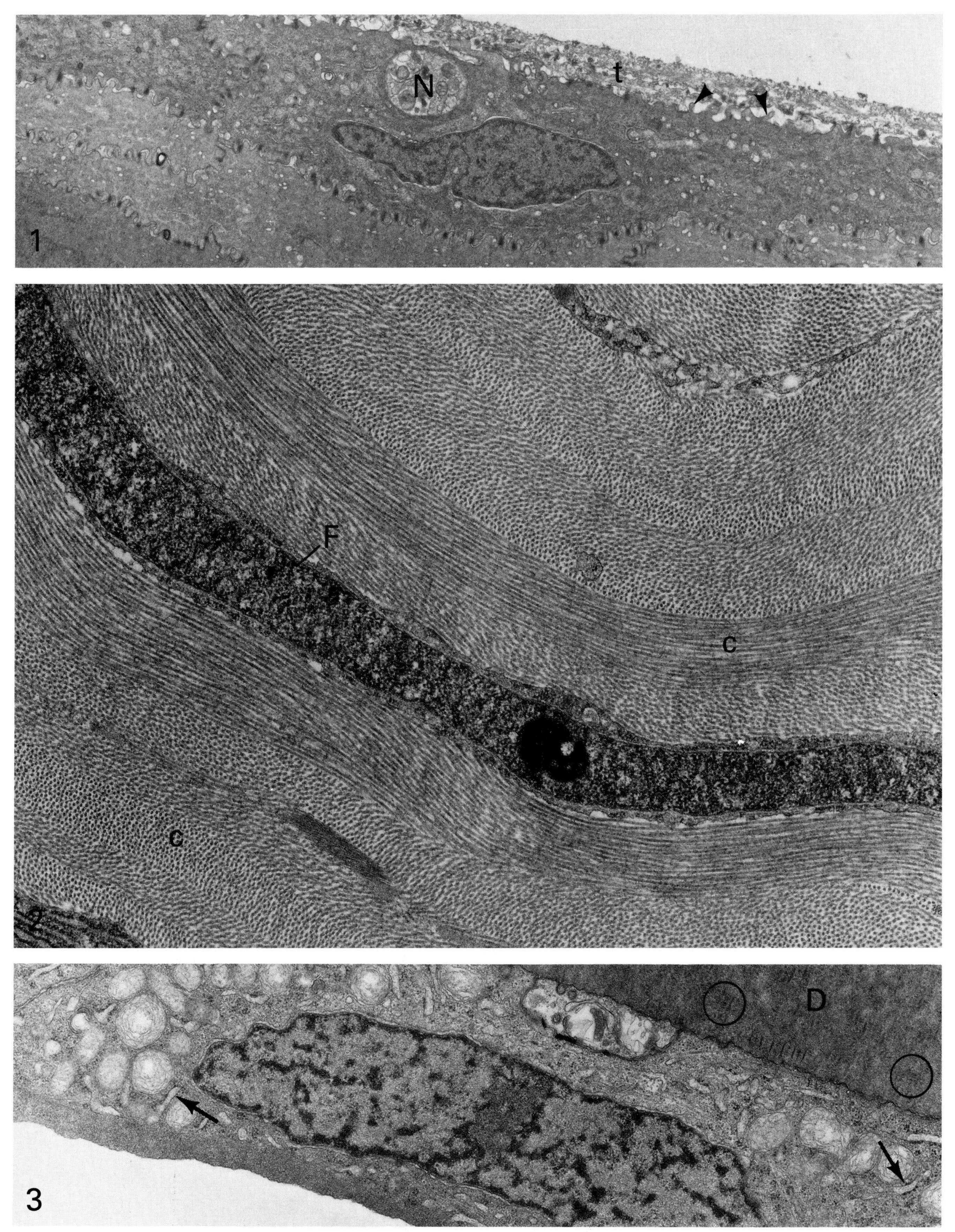

EM 1: 8400× EM 2: 23,000× EM 3: 19,000×

Ear

The sensory regions for hearing (organ of Corti in the cochlea) and equilibrium (macula in the utricle and saccule, for linear acceleration; crista ampullaris in semicircular canals, for angular acceleration) are localized in the inner ear. In each region, **sensory cells** convert a mechanical displacement into a nerve impulse. Bending of a **hair bundle** that projects from the sensory cell apical surface affects the rate of neurotransmitter release at synapses on the basolateral surface. Dendrites of the eighth cranial nerve carry the impulse to the CNS. Hair bundles are tightly associated with an overlying structure (tectorial membrane, otolithic membrane, or cupula) that helps regulate hair bundle displacement.

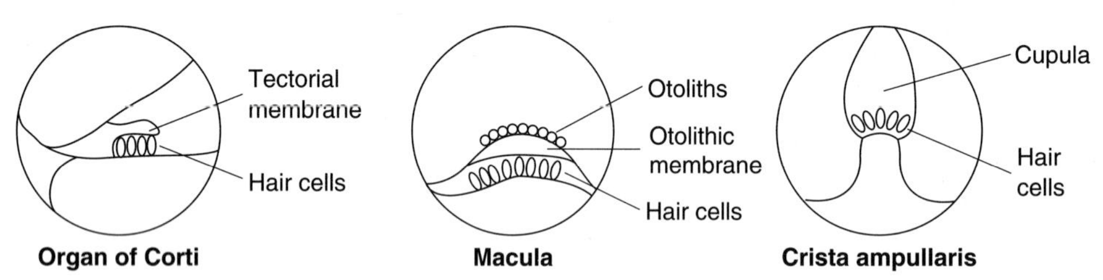

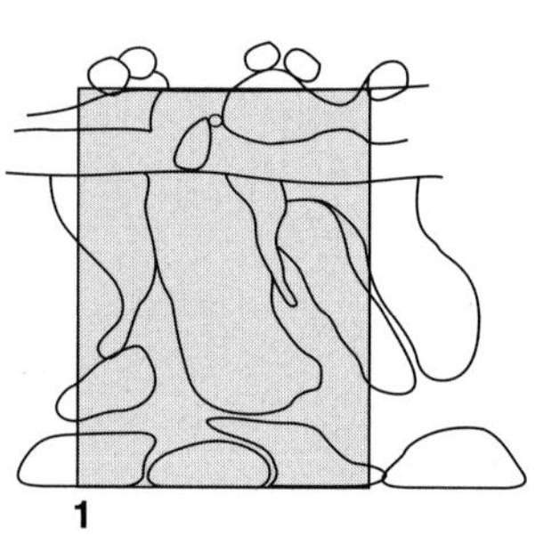

The structure of sensory cells (SC, micrograph 1) is very similar within different classes of vertebrates. A hair bundle (HB, micrograph 1) projects from the sensory cell apical surface in the bullfrog sacculus (micrographs, courtesy of Dr. A. J. Hudspeth and Dr. R. A. Jacobs), as it does in humans. This bundle consists of many **stereocilia** and a single, nonmotile **kinocilium** (arrows, micrograph 1 and inset: apical swelling of kinocilium). The stereocilia increase in length toward the kinocilium (inset). The hair bundle, a stiff unit with individual components attached to one another, is able to pivot at the tapered ends connected to the apical surface. Stability is provided by actin filaments within the stereocilia that extend into the apical cytoplasm to form an unusually dense terminal web (t, micrograph 1). Minute displacements of the bundle toward the kinocilium activate the sensory cells by opening ion channels at the tips of the stereocilia. Cations (primary K^+) enter and the membrane is depolarized. Bending away from the kinocilium closes channels (including those open at rest), thus hyperpolarizing the membrane.

Hair cells are surrounded by **supporting cells** (SU, micrograph 1) that contain microvilli dwarfed in size by the stereocilia of the hair bundle. Supporting cells are tightly attached to hair cells, and zonula occludens junctions effectively separate **endolymph** (bathing the apical surface with a high K^+ concentration) from **perilymph** (surrounding the basolateral surface with a high Na^+ concentration).

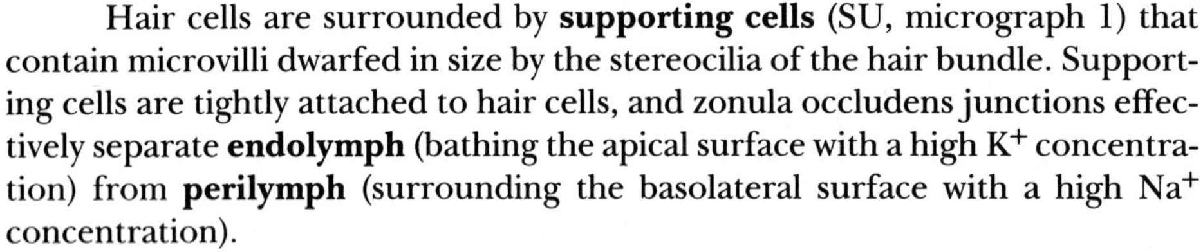

In the saccule (and utricle) an **otolithic membrane** (O, micrograph 1) overlies the sensory region or macula, except directly over the hair bundle (*, micrograph 1). The hair bundle, particularly the kinocilium, is attached laterally to the otolithic membrane. Embedded in this membrane are calcium-containing **otoliths** (not seen in micrograph 1 or inset) of a high specific gravity. Movement of the sensory cells, with head movement, relative to the more stationary otoliths bends the hair bundle.

The mechanical stimulus in the cochlea is **fluid movement.** The amplitude and frequency of waves encode loudness and pitch. In each inner ear sensory region the signal is transduced by sensory cells with essentially the same structure and function.

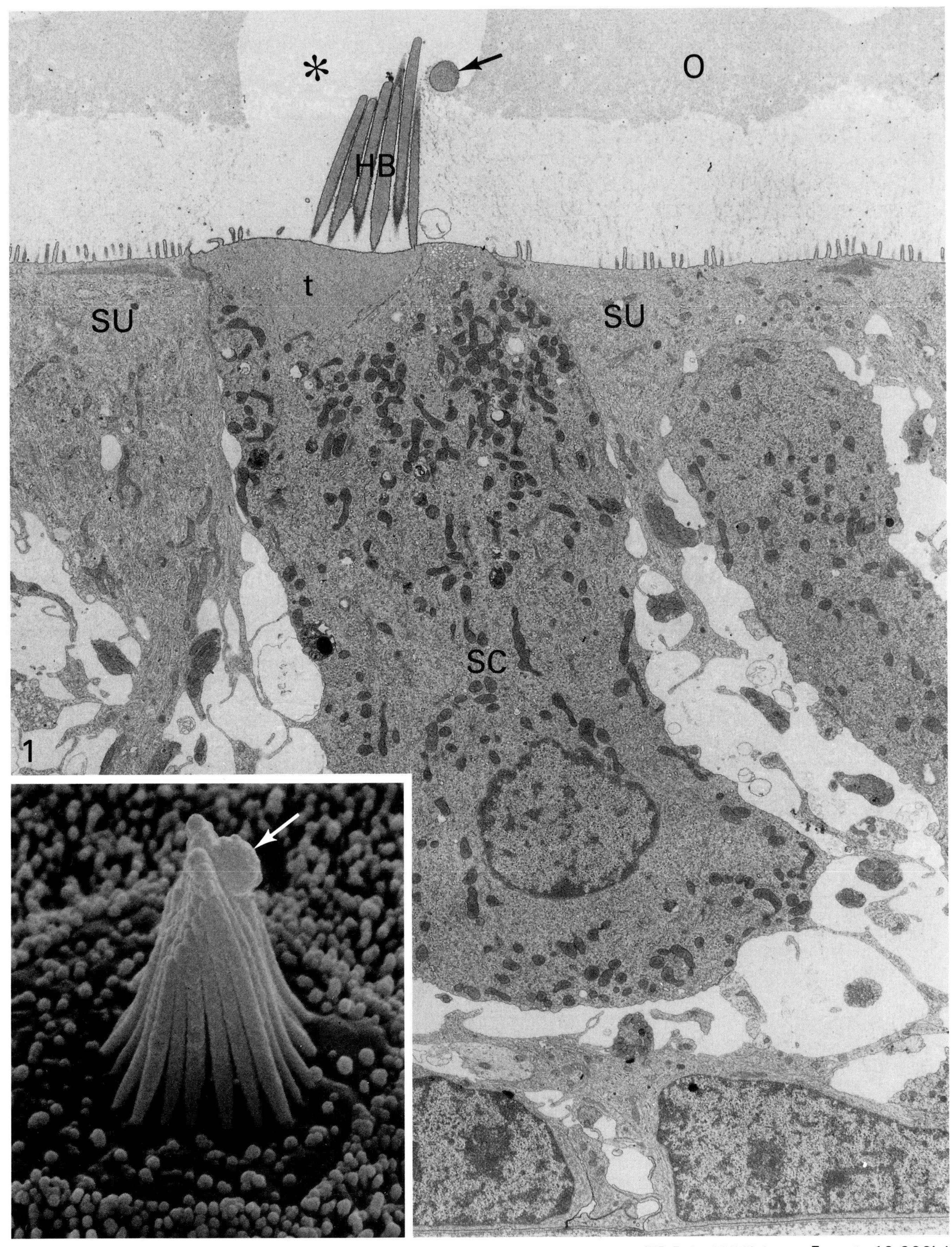

EM 1: 6000× Inset: 10,000×

Olfactory Epithelium

The **olfactory epithelium** is highly specialized to detect small stereochemical differences between odoriferous molecules in extremely low concentrations (parts per trillion). The sensory receptors that accomplish this are **bipolar neurons** whose cell bodies occupy a distinct layer within this pseudostratified epithelium. Axons extend into the underlying connective tissue to the olfactory bulb in the central nervous system (CNS); dendrites project apically and expand to form **olfactory vesicles** (OV, micrograph) on the epithelial surface. Each vesicle contains the basal bodies (b, micrograph) of 5–20 nonmotile cilia that project into the nasal cavity and rest in the overlying secretion.

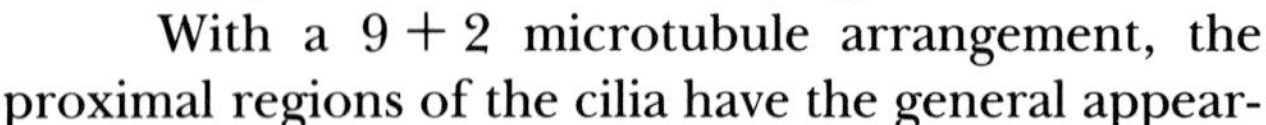

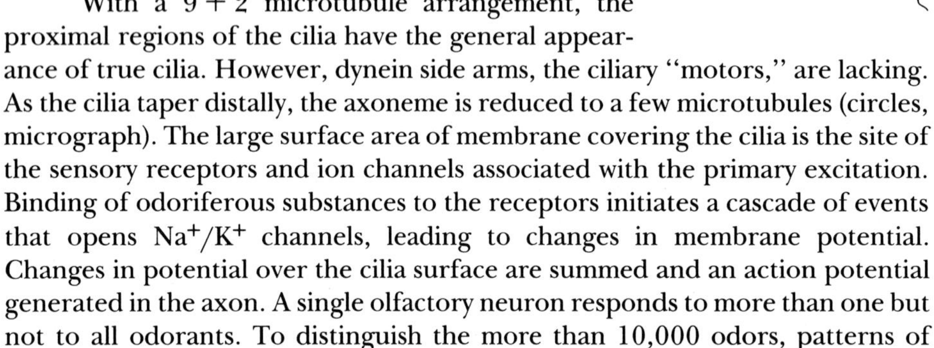

With a 9 + 2 microtubule arrangement, the proximal regions of the cilia have the general appearance of true cilia. However, dynein side arms, the ciliary "motors," are lacking. As the cilia taper distally, the axoneme is reduced to a few microtubules (circles, micrograph). The large surface area of membrane covering the cilia is the site of the sensory receptors and ion channels associated with the primary excitation. Binding of odoriferous substances to the receptors initiates a cascade of events that opens Na^+/K^+ channels, leading to changes in membrane potential. Changes in potential over the cilia surface are summed and an action potential generated in the axon. A single olfactory neuron responds to more than one but not to all odorants. To distinguish the more than 10,000 odors, patterns of neuronal activity are transferred to the brain, where processing occurs.

Each olfactory dendrite is wrapped by and attached to surrounding **supporting cells.** Junctions (arrowheads, micrograph) are frequently seen binding these two cell types together; the olfactory neurons (N, micrograph), recognized by the prominent cytoskeleton of microtubules, are easily distinguished from the supporting cells (S, micrograph). In addition to providing support and a unique environment for the olfactory dendrite, the supporting cells have a direct, essential role in the process of olfaction. The prominent smooth ER (arrows, micrograph) contains the enzymes that remove excess odorous molecules and inactivate them, terminating the sensory signal and thus maintaining the sensitivity of the system. The enzymes, a P-450 detoxifying complex similar to that found in other cells (see Cell, page 22), metabolize the odorants and release the products into the blood.

Both the olfactory neurons and the supporting cells are replaced regularly by the division and differentiation of **basal cells.** Olfactory neurons are the only neurons in the body that are replaced, and the olfactory region is the only site where neurons are directly exposed to an external environment. The nasal cavity is a turbulent area where neurons are easily damaged. As each neuron is replaced, the axon must grow, enter the olfactory bulb, and establish new synapses with CNS neurons. Other axons in the peripheral nervous system are known to regenerate; however, they do not continue to grow after they enter the CNS. The ability of olfactory axons to grow in the CNS is attributed to unique glial cells that wrap the unmyelinated axons from the olfactory epithelium to their region of synapse in the olfactory bulb. These glial cells, which originate from the epithelial basal cells, have characteristics of both Schwann cells and astrocytes.

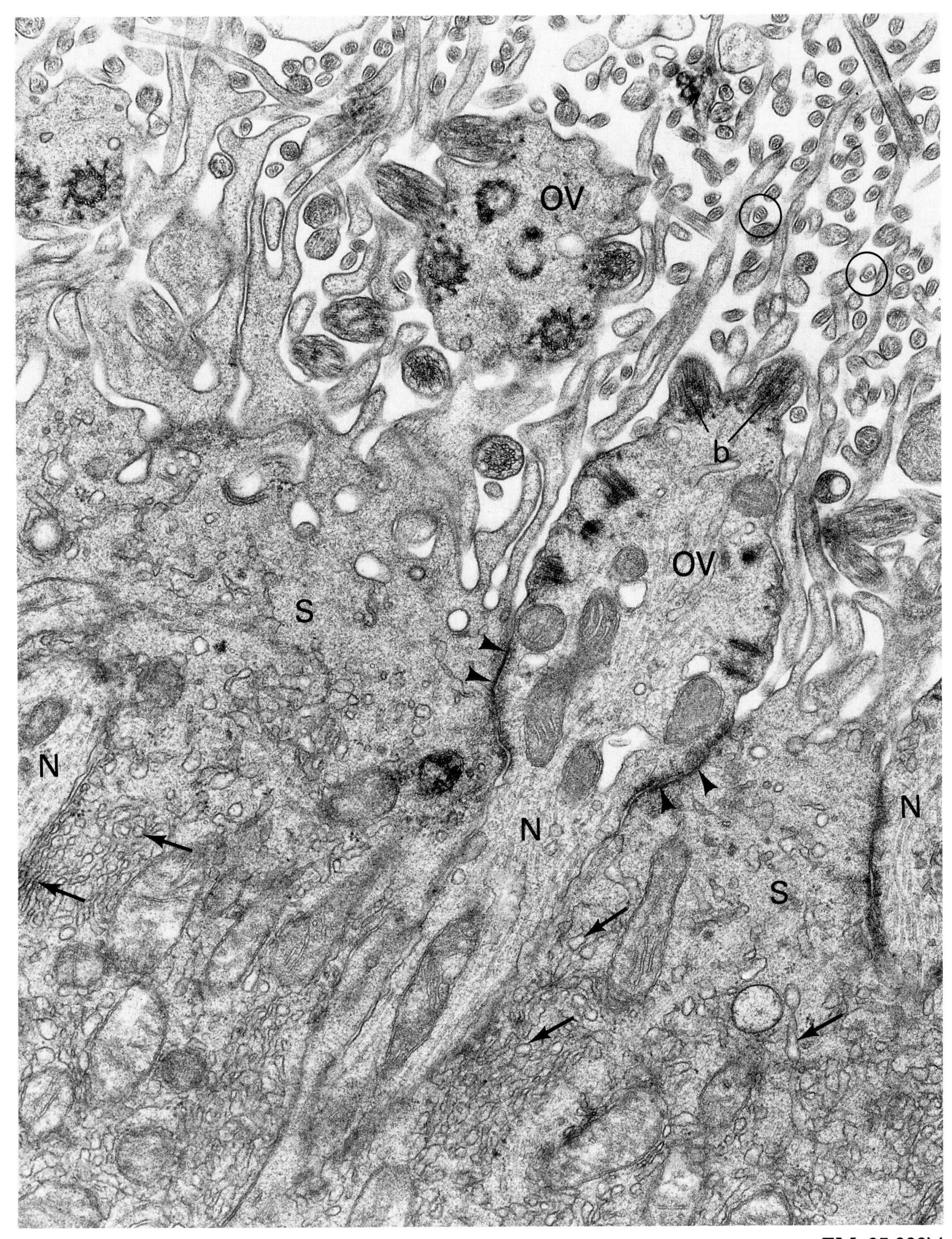

EM: 35,000×

REFERENCES

GENERAL RESOURCES

Cell Biology

Alberts, B., Bray, D., Lewis, J., Raff, M., Roberts, K., and Watson, J. D. 1989. *Molecular Biology of the Cell,* 2nd Ed., Garland Publishing, Inc., New York.

Darnell, J., Lodish, H., and Baltimore, D. 1990. *Molecular Cell Biology,* 2nd Ed., Scientific American Books, Inc.; W. H. Freeman and Company, New York.

Ultrastructure

Fawcett, D. W. 1981. *The Cell,* 2nd Ed., W. B. Saunders Company, Philadelphia.

Porter, K. R. and Bonneville, M. A. 1973. *Fine Structure of Cells and Tissues.* Lea and Febiger, Philadelphia.

Histology

Comprehensive

Cormack, D. H. 1987. *Ham's Histology,* 9th Ed., Lippincott, Philadelphia.

Fawcett, D. W. 1986. *A Textbook of Histology,* 11th Ed., W. B. Saunders Company, Philadelphia.

Weiss, L. 1988. *Cell and Tissue Biology,* 6th Ed., Elsevier Biomedical, New York.

Core

Burkitt, H. G., and Daniels, V. G. 1993. *Functional Histology—A Text and Color Atlas,* 3rd Ed., Churchill Livingstone, New York.

Junquiera, L. C., Carneiro, J., and Kelley, R. O. 1992. *Basic Histology,* 7th Ed., Appleton and Lange Medical Publications, Norwalk, Connecticut.

CHAPTER ONE Cell

Albelda, S. M., and Buck, C. A., 1990. Integrins and other adhesion molecules. *FASEB J.* 4:2868–2880.

Bloemendal, H., and Pieper, F. R., 1989. Intermediate filaments: known structure, unknown function. *Biochimica et Biophysica Acta* 1007:245–253.

Dessev, G. N., 1992. Nuclear envelope structure. *Current Opinion in Cell Biol.* 4:430–435.

Fischer, D., Weisenberger, D., and Scheer, U., 1991. Assigning functions to nucleolar structures. *Chromosoma* 101:133–140.

Forbes, D. J., 1992. Structure and function of the nuclear pore complex. *Annu. Rev. Cell Biol.* 8:495–527.

Gelfand, V. I., and Bershadsky, A .D., 1991. Microtubule dynamics: mechanism, regulation, and function. *Annu. Rev. Cell Biol.* 7:93–116.

Haaf, T., and Schmid, M., 1991. Chromosome topology in mammalian interphase nuclei. *Experimental Cell Res.* 192:325–332.

Hanover, J. A., 1992. The nuclear pore: at the crossroads. *FASEB J.* 6:2288–2295.

Hyman, A. A., and Mitchison, T. J., 1991. Regulation of the direction of chromosome movement. *Cold Spring Harbor Symp. Quant. Biol.* 56:745–750.

Knaus, U. G., Heyworth, P. G., Evans, T., Curnutte, J. T., and Bokoch, G. M., 1991. Regulation of phagocyte oxygen radical production by the GTP-binding protein rac 2. *Science* 254:1512–1515.

Kolega, J., Janson, L. W., and Taylor, D. L., 1991. The role of solation-contraction coupling in regulating stress fiber dynamics in nonmuscle cells. *J. Cell Biol.* 114:993–1003.

Kornberg, R. D., and Lorch, Y., 1992. Chromatin structure and transcription. *Annu. Rev. Cell Biol.* 8:563–587.

Luna, E. J., and Hitt, A. L., 1992. Cytoskeleton–plasma membrane interactions. *Science* 258:955–964.

McIntosh, J. R., and Hering, G. E., 1991. Spindle fiber action and chromosome movement. *Annu. Rev. Cell Biol.* 7:403–426.

Mellman, I., and Simons, K., 1992. The Golgi complex: in vitro veritas? *Cell* 68:829–840.

Nigg, E. A., 1992. Assembly–disassembly of the nuclear lamina. *Current Opinion in Cell Biol.* 4:105–109.

Satterwhite, L. L., and Pollard, T. D., 1992. Cytokinesis. *Current Opinion in Cell Biol.* 4:43–52.

Shimozawa, N., Tsukamoto, T., Suzuki, Y., Orii, T., Shirayoshi, Y., Mori, T., and Fujiki, Y., 1992. A human gene responsible for Zellweger syndrome that affects peroxisome assembly. *Science* 255:1132–1134.

Uyemura, D. G., and Spudich, J. A., 1980. Biochemistry and regulation of nonmuscle actins. In *Biological Regulation and Development,* Vol. 2. Plenum Publishing Corp., New York.

Vale, R. D., 1992. Microtubule motors: many new models off the assembly line. *TIBS: Trends in Biochem. Sci.* 17:300–304.

Wilson, G. N., 1991. Structure–function relationships in the peroxisome: implications for human disease. *Biochemical Medicine & Metabolic Biol.* 46:288–298.

Chapter Two Epithelium

Amano, O., Kataoka, S., and Yamamoto, T. Y., 1991. Turnover of asymmetric unit membranes in the transitional epithelial superficial cells of the rat urinary bladder. *Anatomical Record* 229:9–15.

Badalament, R. A., Franklin, G. L., Page, C. M., Dasani, B. M., Wientjes, M. G., and Drago, J. R., 1992. Enhancement of bacillus Calmette–Guerin attachment to the urothelium by removal of the rabbit bladder mucin layer. *J. of Urology* 147:482–485.

Buxton, R. S., and Magee, A. I., 1992. Structure and interactions of desmosomal and other cadherins. *Seminars in Cell Biology* 3:157–167.

Epstein, E. H., Jr., 1992. Molecular genetics of epidermolysis bullosa. *Science* 256:799–804.

Erle, D. J., and Pytela, R., 1992. How do integrins integrate? The role of cell adhesion receptors in differentiation and development. *Am. J. Respir. Cell Mol. Biol.* 6:459–460.

Gumbiner, B. M., 1992. Epithelial morphogenesis. *Cell* 69:385–387.

Heintzelman, M. B., and Mooseker, M. S., 1992. Assembly of the intestinal brush border cytoskeleton. *Current Topics in Developmental Biology* 26:93–122.

Jones, J. C., and Green, K. J., 1991. Intermediate filament–plasma membrane interactions. *Current Opinion in Cell Biol.* 3:127–132.

Klaunig, J. E., 1991. Alterations in intercellular communication during the stage of promotion. *Proceedings of the Society for Exp. Biology & Medicine* 198:688–692.

Madara, J. L., Parkos, C., Colgan, S., Nusrat, A., Atisook, K., and Kaoutzani, P., 1992. The movement of solutes and cells across tight junctions. *Annals New York Acad. Sci.* 664:47–60.

Matlin, K. S., 1992. W(h)ither default? Sorting and polarization in epithelial cells. *Current Opinion in Cell Biol.* 4:623–628.

Miller, J. M., Wang, W., Balczon, R., and Dentler, W. L., 1990. Ciliary microtubule capping structures contain a mammalian kinetochore antigen. *J. Cell Biol.* 110:703–714.

Molitoris, B. A., and Nelson, W. J., 1990. Alterations in the establishment and maintenance of epithelial cell polarity as a basis for disease processes. *J. of Clinical Investigation* 85:3–9.

Schneeberger, E. E., and Lynch, R. D., 1992. Structure, function, and regulation of cellular tight junctions. *Am. J. Physiol.* 262:L647–L661.

Schroer, T. A., 1991. Association of motor proteins with membranes. *Current Opinion in Cell Biol.* 3:133–137.

Stauffer, K. A., and Unwin, N., 1992. Structure of gap junction channels. *Seminars in Cell Biology* 3:17–20.

Tsukita, S., Tsukita, S., Nagafuchi, A., and Yonemura, S., 1992. Molecular linkage between cadherins and actin filaments in cell–cell adherens junctions. *Current Opinion in Cell Biol.* 4:834–839.

Warner, A., 1992. Gap junctions in development—a perspective. *Seminars in Cell Biology* 3:81–91.

Wheeler, G. N., Parker, A. E., Thomas, C. L., Ataliotis, P., Poynter, D., Arnemann, J., Rutman, A. J., Pidsley, S. C., Watt, F. M., and Rees, D. A., 1991. Desmosomal glycoprotein DGI, a component of intercellular desmosome junctions, is related to the cadherin family of cell adhesion molecules. *Proc. Natl. Acad. Sci. USA* 88:4796–4800.

Witman, G. B., 1992. Axonemal dyneins. *Current Opinion in Cell Biol.* 4:74–79.

Wollner, D. A., Krzeminski, K. A., and Nelson, W. J., 1992. Remodeling the cell surface distribution of membrane proteins during development of epithelial polarity. *J. Cell Biol.* 116:889–899.

Yurchenco, P. D., and Schittny, J. C., 1990. Molecular

architecture of basement membranes *FASEB J.* 4:1577–1590.

CHAPTER THREE Connective Tissue

Boskey, A. L., 1992. Mineral–matrix interactions in bone and cartilage. *Clinical Orthopaedics and Related Research* 281:244–274.

Galli, S. J., 1993. New concepts about the mast cell. *The N. England J. Med.* 328:257–265.

Hardingham, T. E., and Fosang, A. J., 1992. Proteoglycans: many forms and many functions. *FASEB J.* 6:861–870.

Hascall, V. C., Heinegard, D. K., and Wight, T. N., 1991. Proteoglycans: metabolism and pathology. In *Cell Biology of Extracellular Matrix,* 2nd ed., Hay, E. D., ed. Plenum Press, New York. Chapter 5, pp. 149–172.

Haskill, S., Yurochko, A. D., and Isaacs, K. L., 1992. Regulation of macrophage infiltration and activation in sites of chronic inflammation. *Annals New York Acad. Sci.* 664:93–102.

Heinegard, D. and Oldberg, A., 1989. Structure and biology of cartilage and bone matrix noncollagenous macromolecules. *FASEB J.* 3:2042–2051.

Hirsch, J., Fried, S. K., Edens, N. K., and Leibel, R. L., 1989. The fat cell. *Medical Clinics of N. America* 73:83–96.

Jilka, R. L., Hangoc, G., Girasole, G., Passeri, B., Williams, D. C., Abrams, J. S., Boyce, B., Broxmeyer, H., and Manolagas, S.C., 1992. Increased osteoclast development after estrogen loss: mediation by interleukin–6. *Science* 257:88–91.

Katz, H. R., Kaye, R. E., and Austen, K. F., 1991. Mast cell biochemical and functional heterogeneity. *Transplantation Proceedings* 23:2900–2904.

Kuivaniemi, H., Tromp, G., and Prockop, D. J., 1991. Mutations in collagen genes: causes of rare and some common diseases in humans. *FASEB J.* 5:2052–2060.

Lindsay, R., and Cosman, F., 1991. Estrogen, calcium metabolism and the skeleton. *Annals New York Acad. Sci.* 622:307–314.

Mecham, R. P., and Heuser, J. E., 1991. The elastic fiber. In *Cell Biology of Extracellular Matrix,* 2nd Ed., Hay, E. D., Ed. Plenum Press, New York, pp. 79–109 (Chapter 3).

Rappolee, D. A., and Werb, Z., 1992. Macrophage–derived growth factors. *Current Topics in Microbiol. & Immunol.* 181:87–140.

Reiser, K., McCormick, R. J., and Rucker, R. B., 1992. Enzymatic and nonenzymatic cross-linking of collagen and elastin. *FASEB J.* 6:2439–2449.

Stein, G. S., Lian, J. B., Owen, T. A., 1990. Relationship of cell growth to the regulation of tissue-specific gene expression during osteoblast differentiation. *FASEB J.* 4:3111–3123.

Stevens, R. L., 1989. Mast cell proteoglycans. In *Biochemistry of the Acute Allergic Reactions: Fifth International Symp.,* Alan R. Liss, Inc., pp. 131–144.

Suda, T., Takahashi, N., and Martin, T. J., 1992. Modulation of osteoclast differentiation. *Endocrine Revs.* 13:66–80.

Van Der Rest, M., and Garrone, R., 1991. Collagen family of proteins. *FASEB J.* 5:2814–2823.

Weiner, S., and Traub, W., 1992. Bone structure: from angstroms to microns. *FASEB J.* 6:879–885.

CHAPTER FOUR Muscle

Draeger, A., Amos, W. B., Ikebe, M., and Small, J. V., 1990. The cytoskeletal and contractile apparatus of smooth muscle: contraction bands and segmentation of the contractile elements. *J. Cell Biol.* 111:2463–2473.

Dulhunty, A. F., 1989. Feet, bridges, and pillars in triad junctions of mammalian skeletal muscle: their possible relationship to calcium buffers in terminal cisternae and t-tubules and to excitation–contraction coupling. *J. Membrane Biol.* 109:73–83.

Grounds, M. D., 1991. Towards understanding skeletal muscle regeneration. *Path. Res. Pract.* 187:1–22.

Gunning, P., and Hardeman, E., 1991. Multiple mechanisms regulate muscle fiber diversity. *FASEB J.* 5:3064–3070.

Hall, Z. W., and Sanes, J. R., 1993. Synaptic structure and development: the neuromuscular junction. *Cell* 72 (suppl.):99–121.

Kargacin, G. J., Cooke, P. H., Abramson, S. B., and Fay, F. S., 1989. Periodic organization of the contractile apparatus in smooth muscle revealed by the motion of dense bodies in single cells. *J. Cell Biol.* 108:1465–1475.

Lu, M. H., DiLullo, C., Schultheiss, T., Holtzer, S., Murray, J. M., Choi, J., Fischman, D. A., and Holtzer, H., 1992. The vinculin/sacromeric-alpha-actinin/alpha-actin nexus in cultured cardiac myocytes. *J. Cell Biol.* 117:1007–1022.

McMahan, U. J., Horton, S. E., Werle, M. J., Honig, L. S., Kroger, S., Ruegg, M. A., and Escher, G., 1992. Agrin isoforms and their role in synaptogenesis. *Current Opinion in Cell Biol.* 4:869–874.

Morgan, J. P., Perreault, C. L., and Morgan, K. G., 1991. The cellular basis of contraction and relaxation in cardiac and vascular smooth muscle. *American Heart J.* 121:961–968.

Payne, M. R., and Rudnick, S. E., 1989. Regulation of vertebrate striated muscle contraction. *TIBS: Trends in Biochemical Sci.* 14:357–360.

Pette, D., 1992. Fiber transformation and fiber replacement in chronically stimulated muscle. *J. Heart and Lung Transplantation* 11:S299–S305.

Robbins, N., and Nakashiro, S, 1993. Connections among plasticity, regeneration, and aging at the neuromuscular junction. *Advances in Neurology* 59:47–52.

Severs, N. J., 1990. The cardiac gap junction and intercalated disc. *International J. Cardiology* 26:137–173.

Squarzoni, S., Sabatelli, P., Maltarello, M. C., Cataldi, A., Di Primio, R., and Maraldi, N. M., 1992. Localization of dystrophin COOH-terminal domain by the fracture–label technique. *J. Cell Biol.* 118:1401–1409.

Straub, V., Bittner, R. E., Léger, J. J. and Voit, T., 1992. Direct visualization of the dystrophin network on skeletal muscle fiber membrane. *J. Cell Biol.* 119:1183–1191.

Trybus, K. M., 1991. Regulation of smooth muscle myosin. *Cell Motility & the Cytoskeleton* 18:81–85.

Chapter Five Nerve

Allinquant, B., Staugaitis, S. M., U'Urso, D., and Colman, D. R., 1991. The ectopic expression of myelin basic protein isoforms in Shiverer oligodendrocytes: implications for myelinogenesis. *J. Cell Biol.* 113:393–403.

Barres, B. A., 1991. New roles for glia. *J. Neurosci.* 11:3685–3694.

de Waegh, S. M., Lee, V. M., and Brady, S. T., 1992. Local modulation of neurofilament phosphorylation, axonal caliber, and slow axonal transport by myelinating Schwann cells. *Cell* 68:451–463.

Deber, C. M., Reynolds, S. J., 1991. Central nervous system myelin: structure, function, and pathology. *Clin. Biochem.* 24:113–134.

Hinton, G. E., 1992. How neural networks learn from experience. *Scientific American* 267:144–151.

Hirokawa, N., Hisanaga, S.-I., and Shiomura, Y., 1988. MAP2 is a component of crossbridges between microtubules and neurofilaments in the neuronal cytoskeleton: quick-freeze, deep-etch immunoelectron microscopy and reconstitution studies. *J. Neurosci.* 8:2769–2779.

Jessell, T. M., and Kandel, E. R., 1993. Synaptic transmission: a bidirectional and self-modifiable form of cell–cell communication. *Cell 72* Suppl. 10:1–30.

Joe, E. H., and Angelides, K., 1992. Clustering of voltage-dependent sodium channels on axons depends on Schwann cell contact. *Nature* 356:333–335.

Kapfhammer, J. P., and Schwab, M. E., 1992. Modulators of neuronal migration and neurite growth. *Current Opinion in Cell Biol.* 4:863–868.

Kidd, G .J. Hauer, P. E., and Trapp, B. D., 1990. Axons modulate myelin protein messenger RNA levels during central nervous system myelination in vivo. *J. Neurosc. Res.* 26:409–418.

Lisman, J. E., and Goldring, M. A., 1988. Feasibility of long-term storage of graded information by the Ca^{2+}/calmodulin-dependent protein kinase molecules of the postsynaptic density. *Proc. Natl. Acad. Sci. USA* 85:5320–5324.

Naus, C .C. G., Bechberger, J. F., and Paul, D. L., 1991. Gap junction gene expression in human seizure disorder. *Experimental Neurology* 111:198–203.

Nixon, R. A., 1992. Slow axonal transport. *Current Opinion in Cell Biol.* 4:8–14.

Ota, K., Matsui, M., Milford, E. L., Mackin, G. A., Weiner, H. L., and Hafler, D. A., 1990. T-cell recognition of an immuno-dominant myelin basic protein epitope in multiple sclerosis. *Nature* 346:183–187.

Schwab, M. E., 1991. Nerve fibre regeneration after traumatic lesions of the CNS; progress and problems. *Phil. Trans. R. Soc. Lond. B.* 331:303–306.

Schwartz, M., Cohen, A., Stein-Izsak, C., and Belkin, M., 1989. Dichotomy of the glial cell response to axonal injury and regeneration. *FASEB J.* 3:2371–2378.

Smith, R. S., and Snyder, R. E., 1992. Relationship between the rapid axonal transport of newly synthesized proteins and membranous organelles. *Molecular Neurobiology* 6:285–300.

Smith, S. J., 1992. Do astrocytes process neural information? *Prog. in Brain Res.* 94:119–136.

Snyder, S. H., and Bredt, D. S., 1992. Biological roles of nitric oxide. *Scientific American* 266:68–77.

Stevens, C. F., 1993. Quantal release of neurotransmitter and long-term potentiation. *Cell* 72 Suppl. 10:55–64.

Varon, S., Hagg, T., and Manthorpe, M., 1991. Nerve growth factor in CNS repair and regeneration. *Advances Exp. Medicine & Biol.* 296:267–276.

Chapter Six Blood Vessels

Abbott, N. J., Revest, P. A., and Romero, I. A., 1992. Astrocyte–endothelial interaction: physiology and pathology. *Neuropathology and Applied Neurobiology* 18:424–433.

Breitfeld, P. P., Casanova, J. E., Simister, N. E., Ross, S. A., McKinnon, W. C., and Mostov, K. E., 1989. Transepithelial transport of immunoglobulins: a model of protein sorting and transcytosis. *Am. J. Respir. Cell Mol. Biol.* 1:257–262.

Clough, G., 1991. Relationship between microvascular permeability and ultrastructure. *Prog. Biophys. Molec. Biol.* 55:47–69.

D'Amore, P. A., 1992. Capillary growth: a two-cell system. *Seminars in Cancer Biology* 3:49–56.

Davies, P. F., and Tripathi, S. C., 1993. Mechanical stress mechanisms and the cell. An endothelial paradigm. *Circulation Res.* 72:239–245.

Ferns, G. A., Raines, E. W., Sprugel, K. H., Motani, A. S., Reidy, M. A., and Ross, R., 1991. Inhibition of neointimal smooth muscle accumulation after angioplasty by an antibody to PDGF. *Science* 253:1129–1132.

Friden, P. M., Walus, L. R., Watson, P., Doctrow, S. R., Kozarich, J. W., Bäckman, C., Bergman, H., Hoffer, B., Bloom, F., and Granholm, A.-C., 1993. Blood-brain barrier penetration and in vivo activity of an NGF conjugate. *Science* 259:373–376.

Ghitescu, L., and Bendayan, M., 1992. Transendothelial transport of serum albumin: a quantitative immunocytochemical study. *J. Cell Biol.* 117:745–755.

Goldman, R. S., Finkbeiner, S. M., and Smith, S. J., 1991. Endothelin induces a sustained rise in intracellular calcium in hippocampal astrocytes. *Neurosci. Letts.* 123:4–8.

Kolodney, M. S., and Wysolmerski, R. B., 1992. Isometric contraction by fibroblasts and endothelial cells in tissue culture: a quantitative study. *J. Cell Biol.* 117: 73–82.

Kugiyama, K., Kerns, S. A., Morrisett, J. D., Roberts, R., and Henry, P. D., 1990. Impairment of endothelium-dependent arterial relaxation by lysolecithin in modified low-density lipoproteins. *Nature* 344:160–162.

Langille, B. L., Graham, J. J. K., Kim, D., and Gotlieb, A. I., 1991. Dynamics of shear-induced redistribution of F-actin in endothelial cells in vivo. *Arteriosclerosis and Thrombosis* 11:1814–1820.

McEver, R. P., 1992. Leukocyte–endothelial cell interactions. *Current Opinion in Cell Biol.* 4:840–849.

Niimi, N., Noso, N., and Yamamoto, S., 1992. The effect of histamine on cultured endothelial cells. A study of the mechanism of increased vascular permeability. *European J. Pharmacology* 221:325–331.

Palade, G. E., 1988. The microvascular endothelium revisited. In *Endothelial Cell Biology in Health and Disease,* Simionescu, N., and Simionescu, M., Eds. Plenum Press, New York.

Rubanyi, G. M., 1991. Endothelium-derived relaxing and contracting factors. *J. Cellular Biochem.* 46:27–36.

Simionescu, M., and Simionescu, N., 1991. Endothelial transport of macromolecules: transcytosis and endocytosis. A look from cell biology. *Cell Biol. Revs.* 25:1–78.

Smedsrød, B., Pertoft, H., Gustafson, S., and Laurent, T. C., 1990. Scavenger functions of the liver endothelial cell. *Biochem. J.* 266:313–327.

Yednock, T. A., Cannon, C., Fritz, L. C., Sanchez-Madrid, F., Steinman, L., and Karin, N., 1992. Prevention of experimental autoimmune encephalomyelitis by antibodies against $\alpha 4\beta 1$ integrin. *Nature* 356:63–66.

Zilla, P., von Oppell, U., and Deutsch, M., 1993. The endothelium: a key to the future. *J. Card. Surg.* 8: 32–60.

Chapter Seven Blood

Boxer, L. A., and Smolen, J. E., 1988. Neutrophil granule constituents and their release in health and disease. *Hematology/Oncology Clinics of North America* 2:101–134.

Clark, B. R., Gallagher, J. T., and Dexter, T. M., 1992. Cell adhesion in the stromal regulation of haemopoiesis. *Baillière's Clinical Haematology* 5:619–652.

Colman, R. W., 1990. Aggregin: a platelet ADP receptor that mediates activation. *FASEB J.* 4:1425–1435.

Crosier, P. S., and Clark, S. C., 1992. Basic biology of the hematopoietic growth factors. *Seminars in Oncology* 19:349–361.

Dinauer, M. C., and Orkin, S. H., 1992. Chronic granulomatous disease. *Annu. Rev. Medicine* 43:117–124.

Ebbeling, L., Robertson, C., McNicol, A., and Gerrard, J. M., 1992. Rapid ultrastructural changes in the dense tubular system following platelet activation. *Blood* 80:718–723.

Escolar, G., and White, J. G., 1991. The platelet open canalicular system: a final common pathway. *Blood Cells* 17:467–485.

Finley, D., and Chau, V., 1991. Ubiquitination. *Annu. Rev. Cell Biol.* 7:25–69.

Fuhrman, B., Brook, G. J., and Aviram, M., 1992. Proteins derived from platelet alpha granules modulate the uptake of oxidized low density lipoprotein by macrophages. *Biochimica et Biophysica Acta* 1127:15–21.

Handagama, P., Bainton, D. F., Jacques, Y., Conn, M. T., Lazarus, R. A., and Shuman, M. A., 1993. Kistrin, an integrin antagonist, blocks endocytosis of fibrinogen into guinea pig megakaryocyte and platelet α-granules. *J. Clin. Invest.* 91:193–200.

Huang, S., and Terstappen, L. W., 1992. Formation of haematopoietic microenvironment and haematopoietic stem cells from single human bone marrow stem cells. *Nature* 360:745–749.

Ikuta, K., Uchida, N., Friedman, J., and Weissman, I. L., 1992. Lymphocyte development from stem cells. *Annu. Rev. Immunology* 10:759–783.

Kroegel, C., Virchow, J. C., Jr., Kortsik, C., and Matthys, H., 1992. Cytokines, platelet activating factor and eosinophils in asthma. *Respiratory Medicine* 86:375–389.

Lloyd, A. R., and Oppenheim, J. J., 1992. Poly's lament: the neglected role of the polymorphonuclear neutrophil in the afferent limb of the immune response. *Immunology Today* 13:169–172.

Lubbert, M., Oster, W., Ludwig, W. D., Censer, A., Mertelsmann, R., and Herrmann, F., 1992. A switch toward demethylation is associated with the expression of myeloperoxidase in acute myeloblastic and promyelocytic leukemias. *Blood* 80:2066–2073.

Nakamura, Y., Komatsu, N., and Nakauchi, H., 1992. A truncated erythropoietin receptor that fails to prevent programmed cell death of erythroid cells. *Science* 257:1138–1141.

Siczkowski, M., Clarke, D., and Gordon, M. Y., 1992. Binding of primitive hematopoietic progenitor cells to marrow stromal cells involves heparan sulfate. *Blood* 80:912–919.

Stenberg, P. E., and Levin, J., 1989. Mechanisms of platelet production. *Blood Cells* 15:23–47.

Weller, P. F., 1992. Cytokine regulation of eosinophil function. *Clinical Immunology & Immunopathology* 62:S55–S59.

Williams, D. A., Rios, M., Stephens, C., and Patel, V. P., 1991. Fibronectin and VLA-4 in haematopoietic stem cell–microenvironment interactions. *Nature* 352:438–441.

Ziegler-Heitbrock, H. W., 1989. The biology of the monocyte system. *European J. Cell Biol.* 49:1–12.

Zipori, D., 1992. The renewal and differentiation of hemopoietic stem cells. *FASEB J.* 6:2691–2697.

Chapter Eight Immune System

Berek, C., 1992. The development of B cells and the B-cell repertoire in the microenvironment of the germinal center. *Immunological Revs.* 126:5–19.

Butch, A. W., Chung, G. H., Hoffmann, J. W., and Nahm, M. H., 1993. Cytokine expression by germinal center cells. *J. Immunology* 150:39–47.

Chin, Y. H., Cai, J. P., and Xu, X. M., 1991. Tissue-specific homing receptor mediates lymphocyte adhesion to cytokine-stimulated lymph node high endothelial venule cells. *Immunology* 74:478–483.

Goetzl, E. J., and Sreedharan, S. P., 1992. Mediators of communication and adaptation in the neuroendocrine and immune systems. *FASEB J.* 6:2646–2652.

Goodnow, C. C., Adelstein, S., and Basten, A., 1990. The need for central and peripheral tolerance in the B cell repertoire. *Science* 248:1373–1379.

Groom, A. C., Schmidt, E. E., and MacDonald, I. C., 1991. Microcirculatory pathways and blood flow in spleen: new insights from washout kinetics, corrosion casts, and quantitative intravital videomicroscopy. *Scanning Microscopy* 5:159–174.

Isobe, M., Yagita, H., Okumura, K., and Ihara, A., 1992. Specific acceptance of cardiac allograft after treatment with antibodies to ICAM-1 and LFA-1. *Science* 255:1125–1127.

Kopp, W. C., 1990. The immune functions of the spleen. In *The Spleen: Structure, Function and Clinical Significance,* A. J. Bowdler, Ed. Chapman and Hall Medical, London, pp. 103–126.

Kosco, M. H., and Gray, D., 1992. Signals involved in germinal center reactions. *Immunological Revs.* 126: 63–76.

Li, Y., Pezzano, M., Philip, D., Reid, V., and Guyden, J., 1992. Thymic nurse cells exclusively bind and internalize $CD4^+CD8^+$ thymocytes. *Cellular Immunology* 140:495–506.

Low, P. S., 1991. Role of hemoglobin denaturation and band 3 clustering in initiating red cell removal. In *Red Blood Cell Aging,* M. Magnani and A. DeFlora, Eds. Plenum Press, New York, pp. 173–183.

Pardoll, D., and Carrera, A., 1992. Thymic selection. *Current Opinion in Immunology* 4:162–165.

Ramsdell, F., and Fowlkes, B. J., 1990. Clonal deletion versus clonal anergy: the role of the thymus in inducing self tolerance. *Science* 248:1342–1348.

Shimizu, Y., Newman, W., Gopal, T. V., Horgan, K. J., Graber, N., Beall, L. D., van Seventer, G. A., and Shaw, S., 1991. Four molecular pathways of T cell adhesion to endothelial cells: roles of LFA-1, VCAM-1, and ELAM-1 and changes in pathway hierarchy under different activation conditions. *J. Cell Biol.* 113:1203–1212.

Vukmanovic, S., Grandea, A. G., III, Faas, S. J., Knowles, B. B., and Bevan, M. J., 1992. Positive selection of T-lymphocytes induced by intrathymic injection of a thymic epithelial cell line. *Nature* 359:729–732.

Weiss, L., 1990. Mechanisms of splenic clearance of

the blood; a structural overview of the mammalian spleen. In *The Spleen: Structure, Function and Clinical Significance,* A. J. Bowdler, Ed. Chapman and Hall Medical, London, pp. 25–39.

Whitmore, A. C., Prowse, D. M., Haughton, G., and Arnold, L. W., 1991. Ig isotype switching in B lymphocytes. The effect of T cell-derived interleukins, cytokines, cholera toxin, and antigen on isotype switch frequency of a cloned B cell lymphoma. *International Immunology* 3:95–103.

Yagita, H., Nakata, M., Kawasaki, A., Shinkai, Y., and Okumura, K., 1992. Role of perforin in lymphocyte–mediated cytolysis. *Advances in Immunology* 51:215–242.

Chapter Nine Endocrine Glands

Austin, H., Austin, J. M., Jr., Partridge, E. E., Hatch, K. D., and Shingleton, H.M., 1991. Endometrial cancer, obesity, and body fat distribution. *Cancer Res.* 51:568–572.

Bomsel, M., de Paillerets, C., Weintraub, H., and Alfsen, A., 1988. Biochemical and functional characterization of three types of coated vesicles in bovine adrenocortical cells: implication in the intracellular traffic. *Biochemistry* 27:6806–6813.

Burns, R. S., Allen, G. S., and Tulipan, N. B., 1990. Adrenal medullary tissue transplantation in Parkinson's disease: a review. *Advances in Neurology.* 53:571.

Childs, G. V., 1992. Structure–function correlates in the corticotropes of the anterior pituitary. *Frontiers in Neuroendocrinology* 13:271–317.

Childs, G. V., Unabia, G., Tibolt, R., and Lloyd, J. M., 1987. Cytological factors that support nonparallel secretion of luteinizing hormone and follicle-stimulating hormone during the estrous cycle. *Endocrinology* 121:1801–1813.

Csorba, T. R., 1991. Proinsulin: biosynthesis, conversion, assay methods and clinical studies. *Clin. Biochem.* 24:447–454.

Falke, N., 1991. Modulation of oxytocin and vasopressin release at the level of the neurohypophysis. *Prog. in Neurobiol.* 36:465–484.

Gross, D. J., Halban, P. A., Kahn, C. R., Weir, G. C., and Villa-Komaroff, L., 1989. Partial diversion of a mutant proinsulin (B10 aspartic acid) from the regulated to the constitutive secretory pathway in transfected AtT-20 cells. *Proc. Natl. Acad. Sci. USA* 86:4107–4111.

Hartree, A. S., and Renwick, A. G. C., 1992. Molecular structures of glycoprotein hormones and functions of their carbohydrate components. *Biochem. J.* 287:665–679.

Herzog, V., Berndorfer, U., and Saber, Y., 1992. Isolation of insoluble secretory product from bovine thyroid: extracellular storage of thyroglobulin in covalently cross-linked form. *J. Cell Biol.* 118:1071–1083.

Hunter, V. R., Pauly, D. F., Wolkowicz, P. E., McMillin, J. B., and Beckner, M. E., 1990. Mitochondrial adenosine triphosphatase in the oxyphil cells of a renal oncocytoma. *Human Pathology* 21:437–442.

Landström, A. H. S., Andersson, A., and Borg, L. A. H., 1991. Lysosomes and pancreatic islet function: adaptation of β-cell lysosomes to various metabolic demands. *Metabolism* 40:399–405.

Loh, Y. P., Andreasson, K. I., and Birch, N. P., 1991. Intracellular trafficking and processing of pro-opiomelanocortin. *Cell Biophysics* 19:73–83.

MacGregor, R. R., and Bansal, D. D., 1989. Inhibitors of cellular proteolysis cause increased secretion from parathyroid cells. *Biochemical & Biophysical Res. Comm.* 160:1339–1343.

Meeker, R. B., Swanson, D. J., Greenwood, R. S., and Hayward, J. N., 1991. Ultrastructural distribution of glutamate immunoreactivity within neurosecretory endings and pituicytes of the rat neurohypophysis. *Brain Research* 564:181–193.

Navone, F., Di Gioia, G., Jahn, R., Browning, M., Greengard, P., and De Camilli, P., 1989. Microvesicles of the neurohypophysis are biochemically related to small synaptic vesicles of presynaptic nerve terminals. *J. Cell Biol.* 109:3425–3433.

Nishikawa, T., Mikami, K., Saito, Y., Tamura, Y., and Yoshida, S., 1988. Functional differences in cholesterol ester hydrolase and acyl-coenzyme-A/cholesterol acyltransferase between the outer and inner zones of the guinea pig adrenal cortex. *Endocrinology* 122:877–883.

Nordmann, J. J., and Artault, J.-C., 1992. Membrane retrieval following exocytosis in isolated neurosecretory nerve endings. *Neurosci.* 49:201–207.

Ogishima, T., Suzuki, H., Hata, J.-I., Mitani, F., and Ishimura, Y., 1992. Zone-specific expression of aldosterone synthase cytochrome P-450 and cytochrome P-450$11\beta$ in rat adrenal cortex: histochemical basis for the functional zonation. *Endocrinology* 130:2971–2977.

Orci, L., Vassalli, J.-D., and Perrelet, A., 1988. The insulin factory. *Scientific American* 259:85–94.

Pocotte, S. L., Ehrenstein, G., and Fitzpatrick, L. A., 1991. Regulation of parathyroid hormone secretion. *Endocrine Revs.* 12:291–301.

Reetz, A., Solimena, M., Matteoli, M., Folli, F., Takei, K., DeCamilli, P., 1991. GABA and pancreatic beta-cells: colocalization of glutamic acid decarboxylase

(GAD) and GABA with synaptic-like microvesicles suggests their role in GABA storage and secretion. *Embo J.* 10:1275–1284.

Schwartz, J., 1992. At the cutting edge: the forest, the trees and the anterior pituitary. *Molecular & Cellular Endocrinology* 85:C45–C49.

Schwartz, J., and Cherny, R., 1992. Intercellular communication within the anterior pituitary influencing the secretion of hypophysial hormones. *Endocrine Revs.* 13:453–475.

Staats, D. A., Lohr, D. P., and Colby, H. D., 1989. α–Tocopherol depletion eliminates the regional differences in adrenal mitochondrial lipid peroxidation. *Molecular & Cellular Endocrinology* 62:189–195.

Steiner, D. F., and James, D. E., 1992. Cellular and molecular biology of the beta cell. *Diabetologia* 35:S41–S48.

Takasu, N., Ohno, S., Komiya, I., and Yamada, T., 1992. Requirements of follicle structure for thyroid hormone synthesis; cytoskeletons and iodine metabolism in polarized monolayer cells on collagen gel and in double layered, follicle-forming cells. *Endocrinology* 131:1143–1148.

Theodosis, D. T., and Poulain, D. A., 1992. Neuronal-glial and synaptic plasticity of the adult oxytocinergic system: factors and consequences. *Annals New York Acad. Sci.* 652:303–325.

Unsicker, K., Seidl, K., and Hofmann, H. D., 1989. The neuro-endocrine ambiguity of sympathoadrenal cells. *International J. Developmental Neurosci.* 7:413–417.

CHAPTER TEN Skin

Adams, J. C., and Watt, F. M., 1989. Fibronectin inhibits the terminal differentiation of human keratinocytes. *Nature* 340:307–309.

Bell, C. L., Reddy, M. M., and Quinton, P. M., 1992. Reversed anion selectivity in cultured cystic fibrosis sweat duct cells. *Am. J. Physiol.* 262:C32–C38.

Bommannan, D., Potts, R. O., and Guy, R. H., 1990. Examination of stratum corneum barrier function in vivo by infrared spectroscopy. *J. Investigative Dermatology* 95:403–408.

Brod, J., 1991. Characterization and physiological role of epidermal lipids. *International J. Dermatology* 30: 84–90.

Coulombe, P. A., Kopan, R., and Fuchs, E., 1989. Insights into complex programs of differentiation. *J. Cell Biol.* 109:2295–2312.

De Panfilis, G., Manara, G. C., Ferrari, C., Torresani, C., and Rowden, G., 1990. Subsets of keratinocytes and Langerhans' cells express epitopes associated with suppressor-inducer capabilities in resting normal human epidermis. *Immunology* 69:622–625.

Dlugosz, A. A., and Yuspa, S. H., 1993. Coordinate changes in gene expression which mark the spinous to granular cell transition in epidermis are regulated by protein kinase C. *J. Cell Biol.* 120:217–225.

Finch, P. W., Rubin, J. S., Miki, T., Ron, D., Aaronson, S. A., 1989. Human KGF is FGF-related with properties of a paracrine effector of epithelial cell growth. *Science* 245:752–755.

Fuchs, E., and Coulombe, P. A., 1992. Of mice and men: genetic skin diseases of keratin. *Cell* 69:899–902.

Hearing, V. J., and Tsukamoto, K., 1991. Enzymatic control of pigmentation in mammals. *FASEB J.* 5:2902–2909.

Imokawa, G., 1990. Analysis of carbohydrate properties essential for melanogenesis in tyrosinases of cultured malignant melanoma cells by differential carbohydrate processing inhibition. *J. Investigative Dermatology* 95:39–49.

Parent, D., Bernard, B. A., Desbas, C., Heenen, M., and Darmon, M. Y., 1990. Spreading of psoriatic plaques: alteration of epidermal differentiation precedes capillary leakiness and anomalies in vascular morphology. *J. Investigative Dermatology* 95:333–340.

Sato, K., Kang, W. H., Saga, K., and Sato, K. T., 1989. Biology of sweat glands and their disorders. I. Normal sweat gland function. *J. Am. Acad. Dermatology* 20:537–563.

Yamashita, Y., Akaike, N., Wakamori, M., Ikeda, I., and Ogawa, H., 1992. Voltage-dependent currents in isolated single merkel cells of rats. *J. Physiol.* 450:143–162.

CHAPTER ELEVEN Major Exocrine Glands

Adelson, J. W., and Miller, P. E., 1989. Heterogeneity of the exocrine pancreas. *Am. J. Physiol.* 256:G817–G825.

Casanova, J. E., 1992. Transepithelial transport of polymeric immunoglobulins. *Annals New York Acad. Sciences* 664:27–38.

Fielding, C. J., 1992. Lipoprotein receptors, plasma cholesterol metabolism, and the regulation of cellular free cholesterol concentration. *FASEB J.* 6:3162–3168.

Garry, D. J., Garry, M. G., Williams, J. A., Mahoney, W. C., and Sorenson, R. L., 1989. Effects of islet hormones on amylase secretion and localization of somatostatin binding sites. *Am J. Physiol.* 256:G897–G904.

Geiger, S., Geiger, B., Leitner, O., and Marshak, G., 1987. Cytokeratin polypeptides expression in different epithelial elements of human salivary glands. *Virchows Arch. A* 410:403–414.

Schmid, R. M., and Meisler, M. H., 1992. Dietary regulation of pancreatic amylase in transgenic mice mediated by a 126-base-pair DNA fragment. *Am. J. Physiol.* 262:G971–G976.

Singer, J. A., Jennings, L. K., Jackson, C. W., Dockter, M. E., Morrison, M., and Walker, W. S., 1986. Erythrocyte homeostasis: antibody-mediated recognition of the senescent state by macrophages. *Proc. Natl. Acad. Sci. USA* 83:5498–5501.

Chapter Twelve Gastrointestinal Tract

Bhaskar, K. R., Garik, P., Turner, B. S., Bradley, J. D., Bansil, R., Stanley, H. E., and LaMont, J. T., 1992. Viscous fingering of HCl through gastric mucin. *Nature* 360:458–461.

Gleeson, P. A., and Toh, B. -H., 1991. Molecular targets in pernicious anaemia. *Immunology Today* 12:233–238.

Keren, D. F., 1992. Antigen processing in the mucosal immune system. *Immunology* 4:217–226.

Lacasse, J., and Martin, L. H., 1992. Detection of CD1 mRNA in paneth cells of the mouse intestine by in situ hybridization, *J. Histochemistry and Cytochemistry* 40:1527–1534.

Lamont, J. T., 1992. Mucus: the front line of intestinal mucosal defense. *Annals New York Acad. Sci.* 664:190–201.

Louvard, D., Kedinger, M., and Hauri, H. P., 1992. The differentiating intestinal epithelial cell: establishment and maintenance of functions through interactions between cellular structures. *Annu. Rev. Cell Biol.* 8:157–195.

Roth, K. A., Kim, S., and Gordon, J. I., 1992. Immunocytochemical studies suggest two pathways for enteroendocrine cell differentiation in the colon. *Am. J. Physiol.* 263:G174–G180.

Schubert, M. L., and Shamburek, R. D., 1990. Control of acid secretion. *Gastroenterology Clinics of N. Am.* 19:1–25.

Selsted, M. E., Miller, S. I., Henschen, A. H., and Ouellette, A. J., 1992. Enteric defensins: antibiotic peptide components of intestinal host defense. *J. Cell Biol.* 118:929–936.

Tang, L. H., Stock, S. A., Modlin, I. M., and Goldenring, J. R., 1992. Identification of rab2 as a tubulovesicle-membrane-associated protein in rabbit gastric parietal cells. *Biochemical J.* 285:715–719.

Targan, S. R., 1992. The lamina propria: a dynamic, complex mucosal compartment. *Annals New York Acad. Sci.* 664:61–68.

Wilson, D. E., 1991. Role of prostaglandins in gastroduodenal mucosal protection. *J. Clin. Gastroenterol.* 13:(Suppl. 1):S65–S71.

Yeomans, N. D., and Skeljo, M. V., 1991. Repair and healing of established gastric mucosal injury. *J. Clin. Gastroenterol.* 13(Suppl. 1):S37–S41.

Chapter Thirteen Respiratory System

Brandes, M. E., and Finkelstein, J. N., 1990. The production of alveolar macrophage-derived growth-regulating proteins in response to lung injury. *Toxicology Letters* 54:3–22.

Crouch, E., Parghi, D., Kuan, S. F., and Persson, A., 1992. Surfactant protein D: subcellular localization in nonciliated bronchiolar epithelial cells. *Am. J. Physiology* 263:L60–L66.

de Jongste, J. C., Jongejan, R. C., and Kerrebijn, K. F., 1991. Control of airway caliber by autonomic nerves in asthma and in chronic obstructive pulmonary disease. *Am. Rev. Respir. Dis.* 143:1421–1426.

Froh, D., Ballard, P. L., Williams, M. C., Gonzales, J., Goerke, J., Odom, M. W., and Gonzales, L. W., 1990. Lamellar bodies of cultured human fetal lung: content of surfactant protein A (SP-A), surface film formation and structural transformation in vitro. *Biochimica et Biophysica Acta.* 1052:78–89.

Gross, I., 1990. Regulation of fetal lung maturation. *Am. J. Physiol.* 259:L337–L344.

Haagsman, H. P., and van Golde, L. M. G., 1991. Synthesis and assembly of lung surfactant. *Annu. Rev. Physiol.* 53:441–464.

Hawgood, S., and Shiffer, K., 1991. Structures and properties of the surfactant-associated proteins. *Annu. Rev. Physiol.* 53:375–394.

Holian, A., and Scheule, R. K., 1990. Alveolar macrophage biology. *Hospital Practice* 25:49–58.

Jabbour, A. J., Holian, A., and Scheule, R. K., 1991. Lung lining fluid modification of asbestos bioactivity for the alveolar macrophage. *Toxicology and Applied Pharmacology* 110:283–294.

Matthay, M. A., and Wiener-Kronish, J. P., 1990. Intact epithelial barrier function is critical for the resolution of alveolar edema in humans. *Am. Rev. Respir. Dis.* 142:1250–1257.

Plopper, C. G., Macklin, J., Nishio, S. J., Hyde, D. M., and Buckpitt, A. R., 1992. Relationship of cytochrome P-450 activity to Clara cell cytotoxicity. III. Morphometric comparison of changes in the epithelial populations of terminal bronchioles and lobar bonchi in mice, hamsters, and rats after parenteral administration of naphthalene. *Laboratory Investigation* 67:553–565.

Satir, P., and Sleigh, M. A., 1990. The physiology of cilia and mucociliary interactions. *Annu. Rev. Physiol.* 52:137–155.

Verdugo, P., 1990. Goblet cells secretion and mucogenesis. *Annu. Rev. Physiol.* 52:157–176.

West, J. B., and Mathieu-Costello, O., 1992. Strength of the pulmonary blood-gas barrier. *Respiration Physiology* 88:141–148.

Wright, J. R., and Dobbs, L. G., 1991. Regulation of pulmonary surfactant secretion and clearance. *Annu. Rev. Physiol.* 53:395–414.

Chapter Fourteen Kidney

Andrews, P. M., 1989. Shape changes in kidney glomerular podocytes: mechanisms and possible functional significance. In *Cells and Tissues: A Three-Dimensional Approach by Modern Techniques in Microscopy,* Alan R. Liss, Inc., pp. 157–166.

Briggs, J. P., Skøtt, O., and Schnermann, J., 1990. Cellular mechanisms within the juxtaglomerular apparatus. *Am. J. Hypertension* 3:76–80

Christensen, E. I., and Nielsen, S., 1991. Structural and functional features of protein handling in the kidney proximal tubule. *Seminars in Nephrology* 11:414–439.

Deen, P. M., Dempster, J. A., Wieringa, B., Van Os, C. H., 1992. Isolation of a cDNA for rat CH1P28 water channel: high mRNA expression in kidney cortex and inner medulla. *Biochem. & Biophys. Res. Comm.* 188:1267–1273.

Dekan, G., Gabel, C., and Farquhar, M. G., 1991. Sulfate contributes to the negative charge of podocalyxin, the major sialoglycoprotein of the glomerular filtration slits. *Proc. Natl. Acad. Sci. USA* 88:5398–5402.

Desjardins, M., Bendayan, M., 1991. Ontogenesis of glomerular basement membrane: structural and functional properties. *J. Cell Biol.* 113:689–700.

Gupta, S., Rifici, V., Crowley, S., Brownlee, M., Shan, Z., and Schlondorff, D., 1992. Interactions of LDL and modified LDL with mesangial cells and matrix. *Kidney International* 41:1161–1169.

Kahan, B. D., 1989. Cyclosporine. *The New England J. Med.* 321:1725–1738.

Osswald, H., Mühlbauer, B., and Schenk, F., 1991. Adenosine mediates tubuloglomerular feedback response: an element of metabolic control of kidney function. *Kidney International* 39 (Suppl. 32):S128–S131.

Pfeilschifter, J., 1989. Cross-talk between transmembrane signalling systems: a prerequisite for the delicate regulation of glomerular haemodynamics by mesangial cells. *European J. Clinical Investigation* 19:347–361.

Schnabel, E., Anderson, J. M., Farquhar, M. G., 1990. The tight junction protein ZO-1 is concentrated along slit diaphragms of the glomerular epithelium. *J. Cell Biol.* 111:1255–1263.

Schnabel, E., Dekan, G., Miettinen, A., Farquhar, M. G., 1989. Biogenesis of podocalyxin—the major glomerular sialoglycoprotein—in the newborn rat kidney. *European J. Cell Biology* 48:313–326.

Schreiner, G. F., 1992. The mesangial phagocyte and its regulation of contractile cell biology. *J. Amer. Soc. Neph.* 2 (10 Suppl.):S74–S82.

Tsuchiya, K., Wang, W., Giebisch, G., and Welling, P. A., 1992. ATP is a coupling modulator of parallel Na,K-ATPase-K-channel activity in the renal proximal tubule. *Proc. Natl. Acad. Sci. USA* 89:6418–6422.

Walton, H. A., Byrne, J., and Robinson, G. B., 1992. Studies of the permeation properties of glomerular basement membrane: cross-linking renders glomerular basement membrane permeable to protein. *Biochimica et Biophysica Acta* 1138:173–183.

Wilcox, C. S., Welch, W.J., Murad, F., Gross, S. S., Taylor, G., Levi, R., and Schmidt, H. H., 1992. Nitric oxide synthase in macula densa regulates glomerular capillary pressure. *Proc. Natl. Acad. Sci. USA* 89:11993–11997.

Yanagawa, N., 1991. Angiotensin II and proximal tubule sodium transport. *Renal Physiol. & Biochem.* 14:208–215.

Chapter Fifteen Male Reproductive System

Aumüller, G., and Seitz, J., 1990. Protein secretion and secretory processes in male acccessory sex glands. *International Review of Cytology* 121:127–231.

Bearer, E. L., and Friend, D. S., 1990. Morphology of mammalian sperm membranes during differentiation, maturation, and capacitation. *J. Electron Microscopy Technique* 16:281–297.

Braun, R. E., Behringer, R. R., Peschon, J. J., Brinster, R. L., and Palmiter, R.D., 1989. Genetically haploid spermatids are phenotypically diploid. *Nature* 337:373–376.

Caldwell, K. A., and Handel, M. A., 1991. Protamine transcript sharing among postmeiotic spermatids. *Proc. Natl. Acad. Sci. USA* 88:2407–2411.

Carr, D. W., and Acott, T. S., 1989. Intracellular pH regulates bovine sperm motility and protein phosphorylation. *Biology of Reproduction* 41:907–920.

de Kretser, D. M., and Kerr, J. B., 1988. The cytology of the testis. In *The Physiology of Reproduction,* E. Knobil, J. Neill, et al., Eds. Raven Press, Ltd., New York, Vol. 1, pp. 837–932 (Chapter 20).

Douglass, J., Garrett, S. H., and Garrett, J. E., 1991. Differential patterns of regulated gene expression in the adult rat epididymis. *Annals New York Acad. Sci.* 637:384–398.

Fritz, I. B., 1988. Summary of the 5th European workshop on the molecular and cellular endocrinology of the testis. *Molecular and Cellular Endocrinology* 59:147–154.

Grimes, S. R., Jr., 1986. Nuclear proteins in spermatogenesis. *Comp. Biochem. Physiol.* 83B:495–500.

Grove, B. D., Pfeiffer, D. C., Allen, S., Vogl, A. W., 1990. Immunofluorescence localization of vinculin in ectoplasmic ("junctional") specializations of rat sertoli cells. *The Am. J. Anatomy* 188:44–56.

Hecht, N. B., 1990. Regulation of 'haploid expressed genes' in male germ cells. *J. Reprod. Fert.* 88:679–693.

Hedger, M. P., Qin, J.-X., Robertson, D. M., and de Kretser, D. M., 1990. Intragonadal regulation of immune system functions. *Reprod. Fertil. Dev.* 2:263–280.

Jégou, B., 1992. The sertoli cell. *Baillière's Clinical Endocrinology and Metabolism* 6:273–311.

Jones, R., 1989. Membrane remodelling during sperm maturation in the epididymis. *Oxford Revs. Reproductive Biol.* 11:285–337.

Jones, R., Shalgi, R., Hoyland, J., and Phillips, D. M., 1990. Topographical rearrangement of a plasma membrane antigen during capacitation of rat spermatozoa in vitro. *Developmental Biology* 139:349–362.

Linder, C. C., Heckert, L. L., Roberts, K. P., Kim, K. H., and Griswold, M. D., 1991. Expression of receptors during the cycle of the seminiferous epithelium. *Annals New York Acad. Sciences* 637:313–321.

Russell, L. D., Goh, J. C., Rashed, R. M. A., and Vogl, A. W., 1988. The consequences of actin disruption at sertoli ectoplasmic specialization sites facing spermatids after in vivo exposure of rat testis to cytochalasin D. *Biology of Reproduction* 39:105–118.

Stallard, B. J., and Griswold, M. D., 1990. Germ cell regulation of sertoli cell transferrin mRNA levels. *Mol. Endo.* 4:393–401.

Turner, T. T., 1991. Spermatozoa are exposed to a complex microenvironment as they traverse the epididymis. *Annals New York Acad. Sci.* 637:364–383.

Veeramachaneni, D. N., and Amann, R. P., 1991. Endocytosis of androgen-binding protein, clusterin, and transferrin in the efferent ducts and epididymis of the ram. *J. Andrology* 12:288–294.

Verhoeven, G., 1992. Local control systems within the testis. *Baillière's Clinical Endocrinology and Metabolism* 6:313–333.

Yasuzumi, F., Okura, N., Kohata, Y., and Harutsugu, K., 1988. Morphological and functional reconsideration of the cytoplasmic bridges which connect male germ cells in snails. *J. Ultrastructure and Molecular Structure Research* 99:261–271.

Yeung, C. H., Cooper, T. G., Weinbauer, G. F., Bergmann, M., Kleinhans, G., Schulze, H., and Nieschlag, E., 1989. Fluid-phase transcytosis in the primate epididymis in vitro and in vivo. *International J. Andrology* 12:384–394.

CHAPTER SIXTEEN **Female Reproductive System**

Amsterdam, A., Rotmensch, S., and Ben-Ze'ev, A., 1989. Coordinated regulation of morphological and biochemical differentiation in a steroidogenic cell: the granulosa cell model. *TIBS: Trends in Biochemical Sciences* 14:377–382.

Armstrong, D. T., Zhang, X., Vanderhyden, B. C., and Khamsi, F., 1991. Hormonal actions during oocyte maturation influence fertilization and early embryonic development. *Annals New York Acad. Sci.* 626:137–158.

Conley, A. J., and Mason, J. I., 1990. Placental steroid hormones. *Baillière's Clinical Endocrinology and Metabolism* 4:249–272.

Davis, O. K., and Rosenwaks, Z., 1992. Current status of in vitro fertilization and the new reproductive technologies. *Current Opinion in Obstetrics & Gynecology* 4:354–358.

Dockery, P., Li, T. C., Rogers, A. W., Cooke, I. D., and Lenton, E. A., 1988. The ultrastructure of the glandular epithelium in the timed endometrial biopsy. *Human Reproduction* 3:826–834.

Eppig, J. J., 1991. Intercommunication between mammalian oocytes and companion somatic cells. *Bioessays* 13:569–574.

Giovannini, M., Agostoni, C., and Salari, P. -C., 1991. The role of lipids in nutrition during the first months of life. *J. International Medical Research* 19:351–362.

Holt, C., 1992. Structure and stability of bovine casein micelles. *Advances in Protein Chemistry* 43:63–151.

Hultén, M. A., 1990. The origin of aneuploidy: bivalent instability and the maternal age effect in trisomy 21 Down syndrome. *Am J. Medical Genetics,* Suppl. 7:160–161.

Hunt, J. S., and Orr, H. T., 1992. HLA and maternal–fetal recognition. *FASEB J.* 6:2344–2348.

Irianni, F., and Hodgen, G. D., 1992. Mechanism of ovulation. *Endocrinology & Metabolism Clinics of North America* 21:19–38.

Neville, M. C., 1990. The physiological basis of milk secretion. *Annals New York Acad. Sci.* 586:1–11.

Richardson, M. C., Davies, D. W., Watson, R. H., Dunsford, M. L., Inman, C. B., and Masson, G. M., 1992. Cultured human granulosa cells as a model for corpus luteum function: relative roles of gonadotrophin and low density lipoprotein studied under defined culture conditions. *Human Reproduction* 7:12–18.

Roberts, A. J., and Skinner, M. K., 1990. Estrogen regulation of theca cell steroidogenesis and differentiation: thecal cell–granulosa cell interactions. *Endocrinology* 127:2918–2929.

Sayegh, R., and Mastroianni, L., Jr., 1991. Recent advances in our understanding of tubal function. *Annals New York Acad. Sci.* 626:266–275.

Seppälä, M., Julkunen, M., Riittinen, L., and Koistinen, R., 1992. Endometrial proteins: a reappraisal. *Human Reproduction* 7 (Suppl. 1):31–38.

Tabibzadeh, S., 1991. Human endometrium: an active site of cytokine production and action. *Endocrine Reviews* 12:272–290.

Yang, C.-H., and Yanagimachi, R., 1989. Differences betwen mature ovarian and oviductal oocytes: a study using the golden hamster. *Human Reproduction* 4: 63–71.

Chapter Seventeen Sensory Regions

Anholt, R. R. H., 1989. Molecular physiology of olfaction. *Am. J. Physiol.* 257:C1043–C1054.

Bakalyar, H. A., and Reed, R. R., 1990. Identification of a specialized adenyl cyclase that may mediate odorant detection. *Science* 250:1403–1406.

Burchell, B., 1991. Turning on and turning off the sense of smell. *Nature* 350:16–17.

Chuah, M. I., and Au, C., 1991. Olfactory schwann cells are derived from precursor cells in the olfactory epithelium. *J. Neuroscience Research* 29:172–180.

Deigner, P. S., Law, W. C., Cañada, F. J., and Rando, R. R., 1989. Membranes as the energy source in the endergonic transformation of vitamin A to 11-*cis*-retinol. *Science* 244:968–971.

Doucette, R., 1990. Glial influences on axonal growth in the primary olfactory system. *GLIA* 3:433–449.

Garland, D. L., 1991. Ascorbic acid and the eye. *Am. J. Clin. Nutr.* 54:1198S–1202S.

Hudspeth, A. J., 1989. How the ear's works work. *Nature* 341:397–404.

Jacobs, R. A., and Hudspeth, A. J., 1990. Ultrastructural correlates of mechanoelectrical transduction in hair cells of the bullfrog's internal ear. *Cold Spring Harbor Symp. Quant. Biol.* 55:547–561.

Jakus, M. A., 1956. Studies on the cornea. II. The fine structure of Descemet's membrane. *J. Biophysic. and Biochem. Cytol.* 2:243–255.

Jaramillo, F., and Hudspeth, A. J., 1991. Localization of the hair cell's transduction channels at the hair bundle's top by iontophoretic application of a channel blocker. *Neuron* 7:409–420.

Kistler, J., and Bullivant, S., 1989. Structural and molecular biology of the eye lens membranes. *Critical Reviews in Biochem. and Molecular Biol.* 24:151–181.

Komai, Y., and Ushiki, T., 1991. The three-dimensional organization of collagen fibrils in the human cornea and sclera. *Inves. Ophthalmology & Visual Sci.* 32:2244–2258.

Nishida, T., Yasumoto, K., Otori, T., and Desaki, J., 1988. The network structure of corneal fibroblasts in the rat as revealed by scanning electron microscopy. *Inves. Ophthalmology & Visual Sci.* 29:1887–1890.

Pickles, J. O., and Corey, D. P., 1992. Mechanoelectrical transduction by hair cells. *Trends in Neurosciences* 15:254–259.

Scott, J. E., 1992. Supramolecular organization of extracellular matrix glycosaminoglycans, in vitro and in the tissues. *FASEB J.* 6:2639–2645.

Streilein, J. W., and Bradley, D., 1991. Analysis of immunosuppressive properties of iris and ciliary body cells and their secretory products. *Inves. Ophthalmology & Visual Sci.* 32:2700–2710.

Stryer, L., 1987. The molecules of visual excitation. *Scientific American* 255:42–50.

Stryer, L., 1991. Visual excitation and recovery. *J. Biol. Chem.* 266:10711–10714.

Timmers, A. M., Van Groningen-Luyben, D. A. H. M., and De Grip, W. J., 1991. Uptake and isomerization of all-*trans* retinol by isolated bovine retinal pigment epithelial cells: further clues to the visual cycle. *Exp. Eye Res.* 52:129–138.

INDEX

References to primary discussions are in boldface.